Fracture Processes in Concrete, Rock and Ceramics

Volume Two
Macroscopic Modelling
and Structural Engineering

Other RILEM Proceedings available from
E & FN SPON / CHAPMAN & HALL

1 Adhesion between Polymers and Concrete. ISAP 86
Aix-en-Provence, France, 1986
Edited by H.R. Sasse

2 From Materials Science to Construction Materials Engineering
Proceedings of the First International RILEM Congress
Versailles, France, 1987
Edited by J.C. Maso

3 Durability of Geotextiles
St Rémy-lès-Chevreuse, France, 1986

4 Demolition and Reuse of Concrete and Masonry
Tokyo, Japan, 1988
Edited by Y. Kasai

5 Admixtures for Concrete
Improvement of Properties
Barcelona, Spain, 1990
Edited by E. Vázquez

6 Analysis of Concrete Structures by Fracture Mechanics
Abisko, Sweden, 1989
Edited by L. Elfgren and S.P. Shah

7 Vegetable Plants and their Fibres as Building Materials
Salvador, Bahia, Brazil, 1990
Edited by H.S. Sobral

8 Mechanical Tests for Bituminous Mixes
Budapest, Hungary, 1990
Edited by H.W. Fritz and E. Eustacchio

9 Test Quality for Construction, Materials and Structures
St Rémy-lès-Chevreuse, France, 1990
Edited by M. Fickelson

10 Properties of Fresh Concrete
Hanover, Germany, 1990
Edited by H.-J. Wierig

11 Testing during Concrete Construction
Mainz, Germany, 1990
Edited by H.W. Reinhardt

12 Testing of Metals for Structures
Naples, Italy, 1990
Edited by F.M. Mazzolani

13 Fracture Processes in Concrete, Rock and Ceramics
Noordwijk, Netherlands, 1991
Edited by J.G.M. van Mier, J.G. Rots and A. Bakker

14 Quality Control of Concrete Structures
Ghent, Belgium, 1991
Edited by L. Taerwe and H. Lambotte

Publisher's Note
This book has been produced from camera ready copy provided by the individual contributors. This method of production has allowed us to supply finished copies to the delegates at the Symposium.

Fracture Processes in Concrete, Rock and Ceramics

Volume Two

Macroscopic Modelling and Structural Engineering

Proceedings of the International RILEM/ESIS Conference 'Fracture Processes in Brittle Disordered Materials: Concrete, Rock, Ceramics', sponsored by The International Union of Testing and Research Laboratories for Materials and Structures (RILEM) and the European Structural Integrity Society (ESIS), and organized by Delft University of Technology, Delft, The Netherlands.

Noordwijk, The Netherlands June 19–21, 1991

EDITED BY

J.G.M. van Mier

Delft University of Technology, Department of Civil Engineering, Stevin Laboratory

J.G. Rots

Delft University of Technology, Department of Civil Engineering/TNO Building and Construction Research

A. Bakker

Delft University of Technology, Department of Chemical Technology and Materials Science, Materials Laboratory

E & FN SPON

An Imprint of Chapman & Hall

London · New York · Tokyo · Melbourne · Madras

UK	Chapman and Hall, 2–6 Boundary Row, London SE1 8HN
USA	Van Nostrand Reinhold, 115 5th Avenue, New York NY10003
JAPAN	Chapman and Hall Japan, Thomson Publishing Japan, Hirakawacho Nemoto Building, 7F, 1-7-11 Hirakawa-cho, Chiyoda-ku, Tokyo 102
AUSTRALIA	Chapman and Hall Australia, Thomas Nelson Australia, 102 Dodds Street, South Melbourne, Victoria 3205
INDIA	Chapman and Hall India, R. Seshadri, 32 Second Main Road, CIT East, Madras 600 035

First edition 1991

Printed in Great Britain at the
University Press, Cambridge

ISBN 0 419 15870 7 (set) 0 422 31474 4 (USA)
0 419 15860 X (vol. 2)

British Library Cataloguing in Publication Data
Available

Library of Congress Cataloging-in-Publication Data
Available

International Scientific Committee

Local Organizing Committee

J.G.M. van Mier	*Chairman: Delft University of Technology, Stevin Laboratory, The Netherlands*
R. Komen-Zimmerman	*Secretary: Congress Office ASD, Pijnacker, The Netherlands*
J.G. Rots	*Department of Civil Engineering, Delft University of Technology and TNO Building and Construction Research, Rijswijk, The Netherlands*
A. Bakker	*Department of Chemical Technology and Materials Science, Delft University of Technology, The Netherlands*

Sponsors

The organization of this conference was possible due to the generous support of the following sponsors. Their contribution to the success of the conference is greatly appreciated.

Delft University of Technology
TNO Building and Construction Research
Commission of the European Communities
Diana Analysis B.V.
KEMA, Research and Development, Testing and Certification, Consultancy and Engineering Services for the Electric Power Industry
Royal Dutch Association for Brick Manufacturers (KNB)
Ministry of Transport and Public Works, Rijkswaterstaat/ Construction Department
Royal Netherlands Academy of Sciences (KNAW)
Shell Nederland B.V.

Contents

VOLUME TWO
Macroscopic Modelling and Structural Engineering

PART NINE NUMERICAL MACROSCOPIC CRACK MODELS: CONTINUA 599

Preface

This book contains the invited and submitted contributions to the International Conference on 'Fracture Processes in Brittle Disordered Materials: Concrete, Rock, Ceramics', held at Noordwijk, The Netherlands, from June 19–21 1991. The conference was sponsored by RILEM, the International Union of Testing and Research Laboratories for Materials and Structures, (more specifically by Technical Committees 89 FMT 'Fracture Mechanics of Concrete – Test Methods' and 90 FMA 'Fracture Mechanics of Concrete – Applications') and by ESIS, the European Structural Integrity Society (formerly the European Group on Fracture). The primary goal of this conference was to bring together materials scientists, physicists, structural engineers, specialists from fracture mechanics, continuum mechanics and computational mechanics and experimentalists to discuss the advances in fracture research of brittle disordered materials.

The conference has been preceded by a number of conferences starting in 1985, when the first conference on 'Fracture Mechanics of Concrete' was organized by Professor Folker Wittmann at Lausanne. Since then, almost every year follow-up conferences were held: 1987 Houston, 1988 Vienna and 1989 the RILEM workshop in Abisko and the conference in Cardiff. In 1990 a conference was held in Albuquerque, as well as a NATO workshop in Evanston. The series will continue in the future.

The proceedings are split into two parts. The first part (Volume One) contains the contributions related to materials research and materials engineering. The second part (Volume Two) contains the papers related to the development of macroscopic fracture models, methods for the determination of the required macroscopic fracture parameters and applications in structural analysis. The papers, both the invited and submitted contributions, are grouped into thirteen parts.

Knowledge on fracture processes has increased considerably in the past decade, and it seems that materials engineering, i.e. the development of tailor-made materials, comes more and more within reach. Contributions in the first volume explore the micro-world and include observations of microstructural changes (Parts One and Four) as well as analytical and numerical micro-mechanics models for describing the non-linear macroscopic response of materials (Parts Two and Three). Basic to most approaches is the observation that macroscopic fracture properties cannot be measured directly, and depend to a large extent on geometrical and boundary effects. In the micro-models, the heterogeneous material structure is mimicked, and macroscopic non-linearities are analyzed as functions of boundary conditions, geometry and other environmental conditions (e.g. temperature, hygral and loading rate effects). In general two different 'micro'-approaches can be distinguished: (1) micro-mechanics models in which the microstructure is modelled in detail (for example by using finite elements, Part Two); and (2) approaches that depart from regular lattices, where disorder is introduced as fluctuations of strength or stiffness of the constituting bar elements in the lattice (Part Three). Lattice models have received considerable attention in theoretical physics, and many of the ideas developed there seem to be applicable in materials science. An important offspring of micromechanics is the development of new materials. In part Four, fibre reinforced materials are discussed.

As mentioned, the micro-models can be used for validating the macroscopic properties of the material. Using these models, the material behaviour can be studied taking into account realistic boundary conditions and 'real macroscopic fracture properties' can be extracted. For example, manufacturing methods of concretes and ceramics may lead to complex surface crack patterns and thus to different macroscopic properties than expected (Part Five). Next to this, macroscopic fracture properties of brittle disordered materials are size dependent (Part Six). Lattice type models can be used to validate and improve engineering models which are very often a direct fit of experimental data. For engineering purposes the fracture mechanics based models seem accurate enough. However, studies continue to minimize errors.

In order to overcome the difficulties that are met in experiments, researchers try to establish macroscopic fracture parameters needed, for example, in structural analysis, using hybrid approaches (combined numerical or analytical and experimental methods). Results of hybrid studies are reported in several contributions (Part Seven). The parameters derived can be used in numerical models. Two parts (Parts Eight and Nine) of the book are related to the development of numerical macroscopic fracture models. The models are split into two groups: (1) discrete models; and (2) continuum based models. In the first group the discontinuities in the displacement fields that are caused by cracks are modelled explicitly. Growth criteria are either based on conventional stress intensity factors or on cohesive crack models such as those, for example, derived from the hybrid approaches mentioned before. In the second group of models, the localization is modelled using higher order continuum theories. Introducing higher order terms in the continuum formulation introduces an internal length scale in the material, which corresponds to the notion of a crack band. Again, the micromechanics models described in Parts Two and Three may be used for validating the higher order continuum models, specifically for the determination of the internal length scale of the material.

In developing all these models, a strong interation between observation and theory is necessary. Observations become more refined as experimental techniques are improved. Parts Ten to Twelve contain contributions that describe some of the methods used for the determination of macroscopic fracture parameters under various conditions (mode I, mixed mode loading, compression, rate effects, hygral and thermal loading and sustained loading) in relation to the results obtained. Next to this, the development of simple test methods for practical situations is of importance if fracture mechanics is to be used in practice. Engineers should be able to extract the necessary information from these routine tests.

The development of the macroscopic fracture models has come to a point that practical results are starting to emerge, specifically in structural engineering (not least through the activities of RILEM TC 90 FMA). In Part Thirteen a number of applications of fracture mechanics models for structural analysis is included. Code rules for minimum reinforcement, analysis of the pull-out of anchors, assessment of the reliability of structures and analysis of shear capacity of RC beams are possible. The fracture mechanics models developed play a crucial role in rationalizing code rules, but may also be a helpful tool in structural engineering.

It should be realized that progress in the field is possible only through a vivid interaction between experimentalists and theorists, or according to R. Oldershaw: '*Science works well when observation leads theory in the dance of progress, or when the lead alternates between observation and theory on a short time scale.*' The constraints on time are well met in fracture mechanics research of brittle disordered materials. Development of experimental techniques on the one hand and advances in theory and computational techniques on the other hand have shown a good balance in the past decade. The engineering community will take the fruits of these activities, even when this is not visible on the very short time scale of most of their activities.

J.G.M. van Mier, J.G. Rots and A. Bakker
Delft, January 1991

Acknowledgement

The editors would like to acknowledge the continuous support by Mrs Ria Komen-Zimmerman and staff members of Congress Office ASD. Their tireless, efficient and expert work as conference organizer and coordinator is greatly appreciated. In addition to this, the editors would like to thank the members of the international scientific committee for their help in reviewing the abstracts and papers.

PART SEVEN

HYBRID METHODS FOR THE DETERMINATION OF MACROSCOPIC FRACTURE PARAMETERS

42 FRACTURE PROCESS ZONE ASSOCIATED WITH CONCRETE FRACTURE

J.-H. YON, N.M. HAWKINS, Z.-K. GUO
and A.S. KOBAYASHI
University of Washington, Seattle, WA, USA

Abstract
Moire interferometry was used to determine the crack opening displacements (COD) along stably growing and rapidly propagating cracks in concrete fracture specimens. This COD data was used in a hybrid experimental-numerical analysis to determine the constitutive relation governing the fracture process zone (FPZ), the energy release rate of the fracture specimen and the energy dissipation rate in the FPZ. The rising resistance curve was related to the gradual development of the FPZ. FPZ was also identified as the dominant energy dissipation mechanism in concrete.

Keywords:
Concrete Fracture, Static and Dynamic Fracture, Moire Interferometry, Fracture Process Zone, Resistance Curve.

1 Introduction

Since Hillerborg et al (1976) used the concept of a fracture process zone (FPZ) to model the results of direct tension testing of concrete, scores of papers using the FPZ model have been published to date. Most of these papers, including those of two of the authors and their colleagues, assumed as Hillerborg did, a maximum tensile strength criterion for the propagation of an equilibrium crack. One exception to such modelling is the two parameter fracture criterion combining a critical stress intensity factor, K_{IC}, with a critical crack tip opening displacement, $CTOD_C$, and proposed by Jenq and Shah (1985). Most FPZ modelling involves an inverse analysis where the postulated FPZ model is optimized to match measured and computed remote loading parameters, such as the applied load and the load-line displacement. However, such load parameters are not sensitive to small variations in FPZ modelling.

Fracture is a local phenomenon which is governed by crack tip parameter(s), such as the critical stress intensity factor (fracture toughness) of linear elastic fracture mechanics. Any deviation from the theoretically and experimentally proven theorems in fracture requires local theoretical and experimental justifications of such deviation, the FPZ being one of them. Detailed FPZ modelling became possible only with the availability of the crack opening displacement (COD) data obtained using holographic interferometry, Miller et al (1988), or moire

Fracture Processes in Concrete, Rock and Ceramics Edited by J.G.M. van Mier, J.G. Rots and A. Bakker.
 Published by E. & F.N. Spon, 2–6 Boundary Row, London SE1 8HN. ISBN 0 419 15870 7

interferometry, Du et al (1989,1990a). In particular, the data of Miller et al (1988) and Du et al (1990b) showed unambiguously a parabolic crack tip deformation which is indicative of a $r^{-1/2}$ stress singularity at the crack tip. Thus, the equilibrium crack of concrete and its associated FPZ model must be a Griffith crack with a trailing FPZ which retards crack opening and hence reduces the stress intensity factor, K_I. This combined K_I-FPZ model, which is designated here as the singular-FPZ model, was used to replicate numerically, the dynamic fracture responses of displacement-controlled and drop-weight, three-point bend and crack-line wedge-loaded, double cantilever beam (CLWL-DCB) concrete specimens tested by Yon et al (submitted). This inverse analysis showed that the singular-FPZ model also provided a realistic simulation of dynamic fracture of concrete. This success prompted a re-evaluation of the static concrete fracture results which were previously processed without a K_I following Hillerborg's approach.

This paper reviews the experimental evidences, which led to the development of the singular-FPZ model, the procedure used to evaluate that concrete fracture data, and the resultant models.

2 Experimental justification

2.1 Specimens

Figure 1 shows the three-point bend and the CLWL-DCB specimens whose results were used in the static analysis. Only the three-point bend specimens were used in the dynamic analysis involving moire interferometry. High early strength Portland cement, an aggregate of 6.4 mm maximum size and local Seattle sand were used in the mix. All specimens were cast in steel molds. Mechanical properties of the concrete were determined from controlled cylinders as well as small tension bars cut from the broken three point bend and CLWL-DCB specimens.

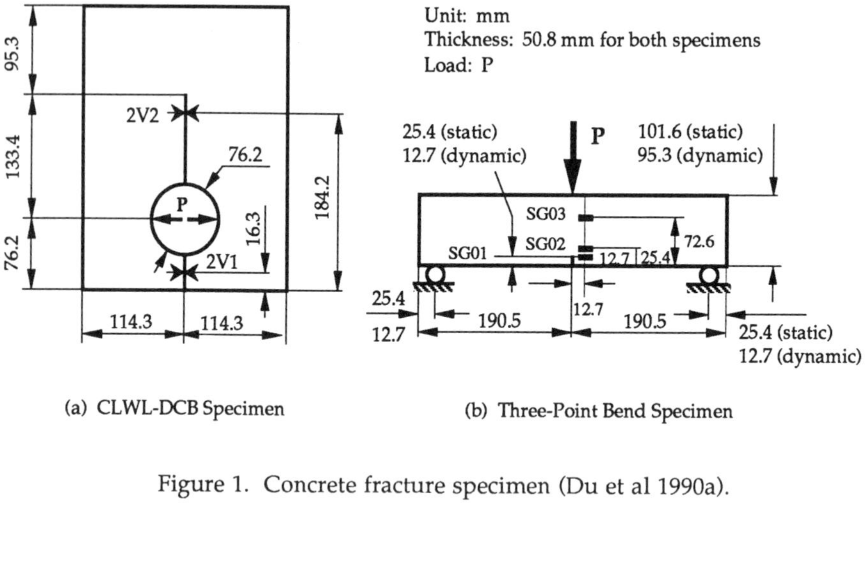

Figure 1. Concrete fracture specimen (Du et al 1990a).

2.2 Experimental procedure

Table 1 shows the test matrix used in this analysis. Further details of the experimental procedures are given in Du et al (1990a, 1990b) and Yon et al (submitted). Static loading was applied in a displacement controlled loading fixture. Dynamic loads were applied by a drop-weight loading system to four three-point bend specimens and by a displacement controlled loading system to three-point bend and CLWL-DCB specimens. For the static and drop-weight tests, moire interferometry was used to measure the displacements, perpendicular to the plane of the crack, in the FPZ and the region surrounding the crack. A fringe multiplication of 2X was achieved by interfering the first order diffracted light beam from the specimen grating of 600 lines/mm with a reference grating of 1200 lines/mm. Moire fringe patterns associated with static crack extension were recorded with a motor driven 35 mm camera. Four moire fringe patterns associated with a rapidly propagating crack were recorded by a specially configured IMACON 790 camera at a rate of 100,000 fps with an exposure time of 2 μs per frame. The development of the FPZ with crack extension was studied by varying the trigger delay time in four specimens with identical loading.

Crack mouth and crack tip opening displacements, $2V_1$ and $2V_2$, respectively, as well as the wedge load, were measured at the locations shown in Figure 1 for the CLWL specimens. Load and load-line displacements were measured in the static, and the displacement controlled dynamically loaded three-point bend specimens. In addition, strains were measured at three locations in the displacement-controlled dynamic tests of the three-point bend and CLWL-DCB specimens.

Table 1. Test Matrix

Specimen	Nos.	Loading	Instrumentation	FPZ
CLWL-DCB	2	Static	Moire interfer.	Nonsingular (Du et al 1990a) Singular (this paper).
3-pt. Bend	2	Static	Moire interfer.	Nonsingular (Du et al 1990a) Singular (this paper).
3-pt. Bend	4	Dynamic Drop wt.	Moire interfer.	Nonsingular and Singular (Yon et al submitted).
CLWL-DCB	9	Dynamic Displ. control	Strain gages	Nonsingular and Singular (Yon et al submitted).
3-pt. Bend	10	Dynamic Displ. control	Strain gage	Nonsingular and Singular (Yon et al submitted).

2.3 Numerical procedure

For static fracture, unlike the previous analyses of Du et al (1989, 1990a), a direct procedure was used to determine the crack closure stress (CCS) versus COD relation. A faired COD curve, which was fitted through the moire interferometry data, was used as the input boundary condition to the finite element models of the statically loaded three-point bend and CLWL-DCB specimens. A similar procedure, minus the fitted COD data, was used in a dynamic analysis of the three-point bend and CLWL-DCB specimens.

The inverse method employed for processing the dynamic data used, as input conditions, the impact load and an assumed CCS versus COD relation. The measured and computed COD's, as well as the measured and computed transient strains at three locations along the crack path, were matched through a tedious iteration scheme in this propagation analysis (Kobayashi, 1979) of the drop-weight loaded, three-point bend specimens. The same procedure, minus the COD measurements, was used to evaluate the results of the displacement-controlled three-point bend and CLWL-DCB specimens. Details of the numerical procedure used in this dynamic analysis are given in Yon et al (submitted).

2.4 Experimental results

Figure 2 shows typical moire fringe patterns from static and dynamic tests of three-point bend specimens.

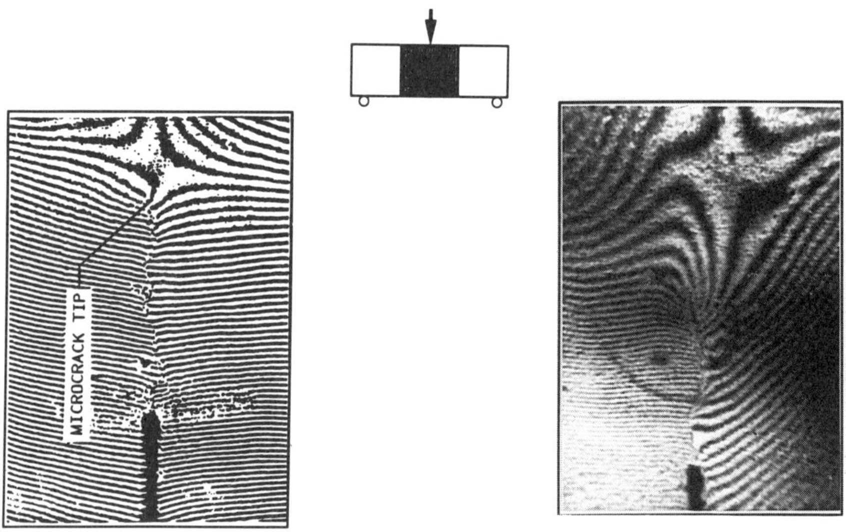

Stable Crack Growth   Dynamic Crack Growth

Figure 2. Moire interferometry fringe patterns for statically and dynamically loaded three-point bend specimens.

Figures 3 and 4 show typical measured and computed COD's for statically loaded CLWL-DCB and three-point bend specimens, respectively. The slight reverse curvature near the crack tip indicates the influence of the FPZ. Figure 5 shows the measured and computed COD's in impact loaded three point bend specimens. At the strain rate of 13.5 sec^{-1}, the influence of the FPZ was overwhelmed by the stress intensity factor at the crack tip resulting in a parabolic shaped profile at that tip.

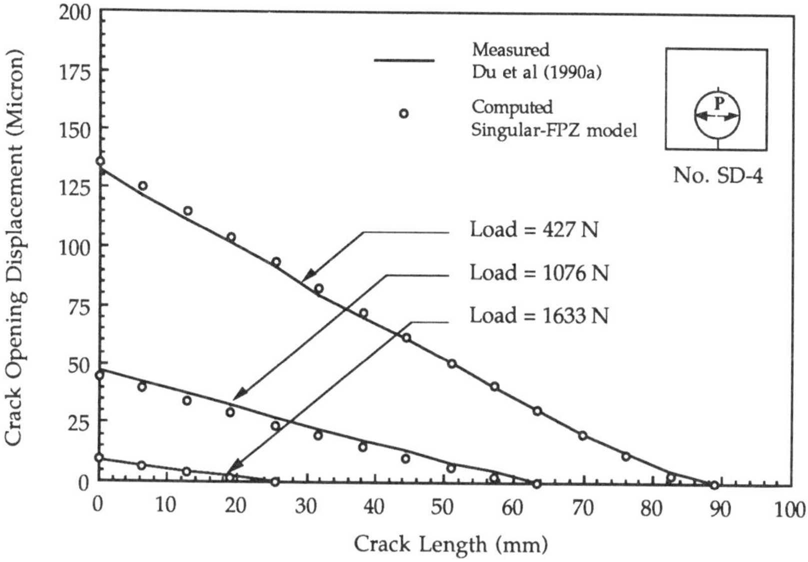

Figure 3. COD in statically loaded specimen CLWL-DCB.

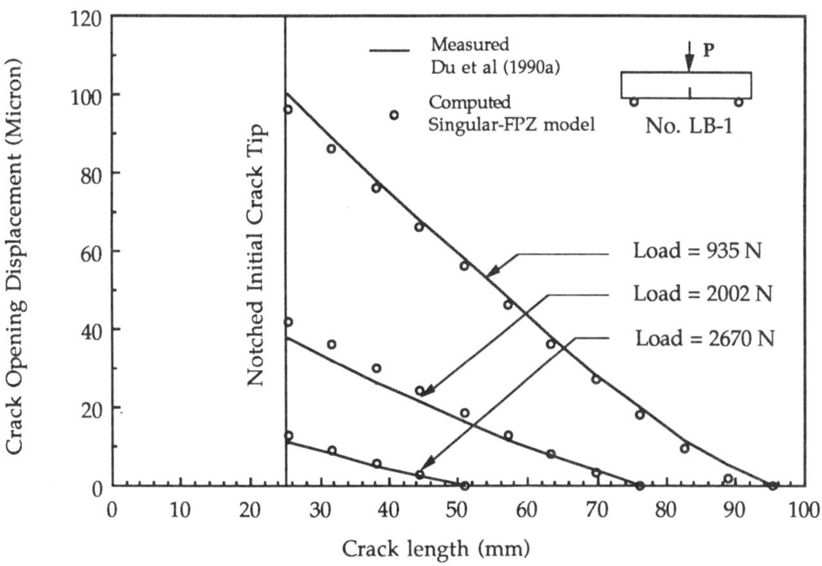

Figure 4. COD in statically loaded three-point bend specimens.

Figure 6 shows a typical measured and computed load versus load-line displacement relation for a statically loaded three-point bend specimen.

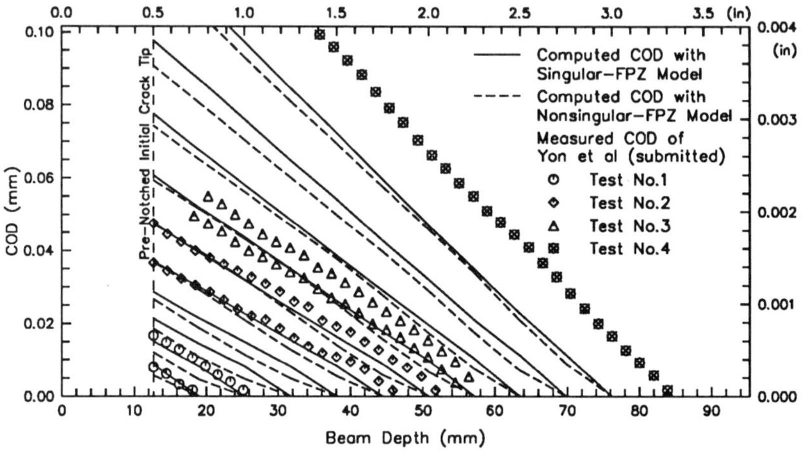

Figure 5. COD in dynamically loaded three-point bend specimens.

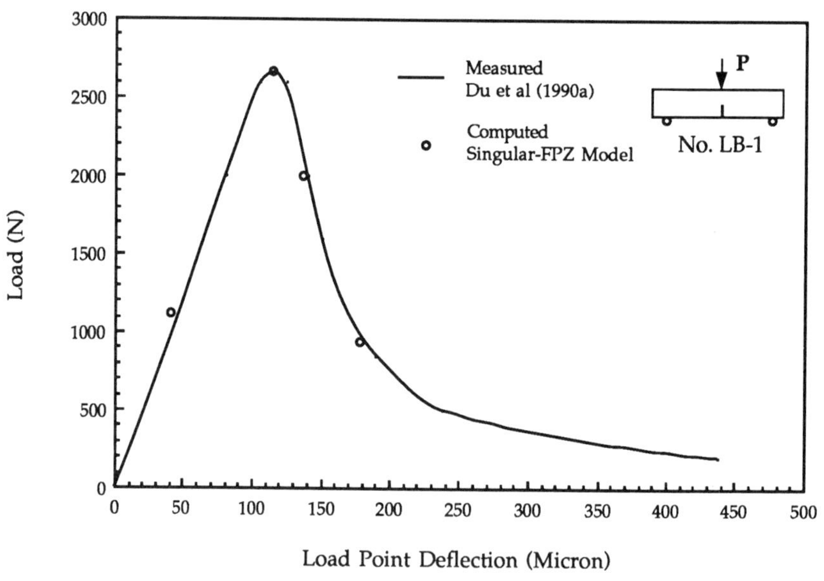

Figure 6. Load versus load-line deflection of statically loaded, three-point bend specimen.

Figure 7 shows the corresponding relations for typical dynamically loaded three-point bend and CLWL-DCB specimens. Only the displacement-controlled dynamically loaded specimens showed the same strain softening relation as that seen in the statically loaded specimen.

Figure 8 shows typical measured and computed strain records for a relatively rapidly loaded test specimen. The three strain gages were bonded at 12.7 mm offsets from the center line of the beam and these gages, identified at SG01, SG02 and SG03 in Fig. 1b, were located at 12.7 mm, 25.4 mm and 76.2 mm from the bottom of the beam, respectively.

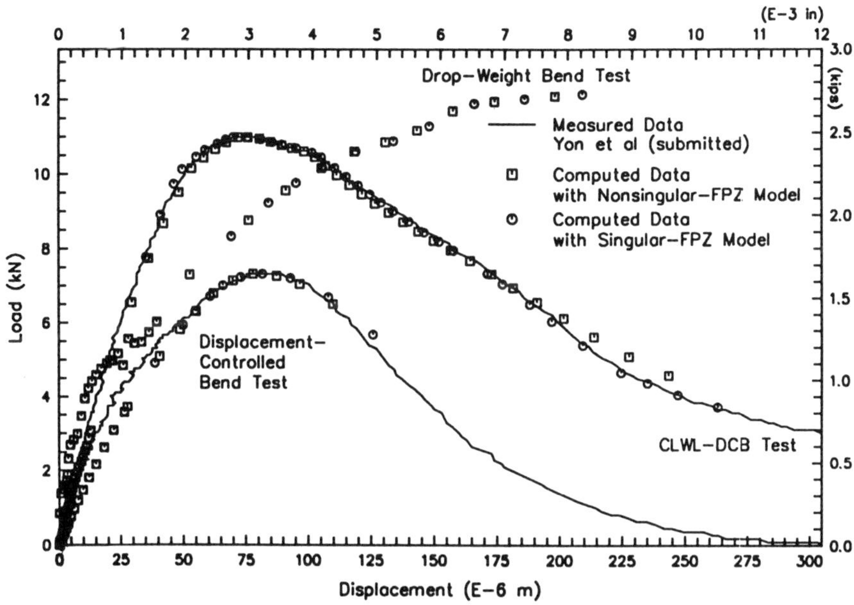

Figure 7. Load versus load-line displacement of dynamically loaded three-point bend and CLWL-DCB specimens

Also shown in Figure 8 is the loading history. Computation was carried out with and without the embedded K_I and the results are designated as the singular-FPZ and the nonsingular-FPZ models, respectively. The computed strains were relatively insensitive to the drastic differences in crack tip modelling, i.e. non-singular versus singular, even though those strains were at widely separated locations along the crack path.

Figure 9 shows the CCS versus COD relations gleaned from the static and dynamic tests of concrete fracture specimens. The maximum strain rates were approximately 12 sec^{-1} for the dynamic CLWL-DCB test and 4 sec^{-1} for the dynamic displacement controlled three-point bend test, respectively. The CCS versus COD relation tends to become stiffer, and the COD at zero CCS tends to decrease with increasing strain rate of loading.

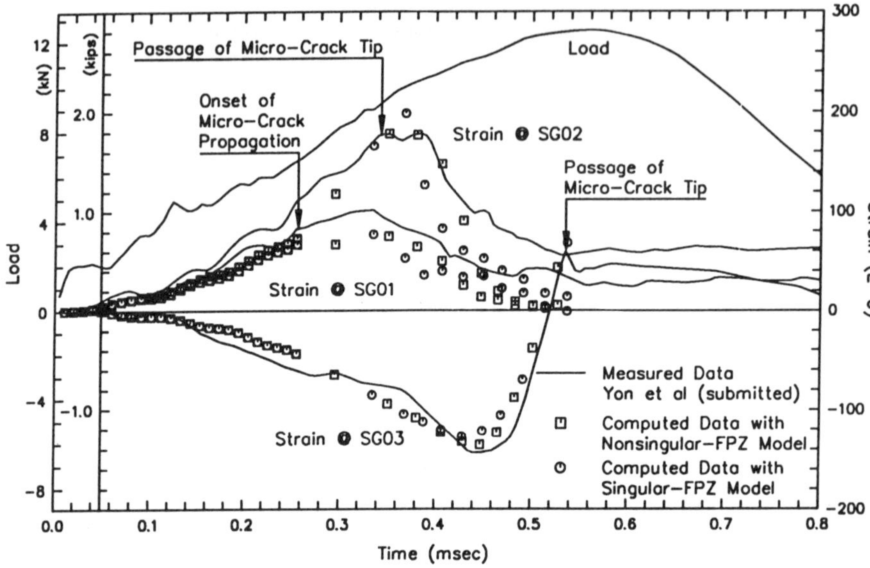

Figure 8. Strain histories for dynamic test on a drop-weight,three-point bend specimen

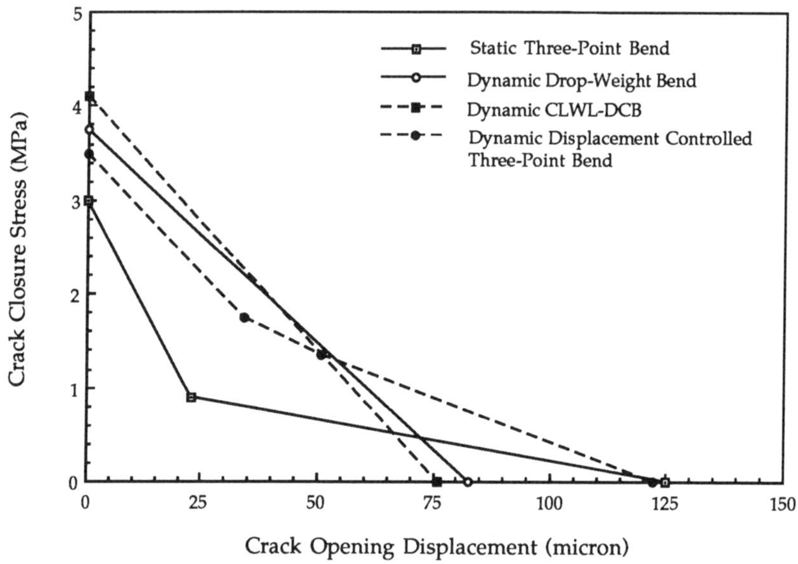

Figure 9. CCS versus COD relation.

Figure 10 shows the increase in energy release rate and energy dissipation rate with increasing crack length for the static and dynamic tests of three-point bend specimens. These rising resistance curves indicate the gradual development of the FPZ with full build-up achieved when the energy rate levels out. The constant differences between the released and dissipated energies for the static and dynamic tests can be attributed to the energy dissipated with micro-cracking in the frontal process zone preceding the extending crack tip.

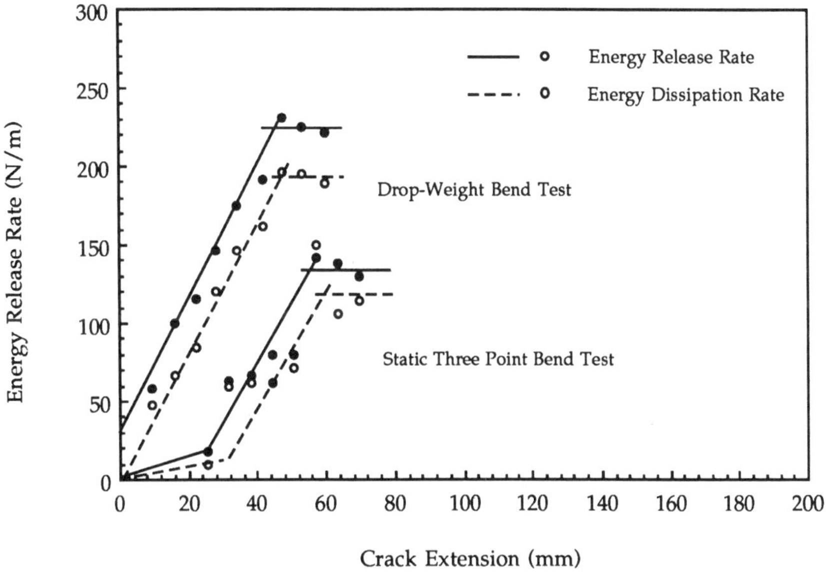

Figure 10. Energy release and dissipation rates for the FPZ in static and drop-weight loaded three-point bend specimen

3 Conclusions

Moire interferometry was used to measure the COD's of statically and dynamically extending cracks in three-point and CLWL-DCB concrete fracture specimens. The COD data was then used in a hybrid experimental-numerical procedure to deduce the CCS versus COD relations which governed the FPZ of the extending crack. The energy release rates and dissipation rates were also obtained. The same procedure minus the COD data was used to deduce the CCS versus COD relations for displacement-controlled, dynamic loading of three-point bend and CLWL-DCB specimens.

Strain, load and COD data are not sensitive to the difference in singular versus non-singular characterization of the FPZ. The COD data in the vicinity of the crack tip, however, are distinctly different for the singular and non-singular FPZ. Only the singular model could realistically model the crack tip profile observed in the tests.

The static and dynamic energy dissipation rates closely followed the related energy release rates thus indicating that the FPZ is the dominant fracture energy dissipation mechanism in this brittle material. In contrast, the energy dissipated at the frontal process zone due to micro-cracking, which is relatively significant during the initial phase of crack extension, is at the most 20 percent of the total release energy for a fully developed FPZ.

4 Acknowledgement

Much of this research was supported by the US Air Force of Scientific Research Grant AFOSR 86-0204. Z.-K. Guo was supported by the National Science Foundation Grant MSM-8922877. The authors wish to thank Dr. Spencer Wu, AFOSR and Dr. Ken Chong, NSF for their continuing support and encouragement.

5 References

Du, J.J., Kobayashi, A.S. and Hawkins, N.M. (1989). Fracture Process Zone of a Concrete Fracture Specimen in Fracture of Concrete and Rock (eds. S.P. Shah and S.E. Swartz), Springer-Verlag, 199-204.

Du, J.J., Kobayashi, A.S. and Hawkins, N.M. (1990a). An Experimental-Numerical Analysis of Fracture Process Zone in Concrete Fracture Specimens. Engineering Fracture Mechanics, 35, 1/2/3, 15-27.

Du, J., Yon, J.H., Hawkins, N.M., Arakawa,K. and Kobayashi, A.S. (1990b). Analysis of The Fracture Process Zone of a Propagating Concrete Crack Using Moire Interferometry, in Micromechanics of Failure of Quasi-Brittle Materials (eds. S.P. Shah, S.E. Swartz and M.L. Wang), 146-155.

Hillerborg, A., Modeer, M. and Petersson, P.E. (1976). Analysis of Crack Formation and Crack Growth in Concrete by Means of Fracture Mechanics and Finite Elements. Cement and Concrete Research, 6, 773-782.

Jenq, Y.-S., and Shah, S.P. (1985). Two Parameter Fracture Model for Concrete. Journal of Engineering Mechanics, ASCE, 110, 1227-1241.

Kobayashi, A.S. (1979). Dynamic Analysis by Dynamic Finite Element Method - Generation and Propagation Analyses in Nonlinear and Dynamic Fracture Mechanics (eds N. Perrone and S.N. Atluri), ASME AMD-35, 19-36.

Miller, R.A., Shah,S.P. and Bjelkhagen, H.I. (1988). Crack Profiles in Mortar Measured by Holographic Interferometry. Experimental Mechanics, 28, 4, 388-394.

Yon, J.-H., Hawkins, N.M. and Kobayashi, A. S. (submitted) Numerical Simulation of Mode I Dynamic Fracture of Concrete. ASCE Journal of Engineering Mechanics.

ASK/cm/11/19/90

43 INTERNAL INSTABILITY IN BRICK ELEMENTS: NUMERICAL AND EXPERIMENTAL RESULTS

I. IORI and G. MIRABELLA ROBERTI
D.I.S., Politecnico di Milano, Italy

Abstract
The topic of the internal instability is now being actively pursued in the field of fracture mechanics. In the present work, following some earlier experimental and theorical researches, the behavior of brick elements is studied mainly with regard to a possible presence of internal instability, or to the sudden and uncontrolled developement of a crack, regardless of the stiffness of the structural elements to wich the brick is connected. By means of an appropriate experimental method (optical interferometry), some point-defined quantities (for example, fracture energy G_f), that are related to the displacement field in the vicinity of the crack tip, are determined.

In the same way, varying the notch height and the brick quality in direct tension and three point bending tests, it has also been possible to outline the role played by the different quantities (besides the fracture energy), that is, the specimen height, the ultimate tension strength and the notch height. An appropriate non-linear numerical analysis based on a cohesive crack model has shown a good agreement with the experimental results, expressed, for example, in terms of the relationship between the applied force F and the total displacement $\bar{u}$ at the specimen boundaries.

Keywords: Crack Propagation, Cohesive Crack Model, Finite Elements, Internal Instability.

1 Introduction

The investigation on the mode of crack propagation (stable vs unstable) is growing in interest in the field of fracture mechanics of brittle materials. This topic can be regarded as an aspect of the so-called internal structural instability, that arises when it is impossible to record the falling branch of a load-displacement curve, despite of the testing-device stiffness.

The question if the crack will develop in a stable or unstable way seems to be the most pressing in the fracture process of concrete-like materials, even if it has been investigated from a lot of time in the more general contest of elastic-plastic stability (Maier, 1968).

The reason is that although some parameters, as the "energetic brittleness number", s_E, could give some good information on the fracture behaviour of a structural element, more investigations are wished when these parameters cannot be derived from the hypothesis of linear fracture mechanics. In fact, in concrete-like materials (and also in the brick, see Cedolin et al.1987) the

Fracture Processes in Concrete, Rock and Ceramics Edited by J.G.M. van Mier, J.G. Rots and A. Bakker.
 Published by E. & F.N. Spon, 2–6 Boundary Row, London SE1 8HN. ISBN 0 419 15870 7

size of the process zone at the crack tip can not be neglected with respect to the structure size. So, in that case, linear fracture mechanics cannot apply (Cedolin et al., 1983, 1987a), and some deeper investigation mainly on the circumstances of internal instability development are needed.

The aim of the present work is to study the problem by means of an appropriate numerical approach, with reference to some particularly meaningful experimental results obtained in the past years by Cedolin et al. (1987); see also Iori and Mirabella (1989).

2 Internal instability in presence of cohesive forces

2.1 Classical solutions in a closed form

Two different constitutive laws are assumed for the uncracked material (Fig. 1a) and for the cracked one (Fig. 1b) in order to represent the mechanical behaviour of concrete-like materials in an effective but also correct way.

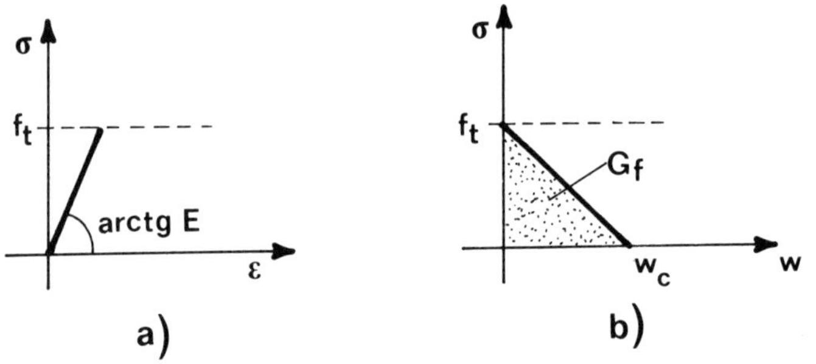

Fig. 1.

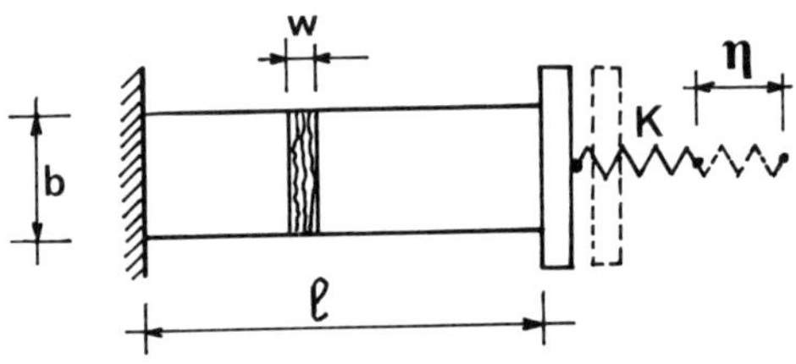

Fig. 2.

In some simple structural geometries, internal instability can be easily investigated by means of closed form solutions. Following Maier (1968), in the specimen of Fig. 2, subjected to uniform tension, internal instability takes place when:

$$(\delta^2 W_p)_{k=\infty} < 0 \tag{1}$$

i.e. when the second variation of the plastic work W_p is less than zero for any virtual growth δw of a single concentrated crack. In that system, the spring with constant k represents for example the testing device stiffness; the expression (1) is obtained taking the value of k to infinity. If the time t is assumed as the ordering variable, the second variation of the plastic work, for a generic value of k, can be written as:

$$(\delta^2 W_p)_k = 1/2\ (\dot{N} - \dot{N}_c)\ \dot{w}\ \delta^2 t) \tag{2}$$

where $N = (f_t - f_t \frac{w}{w_c})\ b\ s,$ $N_c = - \frac{w}{\frac{1}{k} + \frac{l}{Ebs}}.$

being ft the tension strength, E the Young modulus, b the depth of the specimen and s the thickness; the dot represents derivation with respect to the time.

So, assuming $k = \infty$ in (2), the condition (1) of internal instability of the specimen can be written as:

$$[- \frac{ft}{wc} bs - (- \frac{Ebs}{l})]\ \dot{w} < 0 \tag{3}$$

i.e.

$$f_t\ l > E\ w_c \tag{4}$$

Nevertheless, the theory above mentioned is more difficult to apply in structural elements with a more complicated shape, as for example that in Fig. 3a,3b. In that cases the representation of the crack development is more questionable and so is the computation of the crack opening, wich is not anymore uniform along the height of the specimen: a comparison with the results of an accurate experimental analysis is strongly needed (Iori, 1989).

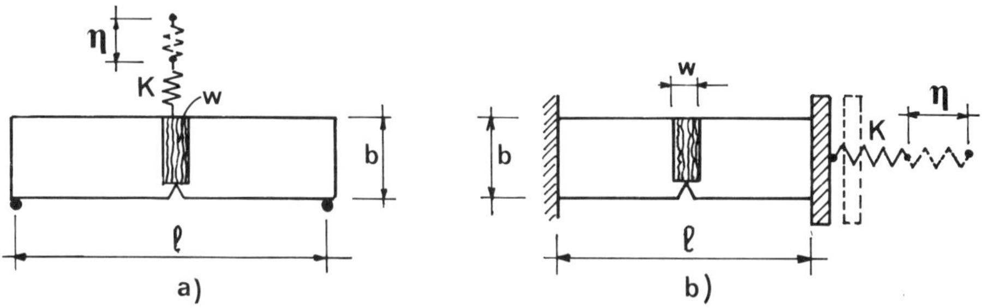

Fig. 3.

2.2 Numerical simulations

Numerical approaches allow one to study structural shapes of any complexity. The choice of the numerical method is tied to the particular performances that are needed in the specific investigation that is pursued. For example, when the crack localization cannot be stated in advance by simple considerations, then an analysis by the boundary elements method seems to be very effective (Cen and Maier, to appear). On the other side, when the crack localization can be recognized very accurately, numerical simulation can be performed by the classical finite elements method, provided that an appropriate fictitious crack model is adopted; that is, the whole process zone could be represented by a single crack.

In the present work a FE computer program was employed, capable of taking into account the non-linear characteristics of the mechanical behaviour as well as the effects of geometrical variation along the loading process (Hibbit et al., 1987). The FE mesh adopted in the analysis is represented in Fig. 4, for one half of the specimen of Fig. 5.

The Fictitious Cohesive Crack Model (Hillerborg, 1976) was adopted: in other words, is assumed that the process zone begin to develop locally only when the maximum principal stress attains the direct tension strength f_t; then the material is yet able to carry some stress, until $w < w_c$, Fig 1b.

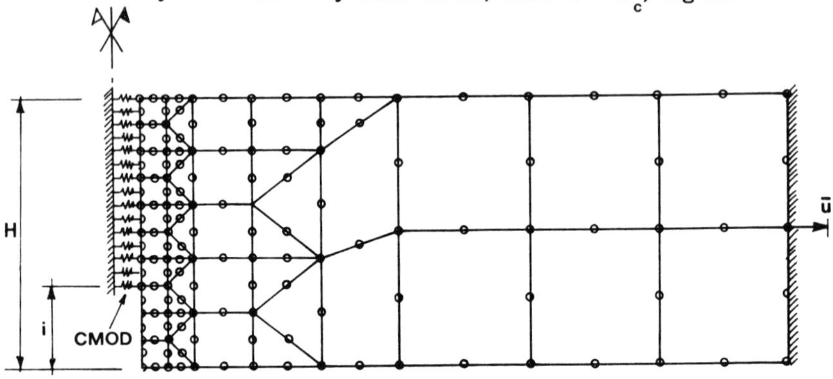

Fig. 4.

The nonlinear relationship between the nodal displacements {u} and the equivalent nodal forces {P} = [K] {u} was studied by means of an incremental tool called "modified Riks method" (Ramm, 1981), wich if needed allows one to follow also the falling branch with positive slope of the system response in the case of internal instability. An "equilibrium path" can be so determined, in the space of the n nodal variables {u} and the load multiplier λ. The increments along the path are calculated, in order to satisfy the equilibrium, taking as unknown both the increments of {u} and of λ.

The process zone was represented by means of elastic boundary elements with non-linear behaviour: a very high initial stiffness (able to impose an uniform state of stress for a constant displacement) is followed by a negative stiffness, that keeps a constant value until the critical value w_c for the crack opening is reached. The specimen heads were subjected to an uniform displacement in the x direction (Fig. 5): the sign of displacement is checked by the computer program automatically, allowing one to find the system equilibrium even when a decrease of loading is associated to a decreasing displacement.

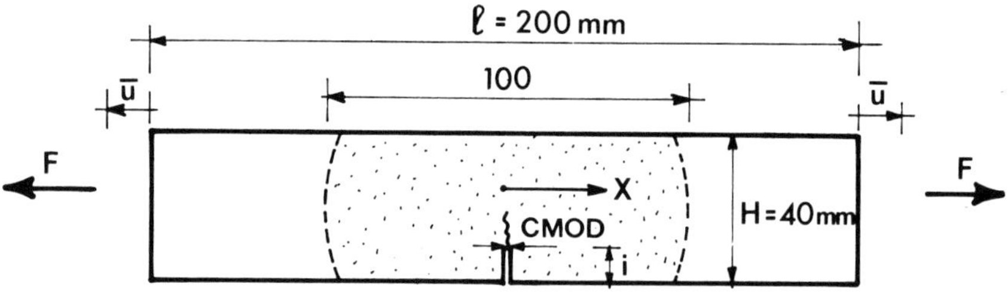

Fig. 5.

Fig. 6.

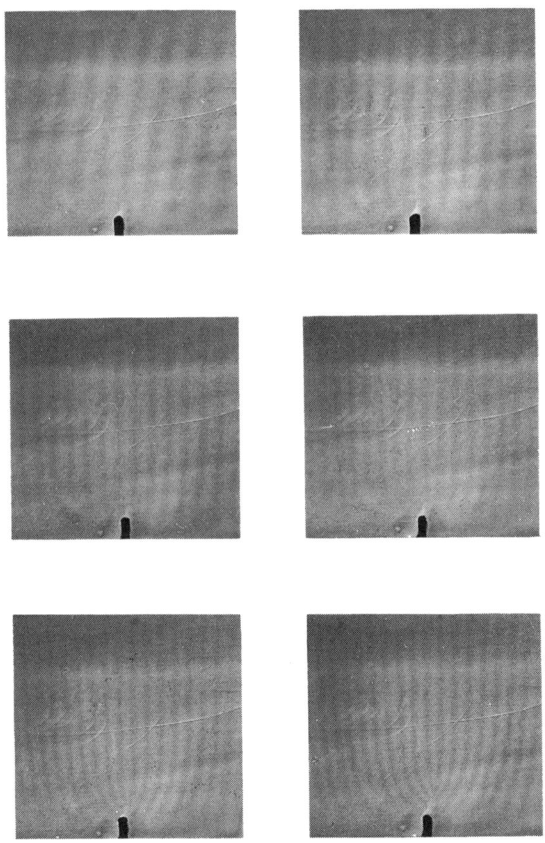

Fig. 7.

3 Experimental and numerical results

For two different types of bricks (named **A** and **B**) the constitutive relationships of Fig. 1 were assumed. The geometrical characteristics of the examined specimens are illustrated in Fig. 5, both for the numerical and experimental approach. For type **B** brick critical values of stress intensity factor were observed, in dependence of the size of the specimen; this does not happen with brick type **A**. So the true specific fracture energy G_f was determined (Cedolin et al., 1987b), by means of size effect law (Bažant, 1984).

That energy was also valued by means of an holographic method by laser light (Iori et al., 1982, and Cedolin et al.,1987a, see also Fig. 6) that has the main scope of studying the local strain development with the crack opening. For example, in Fig. 7 the moiré fringes at different applied load levels **F** are shown for a type **A** brick specimen, with a notch depth of 4 mm (i.e. 0.1 b). Trough the fringe density (Fu-Pen Chiang, 1979), the strain concentration at the crack tip can be recognized very well; for F = 1160 N, a little crack propagation can be noted, revealed by some fringe superposition.

In this way, the experimental behaviour of the specimen is easily determined (Iori and Mirabella,1989), both in terms of the imposed displacement $\bar{u}$ at the boundary and of the crack mouth opening displacement (CMOD) at the notch tip. Moreover, the experimental analysis so carried out allows one to specify the w_c value, that is of basic importance in the determination of the material constitutive relationship (Cedolin et al., 1987b).

For the bricks mentioned above some other direct-tension tests were performed; the basic parameters for constitutive relationship specification are summarized in Tab. 1.

Table 1. Material characteristics

type	A	B	A*	B*
E [MPa]	23200	31200	23200	31200
G_f [N/m]	39.7	17.9	79.4	35.8
f_t [MPa]	5.39	8.93	5.39	8.93
w_c [mm]	0.012	0.004	0.024	0.008

In Fig. 8 the numerical results are reported for two different specimens, made in brick type **A** and **B** with a notch-to-depth ratio of 0.1 and of 0.5 respectively. The applied load is plotted vs imposed displacement $\bar{u}$ and vs the crack opening CMOD; the total external work and the total elastic energy are also plotted vs the imposed displacement $\bar{u}$. In Fig. 8b both the specimens in brick **B** show internal instability, even if the the deeper notch has the effect of diminishing the cumulated energy, and of nearly eliminating the internal instability problem. Fig. 8a, referred to type **A** brick, underlines the basic influence of the fracture energy G_f, that is in this case about two times the type **B** brick value. The internal instability appears only for the notch ratio of 0.1, although in a weaker way; on the other side, the notch ratio 0.5 shows a softening falling branch with negative slope.

These remarks are confirmed in Fig. 8c, if compared with Fig. 8d, by the

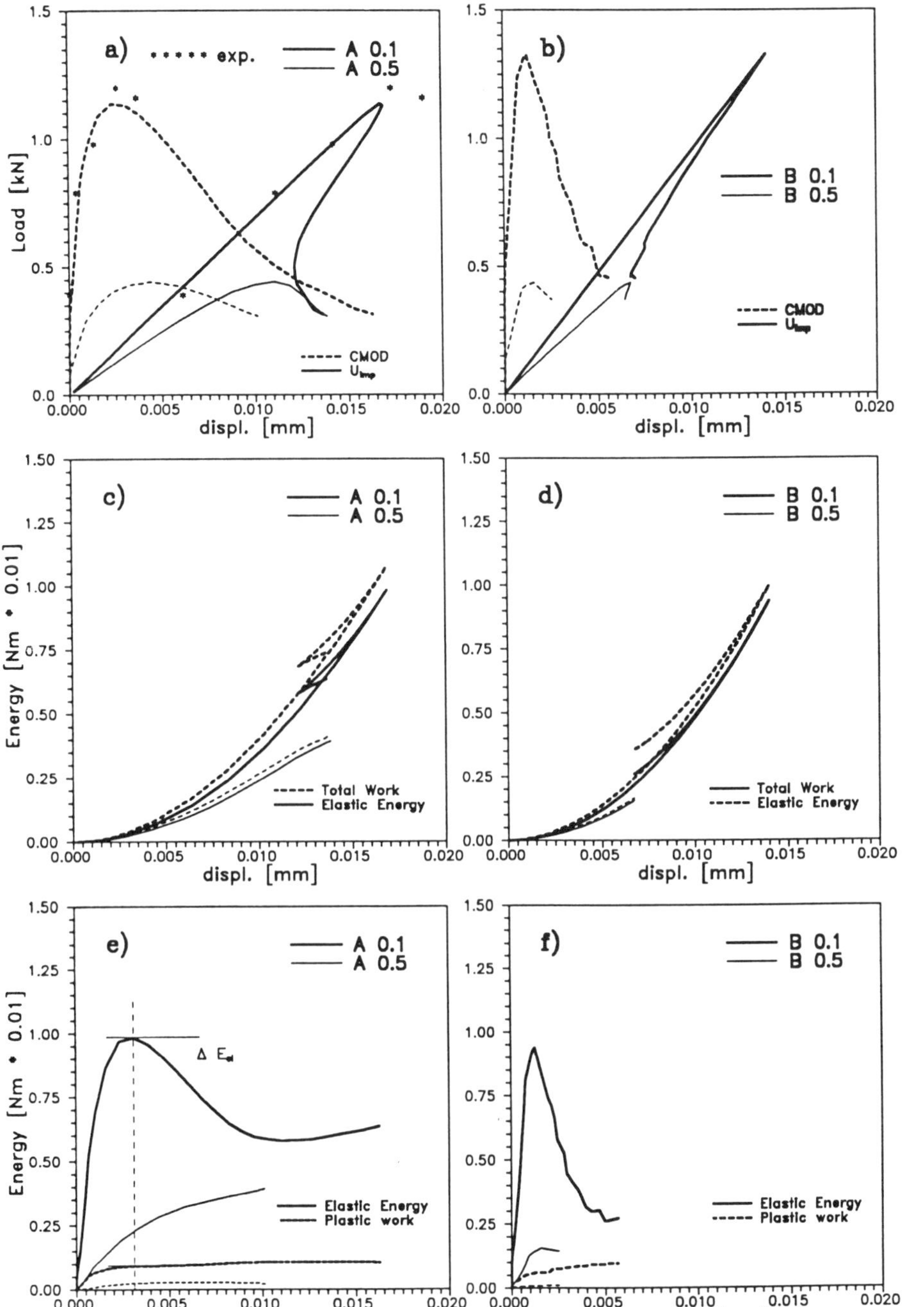

a)
exp.
A 0.1
A 0.5
CMOD
U_{imp}
Load [kN]
displ. [mm]
b)
B 0.1
B 0.5
CMOD
U_{imp}
displ. [mm]
c)
A 0.1
A 0.5
Total Work
Elastic Energy
Energy [Nm * 0.01]
displ. [mm]
d)
B 0.1
B 0.5
Total Work
Elastic Energy
displ. [mm]
e)
A 0.1
A 0.5
Δ E_{el}
Elastic Energy
Plastic work
Energy [Nm * 0.01]
CMOD [mm]
f)
B 0.1
B 0.5
Elastic Energy
Plastic work
CMOD [mm]

Fig. 8.

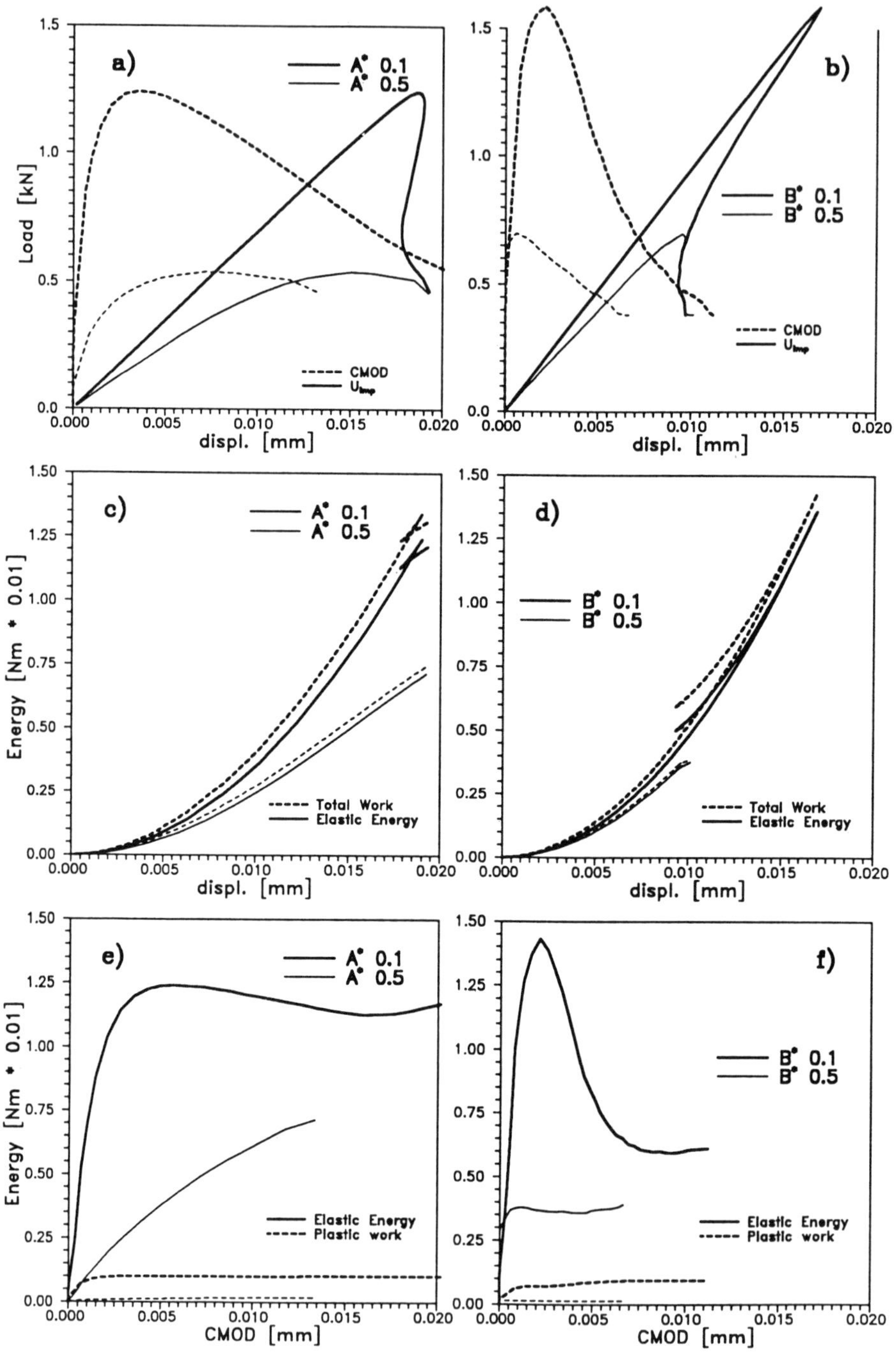
a)
A° 0.1
A° 0.5
Load [kN]
CMOD
U_{imp}
displ. [mm]
b)
B° 0.1
B° 0.5
CMOD
U_{imp}
displ. [mm]
c)
A° 0.1
A° 0.5
Energy [Nm * 0.01]
Total Work
Elastic Energy
displ. [mm]
d)
B° 0.1
B° 0.5
Total Work
Elastic Energy
displ. [mm]
e)
A° 0.1
A° 0.5
Energy [Nm * 0.01]
Elastic Energy
Plastic work
CMOD [mm]
f)
B° 0.1
B° 0.5
Elastic Energy
Plastic work
CMOD [mm]

Fig. 9.

greater amount of energy dissipated in the crack propagation process (i.e. the difference between the total work and the total elastic energy). Another confirmation can be found in Fig. 9, wich is referred to fictitious materials called **A*** and **B*** in Tab. 1, obtained by an increase of the fracture energy, assuming values of w_c double than that of type **A** and **B** bricks. The comparison of Fig 9a, 9b with the analogous Fig. 8a, 8b and similarly the comparison between Fig. 9c, 9d and Fig. 8c, 8d (wich display the total work and the elastic energy vs $\bar{u}$), show the important role played by the specific fracture energy G_f.

The plots in Fig. 8e, 8f and in Fig. 9e, 9f, are particularly meaningful in underlining this aspect. In these figures the elastic cumulated energy and the plastic work (dissipated along the cracking line) are plotted vs the crack opening displacement CMOD. When the internal instability develops it is well apparent that the elastic energy decrease ΔE_{el} is greater than the plastic work increment ΔW_p dissipated along the crack line. When instead the specimen is "internally stable" the elastic energy does not have any release, but is ever growing with the crack opening.

The numerical results here obtained are confirmed by the comparison with the experimental results: for a notch-to-depth ratio of 0.1 and brick type **A**, the numerical results of Fig. 8a are compared with the experimental ones shown in Fig. 7 for the same specimen. This comparison validates the remarks made above, and the capability of representing and defining the complex physical aspects of the problem by the experimental and numerical analysis performed.

4 Conclusions

The principal aim of the analysis performed was the study of crack propagation in notched brick elements subjected to tension and bending. The numerical model, and the particular hypothesis adopted, evidentiates the different role played by the principal mechanical quantities that characterize the problem. Among them, the basic importance of the material fracture energy G_f must be underlined, particularly with reference to the development of the so-called internal instability.

As shown above, the different values of fracture energy modifies the energy balance of the system, mainly for the value of elastic energy release, that is of basic importance in the development of internal instability.

The comparison between the numerical analysis and some earlier experimental works allowed us to individuate a physical phenomenon that have in his whole a wide complexity, both from the numerical and experimental point of view.

5 REFERENCES

BAZANT, Z.P. (1984) Size Effect in Blunt Fracture: Concrete, Rock, Metal, **J. Eng. Mechanics ASCE,** 110.

CARPINTERI, A. (1988), Linear Elastic Fracture Mechanics as a Limit-Case of Strain-Softening Instability, **Meccanica,** 23.

CEDOLIN, L., DEI POLI, S., IORI, I. (1983), Experimental Determination of the Fracture Process Zone in Concrete, **Cement and Concrete Res.,** 13.

CEDOLIN, L., DEI POLI, S., IORI, I. (1987a), Meccanica della frattura di laterizi utilizzati in collaborazione con elementi strutturali in conglome-

rato armato, **Studi e Ricerche**, 9, Politecnico di Milano, Milano.
CEDOLIN, L., DEI POLI, S., IORI, I. (1987b) Tensile Behavior of Concrete, **J. Eng. Mech. ASCE**, 113.
CEN, Z. and MAIER, G.(to appear), Bifurcation and Instabilities in Fracture Mechanics of Cohesive-Softening Media by Boundary Elements.
FU-PEN CHIANG (1979) Moiré Methods of Strain Analysis, in SESA's Manual on **Experimental Stress Analysis**.
HIBBIT, KARLSON and SORESEN (1987) **ABAQUS** - The Theory Manual.
HILLERBORG, A., MODEER, M., PETERSSON, P.E. (1976), Analysis of crack formation and crack growth in concrete by means of fracture mechanics and finite elements, **Cement Concrete Res.**, 6.
IORI I. (1989), Sull'instabilità interna di elementi in conglomerato,in **Atti del Convegno Nazionale AIMETA in ricordo di R.Baldacci e M.Capurso**.
IORI, I., LU, H., MAROZZI, C.A., PIZZINATO, E. (1982) Metodo per la determinazione dei campi di spostamento nei materiali eterogenei a bassa resistenza specifica a trazione, **L' Industria Italiana del Cemento**, 4.
IORI, I., MIRABELLA ROBERTI, G. (1989), Instabilità interna del laterizio: analisi numerico-sperimentale di elementi tenso-inflessi, **Studi e Ricerche**, 11, Politecnico di Milano, Milano.
MAIER G.(1968) Sul comportamento flessionale instabile nelle travi inflesse elastoplastiche, **Rendiconti**, Istituto Lombardo-Accademia di Scienze e Lettere, 102.
RAMM, E. (1981), Strategies for tracing the non-linear response near limit points, in **Nonlinear Finite Elements Analysis in Structural Mechanics**, Wunderlich W., Stein E. and Bathe K.J. Ed., Spriger Verlag, Berlin.
ZIENKIEWICZ, O.C. (1977), **The Finite Element Method**, Mc Graw Hill, Maidenhead.

44 MODE I FRACTURE TESTS AND ANALYSES OF CONCRETE

D.A. KHORDIJK
TNO Building and Construction Research, Rijswijk, The Netherlands
H.W. REINHARDT
Institute of Engineering Materials, Stuttgart University, Germany

Abstract
Based on experimental results of deformation-controlled uniaxial tensile tests, a constitutive model, called "continuous-function-model", for the tensile behaviour of plain concrete was proposed and presented in previous papers. In order to verify the validity of this model for mode I fracture of concrete, four-point bending tests and tensile tests on single-notched plates were performed. The results of these experiments have been simulated by three different numerical techniques, in which the continuous-function-model is used for the material behaviour. In this paper, the experimental results are presented and compared with predictions by the different numerical methods.
Keywords: Concrete, Mode I fracture, Constitutive model, Cyclic behaviour, Experiments, Numerical analyses

1 Introduction

Since it has been recognised that the tensile behaviour of concrete is a leading factor in many failure mechanisms of concrete, it has received very much attention among researchers. As a result, many publications deal with the determination of the tensile behaviour of concrete, especially its softening behaviour. So, also an extensive research programme was performed by the authors. Deformation-controlled uniaxial tensile tests on small concrete bars were performed in the Stevin Laboratory and based on a close inspection of the results, a constitutive model for the stress-crack opening relation of concrete was proposed. The model is called "continuous-function-model" (CFM) and covers the behaviour of a crack under monotonic increasing opening as well as under closing and subsequent re-opening. For the empirical equations of the model, the reader is referred to Hordijk and Reinhardt (1990a).

As usual in research, the model should be verified. A possible way for doing this is to investigate a structure experimentally and to see if its behaviour can be simulated with the model as material input. It is even better to investigate several structures, since the model has only then proven to be a proper description of the real material behaviour, if it can predict the behaviour of all structures in which the behaviour, described in the model, is active. In doing this, however, one should see that the structure under investigation predominantly fails under the mechanism that is modelled. Otherwise,

Fracture Processes in Concrete, Rock and Ceramics Edited by J.G.M. van Mier, J.G. Rots and A. Bakker.
 Published by E. & F.N. Spon, 2–6 Boundary Row, London SE1 8HN. ISBN 0 419 15870 7

there is a possibility that the simulation is blurred by improperly modelled other mechanisms.

In this paper, experimental results of different types of test in which mode I fracture is the leading mechanism, will be discussed and by several simulation techniques, the validity of the "continuous-function-model" will be verified. Here, it is only possible to show some principal results. The total cf the results can be found in Hordijk (1991).

2 Experiments

For verifying the validity of the continuous-function-model, it was chosen to perform four-point bending tests and uniaxial tensile tests on single notched plates. Furthermore, deformation-controlled uniaxial tensile tests were used to determine the material properties: tensile strength f_t, fracture energy G_F and Young's modulus E_o. The bending specimens (500mm*100mm*50mm) were provided with 10, 30 and 50 mm deep notches, respectively. A schematic representation of the test programme, including measured data and type of loading, can be seen in Fig. 1. The size of the specimens was limited due to the fact that the existing tensile loading equipment was applied. Details of the different tests, i.e. test rig, mix composition and curing procedure, can be found in Hordijk (1991).

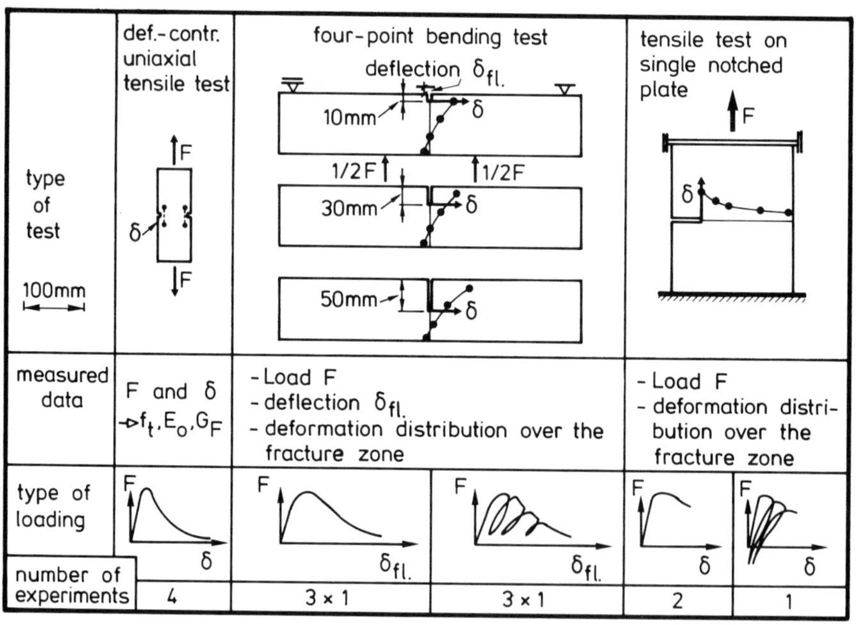

Fig. 1. Experimental programme.

A normalweight type of concrete with a maximum aggregate size of 8 mm was used. The age of the specimens at testing varied between 65 and 86 days. Standard cube compressive and splitting tests were performed at 28 days and 78 days. All specimens and cubes were taken from the same batch. The results obtained from the cube tests and the deformation-controlled uniaxial tensile tests are listed in Table 1. As far as these results are concerned, the following remarks can be made. In tensile tests, a very long tail for the σ-δ relation is generally found. Since this tail affects the absolute value of the fracture energy significantly (see Hordijk and Reinhardt 1990b), G_F was determined by cutting-off this tail via a described procedure. Furthermore, due to the notches and the relatively small specimen dimensions, the applied tensile test is not very suitable for the determination of the Young's modulus. Nevertheless, from the results a value of approximately 40000 MPa could be deduced. Finally, it should be noted that a significant scatter was found for the tensile strength and the fracture energy.

Table 1. Material parameters obtained from cube tests (150 mm^3) and deformation-controlled uniaxial tensile tests.

		28 days	at testing (65-86 days)
from cube tests:			
compressive strength	(MPa)	48.1 (0.8%)*	59.5 (2.6%)
splitting strength	(MPa)	2.96 (8.2%)	3.30 (6.9%)
from tensile tests:			
tensile strength	(MPa)		3.45 (15.2%)
fracture energy**	(J/m^2)		115 (18.7%)
Young's modulus	(MPa)		40000

* coefficient of variation in brackets
** data-points up to 120 μm and linear to an assumed δ_o of 160 μm

The tensile tests in this experimental programme were not used to determine the shape of the stress crack opening relation. Previous investigations showed the relation in the CFM to be a good description for this shape. In this respect, it can also be mentioned that in Hordijk (1991) it is demonstrated that for most types of concrete, the shape of the normalized stress-crack opening relation is almost the same. For the normalization of the curve, stress is divided by the tensile strength and crack opening is divided by a critical crack opening w_c ($=5.14\ G_F/f_t$).

3 Numerical methods

The experiments were simulated by several numerical methods. The most accurate and generally applied method is the simulation by a finite element programme. The four-point bending tests for monotonic, as well as post-peak cyclic loading, were analyzed with the DIANA code (Janssen 1990). The applied FE-idealization can be seen in Fig. 2. A discrete crack concept was applied by defining interface elements at the predefined crack path, which was equal to the axis of symmetry. In accordance with the measurements in the experiments, the deflection was determined by the vertical displacement of point A, while horizontal displacements at points 1 to 4 were used for comparison of deformation distributions over the fracture zone.

Fracture in the bending tests is a local phenomenon. Therefore, it is also possible to model only the fracture zone by a number of layers that are represented by springs and to apply simple equations, known from the bending theory, for the two beam halves. By applying this crude way of modelling, only a rough approximation of the real structural behaviour can be obtained. Nevertheless, it offers a possibility to gain insight in phenomena that may occur in such a test, in a simple and fast way. Based on this method, a so-called "multi-layer model" was developed and used to study the behaviour of the four-point bending beams under monotonic loading (for results see Hordijk 1991). There, also the results of a study on fatigue crack growth and fatigue failure with the same model, are presented.

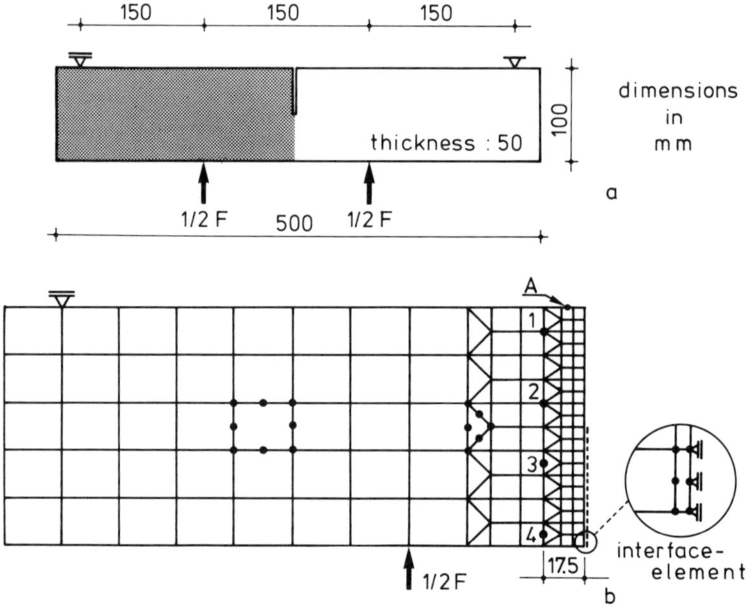

Fig. 2. FE-idealization for the four-point bending tests.

In the above discussed techniques, the experiments are simulated by modelling the specimen and the loading arrangement and by using the appropriate material parameters for the input. In that way, the experimental results and numerical analyses are only linked by the material parameters. If it is only intended to verify the material model, then it is also possible to use a part of the experimental results in the analyses. A method, based on this idea, is applied and is called "hybrid method". Starting point in the analyses is the measured deformation distribution over the fracture zone (see Fig. 3), which is known by the measurements of the four LVDTs. Then the critical cross-sectional area with a height h_c is subdivided in n layers with a height h_i. For each layer, the deformation δ_i can be calculated by linear interpolation between two LVDT-measurements and the corresponding stress σ_i results from the material model. Since in the CFM a stress-crack opening relation is described, an iterative procedure has to be applied in order to find the stress σ_i that belongs to a deformation δ_i for a measuring length of 35 mm. In the bending tests, the resulting stress distribution can be used in two different ways. First of all, it can be investigated whether the resulting normal force approaches zero and secondly, the calculated force F_{calc}, due to the resulting internal moment M, can be compared with the actually applied external load F. In the tests on single notched plates, it is only possible to compare the applied and calculated external load.

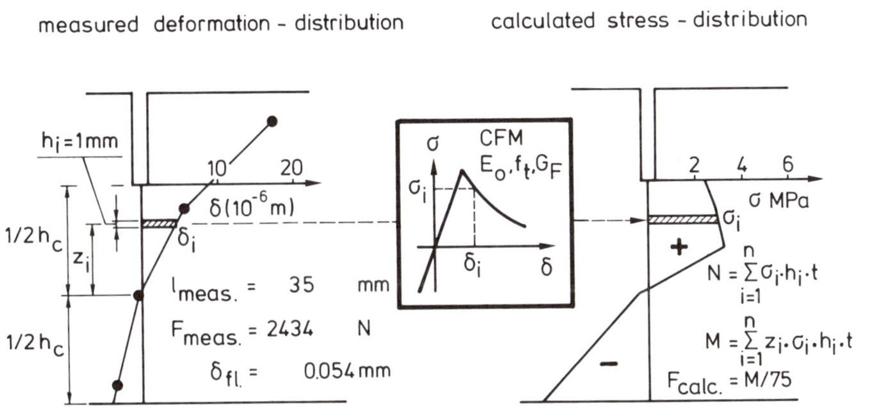

Fig. 3. Principles of the "Hybrid method".

4 Four-point bending tests

4.1 Fracture energy

The bending tests have also been used to calculate a value for the fracture energy. By applying a similar procedure for the tail of the load-deflection curve as done for the tensile tests, average G_F values of 104, 115 and 115 J/m² were found for the beams with 10, 30 and 50 mm notch depth, respectively. The agreement with the value obtained from the tensile tests, 115 J/m², is remarkably good.

4.2 FE-analyses

For the FE-analyses, the input parameters were kept the same for the three different beams. First of all, the input parameters were chosen closely related to the experimental determined values, i.e. tensile strength is 3.3 MPa, Young's modulus is 40000 MPa and fracture energy is 120 J/m². Though, in general, the curves were in rather good agreement with those experimentally determined, the maximum load was slightly overestimated for all three beams. The overestimation was 17, 7 and 5 % for the beams with a 10, 30 and 50 mm deep notch, respectively. Then the input parameters were slightly adjusted in order to obtain an even better prediction. For these analyses, the parameters were chosen as: tensile strength f_t=3.0 MPa, Young's modulus E_o=38000 MPa and fracture energy G_F=125 J/m².

A comparison between experimental and numerical results for the beam with a 30 mm deep notch can be seen in Fig. 4. The prediction of the load-deflection curve up to a deflection of about 0.25 mm is very good. For larger deflections, the load appeared to be underestimated by the FE-analysis. A similar result was also found for the other beams and can be explained by the fact that the long tail of the σ-w relation, as observed in experiments, was not modelled in the continuous-function-model. For most FE analyses, especially

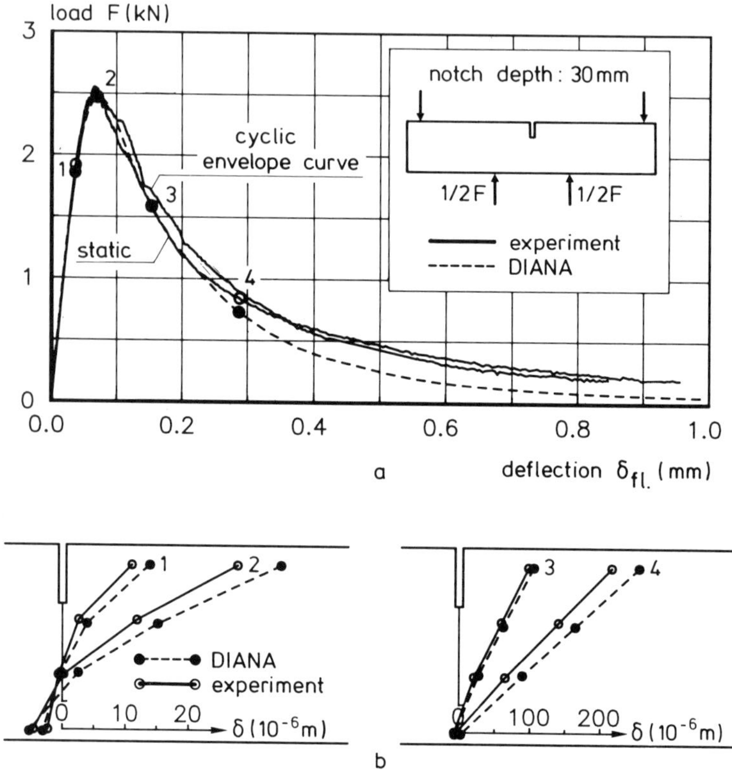

Fig. 4. Comparison between results of experiments and FE-analyses for the beams with a 30 mm notch depth; Load-deflection relations (a) and deformation distributions (b).

those in which it is only intended to predict the peak load, the omission of this part of the curve is not important. In these tests, that were loaded till large crack openings, the influence becomes perceptible. It can be imagined that an increasing stress-transferring capacity at larger crack openings will also lead to an increasing load for larger deflections. By a few analyses with the above mentioned multi-layer model, it was demonstrated that a proper description of the tail of the σ-w relation, indeed results in a load increase for larger deflections. In Fig. 4b, deformation distributions for four points in the F-δ_{fl} relation are plotted. Again the agreement between the measured and calculated results is good. Finally, a result for cyclic loading in the post-peak region is plotted in Fig. 5.

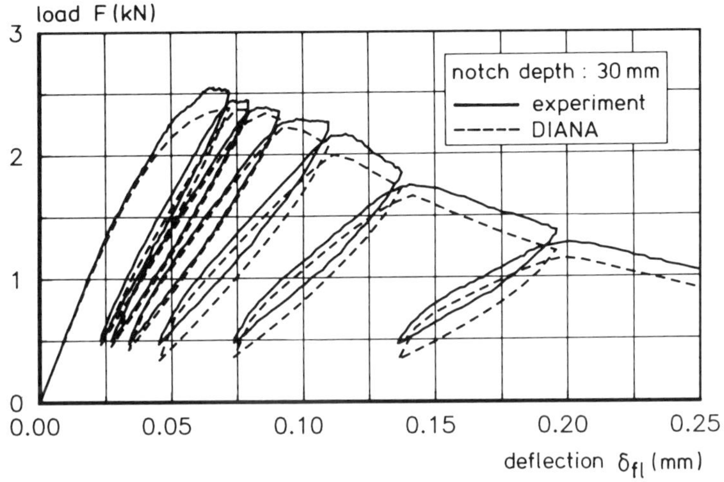

Fig. 5. Prediction of the load-deflection relation of a four-point bending beam under post-peak cyclic loading.

4.3 Hybrid method

In the analyses with the hybrid method, the material properties were chosen equal to those experimentally determined (Table 1) and the cross-sectional area was subdivided in layers of 1 mm (h_i=1 mm). A typical result, which was obtained for the beam with a notch depth of 30 mm, is plotted in Fig. 6. In Fig. 6a it can be seen that the calculated resulting normal force is almost zero up to a deflection of about 0.15 mm, whereafter this value becomes negative. The calculated external load also shows very good agreement with the measured load up to a deflection of 0.15 mm, whereafter the load is overestimated (Fig. 6b). It was surmised that this discrepancy was due to the fact that the applied method is not valid for compression, rather than due to improper tensile modelling. Therefore, the analysis was repeated, while only tensile stresses where taken into account. For the compressive zone, a triangular stress distribution ($N^- = -N^+$) was assumed. In Fig. 7, it can be seen that up to a deflection of 0.2 mm, the calculated load is equal to the measured load, whereafter the calculated load is well below the measured one. The similarity with the results of the FE-analyses is clear. In a second analyses, the stress-crack opening relation in the CFM was extended with a linear relation for larger crack openings. This linear relation fits the experimental results from deformation-controlled uniaxial tensile tests. It can be seen that, by this extension, the calculated load for larger deflections, approaches the measured load rather well.

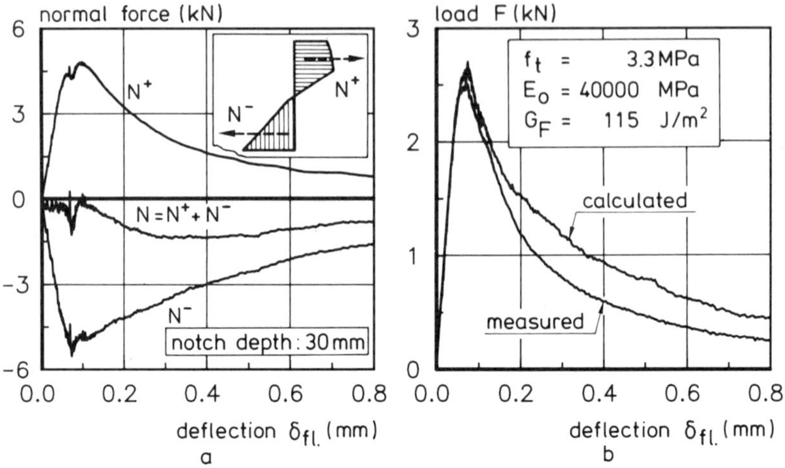

Fig. 6. A typical result for a four-point bending beam obtained with the hybrid method.

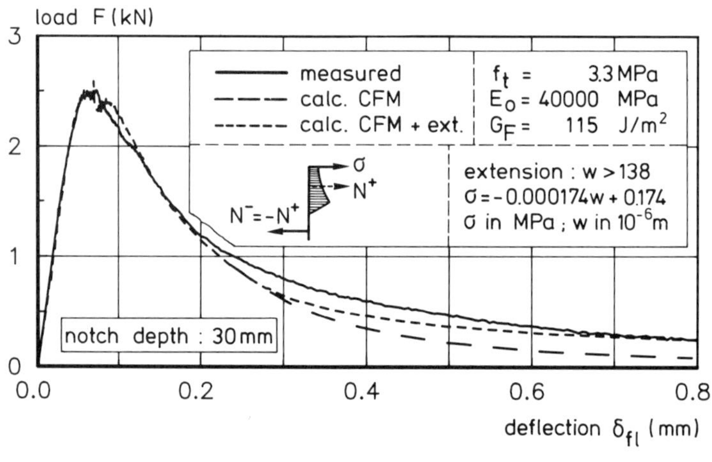

Fig. 7. Results obtained with the hybrid method, while only tensile stresses in the fracture zone have been taken into account.

5 Tensile tests on single notched plates

The tensile experiments on single notched plates were only simulated with the multi-layer model and the hybrid method. The tests were performed in a deformation-controlled way. Nevertheless, it was not possible to load the specimen further than an average deformation (measuring length: 35 mm) equal to about 40 μm, whereafter the test suddenly stopped. This phenomenon could be explained by the existence of a snap-back for the deformation that was used as feed-back signal. Compared to the bending tests, a more significant nonlinearity in deformation distributions occurs in these experiments, especially around peak load (Fig. 8). As a result, compared to uniform crack openings, a much lower peak load is found and the post-peak curve is much more flat. Again the external load could very well be predicted

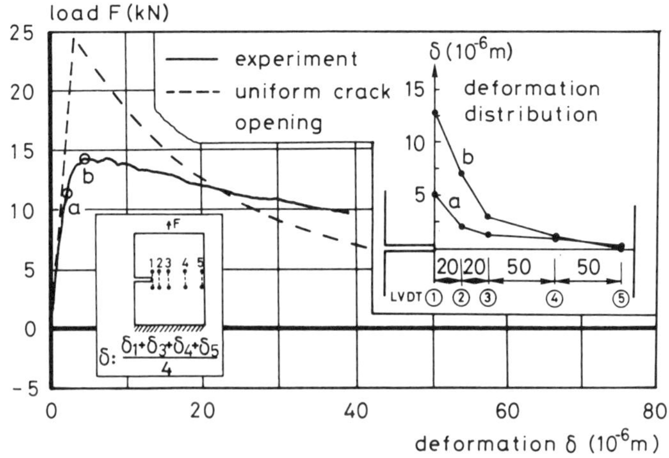

Fig. 8. Experimental results from a tensile test on a single notched plate and the F-δ relation in case of a uniform crack opening.

by the hybrid method (see Fig. 9). Also the agreement between the measured and calculated load during an unloading-reloading loop appeared to be very good. In these analyses, it was not necessary to use the extension for the CFM since the maximum deformations did not exceed a value of 80 μm.

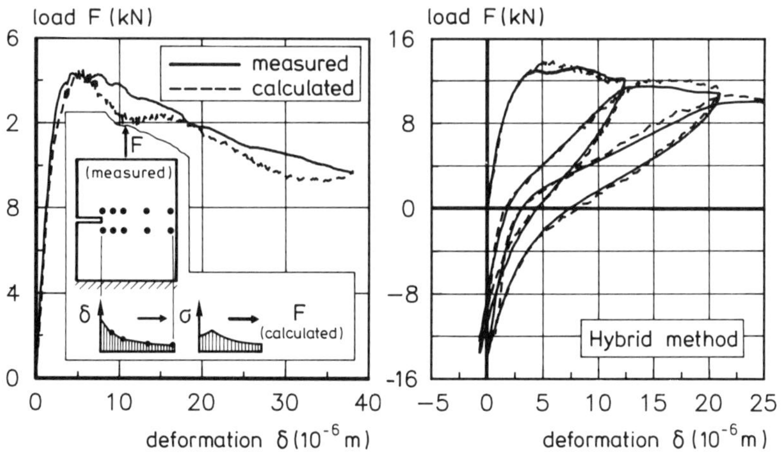

Fig. 9. Comparison between experimental results and predictions with the hybrid method for the tensile tests on single notched plates.

6 Concluding remarks

Meanwhile, many investigators recognize the softening behaviour of concrete to be a material property. Nevertheless, discussions are still going on about the question whether it can also be explained from solely a structural point of view. Results of numerous numerical analyses of experiments have already demonstrated that the experimental results can only be simulated properly, when the softening behaviour of the concrete has been taken into account. The analyses presented in this paper support these findings.

The results of the analyses demonstrated that the continuous-function-model is a proper description for the material behaviour of concrete in the softening zone. Besides the behaviour for monotonic increasing crack openings, also the crack cyclic behaviour is very well described. It should be remarked, however, that this holds true for crack openings smaller than about 120 μm. The results of the bending tests support the existence of a long tail in the stress crack opening relation. The authors believe that this long tail contributes strongly to the generally observed size-dependency of G_F.

So far, the validity of the continuous-function-model has been demonstrated for small scale structures. Now it would be interesting to see if the behaviour of larger specimens, even specimens with dimensions in the order of meters, can also be predicted with the same model. The material behaviour as described in the continuous-function-model is only then a material property if also these specimens can be simulated with it. If such experiments are being performed, however, special attention must be given to phenomena like initial stresses due to differential shrinkage.

7 Acknowledgements

This investigation was supported in part by STW (Netherlands Technology Foundation) and CUR (Netherlands Centre for Civil Engineering, Research, Codes and Specifications). The authors are indebted to A.S Elgersma, who carefully carried out the experiments and to the Computational Mechanics Department of TNO Building and Construction Research, where the DIANA analyses were performed.

8 References

Hordijk, D.A. and Reinhardt, H.W. (1990a) A constitutive model for crack cyclic behaviour of plain concrete, in **Fracture Behaviour and Design of Materials and Structures** (Ed. D. Firrao), Preprint of 8th European Conference on Fracture (ECF8), EMAS, pp. 579-584.

Hordijk, D.A. and Reinhardt, H.W. (1990b) Testing and modelling of plain concrete under mode I loading, in **Micromechanics of failure of quasi-brittle materials** (Eds. S.P. Shah et al.), Elsevier Applied Science, pp. 559-568.

Hordijk, D.A. (1991) Local approach to fatigue failure of concrete, **Doctoral thesis**, Delft University of Technology.

Janssen, J.G. (1990) Mode I fracture of plain concrete under monotonic and cyclic loading; Implementation and evaluation of a constitutive model in DIANA. **Graduate Thesis**, Delft University of Technology, The Netherlands, TNO-IBBC report no. BI-90-110.

45 A COMPARATIVE STRAIN ENERGY DENSITY STUDY FOR VARIOUS FRACTURE SPECIMENS AND A NEW BRITTLENESS NUMBER

A. ASGHARI and B. BARR
School of Engineering, University of Wales, Cardiff, UK

Abstract
The paper reports on a recent study to investigate the strain energy density distribution in a number of fracture test specimens. In particular, a detailed study of the strain energy density in the ligament of three Mode I fracture test specimens is reported. The paper reviews some existing brittleness numbers and then develops a new brittleness number based on the strain energy density calculations. It is shown that the new brittleness number can be used to compare the relative brittleness of different test specimen geometries. In particular, the study shows that the compact compression test specimen geometry results in brittle fracture and that LEFM can be used for such a test geometry.
Keywords: Strain energy density, brittleness number, brittle and ductile fracture.

1 Introduction

This paper reports on a recent study by Asghari (1990) on the strain energy density distribution in a number of fracture test specimens. In particular, a detailed study was carried out on the strain energy density in the uncracked ligaments of test specimens. Three test specimens have been studied in detail - the three-point bend specimen (Fig 1(a)), the four-point bend specimen (Fig 1(b)) and the compact compression test specimen (Fig. 2). In all three test geometries the strain energy density in the ligament is more than that in the other regions. The study by Asghari has also shown that it is possible to estimate accurately failure loads for a range of notch depths in a given test specimen geometry if only one failure load for a given notch depth is known.

One can gain a better understanding of the probable material behaviour during the fracture process (i.e. brittle or ductile fracture) from a consideration of the shape and size of the test specimens. This can be

Fracture Processes in Concrete, Rock and Ceramics Edited by J.G.M. van Mier, J.G. Rots and A. Bakker.
 Published by E. & F.N. Spon, 2–6 Boundary Row, London SE1 8HN. ISBN 0 419 15870 7

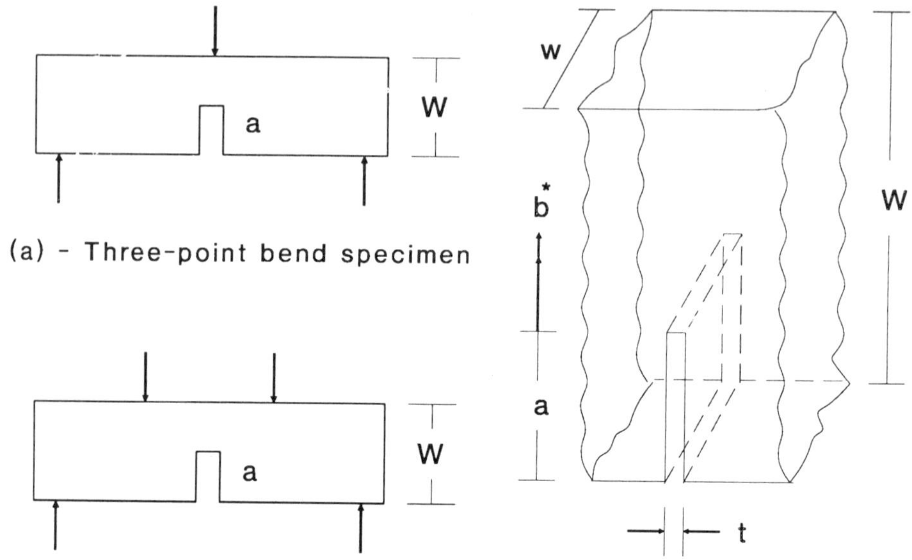

(a) - Three-point bend specimen

(b) - Four-point bend specimen

(c) - Definitions

Fig.1. Three and four-point bend specimens and definitions

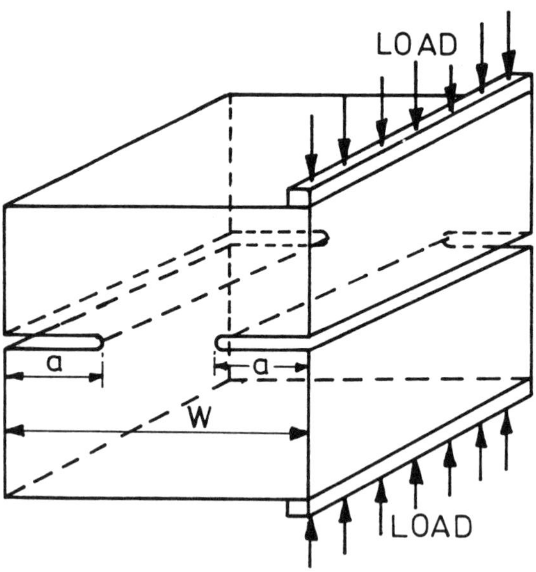

Fig.2. Compact compression test specimen

formulated by means of a brittleness number. Thus the brittleness number can serve as a basic indicator of the type of behaviour (brittle or tough) to be expected. In this sense the brittleness number is analogous to the Reynolds number in Fluid Mechanics. One area of application for the strain energy density study carried out by Asghari has been the development of a brittleness number based on the strain energy density calculations. It is shown that the new brittleness number can be used to compare the relative brittleness of different test specimen geometries.

A number of researchers have already defined various brittleness numbers and a full review cannot be repeated here. By the combination of the fracture energy, G_f, the tensile strength, f_t, and Young's Modulus, E, Hillerborg (1985) introduced a characteristic length denoted l_{ch} which was given by $E.G_f/f_t^2$. Hillerborg proposed that the characteristic length is a material property, which has no physical interpretation, but can be calculated from the values of E, G_f and f_t. Similarly, Carpinteri (1982) characterized the effect of structure size on its brittleness by the non-dimensional ratio of G_f/bf_t where b is the beam depth and G_f and f_t are defined above.

Perhaps it is more meaningful to refer to Hillerborg's and Carpinteri's brittleness numbers as ductility numbers since the larger these numbers the greater will be the ductility. Both researchers have tried to characterise the effect of the size of structures on the fracture behaviour by means of some non-dimensional combination of G_f, f_t and E. Both brittleness numbers can be applied only to geometrically similar test specimens - they cannot be used to compare different geometries. There is a need for brittleness to be determined in a way which will allow comparisons to be made between different geometries as well as a given test specimen geometry.

The method proposed by Bazant and Pfeiffer (1987) overcomes the above difficulties. They proposed that the brittleness number should be defined by the ratio $d/\lambda_o d_a$ where d is a characteristic dimension of test specimen, d_a is the maximum aggregate size and λ_o is evaluated either experimentally or by finite element analysis. If the brittleness number is ≤ 0.1, fracture is ductile and plastic limit analysis applies. On the other hand, if the brittleness number is ≥ 10, fracture is brittle and linear elastic fracture mechanics can be applied to the test geometry.

Bache (1989) has also introduced an interesting fracture characteristic number, B, which is given by the ratio of elastic energy to fracture energy i.e. $(L^3 f_t^2/E)/(L^2 G_f)$ which simplifies to Lf_t^2/EG_f. If the stored elastic energy is high in comparison with the energy required for fracture then the response is brittle.

On the other hand, the material is tough when the stored elastic energy is smaller than the energy needed for crack propagation. The brittleness number proposed here has some similarities with Bache's brittleness number.. The main objective was to develop a new brittleness number which could be used for various test specimen geometries and for all notch depth ratios in each specimen.

2 Strain energy density study

Strain energy density (SED) calculations have been carried out numerically by Asghari (1990) for a range of fracture test specimens. The test specimens considered include beams in three-and four-point bending, compact tension test specimens, compact compression test specimens (i.e. Mode I test specimen); short beam test specimens (primarily Mode II test specimens); four-point shear test specimens and beams with non-central notches (i.e. Mixed-Mode test specimen). For all the above specimens, the geometries were divided into zones and the variation of the strain energy density in each zone was evaluated. In particular the strain energy density variation in the uncracked ligament was studied in detail. Only the results for the Mode I type of test are considered here.

The strain energy density results have been used to investigate three specific fracture areas - crack path prediction, failure load prediction and the development of a brittleness number. This paper deals specifically with the new brittleness number based on SED results.

Most investigations in the past have studied the distribution of energy around the crack tip by using a polar coordinate system. They tended not to study the distribution of energy or strain energy density in a greater area. In the work reported here, by using the Cartesian coordinates, the distribution of energy in a region greater than and ahead of the crack tip has been determined. The best results were obtained from a mesh in which each element was 1mm x 2.5mm.

3 Brittleness Number, b

According to the Gillemot (1976) specific energy concept, failure will occur where the Strain Energy Density, SED, reaches a critical value. This concept is used in the development of the brittleness number, b, reported here. It is assumed that the brittleness number is characterised by an imaginary length, b*, of the ligament length as illustrated in Fig 1(c). A part of the volume of the ligament ahead of the root of the notch is considered (i.e. b* x ligament width (w) x ligament thickness (t)) to possess the critical Strain Energy Density, $(SED)_{cr}$ which is defined as follows:

$$(SED)_{cr} = u/b^*.w.t \tag{1}$$

where u is the total strain energy in the ligament when the applied load becomes the failure load. The volume defined by b*.w.t) represents that volume of the ligament in which the density of energy is critical. (In the experimental work carried out to date, w = 100mm and t = 4mm in all cases).

Equation (1) may be rearranged as follows:

$$b^* = u/(SED)_{cr}.w.t \tag{2}$$

Although b* can be evaluated from Eqn (2), its value is not important since it is only used for comparison purposes. It is assumed that $(SED)_{cr}$ is a material property with a constant value for any given material subjected to identical loading and environmental conditions. If the length of the ligament is (W-a), then the ratio b*/(W-a) is a dimensionless number, b, which can be expressed as follows:

$$b = u/(SED)_{cr}.w.t(W-a) \tag{3}$$

In the above Eqn., (w.t.(W-a)) represents the volume of the ligament and hence u/w.t(W-a) is the average Strain Energy Density, $(SED)_{av}$, in the ligament. Again, this average value is based on the load at failure. Thus Eqn (3) may be written as follows:

$$b = (SED)_{av}/(SED)_{cr} \tag{4}$$

For comparison, the brittleness number of two specimens can be written as follows:

$$b_1 = (SED)_{av1})/(SED)_{cr} \text{ and}$$
$$b_2 = (SED)_{av2}/(SED)_{cr} \tag{5}$$

$$\text{Thus } b_2/b_1 = (SED)_{av2}/(SED)_{av1} \tag{6}$$

The ratio given by Eqn (6) depends on the failure load. In the case where experimental work has been carried out or reported in the literature, it is possible to determine directly the ratio of b_2/b_1. When experimental results are not available the ratio of failure loads for various notch depths can be evaluated from a numerical analysis. Thus for any given geometry the brittleness number for a range of notch depths can be determined without knowing the experimental failure loads. However, if a comparison is to be made between two different test specimen geometries, the relative failure loads must be known.

Three known test specimen geometries have been used in the work reported here. They are the three-point bend specimen, the four-point bend specimen and the compact compression test specimen. The standard geometry was considered to be the three-point bend specimen with a notch depth of 50% of the beam depth - this geometry was assumed to have a brittleness number of unity. All other brittleness numbers have been determined relative to this standard geometry. Considering Eqn (4), as $(SED)_{cr}$ increased the value of b decreases and this corresponds to increased ductility. On the other hand, as b increases the brittleness increases. However, the transition between brittleness and ductility is not clear and further work is required in this area.

4 Results and discussion

The results for the new brittleness number are presented in Table 1 and illustrated graphically in Fig. 3. As

Table 1 Brittleness numbers

Geometry	Notch Depth (mm)	Strain Energy (N-m)	$(SED)_{av}$ J/m^3	Brittleness Number
	50	1.981	119.1	1.000
Three-point	40	2.586	119.2	1.088
bend	30	4.149	154.5	1.496
	20	5.649	174.3	1.782
Four-point	40	3.292	92.4	1.385
bend	30	4.168	77.7	1.503
	20	6.447	77.9	2.034
	50	1.161	350.9	1.465
Compact	45	1.735	346.1	1.752
Compression	40	2.787	396.1	2.345
Specimen	35	3.765	403.5	2.715
	30	6.037	507.5	3.809
	25	9.867	675.7	5.534

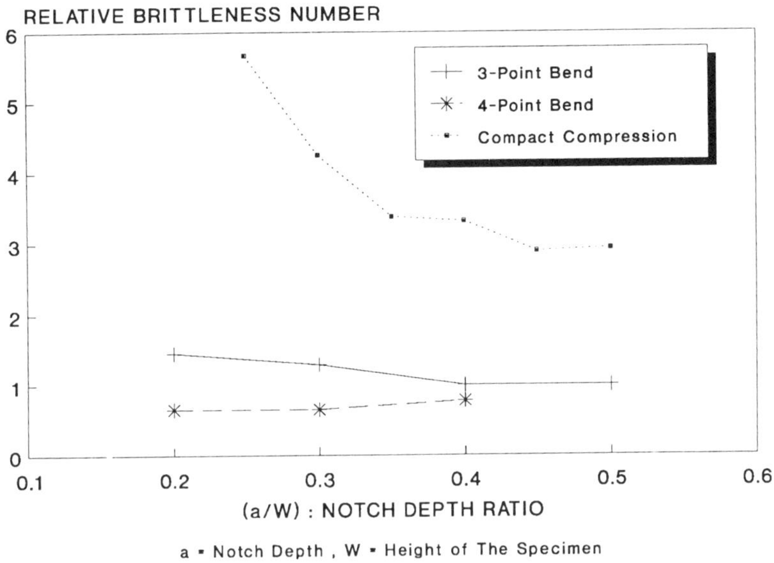

Fig.3. Brittleness number for three fracture test geometries

stated earlier, the standard reference geometry was considered to be the three-point bend specimen illustrated in Fig 1(a) in which the a/W ratio was 0.5. All other brittleness numbers have been evaluated relative to this standard geometry which is assumed to have a brittleness number of unity (Table 1).

The results presented in Table 1 show that the brittleness number increases as the notch depth is reduced (for all three specimens). This is in agreement with the observed performance of fracture test specimens. Smaller notch depths result in increased loads at failure with corresponding increases in the stored strain energy in the specimens and/or loading systems. Thus when crack propagation takes place there is more energy available to drive the cracks and hence shallow notch depths promote a more brittle type of failure.

The brittleness numbers for the compact compression test specimens are significantly greater than the values for the three-point and four-point bend specimens, as illustrated in Fig 3. This is particularly true in the case of the compact compression test specimens with the shallow notch depths. This conclusion is also in agreement with the observed fracture tests, using the compact compression test specimen. Fracture tests on such specimens have produced a brittle type of failure in which the fracture has been sudden, complete and accompanied by a loud noise during the fracture process. Further research is required in this very interesting area of fracture study based on strain energy density considerations and brittleness characteristics.

5 Conclusions

A study of the strain energy density in the ligament of three Mode I fracture test specimens has been carried out. The results have been used to develop a new brittleness number based on the strain energy density calculations. It is shown that the new brittleness number can be used to compare the relative brittleness of different test specimens as well as the effect of notch depth in a given test geometry.

The results show that for all three Mode I test geometries, the brittleness number increased with a reduction in the value of the notch depth. Furthermore, the results show that the compact compression test specimen geometry gives a more brittle type of fracture than the bend tests. These conclusions are in agreement with observations made during fracture test.

6 References

Asghari, A. (1990) Strain energy density and the fracture of concrete, PhD Thesis, University of Wales College of Cardiff.

Bache, H.H. (1989) Brittleness/ductility from deformation and ductility points of view in Chapter 7 of RILEM Report, Fracture Mechanics of Concrete Structures; From theory to applications (editor L. Elfgren), Chapman and Hall pp 202-207.

Bazant, Z.P. and Pfeiffer, P.A. (1987) Determination of Fracture energy from size effect and brittleness number, ACI Materials Journal, pp 463-480.

Carpinteri, A. (1982) Application of fracture mechanics to concrete structures, Journal of the Structural Division, ASCE, 108, pp 833-848.

Gillemot, L.F. (1976) Criterion of crack initiation and spreading, Journal of Engineering Fracture Mechanics, 8, pp 239-253.

Hillerborg, A. (1985) Numerical methods to simulate softening and fracture of concrete. Fracture mechanics of concrete: Structural applications and numerical calculation (eds G.G. Sih et al), Martinus Nijhoff Publishers, pp 141-170.

46 SOFTENING BEHAVIOUR OF CONCRETE IN THREE-POINT BEND TESTS ON SINGLE EDGE NOTCHED BEAMS

Y. SHINOHARA, F. FURUMURA and T. ABE.
The Research Laboratory of Engineering Materials, Tokyo Institute of Technology, Japan

Abstract
An experimental method has been developed to obtain reliable load-displacement curves of plain concrete in three-point bend tests, including the post-peak softening behavior. Three vertical displacements, two crack opening displacements and crack propagation behaviors have been measured to characterize the softening behavior. Twelve test series were performed on plain concrete beams after 28 days of curing under water to study the influence of notch depth, loading rate, maximum aggregate size and water/cement ratio upon the softening behavior, and to investigate how the slow crack propagation in concrete would affect its post-peak softening response. Furthermore, the same tests were conducted on the same beams, one year of age to study the effect of age. Some of the average material properties obtained from tests are summarized and discussed the effect of notch depth, loading rate, maximum aggregate size, water/cement ratio and age. The mode of the deformation for a notched beam is investigated by measured displacements and crack extensions. Finite-element calculations show that the softening behavior of the different specimen-size can be estimated using a bilinear strain softening model with the same tensile strength and fracture energy obtained from tests.
Keywords: Softening Behavior, Crack Propagation, Microscopic, Crack Opening Displacement, Notched Concrete Beam, Bilinear Strain Softening Model.

1 Introduction

It has now been widely accepted that a softening zone is formed at the crack tip in a concrete structure because of the heterogeneity inherent in concrete. The softening is described as a relation between the additional deformation within the fracture zone and the stress which can still be transferred in spite of the damage. The stress distribution in that zone is a function of the deformation and the softening behavior. In order to provide necessary information which applies to nonlinear finite element analysis, stress-crack opening relations should be established by means of deformation controlled tensile or bending tests.

The primary purpose of this investigation is to obtain the reliable post-peak response of concrete subjected to bending, and to investigate how the slow crack propagation in concrete would affect its post-peak softening behavior.

2 Experimental procedures

2.1 Test series and preparation of specimens

The tests to obtain the complete load-displacement response have been conducted on concrete beam specimens in three-point bending(3PB). The

Fracture Processes in Concrete, Rock and Ceramics Edited by J.G.M. van Mier, J.G. Rots and A. Bakker.
 Published by E. & F.N. Spon, 2–6 Boundary Row, London SE1 8HN. ISBN 0 419 15870 7

dimensions of specimens are shown in Fig.1. Twelve series of 3PB specimens were tested at the ages of 28 days and one year. Test conditions for each series are given in Table 1. More than three 3PB specimens were tested for each condition. To facilitate microscopic observations of crack propagation, the relatively low rate(0.005mm/min) was used as a standard rate in tests. Cylinders used in compression tests were 100mm in diameter and 200mm in length. The mix proportions of concrete are given in Table 2. The ordinary Portland cement was used in all mixes. The fine aggregate was a siliceous sand which had a fineness modulus of approximately 2.8 and a specific gravity of 2.44. The coarse aggregate was a semi-rounded sea gravel. The coarse aggregate grading and physical properties are shown in Table 3-1 and Table 3-2 respectively. The average slump was 18cm.

The test beams were cast in 400×100×100mm steel molds. Cylinders used to determine the compressive strength and elastic modulus of each test series were also cast and cured together with the beams. At approximately 24 hours after

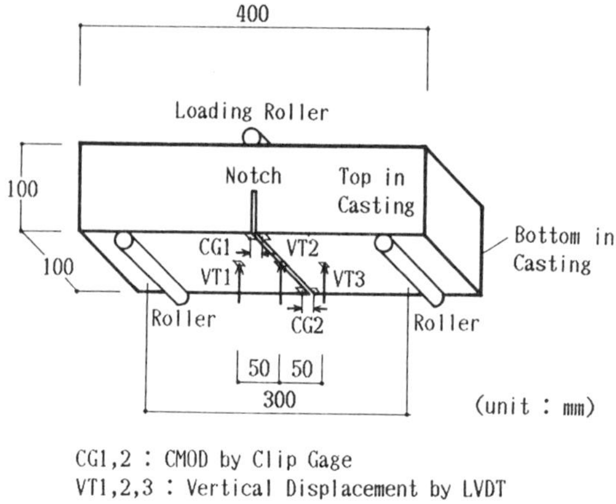

Fig.1. Dimensions of bend specimens and designations of displacements

Table 1. Test series

Series designation	Notch depth (mm)	Loading rate (mm/min)	Max. grain size(mm)	Water/ cement ratio	Remarks
M20-1	10	0.005	20	0.55	Influence
M20-2	30	0.005	20	0.55	of notch
M20-3	50	0.005	20	0.55	depth
M20-4	70	0.005	20	0.55	
M20-5	50	0.05	20	0.55	Influence
M20-6	50	0.5	20	0.55	of rate
M05-1	50	0.005	5	0.55	Influence
M10-1	50	0.005	10	0.55	of max.
M15-1	50	0.005	15	0.55	aggregate
M25-1	50	0.005	25	0.55	size
L20-1	50	0.005	20	0.45	Influence
H20-1	50	0.005	20	0.65	of W/C

Table 2. Concrete composition

Mix designation	Water/cement ratio	Max. grain size (mm)	Unit weight (kg/m^3)			
			Water	Cement	Sand	Coarse aggregate
M20	0.55	20	194	353	689	1054
M05	0.55	5	323	587	1174	----
M10	0.55	10	202	367	700	1020
M15	0.55	15	198	360	703	1037
M25	0.55	25	190	345	700	1071
L20	0.45	20	203	451	603	1054
H20	0.65	20	192	295	766	1037

Table 3-1. Coarse aggregate grading

Sieve nominal size in JIS A 1102 (mm)	25	20	15	10	5	2.5
Percentage passing	100	96	91	55	9	0

Table 3-2. Physical properties of the coarse aggregate

Physical properties	Maximum aggregate size (mm)			
	25	20	15	10
Specific gravity	2.66	2.64	2.64	2.64
Water absorption(%)	0.90	0.90	0.88	0.88
Bulk density(kg/m^3)	1710	1700	1700	1660
Fineness modulus	6.40	6.33	6.30	5.84

casting the specimens were removed from the molds and stored in water maintained at 20°C until the test age. The notches were exactly introduced to the prescribed depth using a diamond wheel saw with 1.2mm thick before carrying out a test.

2.2 Testing procedures

The test set-up is shown in Fig.2. All beam specimens were loaded in a displacement rate controlled by the Instron type testing machine with the capacity of 25ton. To minimize the effect of pin friction, the pins can rotate in slots and are retained by small springs shown in Fig.2. These slots have a width somewhat greater than a pin diameter in order to permit the necessary pin movements. The springs locate the pins accurately at the inner corners of the slots which establish the test span of 300mm at the start of tests.

The loads were measured using a compression load cell with 0.4kg sensitivity. Vertical displacements and crack mouth opening displacements(CMOD) at the positions, as shown in Fig.1 were measured using three linear variable differential transducers(LVDT) with 1/2000mm sensitivity and two clip gages with 1/2400mm sensitivity respectively. The LVDTs were installed in the frame that the influence of any crushing at the supports or under the loading roller was excluded, so that the measured displacements represented the true deflections of the beam (see Fig.3). Slow crack propagations on the both sides of a specimen were observed using two microscopes with digital micrometers. The crack

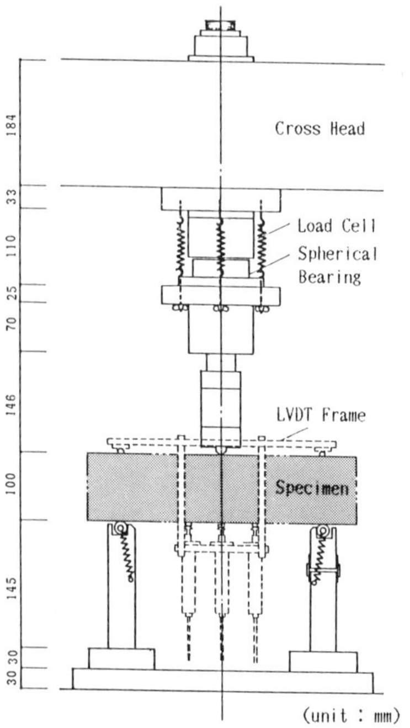

Fig.2. Loading apparatus

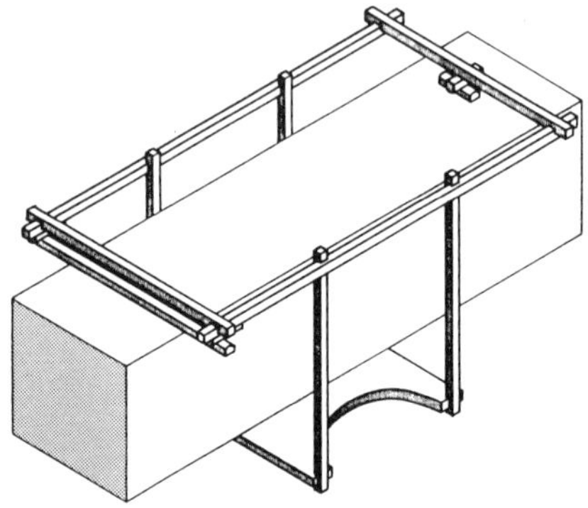

Fig.3. Fixing arrangement of LVDT

lengths at various levels of displacements were measured with a 50× magnification. The total testing time for each specimen was about 120 minutes at the loading rate of 0.005mm/min.

3 Results and discussions

Some of the average material properties obtained from the tests are summarized in Table 4. Values reported therein include the peak load and displacement(L_p and D_p), and fracture energy(G_f) in bend tests, and compressive strength(f'c) and secant modulus at one third of f'c ($E_{1/3}$) in compression tests. The area(U) under the load-load point displacement curve represents the work required to fracture the specimen and the fracture energy(G_f) is defined as

$$G_f = \frac{U}{A} \qquad (1)$$

where A is the uncracked area at the notch (ligament area). The influence of the self-weight of the beam is estimated and included in the calculation of G_f. This method of measuring the fracture energy G_f was the basis for the RILEM recommendation(Ref.5). In this experiment, however, since the beams didn't break except M05-1 series, G_f is calculated using only the area U up to the load point displacement that the notch tip opening displacement reaches 0.6mm

Table 4. Average material properties

Tests	Age	Bend tests			Compression tests	
Series		Peak load L_p (kg)	Disp. at L_p D_p (mm)	Fracture energy G_f (kg/m)	Comp. strength f'c (kg/cm^2)	Secant modulus $E_{1/3}$ (kg/cm^2)
M20-1	4wk	788	0.041	(14.2)	397	2.90×10^5
M20-1	1yr	939	0.034	(15.5)	506	3.43×10^5
M20-2	4wk	471	0.039	10.4	365	2.91×10^5
M20-2	1yr	541	0.037	11.1	495	3.41×10^5
M20-3	4wk	258	0.048	10.0	400	2.93×10^5
M20-3	1yr	272	0.038	10.5	500	3.60×10^5
M20-4	4wk	97.6	0.052	9.5	401	2.91×10^5
M20-4	1yr	105	0.043	8.2	504	3.51×10^5
M20-5	4wk	264	0.042	10.9	417	2.99×10^5
M20-5	1yr	314	0.043	10.8	509	3.46×10^5
M20-6	4wk	291	0.043	10.8	387	2.92×10^5
M20-6	1yr	334	0.046	12.6	486	3.48×10^5
M05-1	4wk	239	0.046	5.7	490	2.42×10^5
M05-1	1yr	254	0.042	5.8	588	2.77×10^5
M10-1	4wk	256	0.048	9.1	421	2.97×10^5
M10-1	1yr	282	0.041	10.2	519	3.51×10^5
M15-1	4wk	256	0.044	9.2	398	3.09×10^5
M15-1	1yr	301	0.048	11.4	470	3.62×10^5
M25-1	4wk	275	0.049	16.3	374	3.13×10^5
M25-1	1yr	303	0.042	12.0	468	3.62×10^5
L20-1	4wk	293	0.044	10.6	513	3.31×10^5
L20-1	1yr	309	0.036	11.3	655	3.75×10^5
H20-1	4wk	219	0.037	8.3	300	2.91×10^5
H20-1	1yr	264	0.041	11.1	388	3.46×10^5

(0.4mm for M05-1 series), assuming that the deformation of beam is the rotation of rigid body as stated later. The calculated values of G_f are in good agreement with those reported by other researchers(Refs.3,4).

Typical load-displacement curves obtained from one year tests are shown in Fig.4. The displacements are designated according to Fig.1. The detail load-displacement curves obtained from 28-day tests have been previously reported(Refs.1,2). Fig.4(a) shows the load-displacement mean curves for four different notch depths. The post-peak resistance and fracture energy increase with decreasing notch depth. However, the fracture energy of 10mm-notched beams is overestimated for the lack of measured points in the descending parts where cracks rapidly propagate. Fig.4(b) shows the load-displacement mean curves for three different loading rates. In Fig.4(c), the load-displacement mean curves for five different maximum aggregate size are shown. The peak load and fracture energy increase with increasing loading rate and maximum aggregate size. The post-peak resistance and fracture energy of mortar specimens (M05-1 series) are considerably lower than those of concrete specimens. In the mortar specimens, the crack extends through the specimen in a relatively straight line. However, for the concrete specimens with larger aggregates, the crack has a

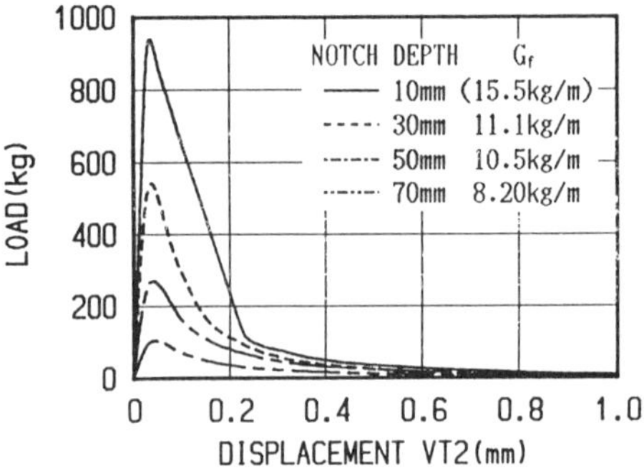

(a) Influence of notch depth

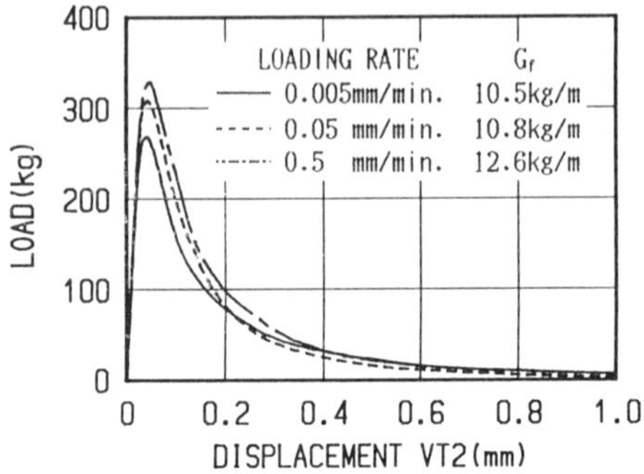

(b) Influence of loading rate

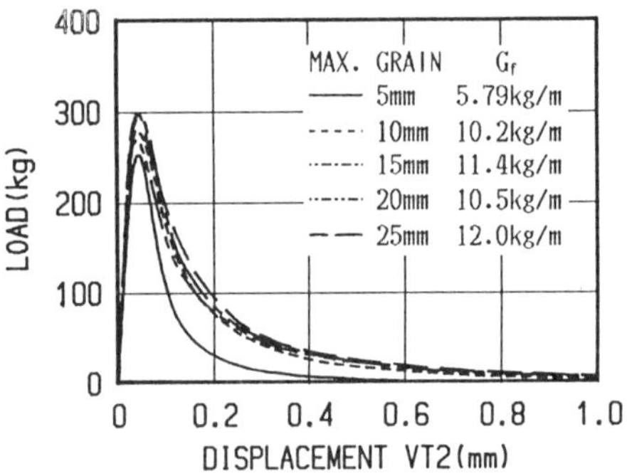

(c) Influence of maximum aggregate size

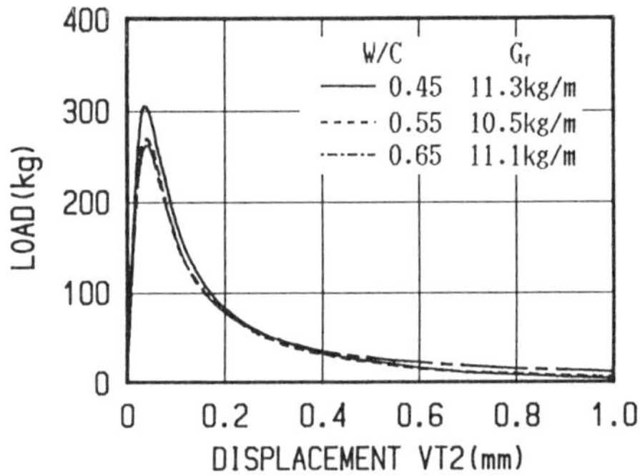

(d) Influence of water/cement ratio

Fig.4. Typical load versus displacement curves

meandering path and tends to run around the aggregates rather than through them because of the crack arresting action of the aggregates. This stabilizes the crack growth and results in a greater curvature of the descending portion of the load-displacement curves. Fig.4(d) shows the load-displacement mean curves for three different water/cement ratios. It can be seen from Fig.4(d) that the fracture energy and the peak load increase with decreasing water/cement ratio. The general tendency is the same as the results obtained from 28-day tests. As can be seen from Table 4, with increase in age, the peak load (L_p) becomes higher but the load-point displacement (D_p) tends to decrease. For one-year specimens, compressive strength increases by more than 20 percent when compared with the corresponding value of 28-day specimens, whereas the fracture energy for some of test series decrease slightly with increasing age. This is probably due to the maturity of hardened cement paste. As the hydration process goes on, the discrimination between aggregates and the matrix is diminished. Thus some cracks may run through the aggregates, leading to more brittle fracture behavior.

From vertical displacements(VT1,2,3) and CMODs(CG1,2) it is possible to derive other parameters of interest by assuming that a notched beam deforms as two rigid halves about a center of rotation. From Fig.5 it is apparent that the

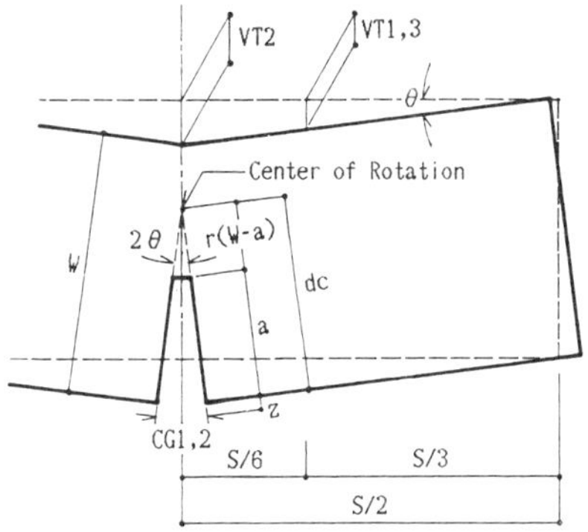

Fig.5. Geometrical relationship between vertical displacements and clip gage displacements

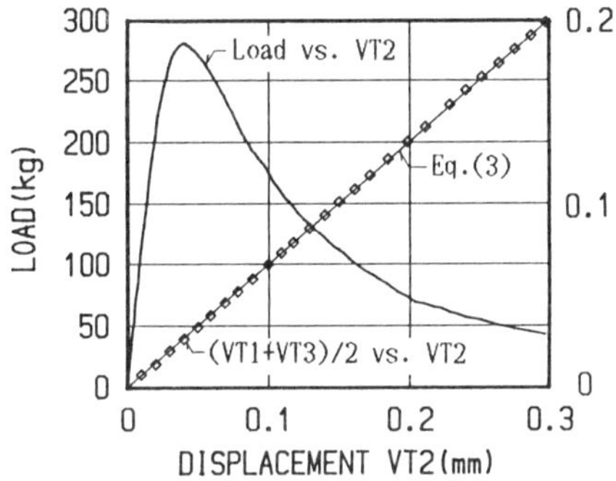

(a) M20-3 series No.5 Specimen (1-year test)

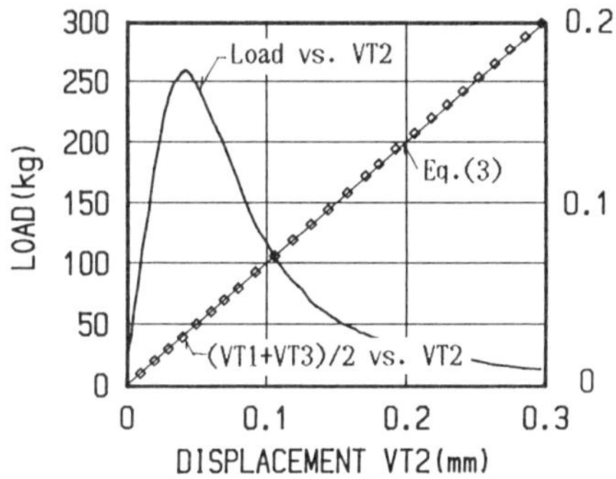

(b) M05-1 series No.3 Specimen (1-year test)

Fig.6. Comparison of (VT1+VT3)/2, obtained from tests and calculated by Eq.(3)

angle of bend θ may be expressed in terms of VT1,2,3 and CG1,2 (assuming small angle) as

$$\theta = \frac{VT2}{S/2} = \frac{(VT1 + VT3)/2}{S/3} = \frac{(CG1 + CG2)/2}{2(r(W - a) + a + z)} \tag{2}$$

where the various terms are defined in Fig.5. The rotation factor(r) is defined as the ratio of the distance between the crack tip and center of rotation to the ligament(W-a). Hence the relationship between vertical displacements is given by

$$(VT1 + VT3)/2 = \frac{2(VT2)}{3} \tag{3}$$

And the distance of the rotation point from the beam bottom (dc) is given by

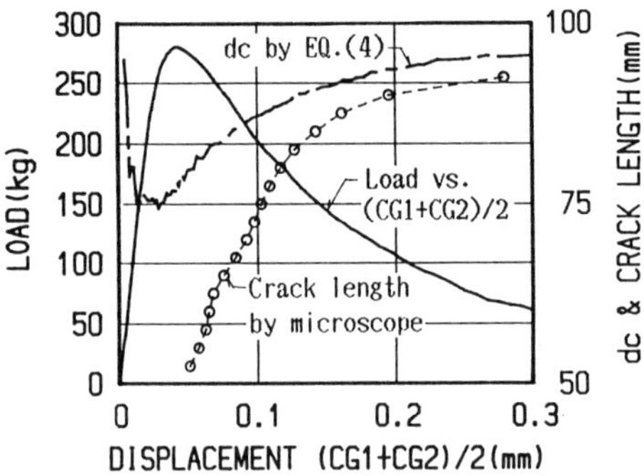

(a) M20-3 series No.5 Specimen (1-year test)

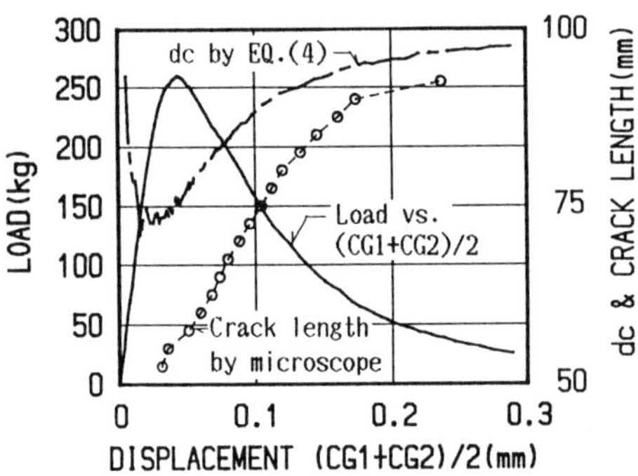

(b) M05-1 series No.3 Specimen (1-year test)

Fig.7. Variation with (CG1+CG2)/2 of dc calculated from Eq.(4) and crack length observed by microscopes

$$dc = r(W - a) + a = \frac{S(CG1 + CG2)/2}{4(VT2)} - z \qquad (4)$$

An evidence that a notched beam deforms as two rigid halves about a center of rotation is obtained from Fig.6 which shows that the linear relationship expressed by Eq.(3) exists between the displacement VT2 and the displacement (VT1+VT3)/2. Fig.7 shows dc calculated by Eq.(4) versus the displacement (CG1+CG2)/2 together with the load-displacement curve. Averaged crack propagations observed during tests are also plotted in Fig.7. At small displacements dc calculated by Eq.(4) becomes considerably large and variable. This is due to the assumption that a notched beam deforms as two rigid halves about a center of rotation. In reality, however, a notched beam behaves elastically for the early stages of deformation, thus giving a smaller VT2. While the displacement (CG1+CG2)/2 increases from 0.015mm to 0.04mm corresponding to the peak load, the center of the rotation stays close to the center of the ligament(r=0.49 to 0.53 for M20-3 and r=0.45 to 0.50 for M05-1). This rotation

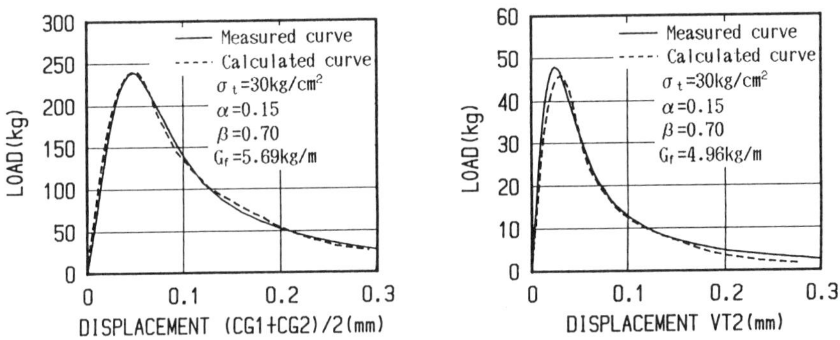

(a) 100×100×400mm specimen (b) 40×40×160mm specimen
Fig.8. Comparison of load versus-displacements, obtained from tests and calculated by FEM. M05-1 series(4-week test)

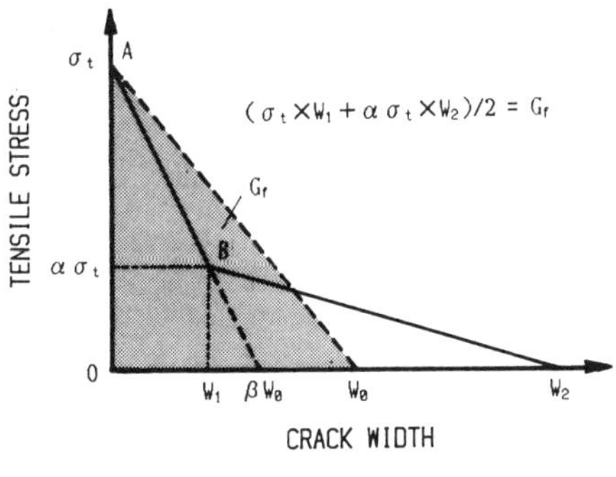

Fig.9. Bilinear strain-softening model used in FEM calculation

factor is somewhat larger than the value of 0.4 assumed in British standard for metal, which means that the latter will underestimate the true crack tip opening displacement(CTOD). Once a significant softening such as microcracks has occurred, the CTOD estimate based on the assumption of a rotational constant becomes considerably less, because the rotational factor increases rapidly with increasing crack extensions as shown in Fig.7.

Fig.8 shows a comparison between the load-displacement curve obtained from tests and the load-displacement curve calculated by finite-element methods using a bilinear strain softening model, as shown in Fig.9. The softening behavior of the different specimen-size can be numerically estimated using the same tensile strength, α and β which characterize the shape of a bilinear strain softening model, and fracture energy obtained from tests.

4 Conclusions

1. The fracture energy increases with increasing loading rate, maximum aggregate size and age, and with decreasing notch depth and water/cement ratio.
2. The post-peak softening behavior of concrete is primarily affected by the meandering path of the main crack and side cracking caused by the crack arresting action of the aggregates. This crack arresting action often masks the influence of loading rate, maximum aggregate size, water/cement ratio and age on the fracture energy.
3. Notched beams deform as two rigid halves about a point located in the ligament above the notch, particularly when the softening behavior becomes significant. The center of the rotation exists near the center of the ligament up to the peak load, and after that moves upward with the slow crack growth.
4. The softening behavior of the different specimen-size can be numerically calculated using a bilinear strain softening model with the same tensile strength and fracture energy obtained tests.

5 References

1) Furumura, F. Ave, T. Shinohara, Y. and Iwaki, H. (1989) Softening behavior of concrete in three-point bend tests on single edge notched beams Part1 Influence of notch depth and loading rate, Report of RLEM, T.I.T., No.14, pp121-133
2) Furumura, F. Ave, T. and Shinohara, Y. (1990) Softening behavior of concrete in three-point bend tests on single edge notched beams Part2 Influence of maximum aggregate size and water/cement ratio, Report of RLEM, T.I.T., No.15, pp147-170
3) Hillerborg, A. (1985) Results of three comparative test series for determining the fracture energy Gf of concrete, Materiaux et Constructions, Vol.18, No.107, pp407-413
4) Nallathambi, P. Karihalloo, B.L. and Heaton, B.S. (1984) Effect of specimen and crack sizes, water/cement ratio and coarse aggregate texture upon fracture toughness of concrete, Magazine of Concrete Research, Vol.36, No.129, pp.227-236
5) RILEM Technical Committee 50-FMC, (1985) RILEM Draft Recommendation, Determination of the fracture energy of mortar and concrete by means of three-point bend tests on notched beams, Materials and Structures, Vol.18, No.106, pp.285-290

47 AN IDENTIFICATION PROCEDURE OF FRACTURE ENERGY IN CONCRETE: MATHEMATICAL MODELLING AND EXPERIMENTAL VERIFICATION

G. ROSATI and C. SCHUMM
D.I.S., Politecnico di Milano, Italy

Abstract
A refined mechanical model for plain concrete elements in bending is presented. In particular, the diffusive phenomena, the strain field evolution and the crack localization are introduced and mathematical relations for both solid and cracked-concrete tensile behavior are proposed and validated vs. experimental results. An identification procedure to determine the fracture energy dissipated both through the microcracked zone and over the macrocrack surface is developed and here based on the results of direct tensile and bending tests.
Keywords: Fracture Energy, Direct Tensile Test, Three-Point Bending Test, Concrete Constitutive Relations, Aggregate Debonding, Notch Sensitivity, Mechanical Model, Identification Procedure.

1 Introduction

The fracture energy dissipated in concrete under uniaxial tensile loading ("opening mode") primarily depends on material constitutive relations and on their governing parameters, Cornellissen et al. (1986) and Mihashi et al. (1988). The most spontaneous way to determine this energy is to carry out direct tensile tests.

Nevertheless, specimen response is highly influenced by its geometry and size, material heterogeneity, boundary and loading conditions; furthermore, the investigated phenomena can develop in a sudden and unstable way and their scale can be small respect to the usual measuring-instrument capacity. As a consequence, experimental outcome can be not representative of the material properties if these affecting factors are not avoided or reduced as much as possible.

The present research efforts are devoted to design proper and simple indirect tensile tests (bending and tensile bending, compact tension,..), where more the reliability than the exactness of the outcome is pursued, in order to make repeatable and codifiable the testing procedure. Due to the numerous interacting and interdependent phenomena involved, these experimental arrangements become a sort of "structural tests", Van Mier and Noouru-Mohamed (1990). In fact, both the co-presence of solid, microcracked, plasticized, fracturing and fractured material and the step-by-step variable loading and geometrical conditions can occur. As a matter of fact, it is impossible to exactly identify each contribution to the global

Fracture Processes in Concrete, Rock and Ceramics Edited by J.G.M. van Mier, J.G. Rots and A. Bakker.
 Published by E. & F.N. Spon, 2–6 Boundary Row, London SE1 8HN. ISBN 0 419 15870 7

specimen response. To correctly study this situation it is necessary to know the local constitutive laws, their governing parameters as functions of the material properties and volumes and surfaces over which they insist; the several structural effects and the consequent stress and strain redistribution; the localized fracturing to adjacent solid region interactions and so on. In order to improve its interpretation, each standard test should be described by a reliable model, helping to estimate the contribution due to any constitutive and structural phenomenon involved.

Referring to the flexural behavior of concrete-like materials, a mechanical-modelling approach to three-point bending test is proposed in this paper and applied to study the softening response of notched and unnotched beams when maximum aggregate size is changed. This model allows a comparison of bending with direct tensile tests and the identification of fracture energy dissipated both through the microcracked zone and over the crack surface as a function of the governing parameters of the material constitutive laws.

2 Concrete in tension

A parabolic stress - strain relation for the ascending branch has been adopted (Fig. 1a):

$$\sigma_{c,\varepsilon} = E_c \varepsilon_c \left(1 - \varepsilon_c/(2\varepsilon_k)\right) \tag{1}$$

where E_c = Young's modulus, ε_k = strain corresponding to zero horizontal slope (virtual strain if $\varepsilon_u < \varepsilon_k$, ε_u = ultimate strain).

The residual strain due to the unloading process and correspondent to a maximum strain ε_c is:

$$\varepsilon_r = \varepsilon_c^2 / 2\varepsilon_k \tag{2}$$

A hyperbolic stress - crack opening relation for the descending branch ("fracturing" concrete), Rosati and Schumm (1990), has been adopted (Fig. 1b):

$$\sigma_{c,w} = \frac{f_{c,w}}{\kappa w/\phi_a + 1}\left(1 - \frac{w}{w_c}\right) \tag{3}$$

where $f_{c,w}$ = maximum stress ("residual" strength), correspondent to zero crack opening w, κ = slope of the hyperbolic curve, ϕ_a = maximum aggregate size, w_c = stress-free crack opening ("critical" opening).

This theoretical relation is based on a similitude criterion of crack opening vs. maximum aggregate size, Giuriani and Rosati (1986).

According to the adopted stress - strain relation (Eq.1), the specific fracture energy dissipated through the diffused microcracked zone in front of a macrocrack tip is:

$$\mathcal{G}'_F = 2\int_0^{\alpha\frac{h}{2}} \frac{1}{2}\left(\int_0^{\varepsilon_{cr}} \sigma_{c,\varepsilon}\, d\varepsilon_c + \int_0^{\varepsilon_u} \sigma_{c,\varepsilon}\, d\varepsilon_c\right) dx = \frac{\alpha h E_c}{2\,\varepsilon_k}\left(\varepsilon_{cr}^3\left(\frac{1}{3} - \frac{\varepsilon_{cr}}{8\varepsilon_k}\right) + \varepsilon_u^3\left(\frac{1}{3} - \frac{\varepsilon_u}{8\varepsilon_k}\right)\right) \tag{4}$$

where ε_{cr} = strain at the microcracking onset and αh = length related to the extension of the microcracked zone and dependent on the boundary conditions. For instance, in three-point bending specimens, assuming a linear distribution of the strain field in the axis direction, the following relation could be adopted:

$$\alpha = \left(\varepsilon_u - \varepsilon_{cr}\right) / \left(\varepsilon_u - \varepsilon_{ref}\right) \tag{5}$$

where ε_{ref} = strain in the reference-section, at a distance from the loaded notched section equal to the section height and at the same level of the notch tip.

According to stress - crack opening relation (Eq.3), the fracture energy dissipated over the macrocrack due to the aggregate debonding phenomena is (Fig. 1b):

$$\mathcal{G}''_F = \int_0^{w_c} \sigma_{c,w} dw = \frac{f_{c,w}}{\kappa/\phi_a} \left(\left(1 + \frac{\phi_a}{\kappa w_c}\right) \ln\left(\kappa w_c/\phi_a + 1\right) - 1 \right) \tag{6}$$

Finally, the global specific fracture energy dissipated is:

$$\mathcal{G}_F = \mathcal{G}'_F + \mathcal{G}''_F \tag{7}$$

3 Concrete in bending

Referring to a beam in bending, its non linear response is due to several constitutive and structural phenomena that mainly develop in a localized region close to the applied load. According to the De Saint Venant principle, this region is a-priori identified as an element extending to a distance h/2 at both sides of the load (Fig. 2). A more accurate extension αh (Eq.5) of the fracturing zone can be a-posteriori determined applying the here proposed mechanical model.

The strain at the notch-tip level in the reference section A (distant h/2 from the loaded notched section B) (Fig. 2) is:

$$\varepsilon_A = \varepsilon_{rA} + \frac{\sigma_{c,\varepsilon}}{E_c} = \frac{\varepsilon^2_{uA}}{2\varepsilon_k} \frac{\left(h - z_A - a\right)^2}{\left(h - z_A\right)^2} + \frac{6\ M}{bh^2 E_c}(1-h/l)\frac{\left(h - z_A - a\right)}{\left(1 - z_A\right)} \tag{8}$$

where a = notch depth, z_A = distance between the neutral axis and the upper edge, ε_{uA} = maximum strain at the lower edge, correspondent to the peak load (maximum $\varepsilon_B = \varepsilon_u$), ε_{rA} = residual strain, equal to zero in the ascending branch of the P-δ curve.

The strain at the notch-tip level in the fracturing section B is:

$$\varepsilon_B = \varepsilon_{rB} + \frac{\sigma_{c,w}}{E_c} = \frac{\varepsilon^2_u}{2\varepsilon_k} + \frac{f_{c,w}}{\kappa w_{max}/\phi_a + 1} \left(1 - \frac{w_{max}}{w_c}\right) \tag{9}$$

where w_{max} = CMOD at the notch tip.

The rotation of section B respect to section A is (Fig. 2a):

$$\varphi = \left(w_{max}/h + \left(\varepsilon_A + \varepsilon_B\right)/2\right)/\left(1 - \zeta_0 - \zeta_1 - a/h\right) \tag{10}$$

where $\zeta_i = z_i/h$ (i = 0, 1, 2), ζ_0 = distance between the neutral and the rotation axis in B, ζ_1 = distance between the neutral axis and the upper edge in B. The crack penetration length is:

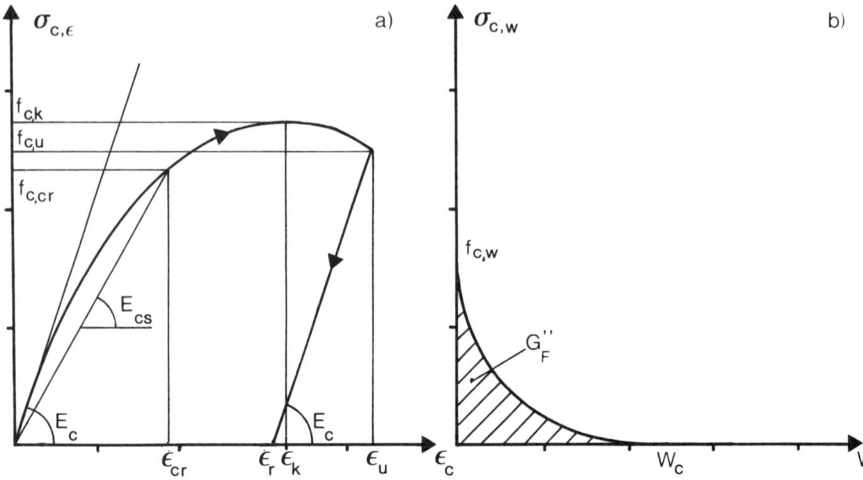

Fig.1. Stress–strain and stress–opening relation for concrete in tension.

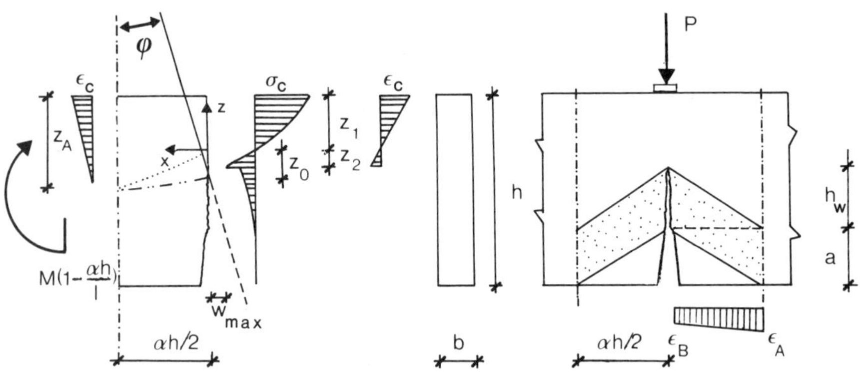

Fig.2. Loaded notched element: stress and strain distribution.

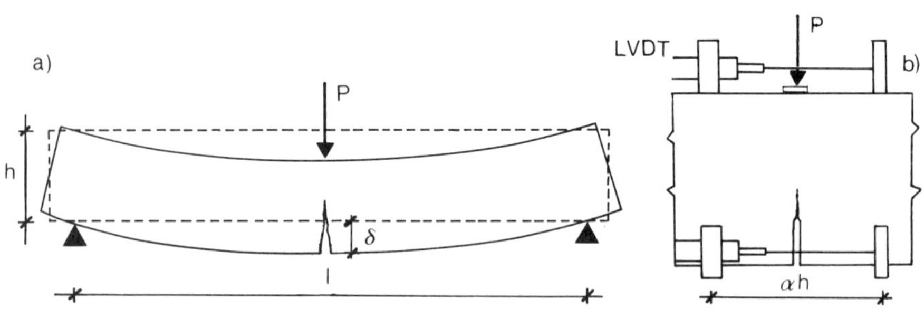

Fig.3. Bending specimen and loaded element instrumentation used.

$$h_W = h - z_1 - z_2 - a \tag{11}$$

where ζ_2 = distance between the neutral axis and the crack tip.

Assuming a linear distribution over the crack, its opening is:

$$w(\zeta) = w_{max}(\zeta - \zeta_2) \,/\, (1 - \zeta_1 - \zeta_2 - a/h) \tag{12}$$

The equilibrium respect to the horizontal direction leads to:

$$N = S_1 E_c \frac{\varphi}{h} + \frac{S_2 E_c}{2\varepsilon_k}\left(\frac{\varphi}{h}\right)^2 + \int_0^{w_c}\int_{z_2}^{h_{Wc}} \sigma_{c,w}\left(w_{max}, z\right)\, b\, dz\, dw_{max} = 0 \tag{13}$$

where S_1 and S_2 = static moments of the solid region in B and h_{Wc} = stress-transfer crack-penetration length:

$$h_{Wc} = h_W\, w_c \,/\, w_{max} \tag{14}$$

The equilibrium respect to the rotation leads to:

$$M = N\left(\frac{h}{2} - z_1\right) + J_1 E_c \frac{\varphi}{h} + \frac{J_2 E_c}{2\varepsilon_k}\left(\frac{\varphi}{h}\right)^2 + \int_0^{w_c}\int_{z_2}^{h_{Wc}} \sigma_{c,w}\left(w_{max}, z\right) bz\, dz dw_{max} = 0 \tag{15}$$

where J_1 and J_2 = moments of inertia of the solid region in B.

The midspan deflection in three point bending test is (Tab. 1):

$$\delta_{tot} = \delta_\gamma + \delta_\varphi + \delta_M + \delta_V + \delta_p \tag{16}$$

Table 1. Contributions to the midspan deflection

contribution to the midspan deflection	non-dimensional midspan deflection δ/h	non-linear structural behavior $E_c \,/\, E_{m,A}$
weight per unit volume	$\frac{5}{384}\,\gamma\left(\frac{l}{h}\right)^4$	1
loaded element rotation	$\frac{\varphi}{4}\frac{l}{h}$	1
bending moment	$\frac{M}{bh^2E_c}\left(\frac{l}{h} - 1\right)^2\left(\frac{h}{l} - 1\right)$	$\frac{1}{(1-\varepsilon_A/2\varepsilon_k)}$
shear effect	$\chi(1+\nu)\frac{2\,M}{bh^2E_c}$	$\frac{1}{(1-\varepsilon_A/2\varepsilon_k)}$
punching effect	$\frac{4\,M}{bh^2E_c}\,\frac{h}{l}\left(\frac{l}{5h} + 5\right)$	1

Referring to a section X along the beam, the mean value of Young's modulus accounting for the non linear response of solid concrete is:

$$E_{m,X} = \frac{1}{2}\left(E_c + E_c\left(1 - \varepsilon_X / \varepsilon_k\right)\right) = E_c\left(1 - \varepsilon_X / 2\varepsilon_k\right) \qquad (17)$$

where ε_X = maximum tensile strain in section X.

4 Identification procedure to determine the fracture energy

The model is applied to a three-step identification procedure.

In the first step a range for each parameter is specified on the basis of direct tensile tests (d.t.t., Figs. 4-6a,b), carried out utilizing a proper mechanical uniaxial testing frame, Giuriani and Rosati (1986). The specimen ends were fixed and the load was applied by variable increments to control the stabilization of deformation and cracking and to maintain the evenly distributed stress state over the fracturing cross-section at each loading step. In the second step three point bending (t.p.b.) specimens were tested on an INSTRON testing machine and their midspan deflection δ (Figs. 3a,4-6c), lower edge elongation e (Figs. 3b,4-6d) and rotation φ (Figs. 3b,4-6e) of the loaded element as functions of the load P were obtained. In the third step these curves were fitted with the theoretically predicted ones by appropriately changing the governing parameters within their first-step-determined range. Finally, fracture energy was determined by integrating the identified constitutive laws (Eq.7 and Tab. 2).

Fracture energy $\mathcal{G}_F^*$ in Tab. 2 is the P-δ curve vs. specimen cross section area ratio. The tested specimens were collected in three series and their material and geometrical properties are indicated in Figs. 4-6. The d.t.t. specimens had the same cross-section area as the t.p.b. specimens and their length was equal to 150mm.

Table 2. Fracture energy for each t.p.b. tested specimen

ϕ_a (mm)	$d\delta/dt$ (μm/sec)	a/h	α	$\mathcal{G}'_F$ (N/mm)	$\mathcal{G}''_F$ (N/mm)	$\mathcal{G}_F$ (N/mm)	$\mathcal{G}^*_F$ (N/mm)
3	0.186	0.0	0.910	0.0097	0.0380	0.0477	0.0502
3	0.186	0.3	0.316	0.0034	0.0380	0.0414	0.0545
7	0.186	0.0	0.625	0.0071	0.0451	0.0522	0.0572
7	0.186	0.3	0.218	0.0025	0.0451	0.0476	0.0632
15	0.093	0.0	1.050	0.0141	0.1327	0.1468	0.1657
15	0.186	0.3	0.367	0.0049	0.0805	0.0854	0.0886

The ultimate strength $f_{c,u}$, the residual strength $f_{c,w}$ and the critical crack opening w_c depend on the water/cement ratio. The following empirical relation for the critical opening, based on the present results and on tests by Mihashi et al. (1988), is proposed:

$$w_c = \frac{\phi_a}{100} + 5\omega\,\frac{w}{c}\left(1 - e^{-2\omega\phi_a}\right) - \omega\left(1 - e^{-4\omega\phi_a}\right) \qquad (18)$$

where ω is a coefficient experimentally determined (ω = 0.15 in the

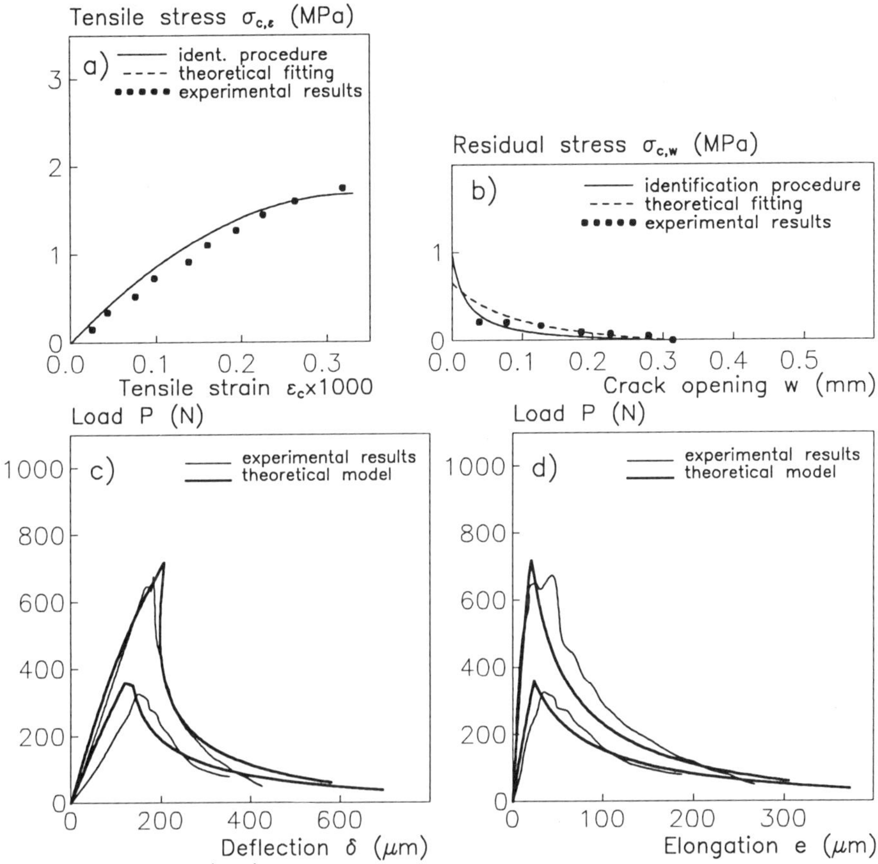

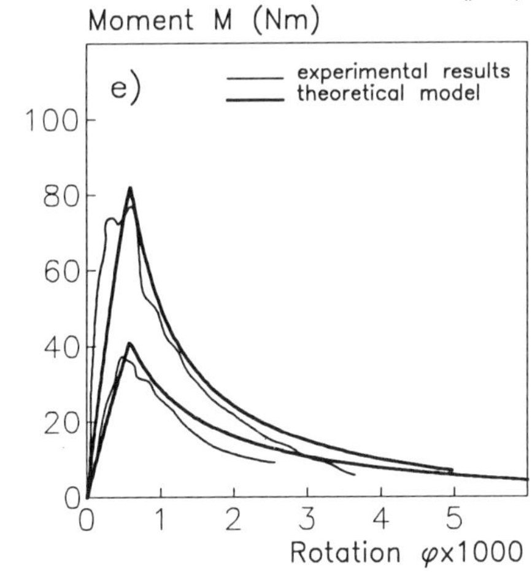

f_{cc} (MPa)	Φ_a (mm)	w/c	b (mm)	h (mm)	l/h	a/h
12.8	3	1.0	50.0	65.3	7	0.3

test	E_c (GPa)	ε_{cr} (%)	$\varepsilon_u=\varepsilon_k$ (%)	$f_{c,u}$ (MPa)
d.t.t.	10	0.024	0.033	1.65
t.p.b.	10	0.024	0.033	1.65

test	$d\delta/dt$ (μm/sec)	$f_{c,w}/f_{c,u}$	κ	w_c (mm)
d.t.t.	–	0.39	30	0.31
t.p.b.	0.186	0.58	150	0.31

Fig.4. Application of the identification procedure to series n.1.

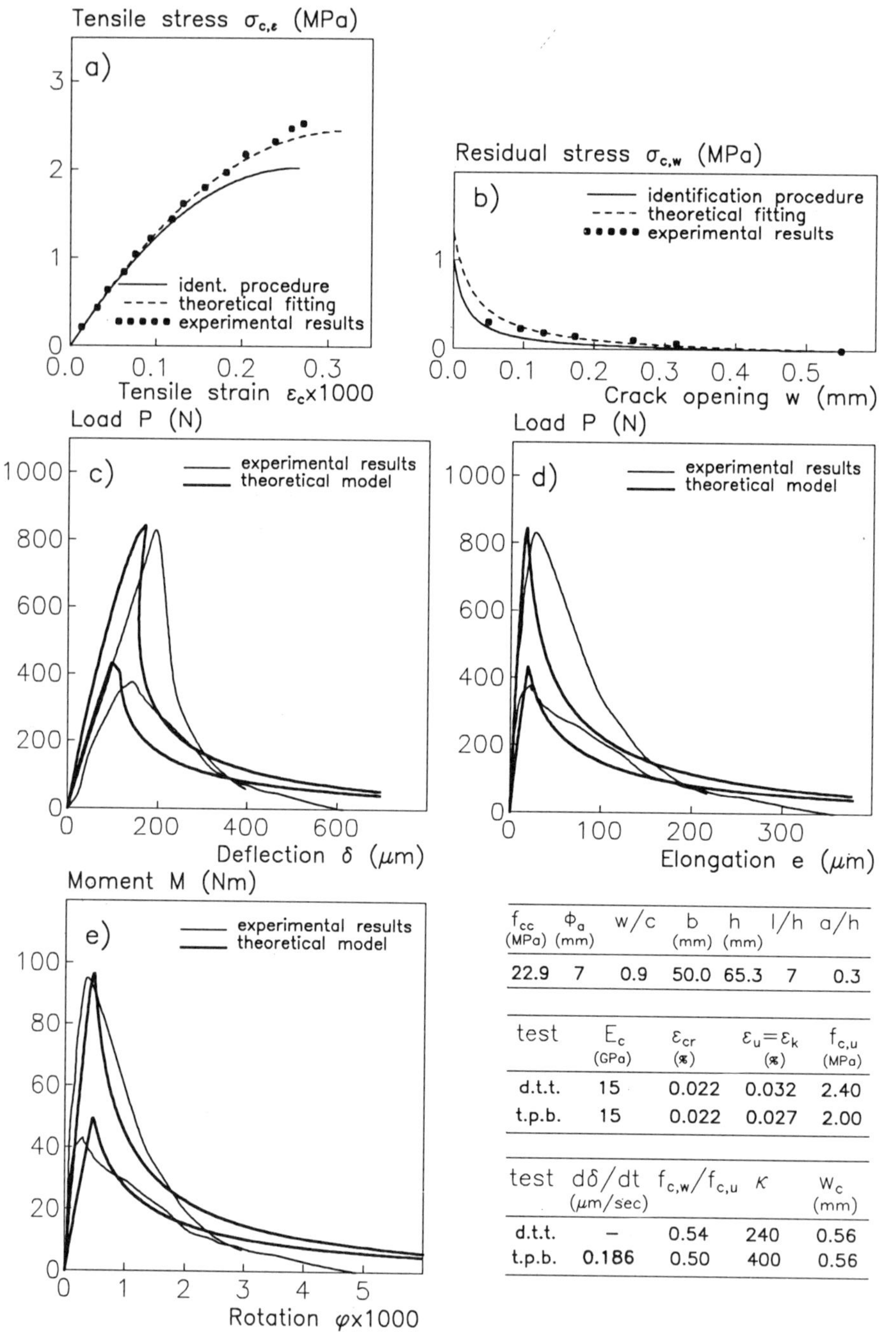

f_{cc} (MPa)	Φ_a (mm)	w/c	b (mm)	h (mm)	l/h	a/h
22.9	7	0.9	50.0	65.3	7	0.3

test	E_c (GPa)	ε_{cr} (‰)	$\varepsilon_u = \varepsilon_k$ (‰)	$f_{c,u}$ (MPa)
d.t.t.	15	0.022	0.032	2.40
t.p.b.	15	0.022	0.027	2.00

test	$d\delta/dt$ (μm/sec)	$f_{c,w}/f_{c,u}$	κ	w_c (mm)
d.t.t.	–	0.54	240	0.56
t.p.b.	0.186	0.50	400	0.56

Fig.5. Application of the identification procedure to series n.2.

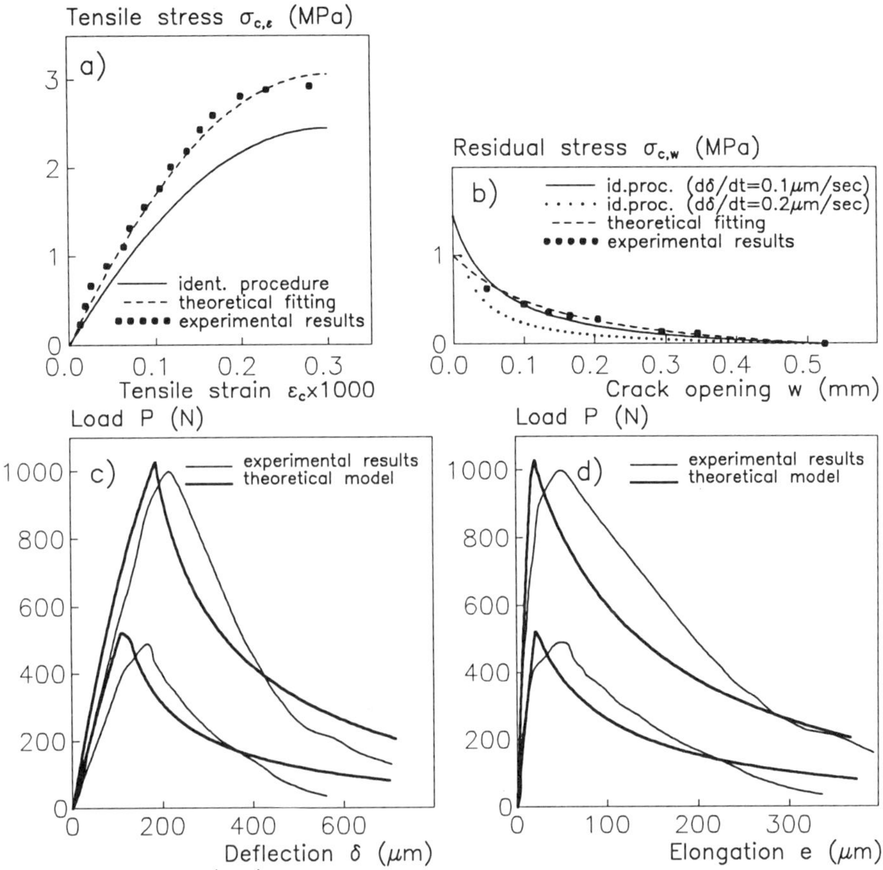

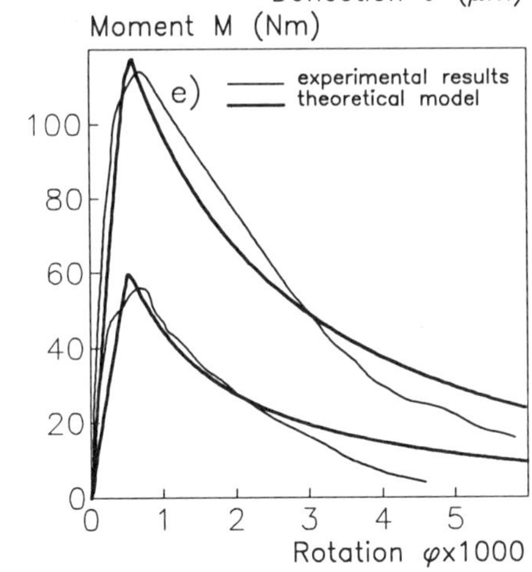

f_{cc} (MPa)	Φ_a (mm)	w/c	b (mm)	h (mm)	l/h	a/h
36.1	15	0.7	50.0	65.3	7	0.3

test	E_c (GPa)	ε_{cr} (‰)	$\varepsilon_u=\varepsilon_k$ (‰)	$f_{c,u}$ (MPa)
d.t.t.	20	0.021	0.030	3.00
t.p.b.	16	0.021	0.030	2.40

test	$d\delta/dt$ (μm/sec)	$f_{c,w}/f_{c,u}$	κ	w_c (mm)
d.t.t.	–	0.33	100	0.52
t.p.b.	0.093	0.60	250	0.52
t.p.b.	0.186	0.60	600	0.52

Fig.6. Application of the identification procedure to series n.3.

present study). In t.p.b. tests of the third series two different values for the deflection rate were adopted (Tab. 2 and Fig. 6).

5 Concluding remarks

The present model predicts the fracture energy dissipation both through the microcracked tip-front and over the surface of a flexural crack. By this approach, it is possible to predict a representative length of the microcracked (αh) and the fracturing (h_{wc}) zone and to detect the material notch-sensitivity.

Fracture energy is mainly governed by aggregate debonding and is quite sensitive to maximum aggregate size.

The slower the deflection rate is, the more stable the fracturing phenomena develop. As a consequence, fracture energy dissipated over the crack surface seems to increase.

The area delimited by a P-δ curve could be representative of the dissipated fracture energy if all the constitutive parameters and the loading rate effects are accurately checked and discussed.

For concrete in bending, the fracture process leads to a strain localization in the loaded element, so that it is possible to fully describe the non linear beam response by moment - rotation relations.

More test data are necessary to support all these observations.

Acknowledgments
The authors wish to thank P. RIVA for his helpful assistance and the D.I.C., University of Brescia, Brescia, Italy, for allowing the use of the INSTRON testing machine.

6 References

Cornellissen, H.A.W., Hordijk, D.A. and Reinhardt, H.W., (1986) Experimental determination of crack softening characteristics of normalweight and lightweight concrete, **Heron**, Delft University of Technology, 31, pp. 45-56.

Giuriani, E. and Rosati, G., (1986) Comportamento di elementi tesi di calcestruzzo in fase fessurata, **Studi e Ricerche**, Politecnico di Milano, 8, pp. 65-82.

Giuriani, E. and Rosati, G., (1987) An analytical model for the study of the crack propagation in plain concrete under Bending, **Studi e Ricerche**, Politecnico di Milano, 9, pp. 107-127.

Mihashi, H. Nomura, N. and Izumi, M., (1988) Influence of internal structure on strain softening and fracture energy of concrete, **International Workshop on Fracture Toughness and Fracture Energy - Test Methods for Concrete and Rock**, Sendai, Japan, pp. 77-89.

Rosati, G. and Schumm, C., (1990) La meccanica della frattura nel calcestruzzo: considerazioni sull'energia specifica di frattura, **Studi e Ricerche**, Politecnico di Milano, 11, pp. 25-59.

Van Mier, J.G.M. and Nooru-Mohamed, M.B., (1990) Geometrical and structural aspects of concrete fracture, **Engineering Fracture Mechanics**, 35, pp. 617-628.

PART EIGHT

NUMERICAL MACROSCOPIC CRACK MODELS: DISCRETE

INVITED PAPER

48 THREE DIMENSIONAL SIMULATION OF HYDRAULICALLY DRIVEN FRACTURE PROPAGATION IN A ROCK MASS

J.O. SOUSA
Universidade Estadual de Campinas, Brazil, presently at Cornell University, Ithaca, NY, USA
E.D. LUTZ and A.R. INGRAFFEA
Cornell University, Ithaca, NY, USA

Abstract
This paper describes numerical simulation of near-wellbore phenomena related to hydraulic fracturing from a perforated wellbore. The objective is to compare numerical results with records of an experiment performed by Behrman and Elbel (1990) in a large triaxial cell under simulated downhole conditions. Comparisons on the geometry of the induced fractures and on the history of fracture evolution are pursued.
Keywords: Hydraulic Fracturing, Fracture Propagation, Perforations.

1 Introduction

Behrman and Elbel(1990) have reported on a sequence of experiments to study the crack shape, pressure history, and fracturing fluid volume in a cased, perforated wellbore. This paper reports on a numerical simulation using FRANSYS† of Experiment 9 in that series.

The type of results expected from the comparison between experimental and computational results consists of:

- Correlation of features of the induced fractures, including:
  - q) coalescence of 8 initial cracks into 4 and, finally, 2 cracks;
  - b) development of classic bi-wing crack shape following coalescence.
  - c) turning of crack orientation to align with the "in situ" stress field, that is, 10 degrees away from the orientation of the perforations;
- Correlation between crack volume and volume of fluid pumped, resulting in characterization of time for each propagation step;
- Correlation between pressure records and crack shape.

† FRANSYS is not related to ANSYS, a product of Swanson Analysis Systems, Inc

Fracture Processes in Concrete, Rock and Ceramics Edited by J.G.M. van Mier, J.G. Rots and A. Bakker.
 Published by E. & F.N. Spon, 2–6 Boundary Row, London SE1 8HN. ISBN 0 419 15870 7

Within the series of experiments done by Behrman and Elbel, the case of 10 degree misalignment between the perforations and the "in situ" stress field is of interest because, among the misalignments tested, it is the largest one to produce the desirable conditions that (a) the cracks initate at the perforations, (b) initial crack growth is along the perforations, and (therefore) (c) the fracturing fluid and proppant pass from the wellbore to the crack by way of the perforation throughout the loading period.

2 Description of the Models

The geometrical and physical data corresponding to the specimen used in the experiment are presented in Figures 2.1 and 2.2, and Tables 2.1 and 2.2. Only the rock and the cement around the casing were simulated, and these were considered as a single material with the rock properties. The existence of the casing was partially taken into account via a reduction in the pressure effectively applied to the rock. The reduced pressure was computed with a simple model of a pressurized steel tube surrounded by rock. The analysis resulted in 27.8% of the pressure internal to the casing being transmitted to the rock. Due to the observed flow of fracturing fluid through the microannulus around the casing near the perforations, 100% of the wellbore pressure was applied to the wellbore walls in the vicinity of the perforations. Linear elastic fracture mechanics behavior was assumed. Poroelastic effects were not taken into account in the simulation.

The pressure in all the cracks was assumed constant, uniformly distributed over the entire crack surface and equal to the wellbore pressure. As the pumping rate was kept constant and equal to 15.0 ml/sec (0.915 cu in/sec) during the experiment, the wellbore pressure was actually variable with respect to time. For the purpose of this simulation, the wellbore pressure was chosen as 5250 psi for all simulations, as an approximated average obtained from the pressure-versus-time records from the experiment.

The geometry of the propagating crack surface is determined according to the stress intensity factors computed for the previous step. The stress intensity factors are used in expressions developed by Hussain et al. (1974) to compute the energy release rate at each point on the crack front as a function of the direction of propagation. The criterion to determine the geometry of the new crack front assumes that:

- The direction of propagation is the one that maximizes the potential energy release rate at each point on the crack front
- The extension of propagation from each point on the crack front is proportional to the energy release rate for the corresponding direction of propagation.

The stress intensity factors were computed with the displacement correlation technique, as described by Sousa et al. (1989).

The structural analyses were performed by the direct boundary element method, using a technique being developed by Lutz and Gray

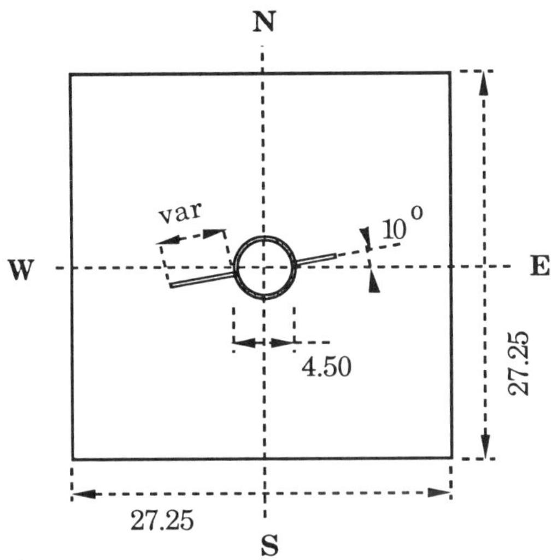

Figure 2.1 - Perf-Frac Experiment - Plan view All Dimensions in Inches

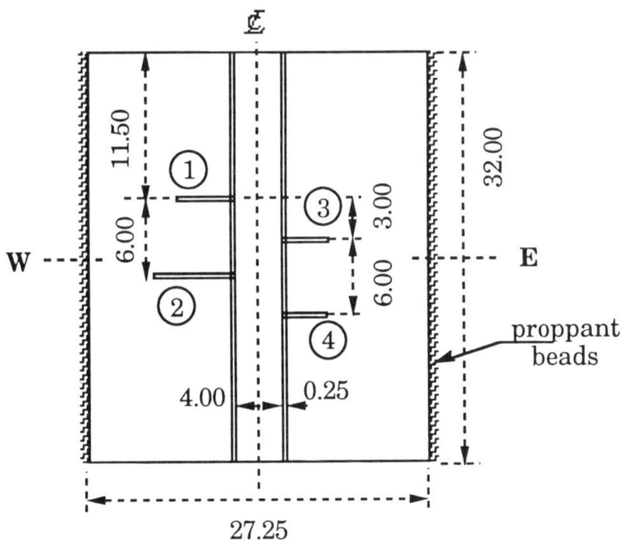

Figure 2.2 - Perf-Frac Experiment - Cross-Sectional View All Dimensions in Inches

(1990). The advantage of this technique is that it does not require that each side of the crack be in a different domain. This is accomplished by replacing the equations corresponding to the elements on one side of each crack by equations related to the surface stresses on the cracks. This corrects the rank deficiency of the primary boundary integral equation for a cracked body. This procedure requires the computation of hypersingular integrals, with special difficulties arising for non-planar cracks and for the intersection of the cracks with surfaces of the object.

A general solution for this problem, including internal and surface cracks of arbitrary shape, was achieved with the use of non-conforming elements on the crack faces and on the external surface framing onto the crack edge. This reduces considerably the work in modeling at each new step of propagation, since there is no need for internal faces of complex geometry ahead of the crack front to subdivide the model into different domains. However this technique has one important drawback: the use of non-conforming elements increases considerably the number of degrees-of-freedom. Although the current implementation of this technique is able to deal with linear and quadratic elements, in this application it was necessary to use only linear elements, with a relatively small number of elements along the crack front, in order to keep the total number of equations within the limits dictated by the available mini-supercomputer.

Table 2.1 - Physical Data

Type of rock: Torrey Buff Sandstone
Fracturing fluid: Brine

E_{rock} = 2,000,000 psi	ν_{rock} = 0.20
K_{Ic} = 622 psi $\sqrt{in}$	
k_v = 0.29 md	k_h = 0.49 md
σ_v = 5460 psi	σ_{NS} = 3180 psi
σ_{EW} = 4100 psi	σ_i = 5250 psi (wellbore pressure)

Table 2.2 - Perf-Frac Experiment - Dimensions of Perforations

perforation number	length [in]	diameter [in]	
		mouth	tip
1	4.10	0.50	0.35
2	5.90	0.40	0.25
3	3.70	0.45	0.30
4	3.40	0.50	0.35

3 Description of the Analyses Performed

Four steps of propagation were analyzed. The geometric models after the first, second and fourth steps are presented in Figures 3.1a through c. The geometric model after the fifth step was predicted, Figure 3.1d, but not analyzed.

3.1 Crack Initiation and First Step of propagation

Two cracks were initiated at the base of each perforation, one above and the other below, following the indication from the contour plot of maximum principal stresses (see contour plot of stress in the direction N-S in Figure 3.2). The cracks were nucleated with a quarter-circular shape of radius one-half inch. The effect of the fluid flowing through the microannulus around the casing was taken into account by applying full pressure on the wellbore walls from 3 inches below the level of the lower perforations to 3 inches above the level of the upper perforations.

The stress intensity factors are presented in Figures 3.3 to 3.5. The abscissae for these plots assume that the crack fronts are oriented from the lower to the upper end; that is, the lower crack front is measured from its intersection with the wellbore towards its intersection with the perforation, and the upper crack front is measured from the perforation to the wellbore.

3.2 Second Step of Propagation

In this step both cracks at each perforation were merged. According to studies conducted by Lamkin et al. (1989), using two dimensional simulations, it is expected that when two propagating cracks coalesce, they first avoid each other and then one intersects the other along its length, not at its tip. Similar behavior is expected in this simulation. However, no attempt to describe this phenomenon was made here, due to the step sizes and mesh refinement being used, which were dictated by the boundary element analysis system and the capacity of the mini-supercomputer used. The alternative chosen was to merge both propagating crack fronts into a single line, and then describe the propagating patch of crack surface by a ruled surface generated by this merged line and the two old crack fronts, Figure 3.1b. This solution, although simplified, describes reasonably well the behavior, since normally only the main branch of the coalescing cracks remains open, while the remaining branches close.

The resulting two cracks at each side of the wellbore were connected by a surface crack patch, created interactively, resulting in a single crack of complex shape, as presented in Figure 3.1b.

The deformed shape in the region of the cracks is presented in a downhole view (Figure 3.6a).

The stress intensity factors are presented in Figures 3.7 to 3.9, together with the values computed for the next two steps. In the region where both crack merged, high values of K_I occurred, as expected, due to the

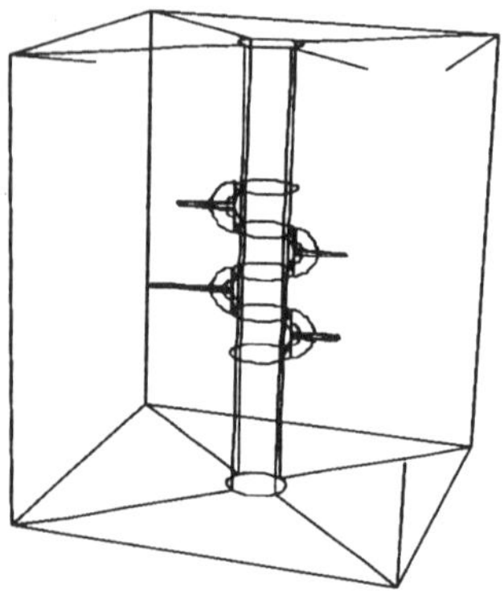

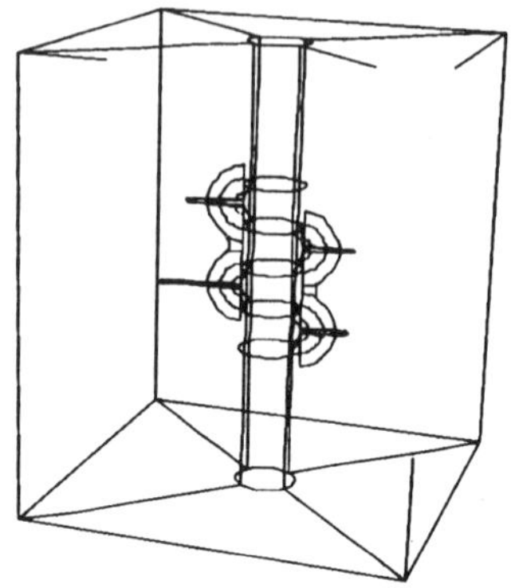

(a) First Propagation Step

(b) Second Propagation Step

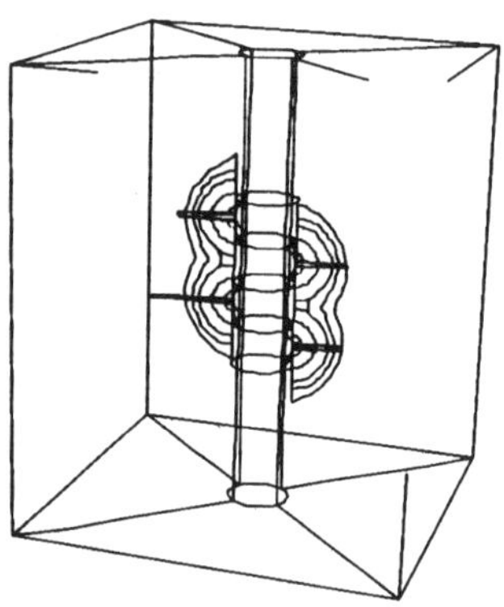

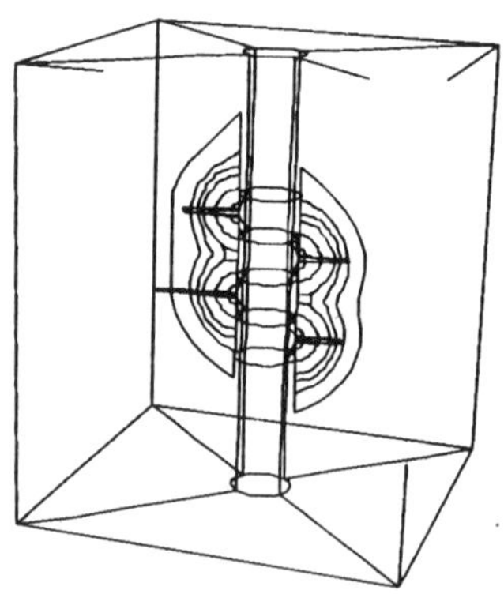

(c) Fourth Propagation Step

(d) Fifth Propagation Step

Figure 3.1 Evolution of Crack Geometry

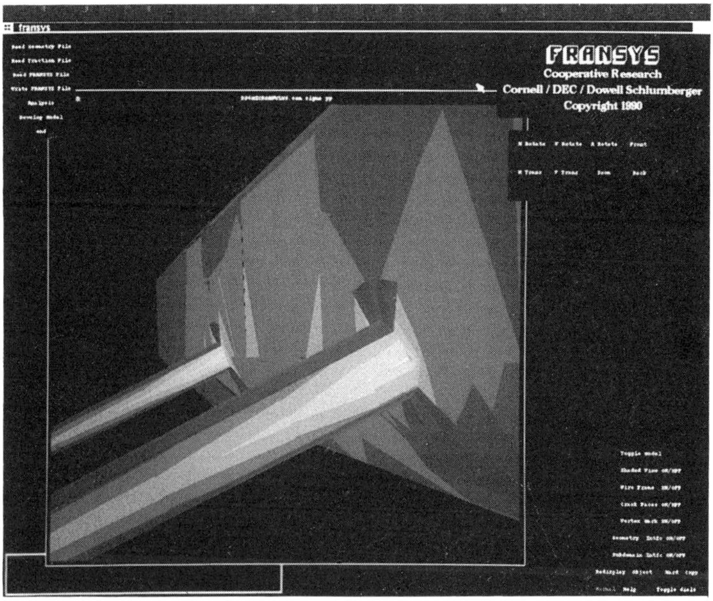

Figure 3.2 - Contour Plot of Stress in the N-S direction in Vicinity of Perforations - σ_i = 5250 psi

re-entrance in the crack front geometry. In the following steps, it was expected that this re-entrance would be progressively reduced.

3.3. Third to Fifth Steps of Propagation

The geometric model and the associated surface mesh for the fourth step are presented on Figures 3.1c and 3.6b. The geometry of the crack fronts at the end of all the previous steps of propagation can be seen on the crack surfaces. The tendency to straighten the re-entering portion of the crack front is clear in the geometry of the second to the fourth steps. In the predicted geometry for the fifth step, Figure 3.1d, this re-entrance in the crack front geometry is almost eliminated, with the crack shape approaching that of a half-penny shaped surface crack. A down hole view of the deformed shape is presented in Figure 3.6c. A side view of the deformed shape is presented in Figure 3.6d.

The predicted evolution of the crack volume is presented in Figure 3.10.

4 Discussion and Summary

4.1 Geometry

The cracks, initiated in pairs at the base of each perforation, changed direction in the first step, aligning with the direction normal to the minimum far field stress. In the second step both cracks at each perforation merged into one. The resulting geometry, with two cracks at each side of the wellbore, is similar to the shapes obtained in the experiment, marked by the red color and carbon black, especially at the East side, as shown in Figures 4.1. The coalescence of the two cracks at each

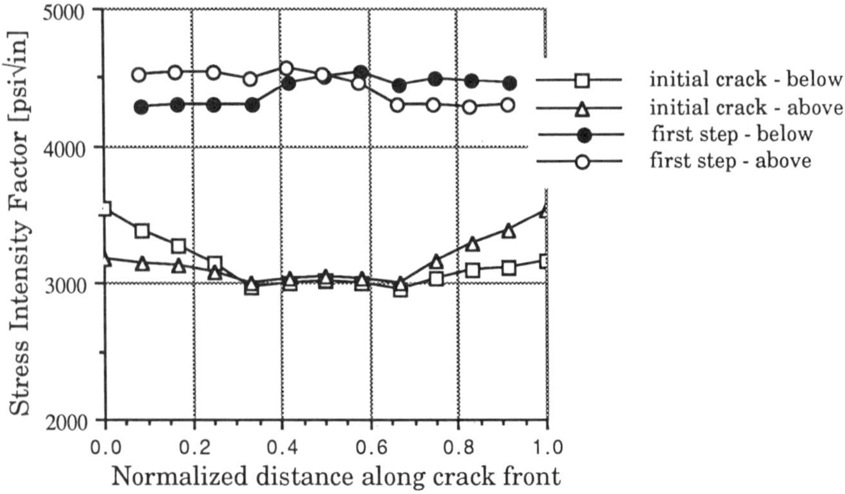

Figure 3.3 Mode I Stress Intensity Factors - σ_i = 5250 psi - Perf. #3

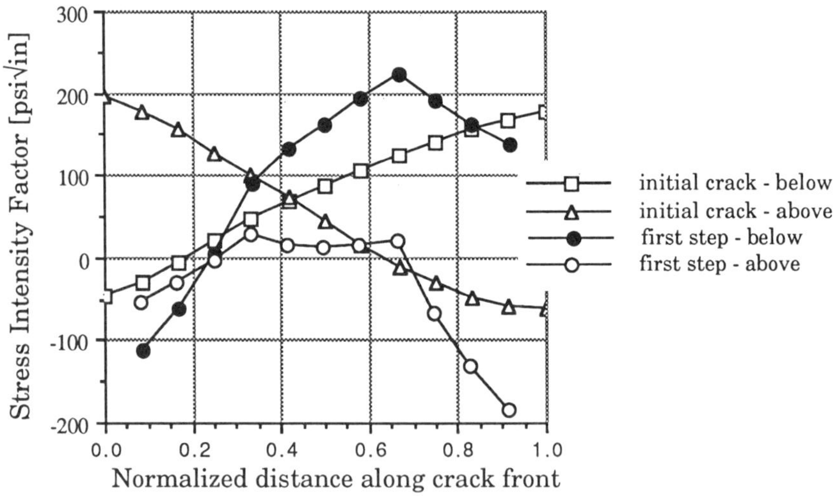

Figure 3.4 Mode II Stress Intensity Factors - σ_i = 5250 psi - Perf. #3

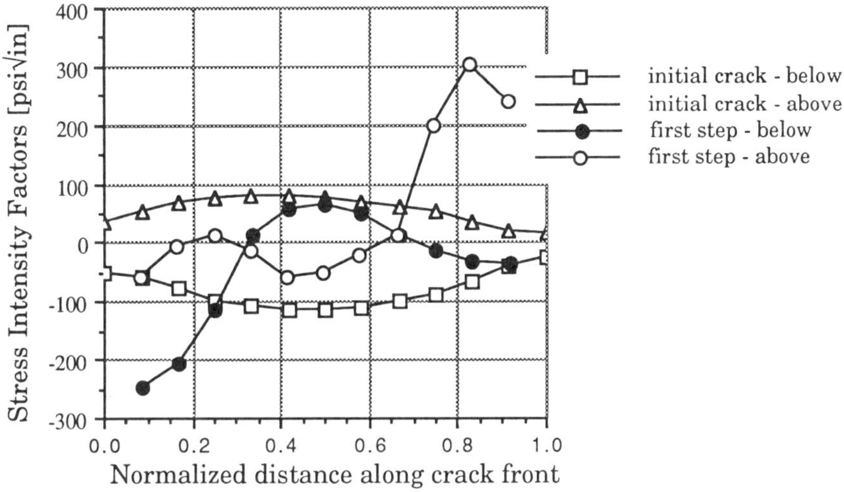

Figure 3.5 Mode III Stress Intensity Factors - σ_i = 5250 psi - Perf. #3

perforation occurred within less than 2 inches from the wellbore walls, as observed in the physical tests. However this conclusion is influenced by the increment sizes chosen for these two steps and should be further investigated in a convergence study.

As mentioned in Section 2.2, the coalescence was modelled by merging the cracks, with no attempt to obtain the detailed history, as in the two-dimensional studies conducted by Lamkin et al. (1989).

The re-entering portion of the crack front, in the region where the cracks from both upper and lower perforations merged, created relatively high values of K_I. This resulted in a progressive reduction of the re-entrance in the following steps. After the fifth step the re-entrance is practically gone, and the crack at each side approaches the semi-circular shape, tending to the classical bi-wing crack. The wing-shape at each side was reached at about 6 inches from the wellbore walls, in agreement with the patterns displayed on Figures 4.1a and b at the West side.

Except for the major change in direction which occurred in the first step and for some oscillations in the orientation of the crack along its intersection with the wellbore, the resulting crack at this stage is practically flat, as observed in the experiment, Figure 4.1c.

4.2 Records of Pressure and Pumping Rates from the Experiment

The computed crack volumes presented in Figure 3.10 lead to the conclusion that the main destination of the pumped fluid was not the induced cracks. A simple model, using just one planar crack of length 7.75 inches at each side of the wellbore was analyzed to obtain an approximate value for the crack volume at a later step of propagation, close to the

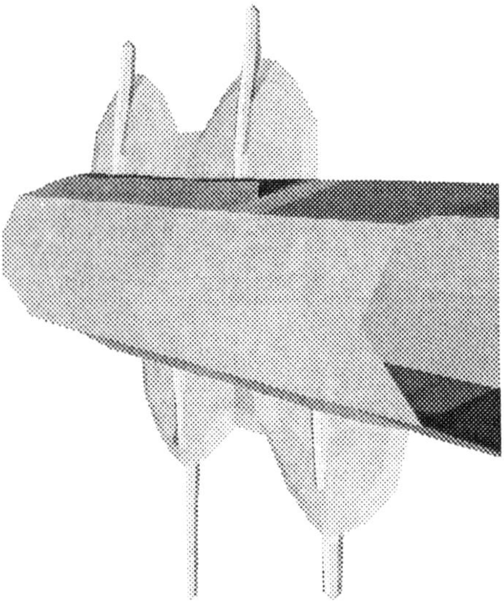

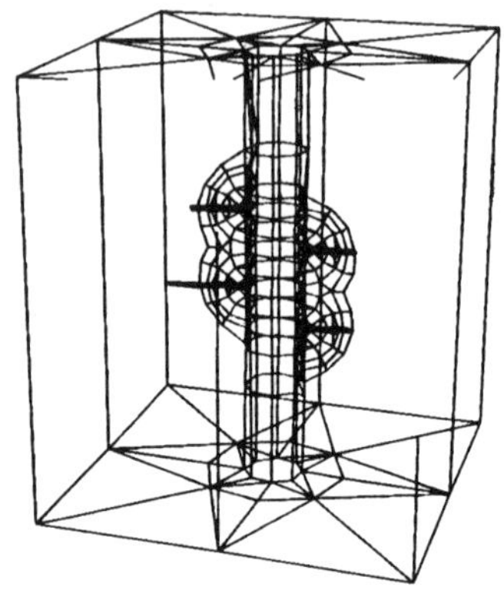

(a) Deformed Shape - 2 nd Step
Magnification Factor: 50

(b) Surface Mesh - 4 th Step

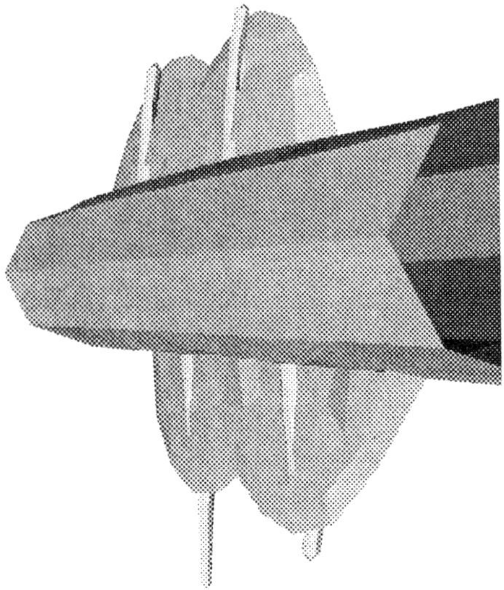

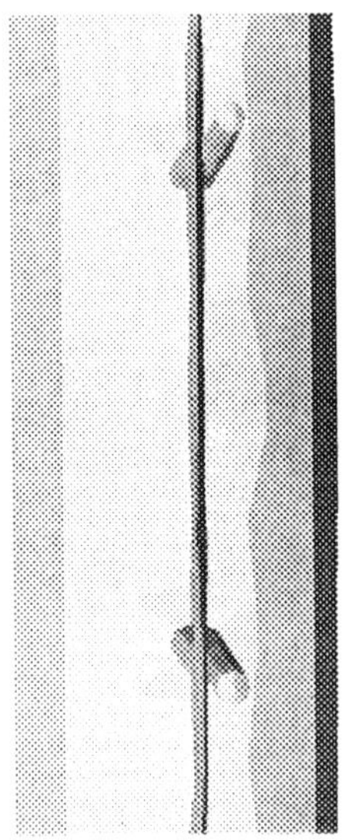

(c) Deformed Shape - 4 th Step
Magnification Factor: 50

(d) Deformed Shape - 4th Step
Magnification Factor: 10

Figure 3.6 Mesh and Deformed Shapes

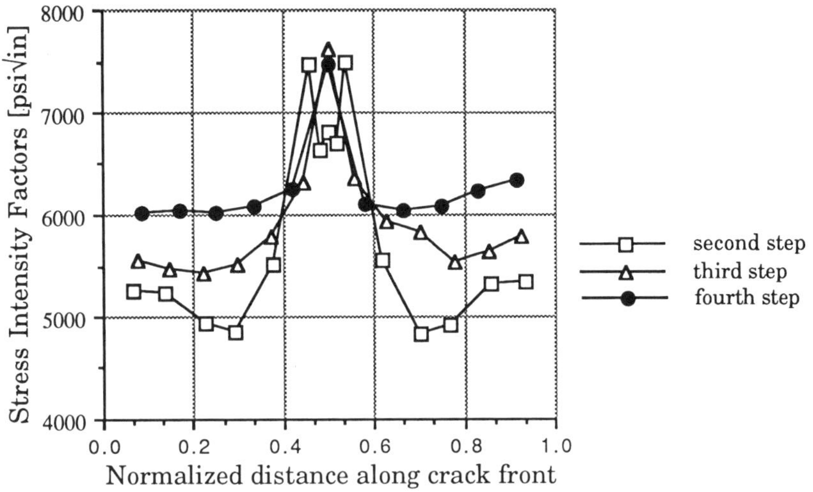

Figure 3.7 - Mode I Stress Intensity Factors - σ_i = 5250 psi - East Side

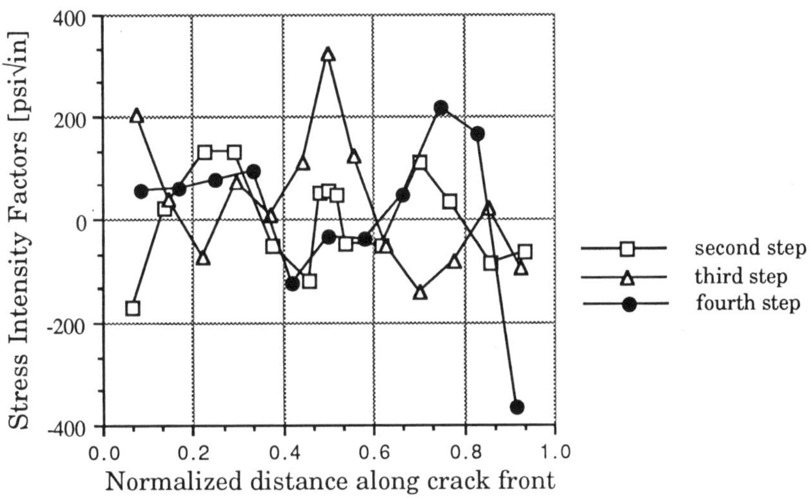

Figure 3.8 - Mode II Stress Intensity Factors - σ_i = 5250 psi - East Side

condition of the block being split. This model was subjected to the same crack pressure assumed for all the simulations in this report: 5250 psi on the entire surface of the crack. The results led to a value of 19.3 cu in (316.3 ml) for a crack area of 496 sq in (3200 cm^2). This computed volume corresponds to only 3.5% of the total volume of fluid, 549.2 cu in (9000 ml), pumped throughout the 600 seconds taken to perform the experiment. Even if the volume of the microannulus, the bottom crack (see Figure 4.1b) and other minor cracks, and the leak off to the matrix are taken into account, it is not possible to account for the destination of about 96.5% of the total volume pumped. The most reasonable explanation is that the fluid leaked through the microannulus and the spurious bottom crack into the beadpack around the specimen (see Figure 2.2). This assumption is supported by beadpack pressure variation during the experiment.

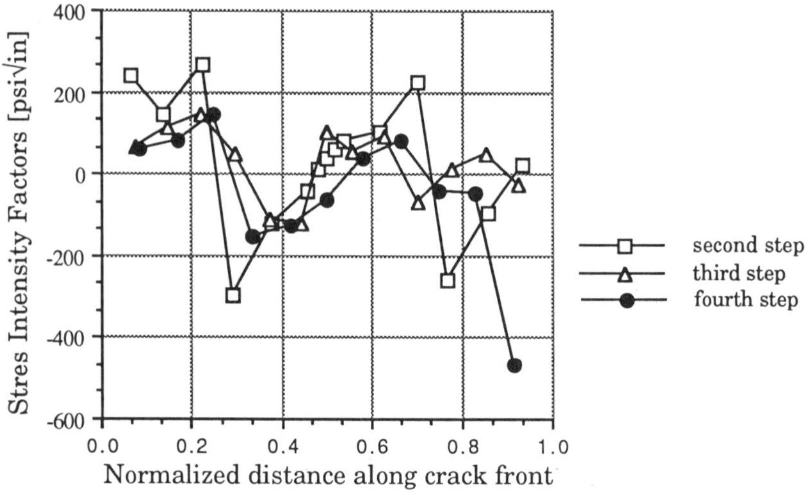

Figure 3.9 - Mode III Stress Intensity Factors - σ_i = 5250 psi - East Side

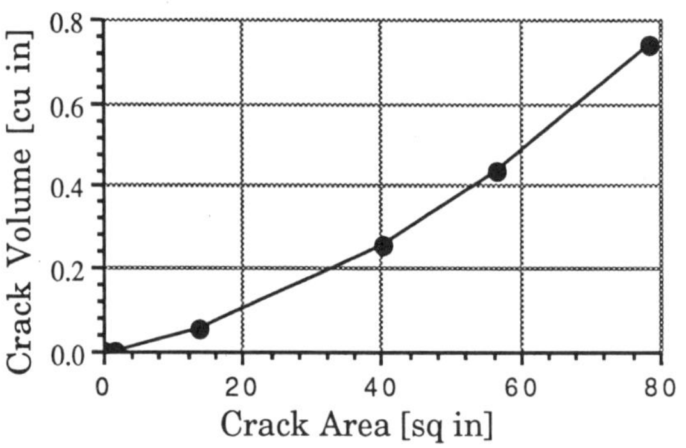

Figure 3.10 - Crack Volume Evolution - σ_i = 5250 psi

(a) North End

(b) South End

(c) Top View - South East Side

(d) Detail - South-East Side

Figure 4.1 Experiment

4.3 Analytic Estimates of Stress Intensity Factors

Because of the assumption of uniform fluid pressure throughout the crack, compute values of K_I are about one order of magnitude higher than the critical stress intensity factor $K_{Ic} = 622$ psi$\sqrt{\text{in}}$ for the rock being fractured. This pressure distribution is known to be inaccurate. It is known from the literature that there is a region in the neighborhood of the crack front that is not reached by the fluid. It is also known that the distribution on the rest of the crack is not constant and depends on the time, fracture conductivity, pressure gradients, etc.

For comparison purposes, consider approximating the crack at the end of the third step of propagation by a simple 16 in. diameter circular crack. If buried in an infinite medium and loaded by a uniform pressure of 2070 psi (the difference between the assumed 5250 psi crack pressure and the 3180 psi far field normal stress), the stress intensity factor in mode I is given (Tada (1977)) by :

$$K_I = \frac{2}{\sqrt{8\,\pi}}\, 2070 \times 8 = 6606 \text{ psi } \sqrt{\text{in}}$$

This is close to the values computed at the end of the third step of propagation (Figure 3.7).

A closer approximation to the real pressure distribution is provided by pressure varying from a maximum at the crack mouth to 0 at the crack front. If the crack pressure has a distribution of the type,

$$\sigma = p \left\{ \sqrt{1 - \frac{r}{a}} \right\}^{\gamma}$$

where
p = maximum value of the crack pressure
a = crack diameter
r = radial coordinate
γ = parameter characterizing the pressure distribution

the result for the circular crack in an infinite medium is

$$K_I = 6606 \frac{1}{\gamma + 1} \text{ psi } \sqrt{\text{in}}$$

Curves of the pressure distribution factor $\left\{ \sqrt{1 - \frac{r}{a}} \right\}^{\gamma}$ are plotted in Figure 4.2 for different values of γ. For γ of 5 or larger, the stress intensity factors are reduced to 1/6 or less of the value obtained with the constant pressure distribution. This is much closer to the fracture toughness of the material.

These estimates show the importance of accurate pressure modeling when deciding whether propagation occurs at each step. This calls for simulation of the coupling between the stress/displacement analysis an analysis of pressure and fluid flow in the crack.

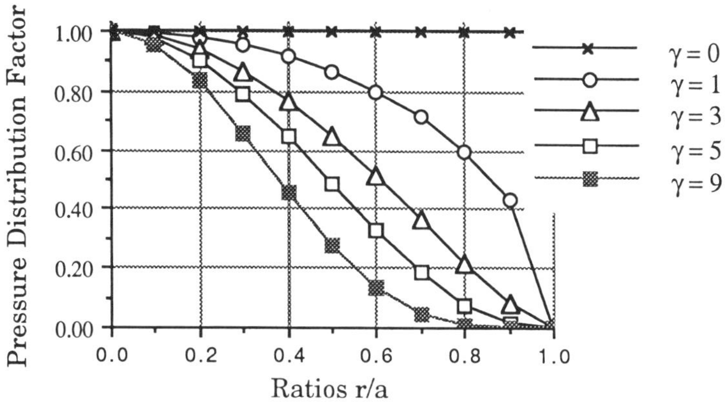

Figure 4.2 - Variation of Pressure Distribution Factor $\left\{ \sqrt{1 - \frac{r}{a}} \right\}^{\gamma}$

4.4 Summary

A computational simulation of the Perf-Frac Experiment 9, Behrman and Elbel (1990), was performed with four complete steps of propagation and the fifth predicted. The results of the simulation were compared to the experimental results, leading to the conclusions:

- The simulation reproduced the geometry of the induced cracks, starting in pairs from the bases of the perforations and coalescing towards a classical bi-wing crack normal to the least far field stress and connected to all perforations.
- Only a small percentage (about 3.5%) of the fluid being pumped, was used to induce the crack. Most of the remaining fluid went, probably, to the beadpack around the specimen via the microannulus and the bottom crack.
- The knowledge of the actual crack pressure distribution is important for the analysis of crack stability. The assumption of constant pressure over the entire crack may lead to values of K_I one order of magnitude higher than required for propagation

The most urgent areas for improvement in the simulation system are:

- Expansion of the limits of the analysis capabilities. The largest model analyzed has 4836 degrees of freedom, requiring 4 hours and 52 minutes of CPU on a FPS/Celerity-500 mini-supercomputer for BEM integration and linear system solution.
- Coupling of the stress/displacement computations with analysis of the fluid pressure and flow in the crack.
- Extension of FRANSYS to support simulation of the fluid flow through the microannulus.
- Extension of FRANSYS to support cracks intersecting each other along a line on a crack surface, and not only by merging crack fronts into a single line, as in the current implementation.

5 Acknowledgements

The authors would like to express their gratitude to the Cornell Program of Computer Graphics for continued support. The detailed information on the experiments was provided by Dr. L. Behrman, from Schlumberger Perforating and Testing Center, whom the authors would like to thank. The collaboration of Paul A. Wawrzynek, David O. Potyondy and Túlio N. Bittencourt is appreciated. This work was partially funded by CAPES (Brazilian Sponsorship Agency), by Dowell-Schlumberger Incorporated, and by the Digital Equipment Corporation.

6 References

Behrman, L. A. and Elbel, J. L., (1990) Effect of Perforations on Fracture Initiation. **65th Annual Technical Conference and Exhibition of the Society of Petroleum Engineers**, SPE 20661, New Orleans, LA, pp 459-472.

Hussain, M. A., Pu, S. L., and Underwood, J. H., (1974), Strain Energy Release Rate for a Crack Under Combined Mode I and Mode II, in **Fracture Analysis, ASTM,** STP 560, pp 2-28.

Lamkin, S. J., Wawrzynek, P. A., and Ingraffea, A. R. (1989) Two-Dimensional Numerical Simulation of Interacting Fractures in Rock, in **Fracture of Concrete and Rock: Recent Developments** (eds S. P. Shah, S. E. Swartz, B. Barr), Elsevier Science Publishing, New York, N. Y., 1989, pp. 121 - 131.

Lutz, E. D., Gray, L. S. and Ingraffea, A. R. (1990) Indirect Evaluation of Surface Stress in the Boundary Element Method. **IABEM-90: Symposium of the International Association for Boundary Elements Method**, (to appear).

Martha, L. F., (1989) Topological and Geometrical Modeling Approach to Numerical Discretization and Arbitrary Fracture Simulation in Three Dimensions. **PhD. Thesis**, Cornell University, Ithaca, NY.

Sousa, J. L.. Martha, L. F, Wawrzynek, P. A. and Ingraffea, A. R (1989) Simulation of Non-Planar Crack Propagation in Structures of Concrete and Rock. **Fracture of Concrete and Rock: Recent Developments** (eds. S. P. Shah, S. E. Swartz, B. Barr), Elsevier Science Publishing, New York, NY, pp. 254 - 264.

Tada, H., Paris, P. and Irwin, G. (1973) **The Stress Analysis of Cracks Handbook**. Del Research Corporation, Hellertown, PA.

49 FRACTURE, PATH BIFURCATIONS AND INSTABILITIES IN ELASTIC-COHESIVE-SOFTENING MODELS: A BOUNDARY ELEMENT APPROACH

G. MAIER
Politecnico di Milano, Italy
Z. CEN
Tsinghua University, Beijing, People's Republic of China
G. NOVATI and R. TAGLIAFERRI
Politecnico di Milano, Italy

Abstract
Response bifurcations and instabilities under load control or displacement control (intrinsic or snap-back), dealt with in the late Sixties for elastic-softening beams, are discussed here in fracturing processes of elastic-cohesive crack models of solids making use of some basic concepts combined with two distinct boundary element methods. These methods appear ideally suited to analyze the fracture propagation and to provide stability criteria.
Keywords: Cohesive-Crack, Fracture, Bifurcations, Instability, Boundary Elements.

1 Introduction

The problem to be discussed in this paper can be described as follows. The fracture process is modelled according to the cohesive-softening crack hypothesis, namely: the stress-strain relationship is linear elastic; however, a discontinuity of normal displacements Δu_n is permitted when the tensile stress attains an ultimate value $\bar{\sigma}$, across a surface or line (called herein "craze") ahead of the crack; the concomitant normal traction p_n is a decreasing function of Δu_n and vanishes when $\Delta u_n \geq \bar{w}$, $\bar{w}$ being another material parameter. As for the shear force p_t along the craze, two alternative hypotheses can be assumed: (a) unlimited p_t with no discontinuity in the tangential displacements ($\Delta u_t = 0$); (b) no shear strength ($p_t = 0$) with admitted displacement jump $\Delta u_t \neq 0$. The latter hypothesis is adopted herein. The above idealization of fracture processes is widely regarded as practically acceptable for concrete-like disordered materials (concrete, rocks, masonry, ceramics) and in some polymeric materials, in which the energy comsumption is mostly concentrated in a 'process zone' ahead of the crack.

Fracture Processes in Concrete, Rock and Ceramics Edited by J.G.M. van Mier, J.G. Rots and A. Bakker.

We mention below related earlier work with no intention to provide a survey or a comprehensive reference list.

Softening, understood as local unstable behaviour described by a constitutive model for materials or structural components, had been discussed with reference to Drucker's postulate for elastoplastic solids (Palmer et al., 1967) and with emphasis on its consequences on stability and rate-solution uniqueness in combination with geometric effects (Maier, 1971b). Flexural softening in beams (primarily reinforced concrete beams) was studied by Rosenblueth et al. (1965) and by Maier et al. (1968, 1971a, 1973). The latter studies led to the following results which anticipated some findings appeared in the literature with reference to continua: constitutive softening implies strain localization (in beams: infinite curvature, i.e. slope jumps) and requires two-law models (for beams: bending moment vs. curvature and vs. relative rotation) unless some "localization arrestor" (then called "contamination length") is introduced; multiplicity of rate solutions, i.e. "equilibrium branching" or bifurcations can be the consequences of constitutive instability; softening may also give rise to overall instability either in load-controlled situations or under displacement control by a rigid loading device ("intrinsic" instability, now called "snap-back" or "catastrophic" instability).

Two-law models (stress vs. strain and vs. Δu_n) have been developed for fracture analysis of concrete by Hillerborg (1976) as "cohesive crack model", alternative to other interpretations resting on elastoplastic softening (and recently damage) constitutive models surveyed e.g. in Bažant (1985) and Bažant and Cedolin (1991). Extensive experimentations corroborate the validity of the cohesive-softening crack model (Cedolin et al., 1987; Biolzi, 1988). Using this model, size effects and instabilities were investigated by Carpinteri and co-workers (1989, 1990). They employ finite element (FE) discretizations and adopt the craze tip advancement as driving variable, like Blandford et al. (1981) in boundary element (BE) analysis of fracture propagation. Cen et al.(1990) applied BE methods to study, in cohesive-crack solids, the same overall consequences of softening (bifurcations and instabilities) pointed out earlier with reference to beam structures (Maier, 1968, 1971a, 1973). This paper presents in a concise form further results in the same direction.

2 Multiplicity of responses and instabilities: some parametric studies

The four-point shear test of concrete or other lapideous materials (fig. 1), already interpreted by Others as a cohesive softening system, is referred to below in order to discuss typical features in the behaviour of such solids and the analysis procedure herein adopted. To within rigid body motions (i.e. for all "intrinsic" quantities), the system exhibits a polar symmetry with respect to point O. This circumstance suggests to compare the results of two strategies for the numerical solution of the fracture propagation problem: (α) the polar symmetry is enforced, with consequent significant savings in the computing effort; (β) the whole system is dealt with, so that one can capture nonsymmetric evolutions. Although the possibility of "branching" of the equilibrium path (i.e. of asymmetric solutions) had been pointed out for elastic-cohesive softening beam structures (Maier et al., 1973), in the literature solids of the type depicted in fig. 1 appear to have often been analyzed only by the approach (α).

Fig. 2 compares graphs of load P vs. displacement u obtained by the approach (α), dashed lines, to those determined by the approach (β),

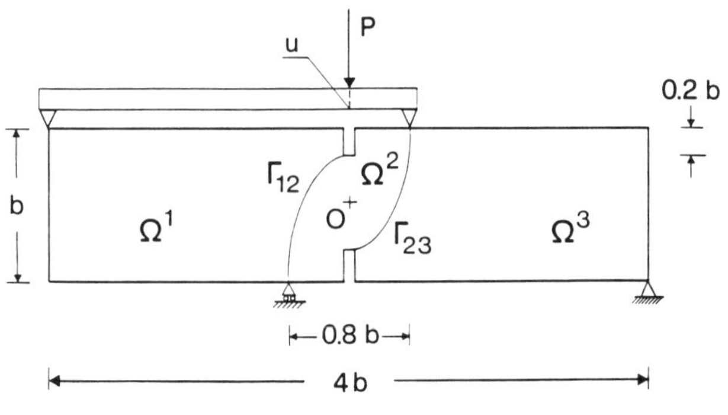

Fig. 1. Four-point shear test: typical configuration and initial subdivision

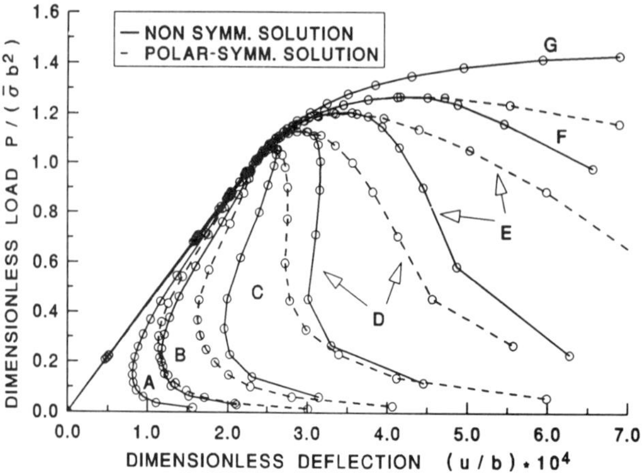

Fig. 2. Polar-symmetric and nonsymmetric time stepping solution for different shear tests.

for various values of the ductility index. This index, which was proposed by Carpinteri (1989), is $\rho = \bar{w}/2b = \mathcal{G}/b\bar{\sigma}$, denoting by $\mathcal{G}$ the dissipated fracture energy per unit surface, i.e. the integral of $p_n \cdot d\Delta u_n$ from 0 to $\bar{w}$. The graphs of fig. 2 were compute for $\rho = 2.5 \cdot 10^{-5}$ (A), $5.0 \cdot 10^{-5}$ (B), $1.25 \cdot 10^{-4}$ (C), $2.5 \cdot 10^{-4}$ (D), $5.0 \cdot 10^{-4}$ (E), $1.0 \cdot 10^{-3}$ (F), $1.0 \cdot 10^{-2}$ (G). It was assumed: (1) a linear decrease of p_n (from the value $\bar{\sigma}$) for increasing Δu_n up to $\bar{w}$; (2) an irreversible, "nonholonomic" p_n vs. Δu_n relationship (i.e. the path-dependence typical of incremental plasticity). The main mechanical circumstance evidenced by fig. 2 is the sharp distinction, for all except the highest numbers ρ, between the symmetric and the asymmetric branch, although the propagation itineraries resulting from the analysis have been kept polar-symmetric. The latter (in a sense "more unstable") path materializes in experiments (Biolzi, 1988) for thermodynamical reasons (Bažant et al. 1991); however, it appears to have been overlooked in earlier numerical analyses.

A well-known circumstance illustrated by fig. 2 is the transition of the overall behaviour from hardening-ductile to "normally" unstable and, finally, to "intrinsically" unstable ("catastrophic" or very brittle), as the elastic energy release rate due to local softening increases with respect to the energy dissipation rate in the deformation process (here as index ρ decreases). This transition was studied for elastic cohesive softening beams (Maier, 1968) and by FE for cohesive-crack twodimensional solids (e.g. Carpinteri, 1989).

The substitution of the irreversibility assumption (2) by a holonomic (reversible, nonlinear elastic) law p_n vs. Δu_n entails some algorithmic and computational simplifications and is mechanically legitimate as long as no local "elastic unloadings" occur. Since such simplification seems to have been adopted in the past, a comparative study is presented in fig. 3 using the same range of ρ values adopted for fig. 2. This shows that, at least for situations like in fig. 1, the holonomy hypothesis leads to significantly different nonsymmetric branches and might lead to quite inaccurate predictions.

The shear test of fig. 1 is referred to also in fig. 4 resulting from analyses carried out at equal geometry b and material parameter $\bar{w}$ (and, hence, at fixed $\rho = 1.5 \cdot 10^{-4}$) for various ultimate tensile stresses $\bar{\sigma}$. Increasing $\bar{w}$ alone provokes, as conflicting consequences, an increase of $\mathcal{G}$ and an increase of the instabilizing effect of softening (steeper p_n vs. Δu_n slope). Fig. 4 shows that the latter consequence prevails,

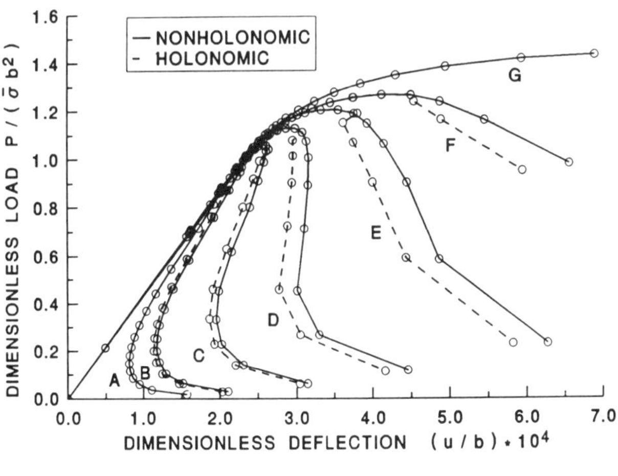

Fig. 3. Holonomic vs. nonholonomic nonsymmetric responses of four-point shear specimen.

in the sense that a significant transition occurs from less brittle (normal stability) to more brittle ("snapback") behaviour, although the ductility ratio ρ is fixed.

The above linearity assumption (1) is rather subjective: in fact experiments can hardly provide the actual behaviour of function $p_n(\Delta u_n)$, but usually determine $\bar{\sigma}$, $\bar{w}$ and possibly $\mathcal{G}$. At equal specific fracture energy, different assumptions as for $p_n(\Delta u_n)$ led to fairly different predictions of overall behaviour for intermediate indices ρ. For paucity of space, this is shown in fig. 5 which concerns the three-point bending test depicted in the inset of fig. 6 and used in later developments.

The analysis method employed for the above numerical results can be outlined as follows with reference to fig. 1 and to the non polar-symmetric case (see Cen et al., 1991, for details). A customary boundary integral equation, centered on Kelvin's fundamental solution, is formulated for each subdomain Ω^r (r=1,2,3) defined by tentative interfaces Γ_{12} and Γ_{23}. Tractions and displacements are modelled on the boundary and the interfaces; the equations are collocated at each node. After imposing the boundary and interface (matching) conditions, a condensation of all non-interface variables is performed. The resulting

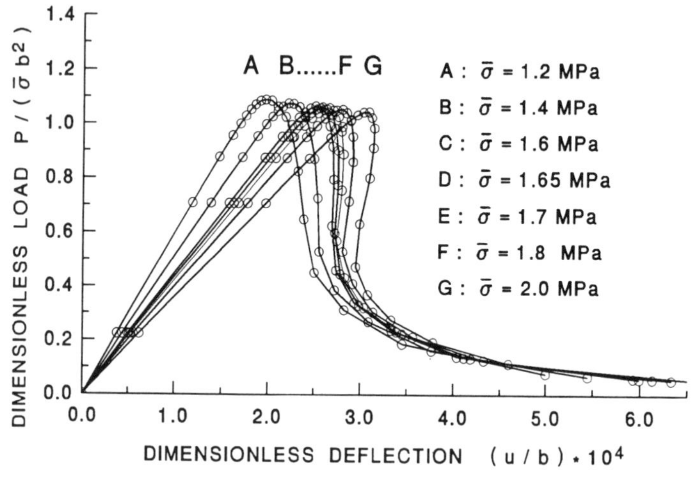

Fig. 4. Analysis of shear tests for varying ultimate tensile stress $\bar{\sigma}$ at equal ductility factor.

linear equations associated to the nonlinear nonholonomic p_n vs. Δu_n relationship are written in finite increments. The step problem is solved iteratively for a given advancement of a craze tip, while the change of the load factor is among the unknowns. Since the fracture propagation itinerary is not known "a priori", before each step in such marching solution the advancement direction is determined (by the max tensile stress criterion) and the interfaces Γ_{12} and Γ_{23} are modified if deviations from the current ones exceed an assigned threshold.

In the specific area of 2D or 3D fracture analyses based on the elastic cohesive model, BE methods like the one developed by the authors, are expected (and do turn out) to be advantageous with respect to FE methods for the following reasons: all nonlinearity is confined to interfaces; remeshing concerns at most interfaces between subdomains and, thus, it is greatly alleviated.

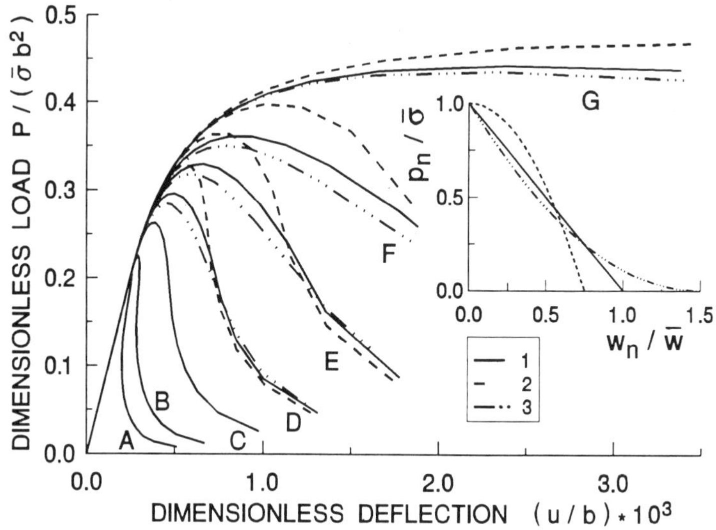

Fig. 5. Three-point test analyses based on various p_n vs. Δu_n behaviors for equal energy fracture $\mathcal{G}$.

3 Instability criteria and a Galerkin BE method

According to a classical general criterion, a system in static equilibrium is stable if and only if $\delta^2 W \geq 0$ for any $\delta \underline{u}(\underline{x})$, where $\delta \underline{u}(\underline{x})$ denotes a field of compatible infinitesimal displacements and $\delta^2 W$ is the second-order work performed by an equilibrium-preserving external agency which promotes $\delta \underline{u}(\underline{x})$. From this criterion the following statement can be proved for the specific solid of concern here: an elastic-cohesive system is stable if, and only if, for any $\delta \Delta u_n$:

$$\delta^2 W^*(\delta \Delta u_n) = -\frac{1}{2} \int_{\Gamma_z} \delta p_n(\delta \Delta u_n)\, \delta \Delta u_n \, d\Gamma + \frac{1}{2} \int_{\Gamma_z} [\delta \Delta u_n]^2 \, h \, d\Gamma \geq 0 \quad (1)$$

where Γ_z is the current craze, h denotes the (negative) instantaneous slope of the p_n vs. Δu_n relationship and δp_n represents the normal traction due to a perturbation $\delta \Delta u_n$ conceived as the only straining effect imposed on the elastic solid.

The linear dependence $\delta p_n(\delta \Delta u_n)$, after suitable discretization of both fields on Γ_z, is expressed by a square matrix $\underline{M}$. This matrix is symmetric and negative semidefinite and the functional $\delta^2 W^*$ becomes a quadratic form of the nodal displacements $\delta \Delta \underline{u}_n$ on Γ_z and preserves its energy meaning, if the BE formulation is carried out by a nonconventional, double integration, Galerkin approach (see in plasticity Maier and Polizzotto, 1989), instead of by the standard approach described earlier.

The distinctive features of the new BE method are: two boundary integral equations are formulated for each subdomain, of which one is based on Kelvin's kernels and the other on Gebbia's kernels (i.e. on the Green's functions of Ω_∞ concerning displacement discontinuities); these equations are enforced in a particular Galerkin weighted-residual fashion, rather than by collocation at nodes, so that double integrations are required in order to generate the coefficient matrices.

As an obvious corollary of the criterion (1) a sufficient (not necessary) condition for overall stability of a discrete model in load-control situations, is represented by the positive semidefiniteness of the matrix in the quadratic form arising from functional (1) through the above new BE discretization. A sufficient and necessary condition is the copositiveness of this matrix. Positive definitess is computationally much easier to assess than copositivity, as it amounts to evaluating the (real) least eigenvalue of a (symmetric) matrix, whose order is equal to the number of nodes in the current craze only.

Fig. 6 shows, for the three-point bending test depicted in the inset, the displacement u_0 at which this minimum eigenvalue vanishes to become negative, for various ductility factors (in abscissae). The displacements corrisponding to load peak $\hat{P}$ are also presented. The differences between the two graphs of fig. 6 indicates that the positive semidefiniteness criterion is sufficient only (in a sense it is "on the safe side") by an amount which increases with the ductility factor ρ. No bifurcation occurs in the cases considered in fig. 6, but it might conveniently be analyzed by the present approach by studying a linear complementarity problem in the opening displacement rates, with a symmetric matrix whose order again equals the number of the current craze nodes. This development is presented elsewhere together with other numerical tests on parametric studies.

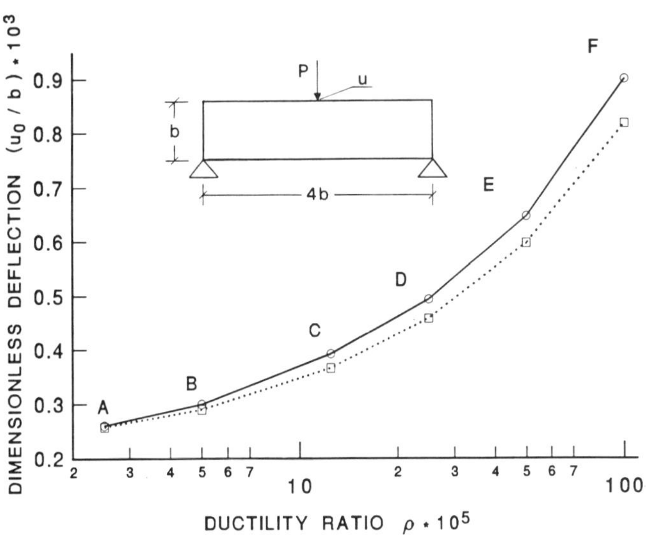

Fig. 6. Instability thresholds for three-point bending test: peak load deflection (solid line) and deflections at vanishing min eigenvalue (dotted line) vs. ductility index.

Acknowledgments. This paper concerns results obtained in a research project financed by G.N.D.T.

5 References

Bažant, Z.P. (1986) Mechanics of distributed cracking. **Appl. Mech. Rev.**, 39, 675-705.

Bažant, Z.P. and Cedolin, L. (1991) **Stability of Structures**, Oxford University Press.

Biolzi, L. (1988) Flexure of beams with elastic-softening behaviour. **SM Archives**, 13, 177-191.

Blandford, G.E., Ingraffea, A.R. and Liggett, J.A. (1981) Two-dimensional stress intensity factor computations using the boundary element method. **Int. J. Num. Meth. Engng.**, 17, 387-404.

Carpinteri, A. (1989) Decrease of apparent tensile and bending strength with specimen size: two different explanations based on fracture mechanics. **Int. J. Solids Structures**, 25, pp. 407-429.

Carpinteri, A. (1990) Cohesive crack tip modelling: size-scale transition from ductile to brittle failure. **Rend. Acc. Lincei,** IX, 1, pp. 59-79.

Cedolin, L., Dei Poli, S. and Iori, I. (1987) Tensile behaviour of concrete. **Proc. ASCE, J. Engng. Mech.**, 113, 431-449.

Cen, Z., Maier, G. and Novati, G. (1990a) On boundary element inelastic analysis in the presence of softening. **Boundary Element Methods in Engineering** (eds. B.S.Annigeri, K.Tseng), Springer, Berlin, 300-309.

Cen, Z. and Maier, G. (1990b) Fracture propagation and instabilities in elastic-cohesive crack models: a boundary element analysis. **ECF 8 Fracture Behaviour and Design of Materials and Structures**, Torino.

Cen, Z. and Maier, G. (1990c) Bifurcations and instabilities in fracture mechanics of cohesive-softening media by boundary elements, to appear

Hillerborg, A., Modeer, M. and Petersson, P.E. (1976) Analysis of crack formation and crack growth in concrete by means of fracture mechanics and finite elements. **Cement and Concrete Research**, 6, 773-782.

Maier, G. (1968) On softening flexural behaviour in elastic-plastic beams (in Italian). **Rendic. Ist. Lombardo, Classe Sci.**, A102, 648-677. English transl. in **Studi e Ricerche**, 8, 1986, Politecnico Milano, 85-117.

Maier, G. (1971a) On structural instability due to strainsoftening, in **Instability of Continuous Systems,** (ed. H. Leipholz), Springer, Berlin, pp.411-417.

Maier, G. (1971b) Incremental plastic analysis in the presence of large displacements and physical instabilizing effects. **Int. J. Solids Structures**, 7, 345-372.

Maier, G. Zavelani Rossi, A. and Dotreppe, J.C. (1973) Equilibrium branching due to flexural softening. **Proc. ASCE, J. Struct. Div.**, 99, N.EM4, 897-906.

Palmer, A.C. Maier, G. and Drucker, D.C. (1967) Normality relations and convexity of yield surfaces for unstable materials or structural elements. **Trans. ASME J. Appl. Mech.**, 464-470.

Rosenblueth, E. Diaz de Cossio, R. (1965) Instability considerations in limit design of concrete beams. in **Flexural Mech. of Reinforced Concrete, ASCE-ACI.**

50 BEM ANALYSIS OF CRACK PROPAGATION BASED ON A DISCRETE CRACK MODEL OF FRACTURE MECHANICS

M. OHTSU
Department of Civil and Environmental Engineering, Kumamoto University, Japan

Abstract
Crack propagation in concrete is investigated by employing the boundary element method (BEM). Applying the discrete crack model, the feasibility of the critical stress intensity factor K_{Ic} and the tensile-softening model is studied in notched-beam specimens of mode I crack propagation . Results show that the tensile-softening model reasonably simulates the load versus crack mouth opening displacement (CMOD) curves, whereas the K_{Ic} criterion recovers an essential feature of the load-CMOD curve.

Mixed-mode crack propagation is studied experimentally by using the acoustic emission (AE) technique. Results of a pull-out test imply that tensile cracks (mode I) precede and shear cracks (mode II) follow the tensile cracks on the final failure surface. Based on these experimental findings, crack propagation in the arbitrary direction of the mixed mode is tested in the notched beams under anti-symmetric loading and is analyzed by employing the maximum circumferential stress criterion and the K_{Ic} criterion. Crack trajectory observed in the test is in reasonable agreement with that of the analysis.
Keywords: BEM Analysis, Discrete Crack Model, K_{Ic}, Tensile-Softening Model, Acoustic Emission.

1 Introduction

Applicability of fracture mechanics to concrete materials is actively investigated to clarify failure of concrete structures. In the analytical approach based on a discrete crack model, the finite element method (FEM) is widely employed (Hillerborg et al., 1976; Ingraffea and Saouma, 1985). The whole domain, in this case, is referred to as homogeneous and elastic, whereas analytical treatments are limited to the boundary of crack surfaces. Both nucleation of a stress-free crack in the linear elastic fracture mechanics (LEFM) and progress of a fracture process zone in nonlinear fracture mechanics (NLFM) correspond to nucleating new boundaries in the numerical analysis. In this respect, FEM is not a smart choice to analyze the discrete crack model, because FEM requires internal nodes and elements to model the domain. In contrast, the boundary element method (BEM) deals with only elements on the boundary and thus leads

Fracture Processes in Concrete, Rock and Ceramics Edited by J.G.M. van Mier, J.G. Rots and A. Bakker.
 Published by E. & F.N. Spon, 2–6 Boundary Row, London SE1 8HN. ISBN 0 419 15870 7

to the privileged applicability to the discrete crack model (Horii, 1989).

In the present paper, the BEM analysis based on the critical stress intensity criterion of LEFM and the tensile-softening model of NLFM is discussed. To check the feasibility of the BEM analysis, notched-beam specimens of concrete are tested and results of load-crack mouth opening displacement (CMOD) curves are compared with those of analytical results in the mode I crack propagation.

To analyze mixed-mode crack propagation in arbitrary directions, mesh rearrangement is necessary. In this case, BEM requires only adding new boundaries. Analytical results based on LEFM are also compared with experimental results after observing the mixed-mode crack propagation by the acoustic emission (AE) in a pull-out test.

2 BEM formulation for discrete crack model

2.1 K_{Ic} criterion

The boundary integral equation in BEM is given by,

$$cu_i(x) = \int_S [G_{ik}(x,y)t_k - T_{ik}(x,y)u_k(y)]dS, \tag{1}$$

where c is the configuration coefficient, which is equal to 1/2 for x located on the smooth boundary. The coefficient is readily obtained from eliminating rigid motion in eq. (1). S is the boundary of the domain. u_k and t_k mean the displacement and the traction on the boundary, respectively. G_{ik} is the Green's function and T_{ik} is the associated traction.

To analyze crack propagation of the Mode I, half domain of concrete specimen is taken into account. An example of mesh division is shown in Fig. 1. The figure shows the case of a coarse

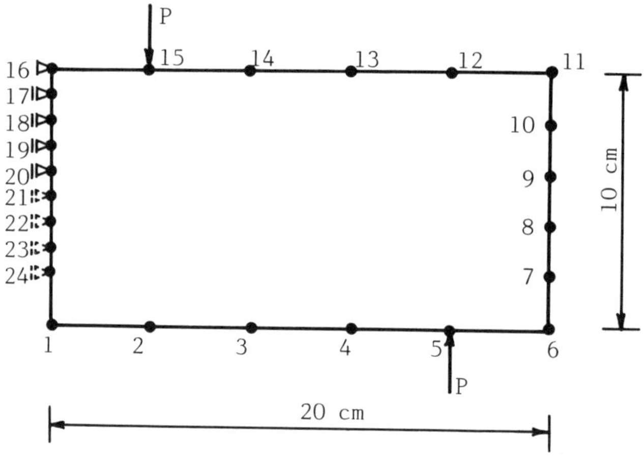

Fig. 1 Mesh arrangement in a BEM model.

mesh division. In practical problems discussed later, the length of the mesh is selected as 2 mm, based on the preliminary survey in respects with the numerical accuracy and the energy absorption.

To analyze eq. (1) numerically, linear elements were employed and the integration of each element was carried out by using the Gaussian 4-point quadrature scheme. The stress intensity factors at a crack tip were determined from Smith's one-point formulae (Smith and Mason, 1982). K_I and K_{II} are determined from the following formulae by using displacement u_i in the x direction and displacement v_i in the y direction, respectively,

$$K_I = (2\pi/L)^{1/2}(2u_i)E/(4 - 4\nu^2), \tag{2}$$

$$K_{II} = (2\pi/L)^{1/2}(2v_i)E/(4 - 4\nu^2). \tag{3}$$

Here L is the distance from the notch tip, E is Young's modulus, and ν is Poisson's ratio. After obtaining these values at five nodes below the notch tip, the stress intensity factors at the tip were evaluated as the averaged values.

In the case of whole domain analysis, new crack surfaces are defined as two coplanar boundaries in eq. (1), which results in a singular problem. Thus, the multi-region approach is proposed by Cruse (1988). This approach is applied to analyzing the mixed-mode crack propagation.

2.2 Tensile-softening model

A tensile-softening model to represent the fracture process zone is shown in Fig. 2. Here, a linear relation is adopted. Note that the stress at zero crack opening is indicated as σ_{AE}, instead of the tensile strength. The treatment on the fracture process zone is completely the same as developed for FEM (Hillerborg, et al., 1976). Cohesive forces obtained from the relation in the figure are applied at the nodes of the fictitious crack boundary in accordance with the crack opening displacements.

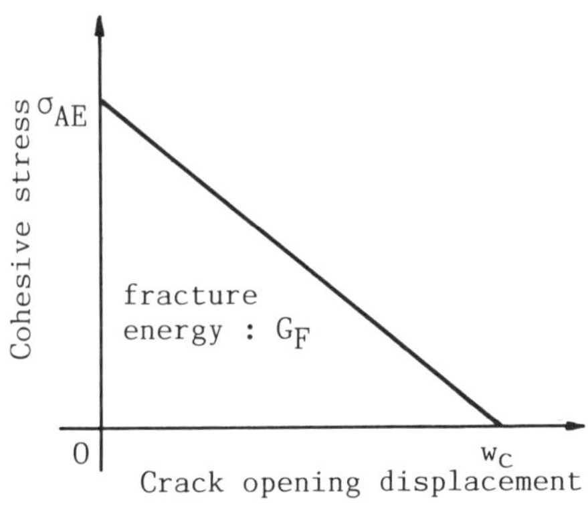

Fig. 2 Tensile-softening model.

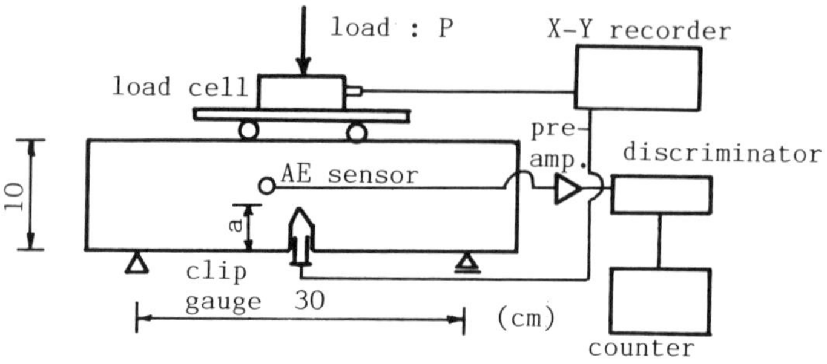

Fig. 3 Experimental set-up of a notched-beam test.

3 Mode I crack

3.1 Notched-beam tests

As material properties of concrete tested, compressive strength 44.2 Mpa, tensile strength 3.78 Mpa, Young's modulus 24.6 Gpa, and Poisson's ratio 0.2 were known. These were determined from three standard specimens of 10 cm diameter and 20 cm height. Notched-beam tests were carried out by employing beam specimens of dimension 10 cm x 10 cm x 40 cm. Plastic plates of 1 mm thickness were inserted to create notches and notch depths were designed as 2 cm, 3 cm, 4 cm, and 5 cm.

An experimental set-up of the notched-beam test is shown in Fig. 3. During four-point bending tests, AE events were detected by an AE sensor attached close to the notch tip. Total amplification was 60 dB gain and the frequency range was 10 kHz to 300 kHz. The threshold level for event counting was 120 mV. The load and the crack mouth opening displacement (CMOD) were measured by using a load cell and a clip gauge, and were recorded by the X-Y recorder.

To identify the load level where the process zone starts to be nucleated, AE activity was measured during experiments. It was observed that total AE-event counts acceleratedly increased at certain load level P_{AE}, which is referred to as the load of the process zone nucleation.

From the beam theory, critical cohesive stress σ_{AE} in Fig. 2 was defined, as follows;

$$\sigma_{AE} = 3P_{AE}/(10 - a)^2, \tag{4}$$

which corresponds to tensile stress at the beam bottom of virtual beam depth (10 - a). The averaged value, σ_{AE}, determined from all notch depths was 3.40 Mpa, which was slightly lower than tensile strength (3.78 Mpa).

To take into account the effect of notch shape, K_{Ic} values were determined, applying load P_{AE} to the BEM model. BEM analysis was performed on the half domain as shown in Fig. 1. Configuration of the notch was modeled according to the shape of an inserted plate. K_{Ic} values determined was equal to 0.54 $Mpam^{0.5}$. It is already reported that the process zone does not have the dominant effect on the determination of K_{Ic} values in the use of P_{AE} (Ohtsu, 1989).

To investigated the feasibility of the tensile-softening model in the region of the fracture process nucleation, the fracture energy, G_F, in Fig. 2 was evaluated from the relation,

$$G_F = (K_{Ic})^2/E. \tag{5}$$

Obtained G_F was 11.85 pam, which is approximately one tenth of ordinary values in concrete. This is because the fracture energy absorbed after the process zone nucleation is neglected. In the present case, the critical crack opening displacement, w_c, in Fig. 2 corresponded 0.00697 mm.

3.2 Comparison with the analysis

After obtaining load versus CMOD curves in experiments, crack propagation of mode I was analyzed. Two procedures to simulate crack propagation were employed. One is the K_{IC} criterion. At a stage, K_I at the notch tip was computed due to unit load and an actual load was determined as K_{Ic}/K_I. Then, the location of the notch tip was advanced up to the next node, removing the constraint at the previous notch tip. The other procedure applied the tensile-softening model determined from the test. The procedure of the numerical analysis is the same as that employed in FEM analysis.

A comparison of analytical results with an experimental results is shown in Fig. 4. It is observed that the load-CMOD curve determined from the tensile-softening model is in good agreement with

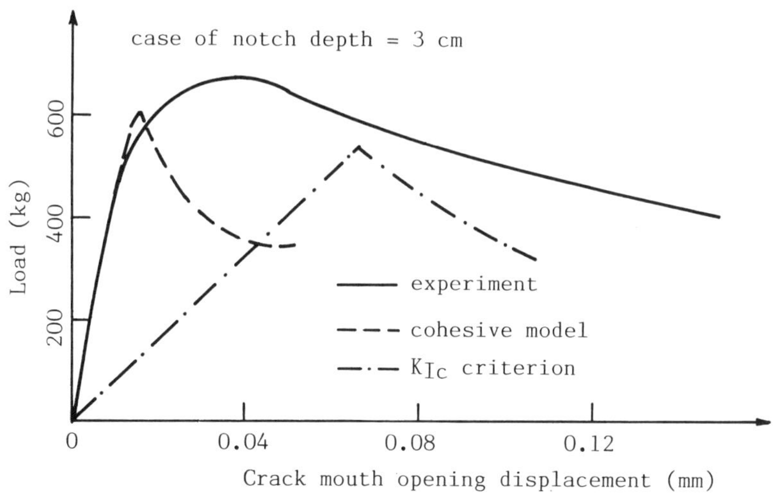

Fig. 4 Load versus CMOD curves in a notched beam.

the experimental curve up to the peak load. The agreement in the post-peak region, however, is not good. This is because the fracture energy, G_F, was determined from the K_{Ic} value and thus underestimated. The load-CMOD curve based on the K_{Ic} criterion recovers an essential feature of the experimental behavior except for the region prior to the peak load. Because the computation on the K_{Ic} criterion is completely the case of linear elastic, it is not easy to simulate nonlinear behavior prior to the peak load.

4 Mixed-mode crack

4.1 AE observation of a pull-out test

It is already clarified that the quantitative waveform analysis of AE can characterize AE sources to identify crack kinematics (Ohtsu and Ono, 1984).

To investigate the mixed-mode crack propagation experimentally, the source inversion procedure based on a SiGMA inversion and a unified decomposition (Ohtsu, 1988b) were applied to AE waveforms in a pull-out test of an anchor bolt from concrete block. A concrete block was of dimension 100 cm x 100 cm x 30 cm. Compressive strength of concrete was 54.3 Mpa and P wave velocity was 4150 m/s. An anchor plate with a nut was embedded 10 cm in the concrete block and was pulled out by a center-hole jack until the frustum of concrete cone was spalled off.

Six AE sensors of 140 kHz resonant frequency were attached to the concrete block. Total amplification of AE signals was 60 dB and the frequency range was 30 kHz to 80 kHz. Reading P-wave travel times, locations of AE sources were determined by the 3D source location procedure developed (Ohtsu and Ono, 1988). Then, moment tensor components were obtained by the SiGMA code, reading amplitudes of first motions. Following the eigenvalue analysis, the unified decomposition was performed and all AE events were classified into either tensile cracks or shear cracks, based on the ratio of shear contribution (Ohtsu, 1988b)

Elevation views of AE sources located, classified, and oriented are shown in Fig. 5. AE sources observed prior to the peak load (stage I) and the post-peak stage (stage II) are plotted at their locations with their types and orientations. In the figure, the final failure surface is drawn by broken lines.

To investigate mechanical characteristics of mixed-mode crack propagation, AE events only detected in the narrow zone of 10 cm width in the plan view, of which zone AE sources are located right on the failure surface in the elevation view, were selected and are plotted. AE source locations of total events are reported elsewhere (Shigeishi and Ohtsu, 1989).

It is observed that directions of tensile cracks (indicated by the direction of arrow symbol) are in remarkable agreement with opening directions of the failure surface, while AE events of shear type are oriented parallel to the failure surface. Here, the direction of shear motion is indicated by orientation of cross symbol.

In the stage I, AE sources are distributed around the anchor bolt. Cracks of both tensile and shear types are clustered. At stage II

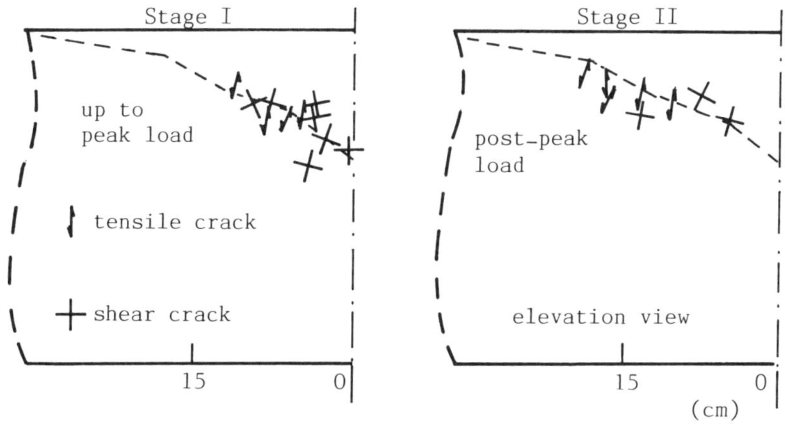

Fig. 5 Crack locations, types, orientations determined by AE in a pull-out test.

shows that tensile cracks are nucleated dominantly and precede, and thus shear cracks follow them. It implies that crack propagation at the post-peak region is dominated by the tensile crack nucleation. After the new stress-free crack boundary is generated, shear cracks trace the final failure surface. The observed trend must be closely associated with the mechanisms of mixed-mode crack propagation. Since the essential feature of the load-CMOD curves in the mode I crack propagation is recovered by the K_{Ic} criterion, experimental results by AE suggest that the same criterion is applicable to the mixed-mode crack propagation.

4.2 Analysis of crack propagation

The case of mixed-mode crack propagation subjected to anti-symmetric loading in a concrete beam was analyzed (Ohtsu, 1988a). An experimental set-up and a BEM model analyzed is shown in Fig. 6. Compressive strength of concrete was 29. 1 Mpa, Young's modulus was 27. 1 Gpa, and Poisson's ratio was 0.2.

Due to the lack of symmetry, the whole domain of the concrete beam is taken into account. By employing the multi-region approach, the region of the left-hand side and that of the right-hand side to the advanced notch are stitched, holding the continuity of displacement and traction as shown in the figure.

The direction of crack growth is determined from the following equation. With reference to the configuration shown in Fig. 6, the angle θ is governed by the criterion of the maximum circumferential stress,

$$K_I \sin\theta + K_{II}(3\cos\theta - 1) = 0. \qquad (6)$$

The stress intensity factors, K_I and K_{II} are computed from eqs. (2) and (3), by substituting the difference of displacements on the coplanar nodes instead of $(2u_i)$ or $(2v_i)$. Based on the results of

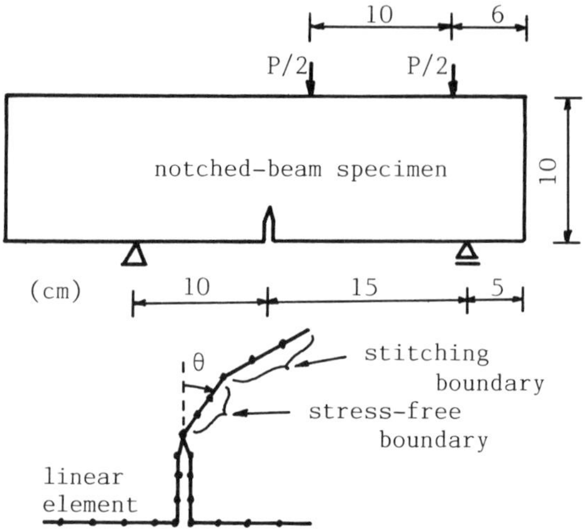

Fig. 6 Experimental set-up in anti-symmetric loading and a two-domain BEM model.

AE observation in the pull-out test, the criterion for crack propagation is based on the critical stress intensity factor of opening mode K_{Ic},

$$\cos\theta/2\ (K_I \cos^2\theta/2 - 3/2\ K_{II} \sin\theta) = K_{Ic}. \tag{7}$$

Here the K_{Ic} value was estimated as 0.65 Mpa from the mode I test, applying the AE observation.

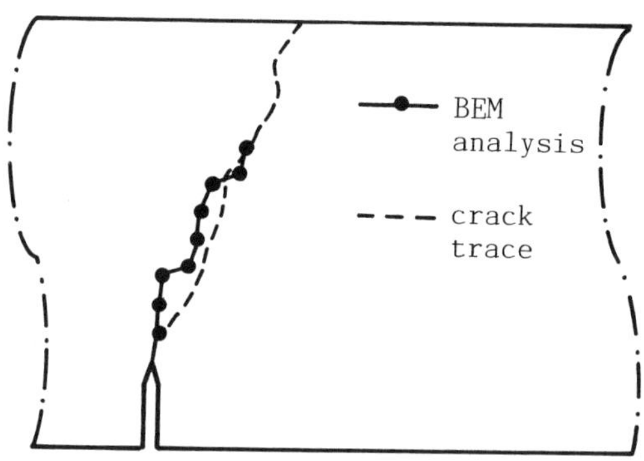

Fig. 7 Comparison of crack trajectories in the BEM analysis and the experiment.

As the similar manner to the mode I crack propagation, the unit load was first applied and then the load level to propagate the crack was evaluated from eq. (7). The orientation of a new crack was determined from eq. (6). Thus, new coplanar stress-free boundaries were added along the orientation of the maximum circumferential stress as shown in Fig. 6. Then, the stitching boundary was created, connecting the notch tip with one node on the top surface.

A predicted crack trajectory by BEM analysis is compared with that of the experiment in Fig. 7. Because the length of stress-free crack element was taken as 1 cm segment, the profile of the trajectory is not smooth. Still, the agreement between the crack trace analyzed and the experimental observation is reasonable. It suggests that the K_{Ic} criterion is available for analyzing the mixed-mode crack propagation.

5 Conclusion

The applicability of BEM analysis to the discrete crack model approach is studied in the crack propagation of the mode I and the mixed mode. To determine the critical stress intensity factor, K_{Ic}, AE observation is applied. Therefore, K_{Ic} values determined are referred to as the load level at the nucleation of fracture process zone. The feasibility of the tensile-softening model is studied, by applying the linear relation and the fracture energy evaluated from K_{Ic}. Results show that the fracture energy was underestimated and the model could simulate up to the peak load of the load-CMOD curve.

The source inversion procedure of the SiGMA inversion and the unified decomposition was applied to AE waveforms detected during the pull-out test of an anchor bolt from concrete block. Crack locations, types, and orientations determined are in good agreement with experimental findings. It suggests that the mixed-mode crack propagation is governed by the nucleation of tensile cracks and the orientation of crack opening. Based on these experimental findings, crack propagation due to anti-symmetric loading was analyzed by the criterion of on the maximum circumferential stress, employing the K_{Ic} value determined from the mode I test. The crack trajectory observed in the test is reasonably predicted by BEM analysis. These results show a great promise of BEM analysis for clarifying the mechanisms of the mixed-mode crack propagation.

6 References

Cruse, T. A. (1988) **Boundary Element Analysis in Computational Fracture Mechanics**, Kluwer Academic Publishers, Dordrecht.

Hillerborg, A., Modeer, M. and Petersson, P. E. (1976) Analysis of crack formation and crack growth in concrete by means of fracture mechanics and finite elements, **Cement and Concrete Res.**, 6, 773-782.

Horii, H. (1989) Models of fracture process zone and a system of fracture mechanics for concrete and rock, in **Fracture Toughness**

and Fracture Energy (eds. H. Mihashi et al.), Balkema, pp. 409-422.

Ingraffea, A. R. and Saouma, V. (1985) Numerical modeling of discrete crack propagation in reinforced and plain concrete, in **Fracture Mechanics of Concrete, Structural Application and Numerical Calculation**, Martinus Nijhoff Publishers, pp. 171-225.

Ohtsu, M. and Ono, K. (1984) A generalized theory of acoustic emission and Green's functions in a half space, **Journal of Acoustic Emission**, 3(1), 27-40.

Ohtsu, M. and Ono, K. (1988), AE source location and orientation determination of tensile cracks from surface observation, **NDT International**, 21(3), 143-150.

Ohtsu, M. (1988a) Crack propagation in concrete: linear elastic fracture mechanics and boundary element method, **Theoretical and Applied Fracture Mechanics**, 9(1), 55-60.

Ohtsu, M. (1988b) Source inversion of acoustic emission waveform, **Proc. JSCE**, 398/I-10, 71-79.

Ohtsu, M. (1989) AE application to fracture mechanics, **Proc. of JCI**, 11(2), 673-678.

Shigeishi, M. and Ohtsu, M. (1989) Application of AE source inversion for concrete structures, **Proc. JCI**, 11(1), 311-316.

Smith, R. N. L. and Mason, J. C. (1982) A boundary element methods for curved crack problems in two dimensions, in **Boundary Element Methods in Engineering**, (ed. C. A. Brebbia), Springer, Berlin, pp. 472-484.

51 SIMULATION OF FRACTURE ZONE DEVELOPMENT IN R.C. MEMBERS

A. SCARPAS
Department of Civil Engineering, Delft University of Technology, The Netherlands

Abstract
A non-linear finite elements code is presented capable to simulate the sequence of physical events leading to the development of a crack in a semi-brittle material like concrete. Two separate phases of crack development are admitted, namely "**micro**" and "**macro**" cracking. The transition from the former to the latter is dictated by means of non-linear fracture mechanics criteria. The Discrete Cracking approach is utilized for the introduction of "macro" cracks in the finite elements mesh. Automatic remeshing allows mixed mode I and II crack propagation to be examined.
Keywords : Finite Element Analysis, Discrete Cracking, Smeared Cracking, Process Zone, Fracture Mechanics.

1 Introduction

Since the first pioneering works on the subject, almost 20 years ago, finite element (F.E.) simulation of cracking remains a point of challenge, dispute and ambiguity among the computational mechanics community.

Responding to our engineering feeling of a crack being a discontinuity in an otherwise continuous medium, in the first attempts of F.E. simulation of concrete cracking, cracks were introduced into the mesh by successively disconnecting the common nodes of adjacent elements. This discrete simulation not only gave its name to the corresponding method but, at the same time, encouraged the adoption of many notions from the pre-existing lacunae of knowledge of linear elastic fracture mechanics [1].

With increased research and understanding of the particularities of the cracking process in semi-brittle materials, doubts were cast, first, on the correctness of utilization of linear elastic fracture mechanics tools for this type of materials and then, on the appropriateness of the discrete simulation itself for granular materials like concrete.

Instead, it appeared that a more distributed simulation would comply better with the nature of the real phenomenon, particularly during the stages of crack initiation. In what is now known as the smeared cracking approach, cracking was simulated by the introduction of axes of material anisotropy in the originally isotropic element. Along these, the

Fracture Processes in Concrete, Rock and Ceramics Edited by J.G.M. van Mier, J.G. Rots and A. Bakker.
 Published by E. & F.N. Spon, 2–6 Boundary Row, London SE1 8HN. ISBN 0 419 15870 7

material stiffness was properly reduced to allow for cracking. Many variations of the basic approach were proposed and techniques for incorporating the non-linear fracture characteristics of concrete cracking were developed [2]. However, sooner rather than later, the inadequacies of the approach were realized as well [3].

After almost two decades of research, it has started to become apparent that none of the above two approaches is ideally suited to realistically simulate on its own the sequence of phenomena that occur during the stages of crack initiation and crack propagation in concrete members.

2 The physical process of concrete cracking

Inspite of extensive international experimental and theoretical research work, the actual cracking mechanism of quasi-brittle materials like concrete is not well understood as yet.

Nevertheless, there is now an abundance of experimental evidence attesting to the fact that, in front of every active crack in concrete, there exists a tensile fracture zone spanning a region, over which, the strains exceed the limiting value corresponding to the tensile strength. Recent experimental evidence by van Mier [4] indicates that within this zone, fracture is manifested in the form of aggregate debonding and overlapping micro-cracks, mainly parallel to each other, with some intact (depending on the degree of straining) grain bridges between them. For illustration purposes this region is portrayed in Fig. 1 as a series of micro-cracks.

In contrast to the macroscopically cracked regions to be discussed in the following, tensile stresses -even though of a diminishing magnitude according to the $\sigma - \varepsilon$ constitutive law of the material, Fig. 1- can still be transmitted through this micro-cracked band which has been termed the "process" zone [2].

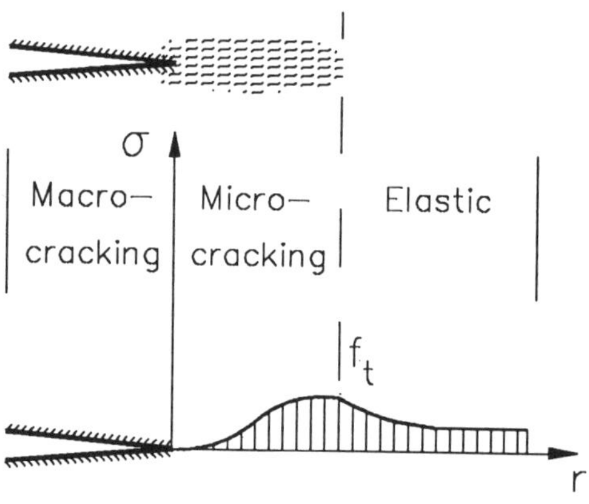

Fig. 1 Concrete cracking

The width of the band (in the tensile stress direction) was originally assumed to be some multiple of the maximum aggregate size, however, the more recent studies of van Mier [4] indicate a width roughly equal to the maximum aggregate size. Its length remains a topic of experimental and theoretical dispute even for the typical types of tests used in laboratory evaluation of material parameters. In actual structural members it is also influenced by the overall state of strains in the cracked region.

It is only after the micro-cracks over some given length of the above zone coalesce into a single crack that a discrete discontinuity develops in the material (complying with our physical sense of a crack corresponding to a physical separation of an initially continuous material). No tensile concrete stresses can be transmitted any longer across this macro-crack, Fig. 1, however, depending on the prevailing state of stress (s.o.s.) significant compressive and/or shear stresses may be possible to be transmitted [6].

3 Constitutive formulation

From the description in Section 2 of the sequence of physical events leading to cracking it is hoped to have become clear that neither of the existing F.E. techniques for simulation of cracking is capable to simulate on its own all stages of cracking in a R.C. member. It appears instead that a combination of the essential characteristics of both approaches in a single methodology may be ideally suited for simulation of the fracture process. In Section 4, the non-linear F.E. code RCLAB will be presented capable of modelling the above described sequence of physical events.

The non-linear path-dependent behaviour of both, the materials and, the processes involved, dictate that an incremental, iterative solution scheme has to be adopted. An incorporated materials library allows use of the program for the study of either fracturing materials like concrete and rock or plastic materials like steel. The program includes restart and element/birth options, as well as very extensive mesh generation and graphics facilities. In the following, those features of the program related to concrete constitutive formulation will be presented so as to serve as a basis for the description of the implemented crack formation and propagation proceedure.

3.1 Concrete idealization

Numerically integrated isoparametric elements are used to model plain concrete. To accomodate the needs of crack propagation and/or mesh refinement, both quadrilateral and triangular elements can be utilized.

For the complete characterization of the two dimensional material behaviour an initial yield surface is defined, Fig. 2. It corresponds to the locus of s.o.s. delimiting the onset of non-linear behaviour. All s.o.s. within this surface are elastic.

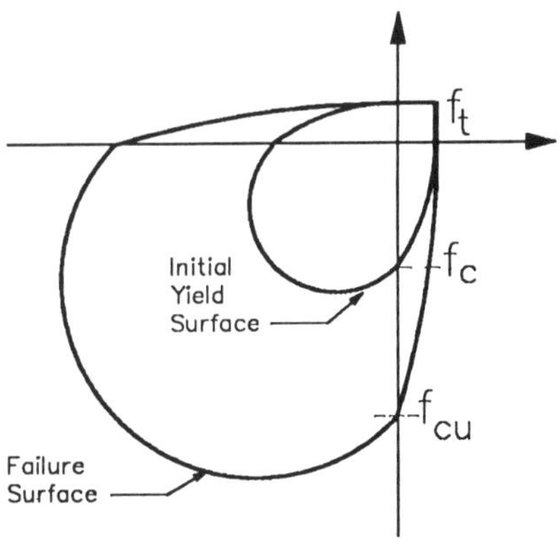

Fig. 2 Biaxial concrete modelling

Following an incremental hardening plasticity formulation, upon further loading, the initial yield surface is assummed to expand isotropically in space. The multitude of resulting loci is known in the literature as "loading surfaces". Expansion of the loading surfaces is accompanied by plastic straining of the material. Associated or non-associated plasticity formulation can be chosen. A Drucker-Prager type surface is utilized for modelling both the initial yield and the subsequent loading surfaces.

Expansion of the loading surfaces is not unlimited. It is restrained by the existence of another, not necessarily similar to the loading surfaces, locus of s.o.s. termed the "failure surface". Physicallly it corresponds to those s.o.s. beyond which degradation of response occurs with increasing strains, either by crushing of the material in the compression-compression (c-c), or, by cracking in the compression-tension (c-t) and tension-tension (t-t) regions. The following criteria can be chosen for the description of the failure surface : Mohr-Coulomb, Drucker-Prager, Chen-Chen and Willam-Warnke's 3-parameter model.

The coincidence of the initial yield surface and the failure surface in the t-t quadrant implies that for purely tension s.o.s. no plastic straining of concrete is allowed to occur untill cracking. On the other hand, for c-t type s.o.s., the limited expansion of the loading surfaces into this region implies that some degree of plastic straining is allowed before cracking.

4 Finite element simulation of cracking

Two separate phases of crack development are admitted in the cracking procedure implemented in the code RCLAB : a) **micro-cracking**, which corresponds physically to the process zone of Fig. 1 and, b) **macro-cracking**, which corresponds physically to the development of a discrete discontinuity in the material.

4.1 Micro-cracking F.E. simulation

During this stage of crack development the material can still be treated elegantly as a continuum even though of an orthotropic nature. The smeared cracking approach is ideally suited for simulating this stage especially if attention is concentrated on response at numerical integration point rather than at element level.

In the RCLAB code, the criterion for initiation of micro-cracking is attainment of the failure surface in the t-t or c-t quadrants, Fig. 2. The micro-cracking plane is defined as being perpendicular to the maximum principal tensile stress at the moment of micro-crack initiation. A fixed crack total strain approach has been utilized in this study.

After many years of experimental research, there appears to be now an almost universal agreement as to the general characteristics of the normal stress σ vs. micro-crack opening w curve. In RCLAB, the F.E. computed normal strains ε in the process region are related to the sum w of the micro-crack opening displacements by means of the smearing assumption $w = h * \varepsilon$, in which h is the width of the process region. In contrast to the usual practise of relating h to the element size, in RCLAB, micro-cracking is represented at the integration point area of influence. This allows for a closer F.E. simulation of the experimental observations of van Mier [4].

A direct consequence of this smearing process is that the amount of energy G_f needed to create one unit of area of a continuous crack can be expressed in terms of F.E. quantities as :

$$G_f = h \cdot \int_0^{\varepsilon_u} \sigma \cdot d\varepsilon = h \cdot g_f \tag{1}$$

As shown in Fig. 3(a), the integral part of Eq. 1 corresponds to the energy released during the softening response of the material.

To the best of the authors knowledge, all currently available experimental research has concentrated on determining the characteristics of the σ-w curve under mainly uniaxial conditions. Yet the c-t regions of response represent the most oftenly encountered s.o.s. within a structural member. Nevertheless, little seems to be known as to the influence of the transverse stress component on G_f.

Even less is known about the response of the micro-cracking region when subjected to shear. Only recently the topic of mixed mode micro-cracking has received some experimental attention by van Mier et al. [9]. The author fully agrees with van Mier's observation that, currently available aggregate interlock models are not applicable for describing the shear response within the process zone.

The procedure advocated in this study allows differentiation between the shear response of the micro-cracks and that of the fully developed macro-crack (refer also to Section 4.2). Intuitively, the degradation of G^{cr} as a function of the micro-crack width has been postulated in the RCLAB implementation to be as shown in Fig. 3(b). Since G_{cr} refers to the process zone micro-cracks, it is natural that the range of its values is higher than normally encountered in aggregate interlock experiments on precracked specimens.

To avoid unlimited increase of shear stress and, also, in order to allow some energy dissipation in mode II, a shear softening formulation has been adopted along the lines of

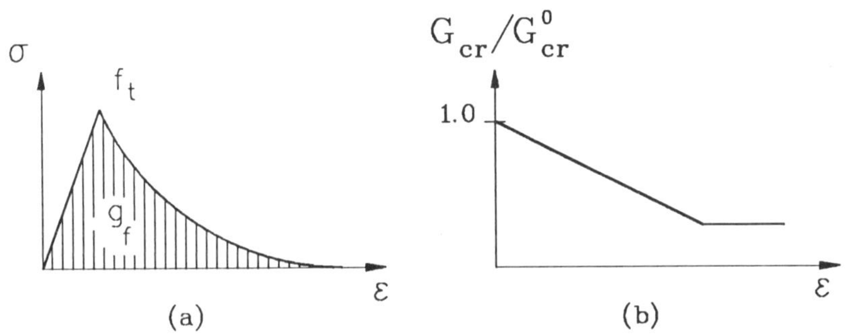

Fig. 3 Constitutive modelling of the micro-cracking zone

Rots et al. [5]. As pointed out above, at present there is very little, if any, experimental evidence as to the actual magnitude of physical quantities like the shear strength on the micro-cracking plane, the mode II fracture energy etc. In this study, the assumptions in [5] have been adopted.

4.2 Macro-cracking F.E. simulation

Macro-cracking consists the final stage of crack development during which the individual micro-cracks, over some length of the process zone, coalesce into a single discrete discontinuity across which no tensile stress can be transferred, Fig. 1.

In the RCLAB implementation, a macro-crack is deemed to have occured only if the necessary for the generation of a continuous crack amount of fracture energy has been expended over some length of the process region. In terms of F.E. quantities, this is equivalent to the requirement that Eq. 1 is fulfilled over a number of integration points.

The discrete cracking approach is utilized as this type of simulation complies with the notion of a macro-crack being a physical separation of the initially continuous material and as, at the same time, allows for realistic inclusion of the shear transfer phenomena that occur at the crack faces.

Discrete cracking procedure

With reference to the F.E. mesh of Fig. 4(a) and assuming that : (a) a discrete crack already exists due to straining in the previous load increments and, (b) Eq. 1 is fulfilled in the region from the current macro-crack tip to node k of the mesh, the algorithm :

(i) determines the node towards which the crack is to be extended, e.g. node k in Fig. 4(a). This node will be labelled in the following as the "new crack tip"

(ii) if necessary (e.g. cracks inclined with respect to the mesh lines), applies local remeshing around the new tip, Fig. 4(b). Cracks may propagate either along element boundaries or through elements.

(iii) releases the elements on either side of the crack propagation axis. This procedure creates the need for generation of additional nodes, Fig. 4(c).

(iv) in problems where it is desired to include in the analysis the force transfer mechanisms at the crack faces, the nodes on either side of the crack line can be joined by means of specially developed "shear transfer" interface elements [6]. Their stiffness characteristics are determined on the basis of available experimental evidence.

(v) after nodal separation, the residual stresses which correspond to the difference between the stresses existing prior to cracking and those that the material can sustain at the same strain level after macro-cracking are computed. An equivalent nodal load vector is formed for each cracked element. Reapplication of this "residual" load vector on the structure simulates the physical process of load redistribution that takes place after cracking.

Internal load redistribution may result to additional crack propagation in the member and

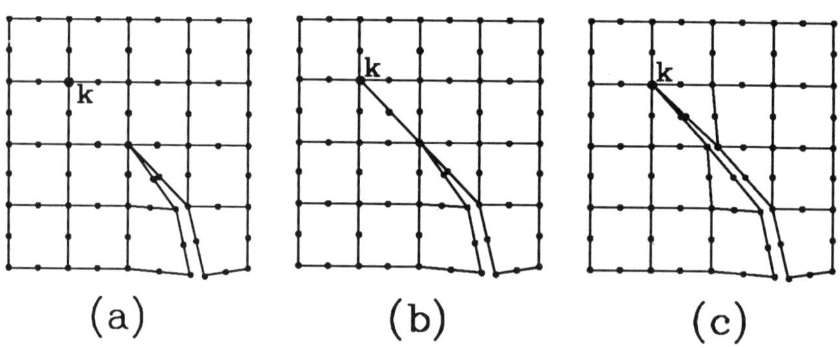

Fig. 4 Steps of discrete crack propagation algorithm

vice-versa. In order to account correctly for these phenomena, the external load incrementation is interrupted automatically until both, stress redistribution and macro-crack opagation have stabilised.

5 Numerical Applications

Due to space limitations, only one example of crack initiation and propagation in a R.C.wall designed and tested by Synge [7] will be presented. The example is typical though of the program's capability to handle cracks under combined Mode I + II conditions.

The length of the wall was 1500 mm, its heigth 3000 mm.and its thickness 100 mm. The concrete compressive strength was 28.0 MPa. The flexural reinforcement consisted of 4-12 mm mild steel bars over a region of 140 mm from both ends of the wall and the same type of bars spaced at 140 mm centers in the web. The shear reinforcement consisted of 16 mm high strength steel bars spaced at 120 mm centers along the height of the wall. Uniform shearing load were imposed on the top edge.

In the F.E. simulation, uniform horizontal displacements were imposed at the top of the wall. The distribution of both, micro-cracking and macro-cracking at the base of the wall for some steps of the analysis is shown in Fig. 5. The gradual propagation of cracking during a given increment is presented in a clock-wise manner.

In agreement with the physical sequence of events portrayed in Fig. 1, macro-cracking developed during Increment 5 only after the development of a relatively long region of micro-cracking at the base of the wall. As mentioned in Section 4.2, discrete cracking progresses through the mesh as long as Eq. 1 is fulfilled. Modelling of the reinforcing bars crossing the discrete crack is achieved by means of interface shear transfer F.E. denoted by dashpots in Fig. 5.

In contrast to the numerical tests of Wang [8], the nature of the proposed procedure allows both, the development and the co-existence of discrete cracks adjacent to regions of micro-cracked numerical integration points. This type of response has been known in

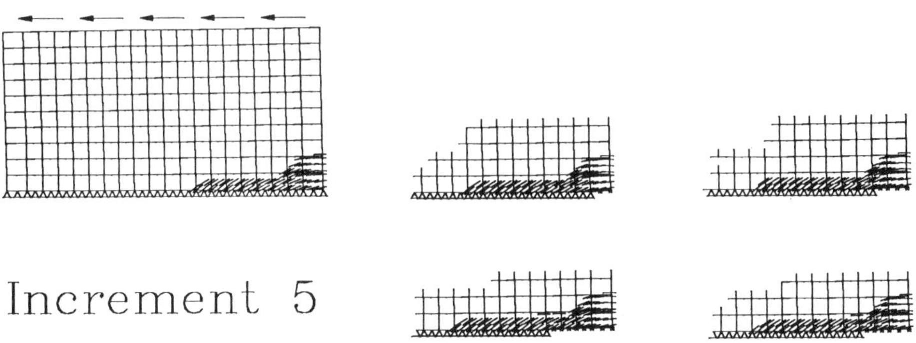

Fig. 5 Micro/macro crack distribution at base

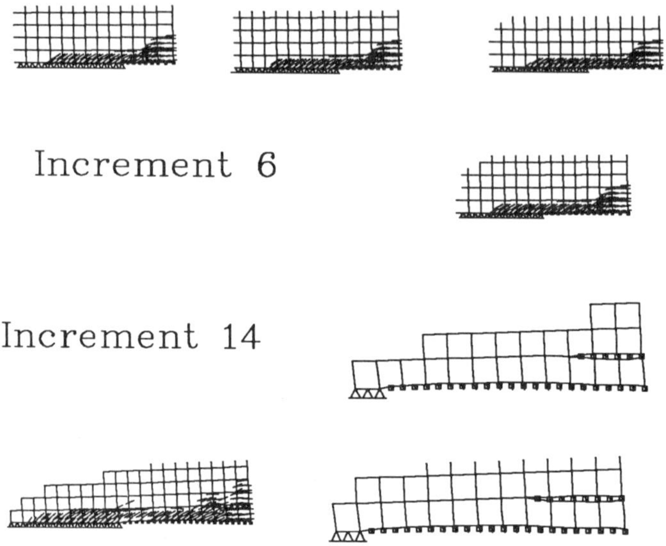

Fig. 5 (cntd.) Micro/macro crack distribution at the base

the past [3], to be difficult to model by the inclusion of discrete cracking elements along predefined crack patterns.In RCLAB, the cracking procedure remained stable even when, in later increments, a second discrete crack developed at the base of the wall. The discrete crack propagation up to Increment 14 is shown in magnification in Fig. 5 together with the extent of micro-cracking that preceeds it.

6 References

1 Saouma V.E. (1981) Finite element analysis of reinforced concrete: A fracture mechanics approach. Ph.D. Dissertation, School of Civil and Environmental Engineering, Cornell University, Ithaca .

2 Bazant Z.P. and Cedolin L. (1979) Blunt crack band propagation in finite element analysis. Journal of the Engineering Mechanics Division, ASCE, Vol. 105, pp. 297-315.

3 Rots J.G. and Blaauwendraad J. (1989) Crack Models for concrete: Discrete or smeared?, Fixed, multi-directional or rotating. HERON, Vol. 34, No. 1.

4 van Mier J. (1990) Mode I fracture of concrete: Discontinuous crack growth and crack interface grain bridging. Research Rept. 25.5-90-7, Faculty of Civil Engineerimg, TU Delft, Delft .

5 Rots J.G. and de Borst R. (1987) Analysis of mixed-mode fracture in concrete. Journal of the Engineering Mechanics Division, ASCE, Vol. 113, pp. 1739-1758.

6 Scarpas A. and Tassios T.P. Reinforced crack interface shear transfer finite element, in **Int. Conf. on Computational Engineering Science**, Patras, Apr. 1991 (accepted for publication).

7 Synge A.J. (1980) Ductility of squat shear walls". Research Rept. 80-8, Dept. of Civil Engineering, Canterbury University, Christchurch .

8 Wang Q.B. (1989) Shear strength of reinforced concrete beams and slabs without shear reinforcement",Research Rept. 25.2-89-2-10, Faculty of Civil Engineerimg, TU Delft, Delft.

9 van Mier J., Nooru-Mohamed and Schlangen E. (1990) Experimental analysis of mixed mode I and II behaviour of concrete", in **Application of Fracture Mechanics**, (eds. Shah S.P. et al.),Chapman and Hall, London, pp. 31-42.

52 THE BIFURCATING CRACK PROBLEM: CONTINUUM AND FRACTURE MECHANICS APPROACHES

G. BAKER and C.A. GRUMMITT
Department of Civil Engineering, University of Queensland, Australia

Abstract
This paper concerns computational aspects of modelling cracks which bifurcate into a T bar in unreinforced concrete. The problem itself is examined, and the difficulties associated with both continuum and fracture mechanics models are discussed, particularly with regard to their ability to predict crack branching. The conclusion is that the use of SIFs must fail and that a stress criterion appears to be necessary. We believe this to be an interesting bench mark for computational fracture procedures.
Keywords: Crack branching, crack direction, continuum models, fracture mechanics.

1 INTRODUCTION

This paper concerns computational aspects of modelling the interesting problem of bifurcating cracks in cementitious materials. The case we consider most closely concerns crack bifurcation near a compression zone in an unreinforced concrete panel (figure 1). When a plain concrete panel supported along its base (a wall) is loaded by a vertical point load on the top surface, static combinations of axial compression and bending are induced. For different eccentricities of the load, different fracture patterns and failure modes can be observed. When the load is applied through the centroid, concrete crushing clearly dominates. At the other eccentricities, between W/6 and W/2 from the centreline, horizontal cracks propagate to near the centreline and then bifurcate into a T bar, and extend vertically in each direction away from the single crack tip.

A second example of this real physical phenomenon has been observed in refractory rings subjected to large thermal gradients (Palmer and Baker, 1990). There, the self equilibriating restraint stresses create large tensions on the inner face of the annulus. Cracks propagate from this face radially, but after a short distance, they meet a field of compressive hoop stress, bifurcate and travel circumferentially causing spalling and failure.

The objective in this article is to examine the ability, and performance, of different computational approaches in modelling the bifurcation of the flexural cracks; we also present a brief study of the phenomenon itself and suggest a simple mechanism to explain the branching. We are interested in both distributed and discrete crack

Fracture Processes in Concrete, Rock and Ceramics Edited by J.G.M. van Mier, J.G. Rots and A. Bakker.
 Published by E. & F.N. Spon, 2–6 Boundary Row, London SE1 8HN. ISBN 0 419 15870 7

formulations. Specifically, we consider standard smeared crack with strength criterion, crack band with energy criterion, and LEFM models. The distributed crack models which use principal stress to determine crack angle and position work well in single crack elements. Crack band models similarly behave well, although there are the difficulties, i.e. singularities, of multiple cracks in elements at the horizontal crack tip. The problem is thus an interesting one for strain localisation models as it is for a fracture mechanics approach where we are, in effect, trying to model micro-damage and then coalescence in bifurcating cracks. We show that traditional criteria for predicting the direction of discrete crack growth are unable to track the branching process.

2 CRACK ARREST NEAR A COMPRESSION FIELD

2.1 Crack Arrest and Bifurcation

As a mode I crack approaches a compression zone ($\sigma_y < 0$) - obviously the crack length influences the flow of vertical stress - the crack will eventually arrest. Thereafter, the stress field around the tip changes dramatically, making simple prediction invalid. At $\theta = \pi/2$, the circumferential stress, i.e. σ_x, becomes tensile, largely because the surface of the crack is free to extend as the panel continues to bend under increasing load. The tip then bifurcates into a T bar, the branches of which propagate in mixed mode: mode I because of the opening displacements caused by further bending, and mode II because of the relative shearing induced by the COD in the original crack.

2.2 Classical Solutions for an Infinite Plate

In fact, there are no classical fracture mechanics solutions for stress intensity for crack bifurcation in a finite panel. Sih and Irwin (1970) give an estimate for stress intensity factor for a symmetrically branched crack in an edge cracked plate, but which is subjected to a uniform tension. However, Paris and Sih (1965) do give solutions for a semi-infinite notch approaching the free surface of a half plane subjected to a combination of far field tension force and bending. We can transform their general forces to the specific case of a compressive load P at eccentricity e (assuming tip stress remains tensile), and relate the size of the uncracked portion to our finite crack length, a, and panel width, W. Then, we find an estimate of stress intensity to be:

$$K_I = \frac{P}{(W - a)^{1/2}}\left(2.16454\left[\frac{2e - a}{W - a}\right] - 0.53694\right) \qquad (1)$$

This will not help to determine the likelihood of bifurcation, but it might indicate the combination of eccentricity and crack length corresponding to crack arrest.

2.3 Finite Element Analysis of a Cracked Panel

Here we present some finite element analyses of a precracked finite panel with the intention of showing the stress distributions around

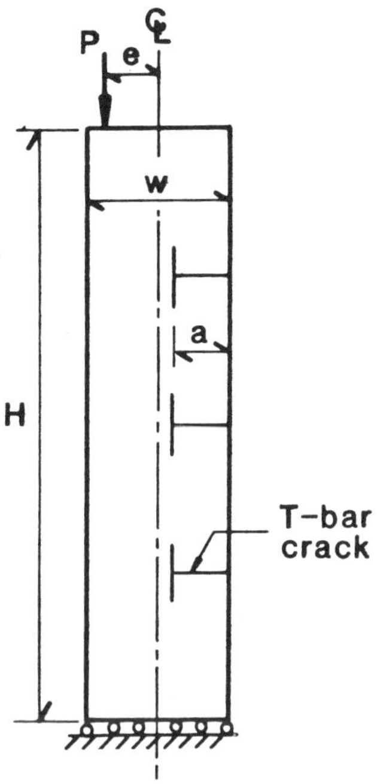

Fig. 1: Crack geometry

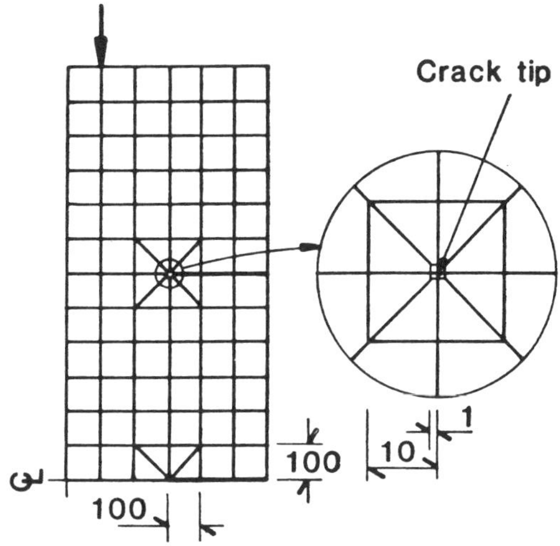

Fig. 2: Mesh layout

the crack tip as it approaches the compression field. In particular, we show how the vertical tensions fade and lateral tensions develop as the crack propagates in mode I. From this, we suggest a simple stress flow which would explain the bifurcation phenomenon. The results are also used in the LEFM section to estimate stress intensity factors, and crack direction based on maximum circumferential stress.

We study a panel with an aspect ratio of 4:1, with three equally spaced flexural cracks, loaded symmetrically with compressive loads; our observation is that while the number of cracks affects the stress disribution generally, it does not significantly affect the flow around the tip of a crack away from the ends. The finite element mesh, of 12dof quadrilaterals, is shown in figure 2. We use a regular grid with concentric refinement of square elements around the crack tip, from 100x100mm down to 10x10mm down to 1x1mm. For brevity here, we study only the case where a reference load of 100kN is placed at eccentricity, e = W/3, but with precracks varying in length from a = 300 to 340mm.

For the case, a = 340mm, we find that the tip just closes under compressive stress, whereas the lateral stress is sustained at around 2 MPa, 1mm from the tip. For a = 330mm, the vertical stress was just tensile, with the lateral stress local to the tip being near 3 MPa, suggesting that the flexural crack would arrest in this vicinity, and may branch under a sufficiently high load. Interestingly, the SIF in (1) correlates remarkably well with the present analyses in predicting the length of crack where mode I propagation would cease:

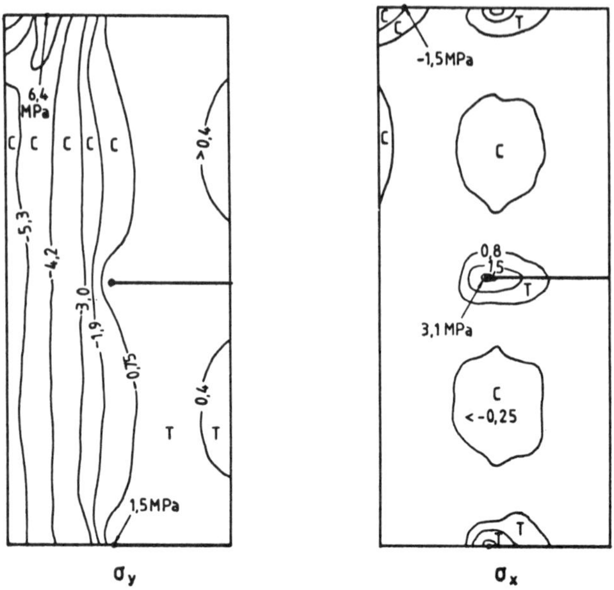

Fig. 3: Contours for cracked panel

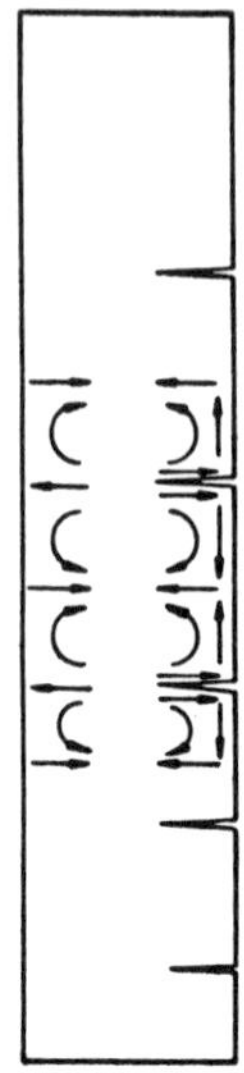

Fig. 4: Stress flow

K_I = 0 and 75 $N/mm^{3/2}$ for a = 335 and 332 mm respectively, which supports the above given a typical K_{IC} of, say, 60. Of course, this view is simplistic. At a=300 mm, tip zone stresses are higher, ≃ 4.5 MPa, and almost hydrostatic, so that arrest and branching might occur between a=300 and 330mm.

Figure 3 shows contour bands of both σ_x and σ_y for a = 330mm. It is interesting to note the extreme lateral tensions near the crack tips and the zones of lateral compression between the cracks. The mechanism which creates the bifurcation thus appears to relate to the stress flow generated in the uncracked regions between the flexural cracks. Figure 4 shows a diagramatic interpretation of this flow. It is the combination of the residual, and tensile, σ_y due to tension stiffening and the fact that the face of the crack is free to extend, thus rotating the panel, which gives rise to the stress flow as shown. These self equilibriating stresses, together with the stress flow around the crack tip, set up transverse couples which induce the vertical cracks emanating from the flexural crack tips.

2.4 Stress Intensity Factors

We evaluate the stress intensities from the calculated nodal displacements predicted near the crack tip in the analysis of the cracked panel problem, using:

$$K_I = COD \left(\frac{2\pi}{r} \right)^{1/2} \frac{E}{4(1-\nu^2)} \qquad (2)$$

$$K_{II} = CSD \left(\frac{2\pi}{r} \right)^{1/2} \frac{E}{4(1-\nu^2)} \qquad (3)$$

for plane strain, where COD and CSD are the crack opening and shearing displacements respectively. In using (2) and (3) to calculate the stress intensity, we find that K_{II} is always zero since the only shearing displacements relate to floating point differences.

For comparison with studies later, we are interested in crack lengths in the range, a=250 to 350mm, with eccentricities e=200,250,300mm. For the value e=W/3, we find K_I reduces from 22 at a=250mm to 0 at 340mm, corresponding to unstable crack growth. For e=250mm, the SIF was essentially constant at ≃60 corresponding to stable growth, but for e=300, stress intensity increased almost linearly from 95 to 125 $N/mm^{3/2}$. Each of the stress intensity curves plotted with a/W has a similar shape: a rising portion, a plateaux and a descending branch, albeit to very different scales. In these results, we see different parts of that function, though there is no simple relationship between SIF, e/W and a/W.

3 CONTINUUM MODELS

3.1 Tension Strength Criterion

The first model we used was a simple smeared crack with a tension strength criterion for crack extension perpendicular to the principal stress angle, based on a reduced integration bilinear quadrilateral element (Baker and Edwards, 1987). The concrete properties were f'_c = 40 MPa, f_t = 3.5 MPa, E_c = 32000 MPa and ε_{ult} = 0.003 in a panel measuring 3000x600x100 mm.

We found a mesh dependence not just in terms of crack tip stress, but the ability to model the branching. With an initial coarse mesh of 30x6 elements, and an eccentricity of e = W/3, there was no evidence of crack bifurcation from first cracking (320kN) to failure. This is partly due to the stiffness of such a coarse mesh: when f_t was reduced to 1.5 MPa, the crack pattern did show evidence of bifurcation. The main effect was that the bifurcation is position sensitive and localised, and that the coarse mesh did not permit sampling at the correct position. Mesh refinement even to 40x8 elements allowed localisation and banding to develop. Cracks initiated at 320kN, bifurcated at 480kN and the panel carried an ultimate load of 680 kN.

We noted also that when using quadratic elements with multiple cracks (at each Gauss point) on a mesh which gave a similar crack refinement to the above, again bifurcation was not captured. Here, the loss of stiffness, which invariably follows from overcracking with a strength criterion, prevented crack localisation. The final mesh dependence occured in the classical shear stiffness of bilinear quadrilaterals. Both selective and full integration predict crack bifurcation. However, with full integration, the pattern was confused, and again less localised. The reason appears to be that the shear stiffness prevents sufficient "shear banding" and localisation along the T bar: we have already noted that the branches propagate in mixed mode and it appears in the blunt crack state that mode II has a significant influence.

3.2 Geometric/Size Effect

There is a marked structural size effect associated with this problem in that the crack pattern, failure mechanism and hence nominal failure stress vary with load eccentricity and aspect ratio; we do not contend that this is the classical size effect where specimen size is the main variable. Figure 5 shows the failure patterns for a panel with an aspect ratio of 5:1 for four different eccentricities. Firstly, it can be seen that as the load eccentricity decreases from W/2 to W/3, the position of bifurcation moves with the eccentricity. In fact, bearing in mind that the cracks are smeared, there is an approximately constant spacing of ≃0.3W between the line of load and vertical cracks. At e = W/4, the force required to bifurcate the bending cracks is sufficiently large so that splitting under the load occurs first. Thus the overall behaviour is changed

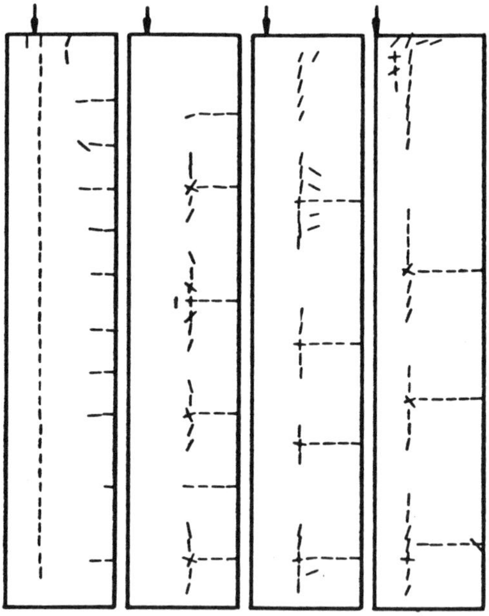

Figure 5: Crack patterns for various Eccentricities

to a panel response and, as eccentricity is further reduced, the response is akin to that of an end block where splitting and crushing completely dominate. We note also that the spacing of flexural cracks depends on eccentricity. In fact, the area of uncracked regions bounded by two T bar cracks increases proportionally with eccentricity. As expected the ultimate load decreased from 900 kN to 180 kN as eccentricity increased from W/4 to W/2.

The obvious size effect to test is the influence on scale for a given eccentricity. The results were repeated for an aspect ratio of 10:3 and e = W/3. Again flexural cracks initiate and bifurcate as before, but the increase in section modulus requires a higher ultimate load with the result that splitting occurs before that load is reached. Bending cracks opened first at 360 kN, and the first splitting occured at 540 kN after bifurcation.

3.3 Crack Band and Damage Models

Crack band models with strain softening across the localised crack within an element should be less mesh dependent. However, because the bifurcation point is critical, there is a certain dependence in requiring a small enough mesh to sample the stresses and strains near the localised branching. For example, Gajer and Dux (1990) use a relatively coarse mesh of quadratic isoparametric elements, with a localised crack band, and one central sampling point.

Of course, damage models based on energy criteria should be more accurate, and so the composite damage model of Bicanic et al.(1985) was used with the smeared crack approach and a strain based fracture criterion. The results were not sufficiently different to the above, and only at first cracking, principally because we are studying crack branching, not straight crack propagation. The only comment is that the technique had difficulty in converging for flat slopes of the softening branch, since in some cases the finite element stress fell artificially inside the constitutive curve, giving reverse release stress; it is tempting to wonder whether Crisfield's remarks (1982) on the subject were as a result of this same problem. An adaptive slope of the softening branch, based on a fracture energy, gave good results, but in this problem, a simple bilinear curve with the slope of the descending branch being the negative of the initial Young's modulus gave the same ultimate load.

4 LINEAR ELASTIC FRACTURE MECHANICS

4.1 Crack Direction

As already stated, the initial crack propagates in pure mode I. The problem, then, is how to calculate the direction for crack extension. Sousa et al. (1989) note that the three common criteria are:

(i) The maximum circumferential stress, which yields the classical relationship:

$$K_I \sin \theta + K_{II}(3\cos \theta - 1) = 0 \qquad (4)$$

(ii) The minimum fracture energy, which yields similar results to (i), but is much more complex.

(iii) The minimum strain energy density.

All three criteria, which use the SIFs, yield $\theta = 0$ for mode I. This is not surprising since they embody Irwin's conviction that cracks propagate in the direction given by $K_{II} = 0$ (Kobayashi, 1975), but it is worth recording. Shephard et al.(1985) use the T-criterion which seeks to maximize the dilatational portion of strain energy:

$$T_v = \left(\frac{1-2\nu}{3Er}\right)\left[K_I^2\cos^2\frac{\theta}{2} - K_I K_{II}\sin\theta + K_{II}^2\sin^2\frac{\theta}{2} \right] \tag{5}$$

For $K_{II} = 0$, this also yields $\theta = 0$.

However, it is possible to evaluate the circumferential stress at a series of points close to the tip and find the maximum from this set of discrete values, thus finding the appropriate crack angle. For example, the results from the analysis given earlier give stresses $\sigma_\theta = 0.4, 2.76, 2.6, 2.4$ for angles $\theta = 0, \pm45^\circ, \pm90^\circ, \pm135^\circ$ respectively, for a crack length a=330, which would suggest that bifurcation would occur rather than straight crack propagation. This process is not necessarily more cumbersome than solving (4), but for normal crack extension, it may well be less accurate because of the reliance on finite element stress rather than displacement.

4.2 Crack Extension and Bifurcation

Notwithstanding the difficulties in determining crack direction, the traditional criteria for assessing crack propagation are also problematic here. It was noted above that, before bifurcation, calculation of intensities using COD and CSD will yield pure mode I fracture, so that the crack will simply arrest near the compression field. Indeed, the precracked analyses yielded the following SIFs using expressions (2) and (3): $K_{II} = 0$, and $K_I = 13.12, 10.25, 7.07, 3.47, 0\ N/mm^{3/2}$ for a = 300,310,320,330,340 mm respectively. Given the continuum prediction of branching at P = 480kN, if we assume a typical value for $K_{IC} = 60$, then the mode I crack would arrest near the centreline for this force, in agreement with the continuum model, although this does not suggest that branching will result.

We need to give consideration to the strains on the surface of the crack near the tip. Indeed, one needs to view the extension along the T as the initiation of two new cracks from the free surface of the mode I portion of the crack. If numerical values of circumferential stress have been found at points a small distance r from the tip, these might be used with the original form of the hoop stress criterion for crack propagation, as given by Erdogan and Sih (1963):

$$\sigma_\theta\,(2\pi r)^{1/2} = K_{IC} \tag{6}$$

For convenience, we would like to calculate the stresses at a specific set of convenient, discrete angles, θ, such as nodes in a uniform rosette or grid, in which case consideration should be given to the use of hybrid elements or a local BEM solution at the tip. From the previous analysis, with elements 1mm square at the tip, the criterion (6) would suggest a crack splitting at a loads of 540 and 870kN for a=300 and 330mm respectively, which exceed the predictions of continuum models. By comparison with the SIFs earlier, it would appear that branching would occur between 300 and 330mm but at a higher load than the continuum models predicted.

We note that the 540kN required for branching and the 480kN required for straight extension from a=300mm agree with the "near hydrostatic" FE stresses at the tip. This suggests that bifurcation is not a linear discrete phenomenon (as seen in the multiple cracks at the tips in figure 5), and that softening in the process zone would allow further rotation and hence branching at a similar load.

4.3 Fracture Mechanics Software: Discrete Crack

The major problem that a bifurcating crack presents with regard to a finite element solution is topological. Even if branching can be predicted, there are real topological problems in an automatic process for remeshing around the two crack tips which emanate from the same point, particularly at the point that bifurcation occurs. We agree with Shephard et al.(1985) that an automatic generator should include the crack geometry as part of the mesh domain description. Our current research in solid modelling suggests that an approach which sees crack tip regions (singular elements or rosettes) as separate geometric/graphics objects with special definitions for the intersection and union with other objects (other mesh areas) is the most structured and robust procedure to cope with these difficulties, particularly in 3D computational fracture.

5 NONLINEAR FRACTURE MECHANICS

It has been known for some time that a nonlinear fracture mechanics approach, with nucleation and softening in the process zone before coalescence, is more appropriate than LEFM in the analysis of concrete structures. The process zone softening will of course affect the lateral stresses around the tip thus influencing the load at which branching occurs. However, whether one adopts the discrete crack formulation as originally introduced by Hillerborg et al.(1976) or a smeared crack formulation, for example Yamaguchi and Chen (1990), the problem is again to define suitable crack orientation criteria. We have already seen that approaches which tend to use principal stress work well.

6 CONCLUDING REMARKS

We have discussed computational aspects of the problem of mode I flexural cracks bifurcating as the crack approaches a compressive field. The smeared approach, whether it uses a strength criterion, a composite damage model, or nonlinear fracture mechanics, easily predicts the branching behaviour because it uses principal stress to determine crack direction. Discrete crack models which use a stress intensity, evaluated on calculated displacements, fail to predict branching since pure mode I propagation cannot deviate.

Our intention is not to criticize LEFM but to point out that crack direction and extension criteria should be considered carefully: bifurcation is a localised, but not a linear discrete phenomenon. In essence, we contend that this interesting phenomenon of bifurcating flexural cracks, which has been observed in experiment, forms a demanding bench mark test for computational fracture models.

7 REFERENCES

Baker, G. and Edwards, A.D. (1987) Experience with a nonlinear FE code for the analysis of cracking in concrete structures. *Civ. Engrg. Trans. I.E.Aust.*, **CE29**(3), 192-201.

Bicanic, N., Pramono, E., Sture, S. and Willam, K.J. (1985) On numerical prediction of concrete fracture localisation. In *Numerical Methods in Engrg.: Theory and Applications* (eds. J.Middleton and G.Pande), A.A.Balkema: Rotterdam, **1**, 385-391.

Crisfield, M.A. (1982) Local instabilities in the non-linear analysis of reinforced concrete beams and slabs. *Proc. Inst. Civil. Engrs.*, **73**(2), 135-145.

Erdogan, F. and Sih, G.C. (1963) On the crack extension in plates under plane loading and transverse shear. *Jnl. Basic Engrg.*, **85**, 519-527.

Gajer, G. and Dux, P.F. (1990) Crack band based model for FEM analysis of concrete structures. *Jnl. Struct. Engrg., ASCE*, **116**(6), 1696-1714.

Hillerborg, A., Modeer, M. and Peterson, P.E. (1976) Analysis of crack formation and crack growth in concrete by means of fracture mechanics and finite elements. *Cem.Conc.Res.*, **6**(6), 773-782.

Kobayashi, A.S. (1975) Some two dimensional problems in fracture mechanics. In *Linear Fracture Mechanics*, (eds. G.C.Sih, R.P.Wei, F.Erdogan), Envo Publishing Co.: Lehigh, 107-128.

Palmer, G.B. and Baker, G. (1990) Analysis of thermally induced fracture of low-cement refractory rings. *Bull. Amer. Ceram. Soc.*, **69**(6), 1035-1038.

Paris, P.C. and Sih, G.C. (1965) Stress analysis of cracks. In *Fracture Toughness Testing and its Applications.* ASTM STP No. 381, 30-83.

Shephard, M.S., Yehia, N.A.B., Burd, G.A. and Weidner, T.J. (1985) Automatic crack propagation tracking. *Comp. Struct.*, **20**, 211-223.

Sih, G.C. and Irwin, G.R. (1970) Dynamic analysis for two-dimensional multiple crack division. *Engrg. Frac. Mech.*, **1**(4), 603-614.

Sousa, J.L., Martha, L.F., Wawrzynek, P.A. and Ingraffea, A.R. (1989) Simulation of non-planar crack propagation in three dimensional structures of concrete and rock. In *Fracture of Concrete and Rock* (eds S.P.Shah, S.E.Swartz, B.Barr), Elsevier Applied Science, 254-264.

Yamaguchi, E. and Chen, W-F. (1990) Cracking model for finite element analysis of concrete materials. *Jnl. Engrg. Mech., ASCE*, **116**(6), 1242-1260.

PART NINE

NUMERICAL MACROSCOPIC CRACK MODELS: CONTINUA

INVITED PAPER

53 CONTINUUM MODELS FOR DISCONTINUOUS MEDIA

R. de BORST
Department of Civil Engineering, Delft University of Technology, and TNO Building and Construction Research, Rijswijk, The Netherlands
H.-B. MUHLHAUS
CSIRO, Division of Geomechanics, Mt Waverley, Victoria, Australia

Abstract
Standard continuum models suffer from pathological mesh dependence when used in fracture analyses of heterogeneous, brittle materials. In this contribution possibilities and limitations will be investigated of two enhancements of conventional continua, namely the Cosserat theory and gradient continuum models, in which the standard continuum description is enriched by adding rotational degrees-of-freedom and strain gradients respectively. Both models incorporate an internal length scale and are therefore, in principle, capable of properly describing discontinuities that arise at a micro-level in the material.
Keywords: Higher-order Continua, Strain-Softening, Fracture, Localization, Finite Element Analysis

1. Introduction

The quintessence of the whole problem of spurious mesh-sensitivity in concrete and rock fracture is the fact that the *force-displacement relations* that are measured in testing devices are simply mapped onto *stress-strain curves* by dividing the force and the elongation by the original load-carrying area and the original length of the specimen respectively. This is done without taking into account the changes in the micro-structure that occur when the material is so heavily damaged as in fracture processes. Mathematically the result is that, when the amount of micro-cracking becomes so severe that, in a macroscopic sense, fracture is observed, the governing field equations lose ellipticity. The mathematical description ceases to be a meaningful representation of the physical reality. The only solution to this problem is the introduction of additional terms in the continuum description which reflect the changes in the micro-structure that occur during fracture processes. The result is that the governing equations remain elliptic also during fracture, which is needed for a proper prediction of the localization process accompanying fracture. For instance, the width of the fracture process zone is contained in the mathematical description and is no longer one element wide as in numerical simulations of fracture in a conventional continuum. When the elements are taken smaller than the width of the fracture process zone, localization occurs in more than one element.

2. Problem statement

The essential shortcomings of the conventional approach as well as the novel features of enhanced continuum models are best demonstrated by the example of a simple bar loaded in

Fracture Processes in Concrete, Rock and Ceramics Edited by J.G.M. van Mier, J.G. Rots and A. Bakker.
 Published by E. & F.N. Spon, 2–6 Boundary Row, London SE1 8HN. ISBN 0 419 15870 7

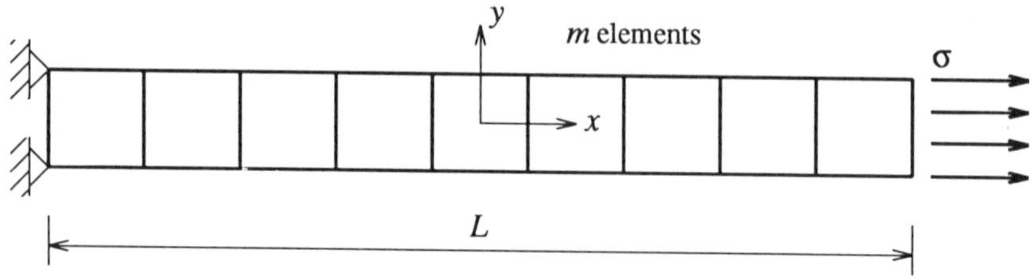

Figure 1. Strain-softening bar subject to uniaxial loading.

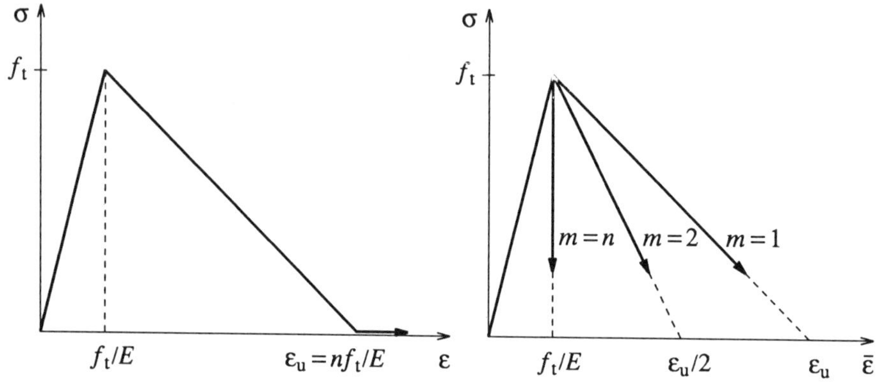

Figure 2. Stress-strain diagram (left) and response of an imperfect bar in terms of an stress-average strain curve (right).

uniaxial tension of Figure 1 (Crisfield 1982, de Borst 1986). Let the bar be divided into m elements. Prior to reaching the tensile strength f_t a linear relation is assumed between the (normal) stress σ and the (normal) strain ε:

$$\sigma = E\varepsilon , \tag{1}$$

with E Young's modulus. After reaching the peak strength a descending slope is defined in this diagram through an affine transformation from the measured load-displacement curve. The result is given in the left part of Figure 2, where ε_u marks the point where the load-carrying capacity is totally exhausted. In the post-peak regime the constitutive model can thus be summarized as:

$$\varepsilon = \varepsilon^e + \varepsilon^i , \tag{2}$$

which constitutes a decomposition of the strain into an elastic part ε^e:

$$\varepsilon^e = E^{-1}\sigma , \tag{3}$$

and a contribution due to inelastic effects (eg, cracking or plastic slip)

$$\varepsilon^i = h^{-1}(\sigma - f_t) , \tag{4}$$

where h plays a similar role for the inelastic strain ε^i that E has for the elastic strains. In case of degrading materials $h < 0$ and h may be termed a softening modulus. In the applications discussed in this article h will be assumed to be a constant, so that linear softening is used. Eq. (4) may also be thought of as an integrated form of the evolution equation for the stress rate after failure:

$$\sigma = f_t + h\,\varepsilon^i \,. \tag{5}$$

Now suppose that one element has a tensile strength that is marginally below that of the other $m-1$ elements. Upon reaching the tensile strength of this element failure will occur. In the other, neighbouring elements the tensile strength is not exceeded and they will unload elastically. The result on the *average* strain of the bar $\bar{\varepsilon}$ is plotted in the right part of Figure 2 for different discretizations of the bar. The results seem to be totally mesh-unobjective, i.e. convergence to a 'true' post-peak failure curve does not seem to occur. In fact, it does occur, because the governing equations predict the failure mechanism to be a line crack with zero thickness. The finite element solution simply tries to capture this line crack, which results in localization in one element, irrespective of the width of this element. The result on the load-average strain curve is obvious: for an infinite number of elements ($m \to \infty$) the post-peak curve doubles back on the original loading curve. A major problem is now that, since the constitutive model has been phrased in terms of a *stress-strain* law and *not* as a *force-displacement* relation, the energy that is dissipated tends to zero upon mesh refinement because the area in which the failure process occurs also becomes zero. From a physical point of view this is unacceptable and we must therefore either rephrase our constitutive model in terms of force-displacement relations, which implies the use of special interface elements (Rots 1988), or enrich the continuum description by adding higher-order terms which can accommodate narrow zones of highly localized deformation quite similar to descriptions for boundary layers in fluids. As an intermediate solution a number of authors (Bazant and Oh 1983, Pietruszczak and Mróz 1981, Willam 1984) have proposed to regard the area under the softening curve in the left part of Figure 2 as a material parameter, namely the fracture energy G_f, and consequently, to make the softening modulus h a function of the element size. For the shape of the load-average strain curve this 'trick' obviously works well, but it leaves the width of the localization zone unspecified.

3. Alternative continuum formulations

To overcome the deficiencies of the conventional continuum models three basically different approaches may be distinguished:

- the use of *non-local* or *gradient* continuum theories,
- the use of *micro-polar* continua, and
- the inclusion of rate-effects in the constitutive model.

Obviously, the last option can only be applied when relatively high loading rates are considered. For (quasi)static loading conditions only the two first-mentioned procedures to restore well-posedness of the partial differential equations can be utilized. In this contribution we shall confine ourselves to problems in which rate effects play no role, and, therefore, we shall only explore the current possibilities and limitations of enhanced continuum models. For a discussion on the role of rate-dependence in regularizing the governing field equations the reader is referred to Sluys and de Borst (1991) and Needleman (1988).

4. Micro-polar continuum models

4.1 Formulation of an elasto-plastic Cosserat continuum

A possible extension of the classical continuum is to allow that, when the elementary area of the insert in Figure 3 is considered, small moments can act on the sides of this rectangle in addition to the normal stresses and the shear stresses. Hence we do not adopt the assumption normally made that the elementary area of Figure 3 can be made so small that the normal stresses are distributed homogeneously over the sides of the elementary area. When we include these so-called *couple-stresses* m_{xz} and m_{yz} the expressions for translational equilibrium are unaltered:

$$\frac{\partial \sigma_{xx}}{\partial x} + \frac{\partial \sigma_{yx}}{\partial y} = 0 \tag{6a}$$

and

$$\frac{\partial \sigma_{xy}}{\partial x} + \frac{\partial \sigma_{yy}}{\partial y} = 0 , \tag{6b}$$

which replicates the result obtained for a classical continuum. For rotational equilibrium, however, we find that

$$\frac{\partial m_{xz}}{\partial x} + \frac{\partial m_{yz}}{\partial y} - (\sigma_{yx} - \sigma_{xy}) = 0 , \tag{6}$$

which shows that the stress tensor is in general only symmetric ($\sigma_{xy} = \sigma_{yx}$) if the couple-stresses m_{xz} and m_{yz} vanish.

Since we have included couple-stresses as additional components in the stress tensor we also need extra strain components, the so-called *micro-curvatures* κ_{xz} and κ_{yz}. To properly define these strain components the rotation ω_z is required as an additional degree-of-freedom, independent of the translations u_x and u_y. Then, kinematics of the so-called unconstrained Cosserat continuum (E. and F. Cosserat 1909), which is but one of the many possible altenatives that exist in the literature, are defined by:

$$\varepsilon_{xx} = \frac{\partial u_x}{\partial x} \tag{7a}$$

and

$$\varepsilon_{yy} = \frac{\partial u_y}{\partial y} . \tag{7b}$$

The shear strains, however, are given a slightly modified form as

$$\varepsilon_{xy} = \frac{\partial u_x}{\partial y} + \omega_z \tag{8a}$$

and

$$\varepsilon_{yx} = \frac{\partial u_y}{\partial x} - \omega_z . \tag{8b}$$

The micro-curvatures are given by:

$$\kappa_{xz} = \frac{\partial \omega_z}{\partial x} \tag{9a}$$

and

$$\kappa_{yz} = \frac{\partial \omega_z}{\partial y} \,. \tag{9b}$$

We next assemble all strain components in the vector $\boldsymbol{\varepsilon}$,

$$\boldsymbol{\varepsilon} = [\, \varepsilon_{xx} \,,\, \varepsilon_{yy} \,,\, \varepsilon_{zz} \,,\, \varepsilon_{xy} \,,\, \varepsilon_{yx} \,,\, \kappa_{xz} l \,,\, \kappa_{yz} l \,]^{\mathrm{T}} \,, \tag{10}$$

with l an additional material parameter. The quintessential point is now that l has the dimension of length. As with the gradient model discussed in a subsequent section we thus have an internal length scale in our continuum description. For this reason we can describe localization of deformation into a band width of a finite size and the important phenomenon of scale effects in failure of brittle materials. In a similar fashion we can assemble all stress components in the vector $\boldsymbol{\sigma}$,

$$\boldsymbol{\sigma} = [\, \sigma_{xx} \,,\, \sigma_{yy} \,,\, \sigma_{zz} \,,\, \sigma_{xy} \,,\, \sigma_{yx} \,,\, m_{xz}/l \,,\, m_{yz}/l \,]^{\mathrm{T}} \,. \tag{11}$$

As normal continuum theory Cosserat plasticity takes its point of departure in an additive decomposition of the strain vector into an elastic contribution $\boldsymbol{\varepsilon}^{\mathrm{e}}$ and an inelastic part $\boldsymbol{\varepsilon}^{\mathrm{i}}$:

$$\boldsymbol{\varepsilon} = \boldsymbol{\varepsilon}^{\mathrm{e}} + \boldsymbol{\varepsilon}^{\mathrm{i}} \,, \tag{12}$$

and we assume that the elastic strain vector is linearly related to the stress vector $\boldsymbol{\sigma}$:

$$\boldsymbol{\sigma} = \mathbf{D}\boldsymbol{\varepsilon}^{\mathrm{e}} \,. \tag{13}$$

$\mathbf{D}$ is the rigidity matrix that contains the elastic moduli (eg, Mühlhaus 1990):

$$\mathbf{D} = \begin{bmatrix} 2\mu a_1 & 2\mu a_2 & 2\mu a_2 & 0 & 0 & 0 & 0 \\ 2\mu a_2 & 2\mu a_1 & 2\mu a_2 & 0 & 0 & 0 & 0 \\ 2\mu a_2 & 2\mu a_2 & 2\mu a_1 & 0 & 0 & 0 & 0 \\ 0 & 0 & 0 & \mu+\mu_c & \mu-\mu_c & 0 & 0 \\ 0 & 0 & 0 & \mu-\mu_c & \mu+\mu_c & 0 & 0 \\ 0 & 0 & 0 & 0 & 0 & 2\mu & 0 \\ 0 & 0 & 0 & 0 & 0 & 0 & 2\mu \end{bmatrix} , \tag{14}$$

with $a_1 = (1-\nu)/(1-2\nu)$ and $a_2 = \nu/(1-2\nu)$. The elastic constants μ and ν have the classical meaning of the shear modulus and Poisson's ratio. μ_c is a rotational shear modulus, completing the total of four material constants, viz. μ, ν, l and μ_c that are needed to describe the elastic behaviour of an isotropic Cosserat continuum under planar deformations. The coefficient 2 has been introduced in the terms D_{66} and D_{77} to arrive at a convenient form of the inelastic constitutive equations (de Borst 1990, 1991). The total (bending) stiffness that sets the relation between the micro-curvatures and the couple-stresses is of course determined by the value of the internal length scale l.

Mühlhaus et al. (1991) have developed a Cosserat model for granular media and have established relationships between the Cosserat elasticities of eq. (14) and the contact stiffnesses,

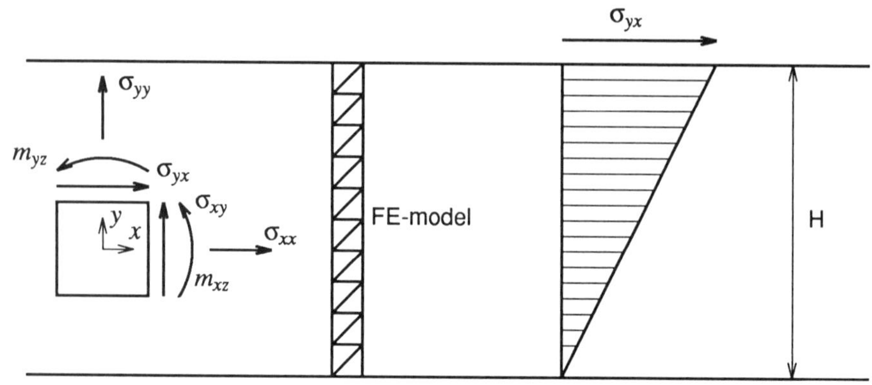

Figure 3. Shear layer: loading conditions and finite element grid (for 20 elements). Insert: stress and couple-stress in a Cosserat continuum.

coordination number and average grain radius of the discrete medium. In another paper (Choi and Mühlhaus 1991) the model has been applied to obtain the analytical solution of simple non-homogeneous shear of an infinite layer and the results have been compared with a corresponding distinct element solution.

In the present treatment we will make a restriction to a von Mises plasticity theory. Accordingly, the yield function f can be written as

$$f = (3J_2)^{1/2} - \bar{\sigma}(\gamma) , \tag{15}$$

with $\bar{\sigma}$ the yield stress which is a function of the hardening parameter γ. J_2 is the second invariant of the deviatoric stresses, which, for a Cosserat continuum, can be generalized as (Mühlhaus and Vardoulakis 1987):

$$J_2 = a_1 s_{ij} s_{ij} + a_2 s_{ij} s_{ji} + a_3 m_{ij} m_{ij} / l^2 , \tag{16}$$

where the summation convention with respect to repeated indices has been adopted. s_{ij} is the deviatoric stress tensor. A generalized expression for the equivalent inelastic strain is (Mühlhaus and Vardoulakis 1987):

$$\dot{\gamma} = [b_1 \dot{e}^{\mathrm{i}}_{ij} \dot{e}^{\mathrm{i}}_{ij} + b_2 \dot{e}^{\mathrm{i}}_{ij} \dot{e}^{\mathrm{i}}_{ji} + b_3 \dot{\kappa}^{\mathrm{i}}_{ij} \dot{\kappa}^{\mathrm{i}}_{ij} l^2]^{1/2} , \tag{17}$$

with $\dot{e}^{\mathrm{i}}_{ij}$ the plastic deviatoric strain-rate tensor. For numerical convenience the choice $a_1 = a_2 = \frac{1}{4}$, $a_3 = \frac{1}{2}$ and $b_1 = b_2 = \frac{1}{3}$, $b_3 = \frac{2}{3}$ has been used (de Borst 1990, 1991). Based on micromechanical considerations Mühlhaus and Vardoulakis (1987) have suggested that other values may be more representative for granular assemblies.

4.2 Localization of deformation: infinite shear layer

The first problem that will be used to scrutinize the characteristic features of a Cosserat plasticity strain-softening model is the shear layer of Figure 3 (de Borst 1990, 1991, Mühlhaus et al. 1991). It is assumed that the shear layer is infinitely long in both the negative and the positive x-direction. The discretization of the shear layer, which has a height $H = 100$ mm, is shown in Figure 3 (for 20 elements).

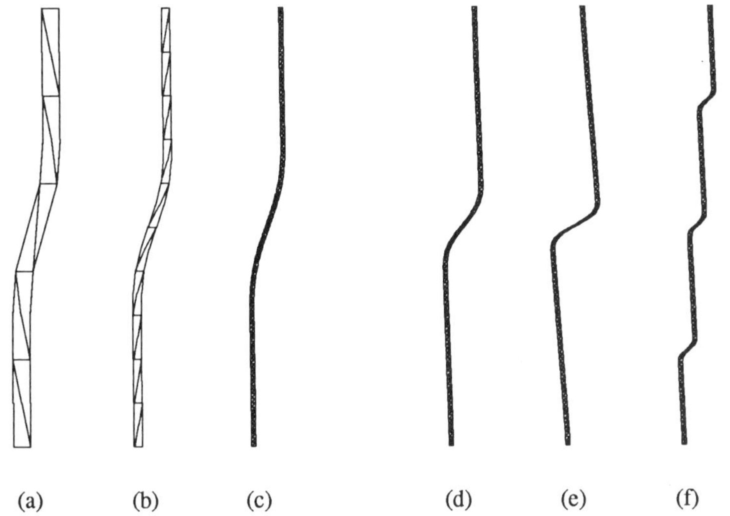

Figure 4. Incremental displacement patterns at a residual load level of $\sigma_{yx}/\bar{\sigma}=0.28$.
$l/H=0.12$: (a) 10 elements; (b) 20 elements; (c) 200 elements.
200 elements: (d) $l/H=0.06$; (e) $l/H=0.03$; (f) $l/H=0.015$.

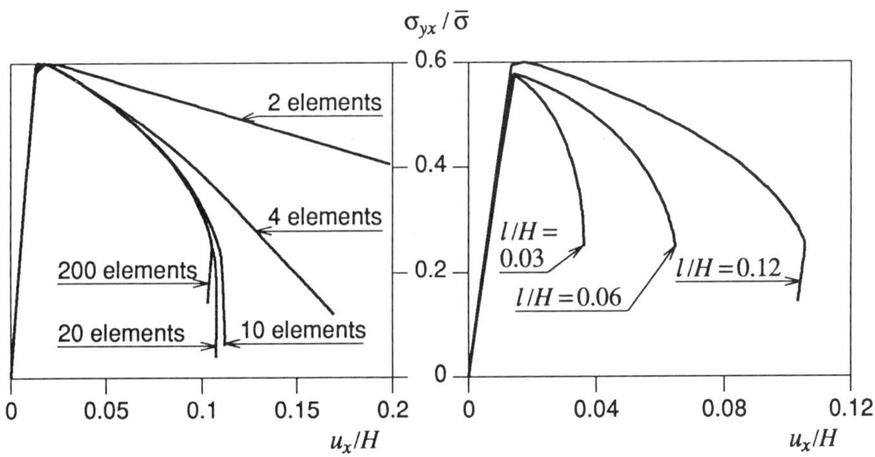

Figure 5. Load-displacement diagrams. Left: different discretizations; right: different l/H ratios.

This problem has intentionally been chosen instead of the tension bar of Figure 1. This is because the rotational degrees-of-freedom in a Cosserat continuum are only activated under shear loadings. In pure tension they do not become active. The micro-curvatures remain zero and so do the couple-stresses. For failure problems in which decohesion plays a prevailing role this is a disadvantage of the Cosserat concept. On the other hand, brittle failures under compressive loading often involve a highly localized band of intense shearing. For analyzing these types of localization modes the Cosserat concept seems a most natural framework.

The problem has been analysed using two-dimensional, six-noded triangles with three degrees-of-freedom per node: two translations (u_x, u_y) and one rotation (ω_z). The problem is one-dimensional and could have been analysed using line elements. Use of two-dimensional elements requires that linear dependence relations are added for the displacements in the x-direction and for the rotation about the z-axis. In the y-direction all displacements are prevented (isochoric motion). At the bottom of the shear layer all displacements and rotations are kept zero ($u_x = 0$, $u_y = 0$, $\omega_z = 0$). The upper boundary is subjected to a shear force σ_{yx} (per unit area), which has been controlled using a standard arc-length technique, eg, de Borst (1988). Furthermore the condition $\omega_z = 0$ has been imposed at the upper boundary.

The shear moduli have been chosen as: $\mu = 4000$ N/mm^2 and $\mu_c = 2000$ N/mm^2. In the reference calculations the internal length parameter has been assigned the value $l = 12$ mm. The initial yield strength has been taken equal to $\bar{\sigma} = 100$ N/mm^2. A linear softening diagram has been used with a hardening (softening) modulus $h = -0.125\,\mu$.

While an imperfection is necessary to trigger localization in a conventional strain-softening solid, this is not the case for Cosserat elasto-plasticity. For a Cosserat model a non-homogeneous strain state is obtained already in the elastic regime. A *boundary layer* of approximately a height l develops at the upper and lower parts of the shear layer and the strains are not uniform over the height of the shear layer (Schaefer 1962). As a consequence the localization zone develops smoothly and gradually in the middle of the shear layer. When the discretization is refined to such an extent that more than one set of two triangular elements are placed over the localization zone, the width of the numerically predicted localization zone becomes constant (Figure 4). This is the case when the shear layer is subdivided into more than 10 elements. Figure 5 shows that also the load-displacement curve then converges to a unique, physically meaningful solution with a finite energy consumption.

As discussed previously a most prominent feature of higher-order continuum models including the Cosserat continuum is the presence of an internal length scale. On one hand this internal length scale causes localization zones to have a finite width and not to collapse into genuine discontinuities. On the other hand it introduces a structural size effect. To demonstrate this in a quantitative sense the shear layer of Figure 3 has been analysed for various ratios l/H, l being the internal length scale and H being the height of the layer. For larger structures or for smaller values of the internal length parameter l, the ratio l/H diminishes. Physically, one would then expect a more brittle structural behaviour. This is confirmed in the calculations, in which l/H was varied between 0.03 and 0.12 (Figure 5). Also the size of the localization zone shrinks compared with the total height of the layer if the ratio l/H is reduced (Figure 4).

A remarkable observation in Figure 4 is the fact that for $l/H = 0.015$ *three* localization zones are computed. Intuitively, one would associate this result with patterning and periodicity. Nevertheless, such behaviour is not contained in the Cosserat continuum model. It is simply a numerical artefact, caused by the fact that for smaller l/H-ratios the stress distribution in the elastic regime becomes more and more homogeneous. The strong maximum of the value of J_2 of the shear layer, which causes the initiation of yielding and the ensuing development of the localization zone in the centre, becomes weaker and, owing to the fact that we deal with finite load increments in a numerical analysis, the yield condition is violated also elsewhere in the layer. These points then also act as nuclei for the formation of shear bands. However, it is emphasized that, although the solution corresponds to a sequence of perfectly converged equilibrium states, this solution is *not* the most critical solution, as a solution that involves a

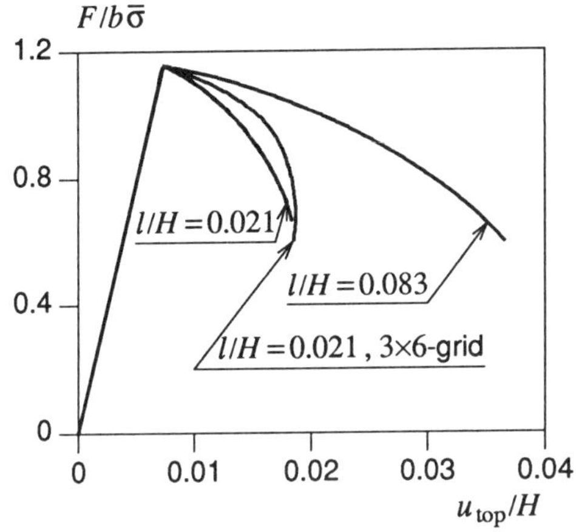

Figure 6. Load-displacement diagram for a biaxial compression test (Cosserat continuum).

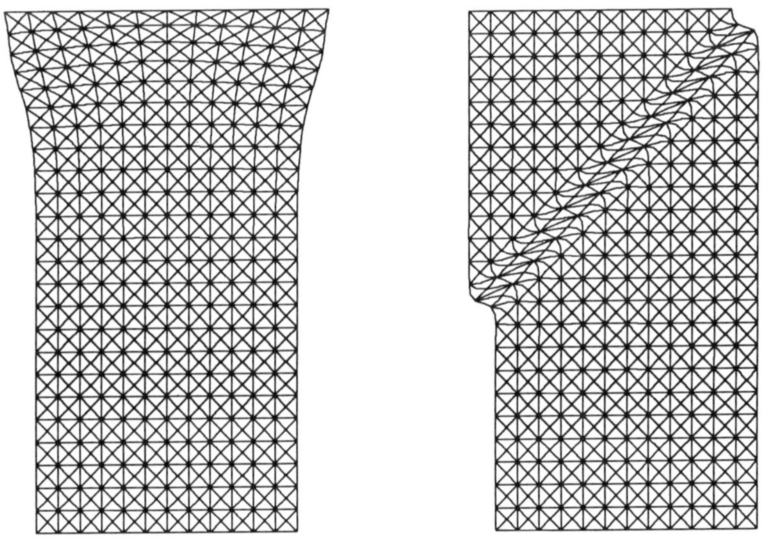

Figure 7. Incremental displacement patterns at failure. Left: $l/H = 0.0833$; right: $l/H = 0.0052$.

single localization zone exists and is associated with a load-displacement path that is lower than the equilibrium path computed for the profile with the three distinct localization zones.

4.3 Localization of deformation: biaxial test

To investigate whether the excellent, mesh-independent behaviour of the Cosserat model also holds for strain-softening calculations in truly two-dimensional finite element configurations a biaxial compression test under plane-strain conditions was simulated. For a specimen with a height $H = 120$ mm and a width $b = 60$ mm analyses have been carried out for three different values of the characteristic length ($l = 10$ mm, $l = 2.5$ mm, and $l = 0.625$ mm), while for one value of the length parameter ($l = 2.5$ mm) calculations have been made for three different meshes (72, 288 and 1152 six-noded triangular elements respectively).

In these analyses the elasticity parameters have been assigned the following values: classical shear modulus: $\mu = 4000$ N/mm^2, Poisson's ratio $\nu = 0.49$, Cosserat rotational shear modulus $\mu_c = 2000$ N/mm^2. A linear softening model has been used with a hardening modulus $h = -0.1\,\mu$ and a value of $\bar{\sigma} = 100$ N/mm^2 for the virgin yield strength.

The translational degrees-of-freedom at both sides of the specimen were free, while the upper and lower boundaries were perfectly smooth. The upper boundary was subjected to a vertical force F, which has been controlled using indirect displacement control (de Borst 1986). At all boundaries the micro-rotations ω_z were free (natural boundary conditions). Since pure compression does not result in any activation of the rotational degrees-of-freedom and since it causes the specimen to deform homogeneously also in the post-peak regime, a material imperfection was used to trigger localization. The results of these analyses are shown in Figures 6 and 7. Figure 6 shows that the Cosserat model results in a global, structural behaviour that is fully mesh-objective. As with the case of the infinitely long shear layer the role of the internal length in governing the brittleness and setting the width of the localization zone comes out nicely (Figures 6 and 7).

5. Non-local and gradient continuum models

Another way to depart from the concept of 'simple materials' (Noll 1958) used hitherto for the numerical modelling of degrading materials is to modify eq. (5) which states that the residual stress beyond peak load is exclusively a function of the inelastic strain in the material point itself. Instead, one could write:

$$\sigma = f_t + h\bar{\varepsilon}^i , \tag{18}$$

where $\bar{\varepsilon}^i$ is defined by the integral expression

$$\bar{\varepsilon}^i = \int_{-L/2}^{L/2} g(s)\varepsilon^i(x+s)\mathrm{d}s , \tag{19}$$

with L the length of the bar (Figure 1) and $g(s)$ a weighting function. While this expression can be generalized to three dimensions (eg, Bazant *et al.* 1984, Pijaudier-Cabot and Bazant 1987, Mühlhaus and Aifantis 1991), this will not be done here to preserve transparency of the treatment. A possible choice for $g(s)$ that meets the intuitive notion that the non-local character of a constitutive law as eq. (19) should fade away rapidly for larger distances is the error function. Then eq. (19) changes into

$$\bar{\varepsilon}^i = \int_{-L/2}^{L/2} \frac{l}{2\sqrt{\pi}} \exp(-l^2 s^2/4)\varepsilon^i(x+s)\mathrm{d}s , \tag{20}$$

where l is a characteristic material parameter with the dimension of length that can be related to the size of the softening zone. When $\varepsilon^i(x+s)$ is developed in a Taylor series around $s = 0$, we obtain instead of eq. (20):

$$\bar{\varepsilon}^{\mathrm{i}} = \int_{-L/2}^{L/2} \frac{l}{2\sqrt{\pi}} \exp(-l^2 s^2/4) \varepsilon^{\mathrm{i}}(x) \mathrm{d}s + \int_{-L/2}^{L/2} \frac{ls}{2\sqrt{\pi}} \exp(-l^2 s^2/4) \frac{\mathrm{d}\varepsilon^{\mathrm{i}}(x)}{\mathrm{d}x} \mathrm{d}s$$

$$+ \int_{-L/2}^{L/2} \frac{ls^2}{4\sqrt{\pi}} \exp(-l^2 s^2/4) \frac{\mathrm{d}^2\varepsilon^{\mathrm{i}}(x)}{\mathrm{d}x^2} \mathrm{d}s + \ldots\ldots\ldots\ldots\ldots\ldots \quad . \tag{21}$$

If it is assumed that $L \gg l$ (which implies that the structural size is several orders of magnitude bigger than the size of the localization zone) and if higher-order terms are neglected the integrals of eq. (21) can be evaluated analytically. The result is

$$\bar{\varepsilon}^{\mathrm{i}} = \varepsilon^{\mathrm{i}} + \frac{1}{l^2} \frac{\mathrm{d}^2\varepsilon^{\mathrm{i}}}{\mathrm{d}x^2} \ . \tag{22}$$

Note that the second term on the right-hand of eq. (21) cancels in the integration process. More generally all odd derivatives of ε^{i} with respect to x cancel upon integration over the entire bar. Extension of eq. (5) to include just the first gradient of the inelastic strain (Schreyer and Chen 1986) therefore seems less natural. It is furthermore noted that extension to higher-order derivatives than included in eq. (22) is possible. Zbib and Aifantis (1988) and Mühlhaus and Aifantis (1991) for instance also consider the fourth derivative of ε^{i} with respect to x. With expression (22) eq. (18) can be replaced by:

$$\sigma = f_{\mathrm{t}} + h\,\varepsilon^{\mathrm{i}} + \frac{h}{l^2} \frac{\mathrm{d}^2\varepsilon^{\mathrm{i}}}{\mathrm{d}x^2} \ . \tag{23}$$

In sum, gradient models as defined in eq. (23) can be derived from non-local models by expanding the kernel of the integral employed in the averaging procedure for the inelastic strains. However, it is important to note that the gradient term in eq. (23) acts as a stabilizer and garantuees ellipticity of the governing partial differential equations only if $h < 0$ from the onset of plastic deformation. To show this we calculate the rate of work corresponding to the gradient term:

$$\frac{1}{2} \int_{-L/2}^{L/2} \frac{h}{l^2} \frac{\mathrm{d}^2\dot{\varepsilon}^{\mathrm{i}}}{\mathrm{d}x^2} \dot{\varepsilon}^{\mathrm{i}} \mathrm{d}x = -\frac{1}{2} \int_{-L/2}^{L/2} \frac{h}{l^2} \left(\frac{\mathrm{d}\dot{\varepsilon}^{\mathrm{i}}}{\mathrm{d}x} \right)^2 \mathrm{d}x \ , \tag{24}$$

whereby we have assumed the variational boundary condition $\mathrm{d}\dot{\varepsilon}^{\mathrm{i}}/\mathrm{d}x = 0$ at $x = \pm L/2$ as suggested by Mülhaus and Aifantis (1991). For the gradient term to act as a stabilizer expression (24) has to be positive definite and this is the case if and only if $h < 0$. This implies that the gradient term as introduced in eq. (23) would indeed be stabilizing in the softening regime, but would be destabilizing in case of hardening, which would be neither physically meaningful, nor computationally desirable. Motivated by recent micromechanical models (Walgraef and Aifantis 1985, Kubin and Lépinoux 1988, Kratochvil 1988) we shall henceforth use the more general expression

$$\sigma = f_{\mathrm{t}} + h\,\varepsilon^{\mathrm{i}} - \frac{c}{l^2} \frac{\mathrm{d}^2\varepsilon^{\mathrm{i}}}{\mathrm{d}x^2} \ , \tag{25}$$

instead of eq. (23), where $c > 0$ is a material constant.

When compared to the original non-local formulation as expressed through eqs. (18) and (19) the gradient model (25) has two important advantages:

- It is computationally much more efficient, as for the gradient models an efficient algorithm can be developed that satisfies the evolution equation for the inelastic strains in a distributed sense (de Borst and Mühlhaus 1991). This is exactly the requirement imposed on non-local or gradient models (Simo 1989).
- The additional boundary conditions on the inelastic strains which have to be enforced in non-local and gradient continuum models can be solved uniquely and elegantly for gradient models. Indeed, for gradient plasticity models it is even possible to deduce these non-standard boundary conditions from a variational principle (Mühlhaus and Aifantis 1991, de Borst and Mühlhaus 1991).

5.1 Some salient characteristics of gradient models

The properties of non-local and gradient continuum models can most lucidly be brought out by constructing analytical solutions. For the bar of Figure 1 and considering the 2nd grade material as defined by eq. (25) a closed-form solution exists which shows that, when a steady-state solution has developed, the width of the fracture process zone is given by (de Borst and Mühlhaus 1991)

$$w = 2\pi l \,, \tag{26}$$

while the slope in the diagram in which the stress in the bar σ *vs* the displacement of the end of the bar u is given, reads:

$$\frac{\dot{u}/L}{\dot{\sigma}} = \frac{1}{E} + \frac{2\pi l}{Lh} \,. \tag{27}$$

It is emphasized that the solution holds only for a steady-state solution with a constant softening modulus h. If ε_u has been exceeded somewhere in the bar, h effectively becomes zero there and the solution ceases to be valid.

5.2 Size effect and gradient models

A characteristic effect that is observed in testing of brittle solids, is the size effect: a bigger but otherwise identical structure has a lower failure load than a smaller specimen. The failure of classical continuum models to predict this property is attributable to the absence of an internal length scale. The numerical simulations for the Cosserat continuum in the previous section showed that the presence of an internal length scale indeed introduced a size effect under compressive/shear loadings. From eq. (27) it is evident that the incorporation of gradients in a continuum model results in a size effect that also manifests itself in tension, since the ratio l/L, i.e. the quotient of the internal length scale and the structural size, basically governs the post-peak behaviour. A smaller value for l or a larger value for L both result in a more brittle post-peak load-displacement path. Indeed, for small values of l ($l/L < -h/(2\pi E)$) we even obtain so-called snap-back behaviour. This implies that under quasi-static loadings not only the load, but also the displacement of the right end of the bar decreases. Experiments can no longer be kept stable under displacement control. Owing to the fact that the localization zone cannot absorb the elastic energy released in the unloading remaining parts of the bar, the observed failure of the specimen is of a highly explosive character.

As another salient feature of the gradient model we observe from eq. (26) that the width of the fracture process zone w is directly proportional to the internal length scale l: if l is taken smaller, w is reduced proportionally.

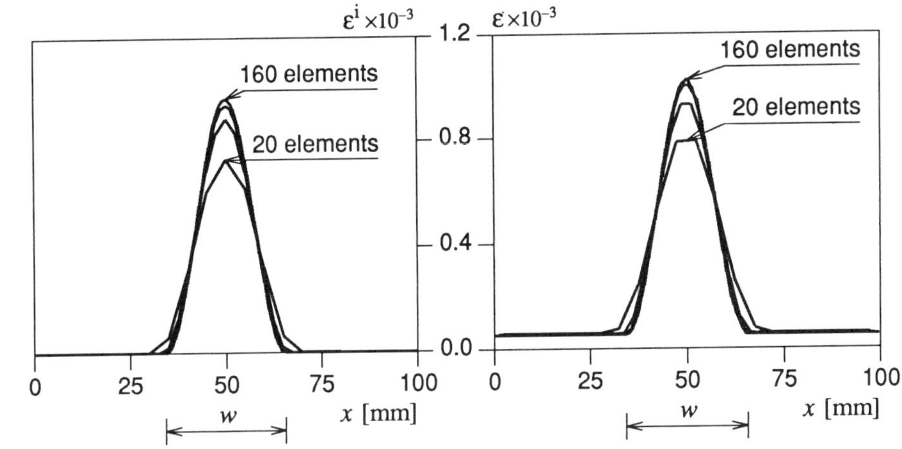

Figure 8. Plastic (left) and total (right) strain distribution along the axis of a tension bar with an imperfection at the centre for different discretizations ($l/L=0.05$ and $u=0.02$).

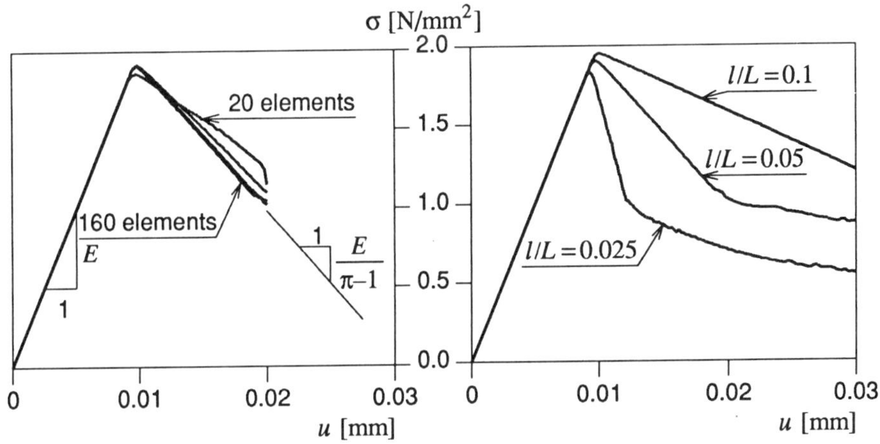

Figure 9. Axial stress *vs* axial displacement of right end of a tension bar.
Left: results for $l/L = 0.05$ and different discretizations (20, 40, 80 and 160 elements).
Right: results for different ratios of l/L (160 elements).

5.3 Numerical solutions for gradient models

As an example a bar with a length $L=100\,\text{mm}$ is taken. A Young's modulus $E=20000\,\text{N/mm}^2$ is used for the entire bar. The initial tensile strength is $f_t=2\,\text{N/mm}^2$ except for the elements which are between $x=45$ mm and $x=55$ mm. For these elements the initial tensile strength has been reduced by 10%. After reaching the peak strength a softening modulus $h=-0.1\,E$ is used for all elements, including the weakened elements. When the ultimate tensile strain $\varepsilon_u=10^{-3}$ (Figure 2) has been reached at which the residual load-carrying capacity has been reduced to zero the stress due to local softening is kept at that value, so that the softening modulus also becomes zero. For all values of ε^i we take $c = 0.1\,E$.

The reference solution has been carried out for a value of the internal length parameter of $l = 5$ mm. The numerical solutions show that upon mesh refinement the width of the localization zone and the strain profile over this zone rapidly converge towards a unique value (Figure 8). While the strain profiles of the solution for 20 elements still deviate the solutions for 40, 80 and 160 elements are very similar.

Figure 9 shows the convergence of the load-displacement curves. Again the solution for 20 elements is still slightly stiff, but the solutions for the discretizations with 40, 80 and 160 elements agree well, not only mutually, but also with the analytical solution. For the present geometry and set of material parameters ($l = 5$ mm) eq. (27) gives: $\dot{u} / (L\,\dot{\sigma}) = (1 - \pi)/E$. Especially the results for the meshes with 80 and 160 elements match the analytical solution perfectly.

All numerical solutions correspond to perfectly converged equilibrium states (force norm 10^{-6}). When the steady-state solution has developed the problem is linear until the strain ε_u is reached in the centre of the bar. Indeed, in this part of the localization process only one iteration with a tangent operator is needed to achieve equilibrium.

For the mesh with 160 elements the problem has also been analysed for $l = 2.5$ mm and $l = 10$ mm. It is interesting to observe from Figure 9 that, for a residual stress level $\sigma \approx 1$ N/mm^2, the load-displacement curve for $l = 2.5$ mm and $l = 5$ mm show a rather sharp cusp. At this point the centre two integration points exit from the softening part of the local stress-strain diagram. All residual load-carrying capacity in these integration points is then due to the gradient effect. The zone in which all load-carrying capacity is due to the gradient effect gradually grows from this point onwards until a new steady-state situation is attained where the residual stress level becomes constant. Indeed, when the material parameter c that precedes the gradient term, is assigned a constant, positive value, but if h is set equal to zero, eq. (27) predicts that $\dot{\sigma} \to 0$ and the stress-displacement curve converges towards a horizontal asymptote. To avoid that even in pure tension a steady-state solution is reached at a non-zero residual stress level c would have to be made a function of ε^i: $c = c(\varepsilon^i)$.

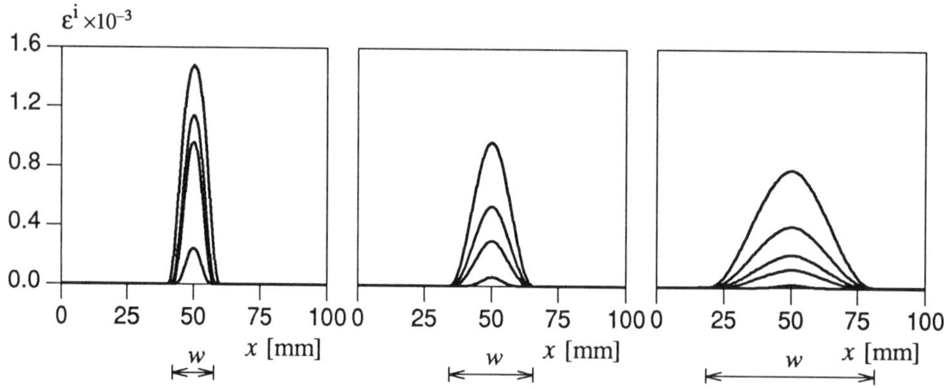

Figure 10. Evolution of the plastic strain distribution along the axis of a tension bar. Left: $l/L = 0.025$; centre: $l/L = 0.05$; right: $l/L = 0.1$.

Figure 10 illustrates the gradual growth of the localization zone for a division of the bar into 160 elements. For $l=2.5$ mm, $l=5$ mm and $l=10$ mm the plastic strain distribution has been plotted for $u=0.01$, 0.0125, 0.015 and 0.02 mm respectively. In Figure 10c the plastic strain distribution is also shown for $u=0.03$ mm ($l/L=0.1$). The numerical solutions show that, starting from the initial imperfect elements, a localization zone grows gradually until the analytically predicted widths are reached.

5.4 Relation to fracture energy based approaches

It was already alluded to in the beginning of this article that the first step to model fracture and failure in a smeared concept that goes beyond the classical idea of a simple solid (Noll 1958) is to regard the area under the stress-displacement curve as a material parameter. This parameter is commonly called the fracture energy and is defined by:

$$G_f=\int\sigma\,du=\int\sigma\,\varepsilon(s)\,ds\,. \tag{28}$$

Carrying out the integration for a linear softening diagram and assuming that the strains are constant over the band width (an assumption commonly made in numerical analyses) we arrive at the following relation between the strain ε_u at which the residual strength is exhausted, and the fracture energy G_f:

$$\varepsilon_u=\frac{2G_f}{f_t w}\,, \tag{29}$$

with w the width of the localization zone. The softening modulus h is thus given by:

$$h=\frac{wf_t^2}{2G_f}\,.$$

We shall now carry out an analysis for the tension bar of Figure 1 and give one element a tensile strength marginally below the other elements. Beyond the peak stress the tangential relation between the stress σ in the bar and the displacement u then reads:

$$\frac{\dot{u}/L}{\dot{\sigma}}=\frac{1}{E}+\frac{1}{mh}\,. \tag{30}$$

Since in a conventional continuum localization always occurs in one element, i.e. $w=L/m$, and using eq. (29) we can rewrite eq. (30) as

$$\frac{\dot{u}/L}{\dot{\sigma}}=\frac{1}{E}+\frac{2G_f}{Lf_t^2}\,. \tag{31}$$

Comparing expression (27), which resulted from a one-dimensional gradient model, and eq. (31) shows that the fracture-energy based model can give the same result as the gradient model, in the sense that the same slope in the stress-displacement diagram is computed, when

$$G_f=\frac{\pi l f_t^2}{h}\,. \tag{32}$$

It is emphasized that in eq. (32) h is the softening modulus in the gradient model and not the size-dependent softening modulus in the fracture energy model. When we substitute the tensile strength f_t and the softening modulus h used in the numerical analyses and take $l=10$ mm we obtain $G_f=0.0628$ N/mm, which, for a linear softening diagram, is a commonly accepted value for normal-weight concrete. Also the width $w = 62.8$ mm which follows from eq. (26) reasonably agrees with data reported in the literature which suggest that $w \approx 3d_a$ (Bazant and Pijaudier-Cabot 1989). Of course, many crude assumptions have been made in deriving this correspondence, eg, the adopted linear shape of the softening curve and, in the G_f-model, the homogeneous distribution of the strains over the localization zone. The above coarse analysis nevertheless gives an indication on the range of values for the length parameter l in the gradient model.

6. Discussion

In this paper an overview has been given of various possibilities to numerically analyse failure processes in heterogeneous materials. If one adheres to the concept of a continuum at a meso-level three approaches seem potentially useful: (i) non-local or gradient models, (ii) micro-polar (Cosserat) continua, and (iii) inclusion of rate-dependence. Obviously the latter possibility is only suited for relatively rapid loadings. For quasi-static loading conditions the addition of rotational degrees-of-freedom (micro-polar continuum models) has a firm physical basis for processes in which frictional slip is the most prominent failure mechanism. In this case micro-mechanical considerations can be set up in a very natural fashion that result in constitutive equations for such an enhanced continuum (eg, Mühlhaus and Vardoulakis 1987). The Cosserat continuum model is also appealing for the engineer, since it introduces, albeit at a meso-level instead of at a macro-level, notions as bending stiffnesses, moments and curvatures that are familiar. For example, a one-dimensional Cosserat continuum bears much resemblance with Timoshenko beam theory.

A disadvantage of the Cosserat continuum is that the rotational degrees-of-freedom are only activated under shear loadings. In pure (mode-I) tension rotational degrees-of-freedom do not become active, the micro-curvatures are zero and so are the couple-stresses. Although pure mode-I problems are of academic rather than of practical interest, and although all external actions on a structure generally result in a combination of mode-I (tension) and mode-II (shear) behaviour, numerical experiments suggest that, when decohesion is the predominant failure mode rather than frictional slip, the Cosserat effect is generally too weak to preserve ellipticity of the boundary value problem. It thus seems that the Cosserat continuum is primarily useful for compressive loadings, where the failure mode nearly always consists of a localization zone of a very high shear intensity. At a micro-level this high shear-strain intensity can either be caused by interparticle slip as in sands (eg, Vardoulakis and Graf 1985), or by severe micro-cracking as in concrete and rock (eg, van Mier 1984).

For tensile loadings where decohesion is the main cause of structural failure non-local models (eg, Bazant *et al* 1984) are very effective in keeping the boundary value problem elliptic. A definite disadvantage of current non-local formulations is that they are at odds with existing numerical strategies (Simo 1989). This drawback can be obviated by expanding the kernel of the integral in the non-local formulation into a Taylor series. The gradient models which then ensue are much more amenable to an efficient numerical implementation (de Borst and Mühlhaus 1991), while preserving the favourable property that they contain an internal length scale. The localization zone thus has a finite width and the size effect can be described in a proper fashion. This has been demonstrated analytically as well as numerically.

7. Acknowledgements

Financial support from CUR-committee A30 "Concrete Mechanics" and from the Commission of the European Communities under contract BREU-0070 to the first author is gratefully acknowledged.

8. References

Bazant, Z.P. and Oh, B. (1983) Crack band theory for fracture of concrete. *RILEM Materials and Structures,* 16, 155-177.

Bazant, Z.P., Belytschko, T.B. and Chang, T.-P. (1984) Continuum theory for strain-softening. *ASCE J. Eng. Mech. Div.,* 110, 1666-1692.

Bazant, Z.P. and Pijaudier-Cabot, G. (1989) Measurement of characteristic length of nonlocal continuum. *ASCE J. Eng. Mech.,* 115, 755-767.

de Borst, R. (1986) *Non-Linear Analysis of Frictional Materials.* Dissertation, Delft University of Technology, Delft.

de Borst, R. (1990) Simulation of localisation using Cosserat theory, in *Proc. 2nd Int. Conf. Computer Aided Analysis and Design of Concrete Structures,* (eds N. Bićanić and H.A. Mang), Pineridge Press, Swansea, pp. 931-944.

de Borst, R. (1991) Simulation of strain localisation: A reappraisal of the Cosserat continuum. *Engineering Computations,* in press.

de Borst, R. and Mühlhaus, H.-B. (1991) Gradient-dependent plasticity: formulation and algorithmic aspects. Submitted for publication.

Choi, S.K. and Mühlhaus, H.-B. (1991) Distinct elements vs Cosserat theory: a comparison for the case of an infinite shear layer, in Proc. Seventh Conf. Int. Assoc. Comp. Methods and Advances in Geomechanics, (eds G. Beer and J.R. Booker), Balkema, Rotterdam and Boston.

Cosserat, E. and Cosserat, F. (1909) *Theórie des Corps Deformables.* Herman et fils, Paris.

Crisfield, M.A. (1982) Local instabilities in the non-linear analysis of reinforced concrete beams and slabs. *Proc. Instn. Civ. Engrs,* Part 2, 73, 135-145.

Kubin, L.P. and Lépinoux, J. (1988) The dynamic organization of dislocation structures, in *Proc. Eight Int. Conf. on Strength of Metals and Alloys,* Tampere, Finland, pp. 35-59. 24, 581-597.

Kratochvil, J. (1988) Dislocation pattern formation in metals. *Revue Phys. Appl.,* 23, 419-429.

van Mier, J.G.M. (1984) *Strain-Softening of Concrete under Multiaxial Loading Conditions.* Dissertation, Eindhoven University of Technology, Eindhoven.

Mühlhaus, H.-B. and Vardoulakis, I. (1987) The thickness of shear bands in granular materials. *Geotechnique,* 37, 271-283.

Mühlhaus, H.-B. (1990) Continuum models for layered and blocky rock, in *Comprehensive Rock Engineering, Vol. 2: Analysis & Design Methods,* (eds J.A. Hudson, E.T. Brown, C. Fairhurst and E. Hoek), Pergamon Press, Oxford.

Mühlhaus, H.-B., de Borst, R. and Aifantis, E.C. (1991) Constitutive models and numerical analyses for inelastic materials with microstructure, in *Proc. Seventh Conf. Int. Assoc. Comp. Methods and Advances in Geomechanics,* (eds G. Beer and J.R. Booker), Balkema, Rotterdam and Boston.

Mühlhaus, H.-B. and Aifantis, E.C. (1991) A variational principle for gradient plasticity. *Int. J. Solids Structures,* in press.

Needleman, A. (1988) Material rate dependence and mesh sensitivity in localization problems. *Comp. Methods Appl. Mech. Engng.*, 67, 69-86.
Noll, W. (1958) A mathematical theory of the mechanical behaviour of continuous media. *Arch. Rat. Meth. Anal.*, 2, 197-226.
Pietruszczak, S. and Mróz, Z. (1981) Finite element analysis of deformation of strain softening materials. *Int. J. Num. Meth. Eng.*, 17, 327-334.
Pijaudier-Cabot, G. and Bazant, Z.P. (1987) Nonlocal damage theory. *ASCE J. Eng. Mech.*, 113, 1512-1533.
Rots, J.G. (1988) *Computational Modeling of Concrete Fracture.* Dissertation, Delft University of Technology, Delft.
Schaefer, H. (1962) Versuch einer Elastizitätstheorie des zweidimensionalen ebenen Cosserat-Kontinuums, in: *Miszellaneen der Angewandten Mechanik,* (ed M. Schäfer), Akademie-Verlag, Berlin, pp. 277-292.
Schreyer, H.L. and Chen, Z. (1986) One-dimensional softening with localization, *J. Appl. Mech.* 53, 791-979.
Simo, J.C. (1989) Strain softening and dissipation: a unification of approaches, in *Cracking and Damage: Strain Localization and Size Effect* (eds J. Mazars and Z.P. Bazant), Elsevier Applied Science, London and New York, pp. 440-461.
Sluys, L.J. and de Borst, R. (1991) Solution methods for localization in fracture dynamics, this volume.
Vardoulakis, I. and Graf, B. (1985) Calibration of constitutive models for granular materials using data from biaxial experiments, *Geotechnique*, 35, 299-317.
Walgraef, D. and Aifantis, E.C. (1985) On the formation and stability of dislocation patterns, I-III. *Int. J. Engng. Science,* 12, 1351-1372.
Willam, K.J. (1984) Experimental and computational aspects of concrete fracture, in *Proc. Int. Conf. Computer Aided Analysis and Design of Concrete Structures,* (eds F. Damjanić et al.), Pineridge Press, Swansea, Part 1, pp. 33-70.
Zbib, H.M. and Aifantis, E.C. (1988) On the localization and postlocalization behavior of plastic deformation, I,II,III, *Res Mechanica,* 23, 261-277, 279-292, 293-305.

54 SMEARED ANALYSIS OF CONCRETE FRACTURE USING A MICROPLANE-BASED MULTICRACK MODEL WITH STATIC CONSTRAINT

I. CAROL and P.C. PRAT
Universitat Politècnica de Catalunya, Barcelona, Spain

Abstract
The smeared analysis of cracking and fracture requires constitutive models representing both the behavior of cracks and continuum between cracks simultaneously. In this paper, a new model is presented that is intended to be used in that context. In a similar way to the microplane-type models, the fundamental constitutive laws —in this case the cracking laws— are established in terms of stresses and strains on a plane of generic orientation. The overall behavior of the model is derived by combining cracking with the behavior of the continuum between cracks by means of a static constraint (tractions on a crack plane are equal to the projection of the stress tensor on that plane) and an additive combination of strains. Emphasis has been given to a consistent definition of cracking modes other than tensile Mode I, since it seems important that a crack can start and develop also under shear-compression conditions. To this end, we introduce an asymptotic Mode II (Mode IIa) of cracking under shear and very high compression, which is completely independent of the tensile Mode I. Two of the parameters of the model are the fracture energies (g_f^I and g_f^{IIa}) themselves, thus simplifying the task of parameter adjustment with element size. Numerical results of both constitutive and finite element analysis show good qualitative agreement with experimental data.
Keywords: Fracture, Fracture Energy, Concrete, Micromechanics, Microplane Models, Finite Elements.

1 Introduction

The "smeared approach" to the finite element analysis of cracking and fracture is nowadays a widely used method for modeling these phenomena without need to implement complicated and expensive techniques (such as remeshing). In this approach, an equivalent continuum is assumed for the cracked medium, and the behavior of both cracks and material between cracks is assumed to follow the material laws of the equivalent continuum. However, some characteristics needed for these laws in order to represent the actual behavior of cracked solids (e.g. softening and non-associativeness) introduce well known additional difficulties in the calculations, like inobjectivity of the results with respect to the mesh size, etc. (Bažant & Cedolin, 1979; Bažant & Oh, 1983; de Borst, 1984; Darwin, 1985).

Fracture Processes in Concrete, Rock and Ceramics Edited by J.G.M. van Mier, J.G. Rots and A. Bakker.
 Published by E. & F.N. Spon, 2–6 Boundary Row, London SE1 8HN. ISBN 0 419 15870 7

A good constitutive model for the smeared analysis of cracking and fracture should provide at least the following features: (a) a clear definition of the fracture modes and the corresponding fracture energies; (b) knowledge *a priori* (by means of the parameters), of the energy spent for cracking under the basic modes; (c) cracks can start and develop under any mode or combination of modes and not only in Mode I; (d) possibility to combine the cracking model with a model for the continuum medium; and (e) cracks with different directions can exist simultaneously. On the basis of these considerations, a new model is presented in this paper within the framework of the microplane-type models. The fundamental material laws are established first on a plane of generic orientation, and then the tensorial formulation is obtained as a combination of the behavior on all possible microplanes in space (Bažant & Prat, 1988; Carol et al., 1990). The present approach, however, presents a deviation from the microplane formulation: the model is divided in two parts, one for cracking (for which we use the microplane environment) and the other for the continuum between cracks (for which any available continuum medium model can be used). In the next sections, a description of the formulation is given, together with numerical results for both constitutive ("single point") and finite element analysis used to verify the model More details of the formulation and test results for mixed mode cracking at constitutive level can be found in Carol & Prat (1990).

2 Cracking model

2.1 Model for a single crack

The laws for a single crack are defined in terms of the normal (σ) and shear (τ) stresses applied on the crack plane. The two corresponding strain components are ε^{cr} and γ^{cr}. The crack is assumed to start when the stresses on the σ–τ space reach the hyperbolic "cracking surface" $F(\sigma,\tau)$=0 shown in Fig. 1a, (tensile stresses are positive). The "cracking function" F is defined by the following equation:

$$F = \tau^2 - (c - \sigma \tan\phi)^2 + (c - \chi \tan\phi)^2 \tag{1}$$

where $\tan\phi$, c and χ are model parameters. From Fig. 1a one can distinguish two limiting situations: (a) cracking occurs under pure tension with zero shear stresses (Mode I), and (b) cracking occurs under shear and very high compression. In the latter case, the hyperbola approaches its asymptote, which in fact represents a Mohr–Coulomb criterion. We call this second mode the "asymptotic Mode II" or Mode IIa of cracking. The hyperbola used provides smooth transition for crack formation (mixed mode cracking) between these two limiting states.

As cracking progresses, the cracking surface shrinks so that the current stress point always remains on the surface. The initial shape of the cracking surface is represented by the curve labeled "0" in Fig. 1b, while its final shape depends on the mode of cracking. In pure tension (Mode I), the crack develops following the weakest path between aggregate and cement (Fig. 2a). The only kinematic condition once the crack is completely formed, is that both sides of the crack can separate from each other. Because some interface roughness is still present, the final surface is another hyperbola (curve "1" in Fig. 1b). On the other hand, Mode IIa would be a limiting case with very high compressive and shear stresses, where shear displacement should be kinematically admissible and no dilatancy allowed. According to these restrictions, the only possible path for the crack is a straight

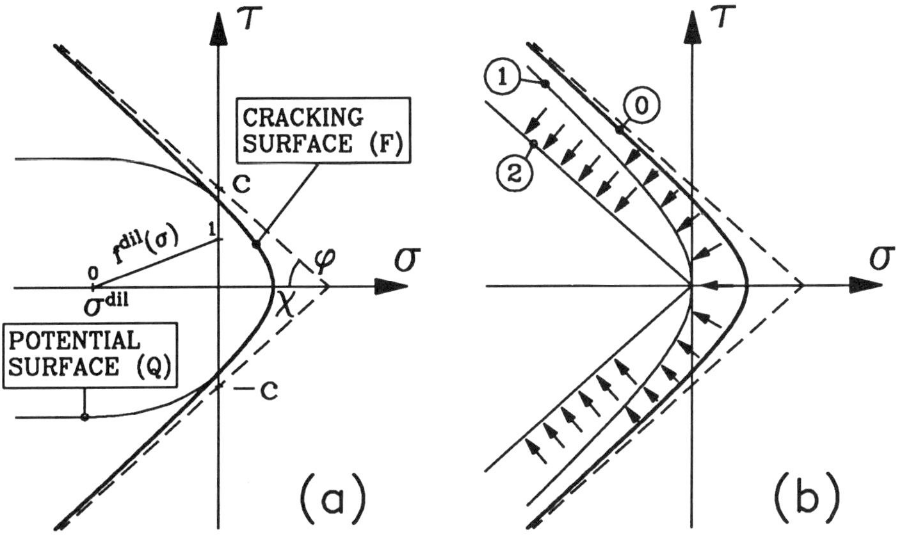

Fig. 1 *(a) Hyperbolic cracking surface; (b) Evolution of the cracking surface.*

line cutting through both cement paste and aggregate (Fig. 2b), and the corresponding surface is defined by a pair of straight lines representing a pure frictional behavior (curve "2" in Fig. 1b).

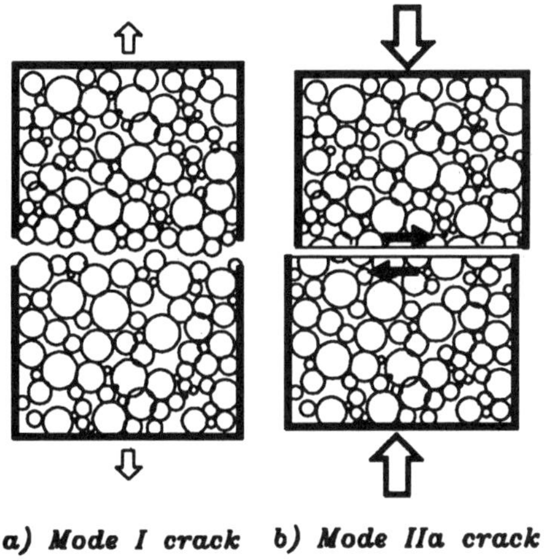

Fig. 2 *Crack paths in the microstructure*

We consider the evolution of the cracking suface for both modes as a single process, in which the evolution of parameters c and χ ($\tan\phi$ is assumed to remain constant) is controlled by a single internal variable w^{cr} (i,e. the work spent on the formation of the crack) defined as:

$$\begin{aligned} \mathrm{d}w^{cr} &= \sigma \mathrm{d}\varepsilon^{cr} + \tau \mathrm{d}\gamma^{cr} && \text{if } \sigma > 0 \\ \mathrm{d}w^{cr} &= (|\tau| - |\sigma| \tan\phi)|\mathrm{d}\gamma^{cr}| && \text{if } \sigma < 0 \end{aligned} \tag{2}$$

Note that in compression the frictional work is not taken into account for w^{cr} as already suggested elsewhere (Willam, 1984; Willam & Sture, 1985). The parameters c and χ are assumed to decrease with w^{cr} from their initial values (c_0 and $\chi_0 = f_t'$) to a zero value at $w^{cr} = g_f^I$ and $w^{cr} = g_f^{IIa}$ respectively. The variation is assumed linear in terms of the intermediate scaling function $\mathcal{S}$:

$$\mathcal{S}(\xi) = \frac{e^{-\alpha\xi}}{1 + (e^{-\alpha} - 1)\xi} \tag{3}$$

where $\xi = w^{cr}/g_f^I$ and $\alpha = \alpha_\chi$ for parameter χ, and $\xi = w^{cr}/g_f^{IIa}$ and $\alpha = \alpha_c$ for parameter c. This definition provides a family of descending curves for c and χ depending on the parameters α_χ and α_c (Fig. 4). A linear function is obtained if $\alpha = 0$, i.e. $\mathcal{S}(\xi) = \xi$.

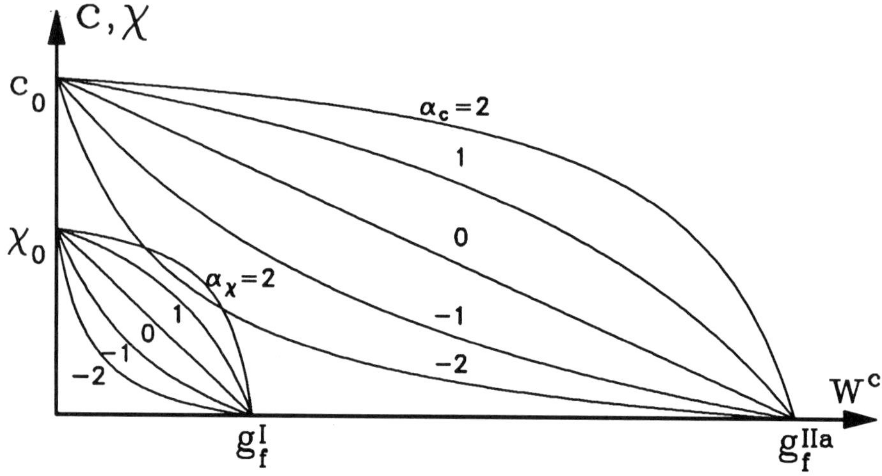

Fig. 3 *Evolution of the cracking surface parameters with* w^{cr}

To calculate the direction of the crack strain increments, we assume that they follow a path perpendicular to the "potential surface" Q=0 shown in Fig. 1a. In tension, this surface coincides with the cracking surface, i.e. the crack strains follow an associated rule. In compression, the rule is non–associated so that for high compression ($|\sigma| > |\sigma^{\mathrm{dil}}|$ the crack shows no dilatancy. To this end, the horizontal component of the vector normal to the cracking surface is reduced by the factor f^{dil}, which is defined by a decreasing function of the compressive stress also represented in Fig. 1a. The equations defining Q are:

$$\frac{\partial Q}{\partial \sigma} = 2\tan\phi(c - \sigma\tan\phi) f^{\mathrm{dil}}, \qquad \frac{\partial Q}{\partial \tau} = 2\tau \tag{4}$$

2.2 Formulation of the macroscopic model

To formulate a consistent macroscopic model including both continuum and cracks, a static constraint between macroscopic and crack plane levels has been assumed (i.e. the stresses acting on a crack plane —the "microplane"— are assumed to be equal to the resolved components of the macroscopic stress tensor). According to the static constraint, the stresses $\underset{\sim}{s}_i = [\sigma, \tau]^t$ on a given plane of orientation $\underset{\sim}{n}$, can be expressed in terms of the macroscopic stress tensor $\underset{\sim}{\sigma}$ as:

$$\underset{\sim}{s}_i = \underset{\sim}{N}_i^t \underset{\sim}{\sigma}, \qquad \underset{\sim}{N}_i = \begin{bmatrix} \cos^2\theta & -\cos\theta\sin\theta \\ \sin^2\theta & \cos\theta\sin\theta \\ 2\cos\theta\sin\theta & \cos^2\theta - \sin^2\theta \end{bmatrix} \tag{5}$$

where θ is the angle between the normal to the crack plane i and the x axis. The total macroscopic strain tensor is obtained from the contributions of the continuum and crack components:

$$\underset{\sim}{\varepsilon} = \underset{\sim}{\varepsilon}^{co} + \sum_{i=1}^{N_c} \underset{\sim}{\varepsilon}_i^{cr}, \qquad \underset{\sim}{\varepsilon}_i^{cr} = \underset{\sim}{N}_i \underset{\sim}{e}_i^{cr} \tag{6}$$

where N_c is the total number of active cracks and $\underset{\sim}{e}_i^{cr} = [\varepsilon^{cr}, \gamma^{cr}]^t$ is the crack strain vector in local coordinates. This kind of approach has already been used with similar purposes by other authors (Pietruszack & Mróz, 1981; de Borst & Nauta, 1984; Rots & de Borst, 1987). In some of these works devoted to cracking, however, different hypothesis concerning the conditions for crack formation and evolution led to a different formulation which, although less expensive in computational terms, in fact restricted the possibilities of crack formation to the case of pure tensile Mode I.

The direction of the crack strain increment is given by the plastic potential:

$$\mathrm{d}e_i^{cr} = \left[\frac{\partial Q}{\partial \underset{\sim}{s}}\right]_i \mathrm{d}\lambda_i \tag{7}$$

where $\mathrm{d}\lambda_i$ is a scalar factor. Note that in the present derivation repetition of subindex i does not imply summation, unless otherwise indicated.

As in classical plasticity, a macroscopic stress increment $\mathrm{d}\underset{\sim}{\sigma}$ is related exclusively to the continuum part of the strain tensor increment $\mathrm{d}\underset{\sim}{\varepsilon}^{co}$:

$$\mathrm{d}\underset{\sim}{\sigma} = \underset{\sim}{D}^{co}\mathrm{d}\underset{\sim}{\varepsilon}^{co} \tag{7}$$

From this point, a classical derivation leads to the final incremental relationship between macroscopic stress and strain tensor increments:

$$\mathrm{d}\underset{\sim}{\sigma} = \underset{\sim}{D}^{crco}\mathrm{d}\underset{\sim}{\varepsilon}; \qquad \underset{\sim}{D}^{crco} = \underset{\sim}{D}^{co}\left(\underset{\sim}{I} - \sum_{i=1}^{N_c} \underset{\sim}{N}_i \left[\frac{\partial Q}{\partial \underset{\sim}{s}}\right]_i \left[\frac{\partial \lambda_i}{\partial \underset{\sim}{\varepsilon}}\right]^t\right) \tag{8}$$

where $\underset{\sim}{D}^{crco}$ is the tangent stiffness matrix of the continuum-plus-cracks equivalent medium. The vectors $\partial\lambda_i/\partial\underset{\sim}{\varepsilon}$ are the solution of an algebraic system of equations, which is obtained by enforcing the consistency condition for the cracking function F of every active crack.

2.3 Numerical implementation

The proposed cracking model has been implemented with linear elasticity for the continuum between cracks. It has a total of 10 parameters: 2 (E and ν) for the continuum model, and 8 ($\tan\phi$, c_0, χ_0, g_f^I, g_f^{IIa}, σ^{dil}, α_χ and α_c) for the crack model. A classical incremental plasticity procedure is used to integrate Eqs. 8. However, these equations are valid for a known set of active cracks but they tell nothing about which are the active cracks, and if some of them stop being active or start opening during a prescribed step. Establishing a correct scheme to check crack opening and closing is also an important part of the model implementation. As in previous microplane formulations, a fixed number of planes of pre-established orientation have been considered. 12 sample directions have been used in this work, distributed at constant intervals over the upper half hemicircle and therefore, the same number of 12 internal variables (w^{cr} for each direction) need to be stored and updated during computations.

3 Examples of application to Mode I tests

Two types of results are presented: results at the constitutive equation level and results for the boundary value problem (finite element analysis) level. The same set of constitutive subroutines are used for both types of analysis, although they are called either from a "single point" main program or from a finite element program, according to the scheme reported by Carol et al. (1990). Although the model is intended to cover all cracking modes, the examples presented in this paper refer to tensile Mode I exclusively, which is the objective of the current stage of our investigation. Additional results at constitutive level showing the influence of various parameters on mixed mode cracking can be found in Carol & Prat (1990).

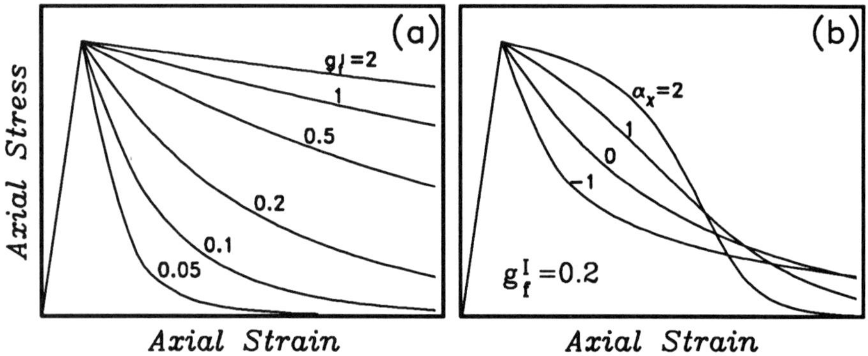

Fig. 4 *Mode I cracking with different values of parameters: (a) g_f^I and (b) α_χ*

Fig. 4a shows the influence of parameter g_f^I on the stress–strain relation obtained for a pure tensile loading test. In the way the model has been defined, the area under each curve of the diagram is equal to the corresponding value of the parameter g_f^I. Fig. 5b shows the influence of parameter α_χ on the same relation. As expected, the shape of the softening law for χ (Fig. 3) is somehow reproduced in the σ–ε diagram: negative values of

α_χ give σ–ε curves with a sharp drop at the begining and a long asymptote reaching the x–axis slowly, while positive values give slow decreasing curves at the begining, then a sharp drop to zero (step–shaped variation). All curves have been obtained with the same values of the remainig parameters, thus having the same area under the curve (value of the parameter g_f^I).

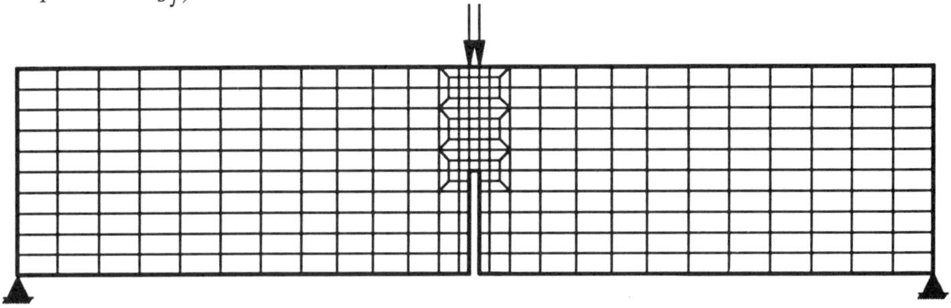

Fig. 5 *Finite element mesh of Kormeling's beam*

In Figs. 5–6, the results of the finite element analysis of a plain concrete beam tested under three-point bent conditions by Kormeling & Reinhard (1983), and analyzed by Rots (1988) are presented. In Fig. 5, the finite element mesh is depicted. For the sake of comparison, the number and order of the finite elements are the same used by Rots (1988), including the special integration rule with a single integration point in the horizontal direction for the elements ahead of the notch, where the crack is to propagate. In Fig. 6, the results of the analysis, (load–displacement curve at the top midpoint of the beam) for three different values of parameter α_χ are shown, together with experimental results (shadowed zone), and the curves obtained by Rots (1988) for various shapes of the tensile σ–ε^{cr} descending curve (the basic feature of that model, with several possible alternative definitions).

Fig. 6 shows that variations of parameter α_χ influence the load–displacement curve with a similar trend as it does on the σ–ε diagram, but in this case the value of the peak load is also affected, and not only the shape of the descending portion of the curve.

4 Summary and concluding remaks

A new constitutive model for the analysis of concrete cracking in the context of a smeared finite element analysis has been presented. In this model, based on the microplane concept with static constraint, the laws for crack formation and evolution are established on a plane of generic orientation in terms of the stresses on that plane, which are assumed to be the resolved components of the macroscopic stress tensor. The behavior of the crack is governed by a hyperbolic cracking surface. This surface shows two limiting situations for cracking: the classical Mode I in pure tension, and a newly introduced asymptotic Mode IIa for very high shear–compression stresses with an independent g_f^{IIa} fracture energy in which friction is not taken into account. Between these two limit states, the hyperbola provides smooth transition so that a crack can start and develop under tension, shear–tension or shear–compression stress states. Both Mode I and Mode IIa fracture energies are parameters of the model. This makes the model adequate for smeared crack

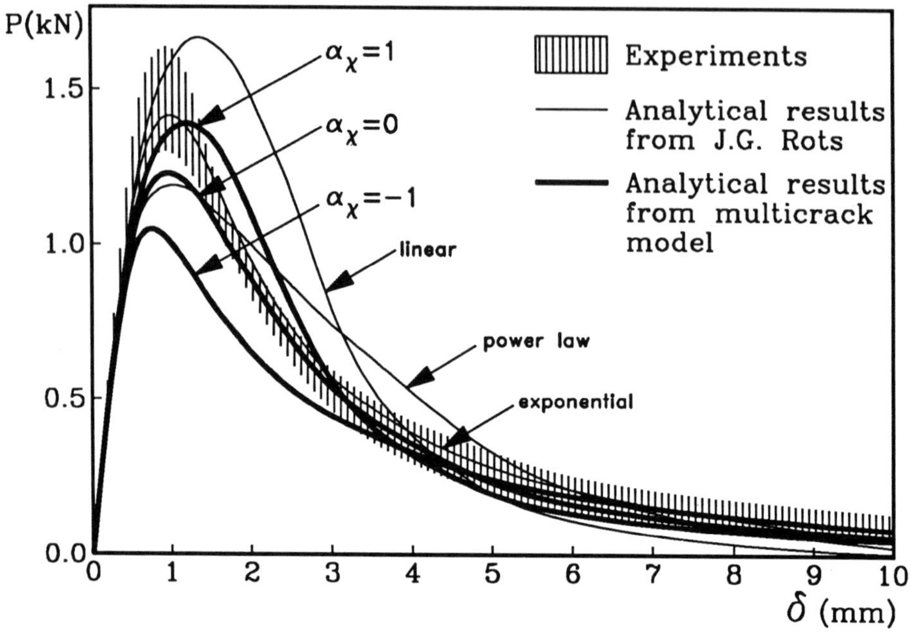

Fig. 6 *Force–displacement diagram for the three-point bent tests*

analysis in which automatic adjustement of parameters is performed at each integration point.

Although to the knowledge of the authors not yet supported by experimental evidence, the fracture Mode IIa provides new possibilities of theoretical interpretation and model development. The proposed definition of Mode IIa is an alternative to other definitions of a second mode of fracture (and its associated energy parameter) that can be found in the existing literature (de Borst & Nauta, 1984; Rots & de Borst, 1987; Oliver et al., 1990). Compared to the latter possibilities, the proposed approach shows some important advantages: it has a simple and clear definition in terms of the stresses on the crack plane, it is a "true" shear mode (the crack can start under shear–tension or shear–compression stress states and not necesarily under pure tension), and it is defined completely uncoupled from the tensile Mode I, with independent energy parameters. In this context, the existing shear fracture tests under low or null confinement stresses (Arrea & Ingraffea, 1981; Bažant & Pfeiffer, 1986) would represent combinations of Mode I and Mode IIa. New series of shear tests with increasing confining stress normal to the expected crack plane seem inevitable to clarify the adequacy of the model proposed.

The formulation of an overall macroscopic model including cracks in various directions and the possibility of using any existing continuum model for the intercrack material has also been presented. Its structure is of a multisurface non-associated plasticity type, from which the formulation of the tangent stiffness matrix and the step-by-step integration techniques have been borrowed. A number of pre-established sample directions over the hemicircle are checked sistematically, and crack opening and arrest can be detected. The model has been implemented in a computer subroutine which performs strain-to-stress calculations and can be used in conjunction with either a "single-point" program or a finite element code.

Numerical results for both constitutive and finite element analysis show good qualitative agreement with experimental results of Mode I tests, as well as with other numerical results available in the literature. Numerical and experimental developments are currently under way in order to verify the hypothesis of Mode IIa of fracture, and to extend the application of the model to the finite element analysis of mixed mode fracture problems.

5 Acknowledgements

Partial financial support from the spanish CICYT under research grant PB87–0861 is gratefully acknowledged.

6 References

Arrea, M. & Ingraffea, A.R. (1981) *"Mixed–mode crack propagation in mortar and concrete."* Rep. 81–13. Dept. of Struct. Engrg., Cornell Univ., Ithaca, New York.

Bažant, Z.P. & Cedolin, L. (1979) "Blunt crack band propagation in finite element analysis." *J. Engrg. Mech.* ASCE **105**(2), 287–315.

Bažant, Z.P. & Oh., B.H. (1983) "Crack band theory for fracture of concrete." *Materials and Structures* RILEM, **16**, 155–177.

Bažant, Z.P. & Pfeiffer, P.A. (1986) "Shear fracture tests of concrete." *Materials and Structures* RILEM, **19**, 111–121.

Bažant, Z.P. & Prat, P.C. (1988) "Microplane model for brittle–plastic material. I: Theory & II: Verification." *J. Engrg. Mech.* ASCE, **114**(10), 1672–1702.

de Borst, R. (1984) "Application of advanced solution techniques to concrete cracking and non–associated plasticity." *Proc. Int. Conf. on Numerical Methods for Nonlinear Problems.* C. Taylor et al. (eds.) Pineridge Press (U.K.)

de Borst, R. & Nauta, P. (1984) "Smeared crack analysis of reinforced concrete beams and slabs failing in shear." *Proc. Computer aided analysis and design of concrete structures.* F. Damjanić et al. (eds.) Pineridge Press (U.K.)

Carol, I., Bažant, Z.P. & Prat, P.C. (1990) *"New explicit microplane model for concrete: theoretical aspects and unified implementation for constitutive verification and F.E. analysis."* Report, E.T.S. Camins, Canals i Ports – U.P.C., Gran Capità, s/n, E-08034 Barcelona, Spain.

Darwin, D. (1985) "Concrete crack propagation—Study of model parameters." *Proc. Finite element analysis of reinforced concrete.* C. Meyer et al. (eds.), ASCE (New York), 184–203.

Kormeling, H.A. & Reinhard, H.W. (1983) *Determination of the fracture energy of normal concrete and epoxy modified concrete.* Report 5-83-18, Stevin Lab., Technische Universiteit Delft.

Oliver, J., Cervera, M., Oller, S. & Lubliner, J. (1990) "Isotropic damage model and smeared analysis of concrete." *Proc. 2nd. Int. Conf. on Computer aided analysis and design of concrete structures "Sci–C 1990."* Bićanić & Mang (eds.), Pineridge Press (U.K.), Vol. 2, 945–958.

Pietruszack, S.T. & Mróz, Z. (1981) "Finite element analysis of deformation of stran–softening materials." *Int. J. of Numerical Methods in Engineering.* **17**, 327–334.

Rots, J.G. & de Borst, R. (1987) "Analysis of mixed–mode fracture in concrete." *J. Engrg. Mech.* ASCE, **113**(11), 1739–1758.

Rots, J.G. (1988) *Computational Modeling of Concrete Fracture.* Ph.D. Thesis presented to the Technische Universiteit Delft, The Netherlands.

Willam, K.J. (1984) "Experimental and computational aspects of concrete fracture." *Proc. Computer aided analysis and design of concrete structures.* F. Damjanić et al. (eds.), Pineridge Press (U.K.)

Willam, K.J. & Sture, S. (1985) "A composite fracture model for localized failure in cementitious materials." *Proc. 2nd. Symp. on the Interaction of Nonnuclear Munitions with Structures*, Panama City Beach, Florida.

55 A COMPARISON OF DIFFERENT CRACK MODELS APPLIED TO PLAIN AND REINFORCED CONCRETE

P.H. FEENSTRA
Department of Civil Engineering, Delft University of Technology, The Netherlands
R. de BORST and J.G. ROTS
Department of Civil Engineering, Delft University of Technology, and TNO Building and Construction Research, Rijswijk, The Netherlands

Abstract
The paper describes two crack models embedded in a total strain concept, the fixed orthogonal crack model and the coaxial rotating crack model. These two crack models are compared with the corresponding crack models embedded in a decomposed strain concept. Further, a model for reinforced concrete based on a total strain concept is presented. The salient feature in this model is the stress decomposition into a stress contribution of the concrete with a strain softening model, a stress contribution of the reinforcing steel with a elasto-plastic model and a stress contribution owing to the interaction between concrete and reinforcement. The model capabilities are demonstrated in a simulation of a unidirectionally reinforced concrete panel.
Keywords : Rotating Crack Model, Fixed Crack Model, Reinforced Concrete Panels, Tension-stiffening, Tension-softening, Bond.

1. Introduction

Fracture of concrete can be modeled following several approaches. One of the most promising approaches for implementation in a finite element program is the phenomenological approach, in which fracture is described within a continuum. This approach is known as the smeared crack model. Generally an elastic-softening behavior is assumed in which a gradual decrease of the stress occurs after the crack is initiated. These constitutive models can be embedded in different concepts. In this study two concepts will be used, the decomposed strain concept and the total strain concept. The first concept is based on a strain decomposition into an elastic part and an inelastic part: $\boldsymbol{\varepsilon} = \boldsymbol{\varepsilon}^{el} + \boldsymbol{\varepsilon}^{cr}$. This model has been treated comprehensively in the work of de Borst and Nauta (1985) and Rots (1988), and will not be treated in this paper. The second concept is the total formulation in which the stress is assumed to be a function of the total strain. The constitutive behavior of models embedded in both concepts is highly dependent upon the modeling of the shear stress-strain relation. This has resulted in different crack models, the fixed smeared crack model and the rotating smeared crack model. In this study the fixed and the rotating crack model embedded in a total strain concept will be treated.

The practical application of fracture models for concrete generally concerns reinforced concrete panels. The major characteristics of reinforced concrete are the distributed fracture in the neighborhood of the reinforcing bars and the localized fracture some distance away from the reinforcement. Especially in unidirectionally reinforced concrete panels this phenomenon becomes very challenging when cracks occur which are inclined at some angle with the reinforcement. This study deals with the gradual transition between the localized fracture in the unreinforced direction and the distributed fracture in the reinforced direction.

Fracture Processes in Concrete, Rock and Ceramics Edited by J.G.M. van Mier, J.G. Rots and A. Bakker.
 Published by E. & F.N. Spon, 2–6 Boundary Row, London SE1 8HN. ISBN 0 419 15870 7

2. Crack models in a total strain concept

In smeared crack models a crack is conceived to be distributed over the area of an element represented by an integration point. With this approach the cracked material is considered as a continuum, for which the notions of stress and strain are defined. This means that the constitutive relation can be described in terms of stress-strain relations. Commonly, and this has also been done in this study, the constitutive relation in the uncracked state is restricted to linear-elasticity. When a tension cut-off criterion is violated, this linear-elastic relation is replaced by an orthotropic stress-strain law with the axes of orthotropy in accordance with the directions of the principal stresses. The explicit application of the orthotropic elasticity theory to cracked concrete is not straightforward. Indeed, the concept of a fully orthotropic material has not yet been applied to describe smeared cracking. Instead, it is assumed that cracking only influences the diagonal terms in the compliance matrix (Bažant and Oh, 1983). Then the total stress-strain relation in a plane-stress configuration is given by the following stiffness relation in the crack coordinate system n,s :

$$\sigma_{n,s} = D_{n,s}\,\varepsilon_{n,s} \tag{1}$$

with

$$D_{n,s} = \begin{bmatrix} \dfrac{\alpha_n E}{1-\alpha_n\alpha_s\nu^2} & \dfrac{\alpha_s\alpha_n E}{1-\alpha_n\alpha_s\nu^2} & 0 \\ \dfrac{\alpha_n\alpha_s E}{1-\alpha_n\alpha_s\nu^2} & \dfrac{\alpha_s E}{1-\alpha_n\alpha_s\nu^2} & 0 \\ 0 & 0 & \beta_s G \end{bmatrix} \tag{2}$$

As a first approximation it is assumed that the reduction factors α_n, α_s for the elastic Young's modulus E are only functions of the tensile strains in the considered directions, i.e. $\alpha_n = \alpha_n(\varepsilon_{nn})$, $\alpha_s = \alpha_s(\varepsilon_{ss})$ and $\beta_s = \beta_s(\gamma_{ns})$.

The constitutive relation, eq.(1), has been expressed in the local coordinate system. Using the strain transformation matrix $\mathbf{T}(\phi)$ the relation in the global coordinate system is given by:

$$\sigma_{x,y} = [\,T(\phi)^T D_{n,s}\, T(\phi)]\,\varepsilon_{x,y} \tag{3}$$

in which the superscript T denotes the transpose and ϕ is the angle between the local crack coordinate system and the global coordinate system. The tangential stiffness matrix is always needed in a non-linear analysis which employs a Newton-Raphson type iterative procedure. These stiffness moduli may be readily obtained by differentiating:

$$\sigma_{x,y} = T^T(\phi)\,\sigma_{n,s} \tag{4}$$

which results in:

$$\dot{\sigma}_{x,y} = {}^tD_{x,y}\,\dot{\varepsilon}_{x,y} \tag{5}$$

with ${}^tD_{x,y}$ the tangent stiffness matrix

$${}^tD_{x,y} = \left[\, T^T(\phi)\,\frac{\partial\sigma_{n,s}}{\partial\varepsilon_{n,s}}\,T(\phi) + \frac{\partial T^T(\phi)}{\partial\phi}\,\sigma_{n,s}\,\frac{\partial\phi^T}{\partial\varepsilon_{x,y}} \right] \tag{6}$$

2.1 Fixed crack model

In the fixed crack model the principal axes of orthotropy are kept fixed during the post-cracking phase. This results in the following constitutive model:

$$\sigma_{x,y} = [\, \mathbf{T}^T(\phi_0)\, \mathbf{D}_{n,s}\, \mathbf{T}(\phi_0)]\, \varepsilon_{x,y} \tag{7}$$

in which ϕ_0 the angle between the global coordinate system and the crack coordinate system. Because $\partial \mathbf{T}^T / \partial \phi = \mathbf{0}$, the tangential stiffness matrix reads explicitly

$$^t\mathbf{D}_{x,y} = \mathbf{T}^T(\phi_0) \begin{bmatrix} \partial\sigma_{nn} / \partial\varepsilon_{nn} & \partial\sigma_{nn} / \partial\varepsilon_{ss} & 0 \\ \partial\sigma_{ss} / \partial\varepsilon_{nn} & \partial\sigma_{ss} / \partial\varepsilon_{ss} & 0 \\ 0 & 0 & \partial\sigma_{ns} / \partial\gamma_{ns} \end{bmatrix} \mathbf{T}(\phi_0) \tag{8}$$

This model has also been used by Rots, Nauta, Kusters and Blaauwendraad (1985) as an extension of the fixed crack model proposed by Bažant and Oh (1983). A more fundamental treatment of the fixed crack model is given in the inspirational work of Willam, Pramono and Sture (1987).

2.2 Rotating crack model

The rotating crack model differs notably from the fixed crack model in the post-cracking phase. The principal axes of orthotropy are not kept constant, but rotate coaxially with the principal strains during crack propagation. The constitutive model is now defined by the full expression of eq.(3) with $\mathbf{T}(\phi)$ a function of the global strains. The tangential stiffness matrix $^t\mathbf{D}_{x,y}$, is more complicated because of the introduction of the spin of the principal stress axes and is given by the full expression of eq.(6). The first term of the tangential stiffness matrix is equal to the tangential stiffness matrix that follows from the fixed crack approach, and can be considered as the material tangent stiffness. The second term, due to the spin of the principal axes, is a function of the stresses and the angle ϕ between global coordinate system and crack coordinate system. The rotating crack model does not involve an independent shear-retention factor β, but the coaxiality condition results in a shear stiffness that is associated with the rotation of the principal axes. Many authors, e.g. Bažant (1983), Crisfield and Wills (1989) and Willam, Pramono and Sture (1987), have proved that in a two-dimensional configuration $^t\mathbf{D}_{x,y}$ can then be written as:

$$^t\mathbf{D}_{x,y} = \mathbf{T}^T(\phi) \begin{bmatrix} \partial\sigma_{nn} / \partial\varepsilon_{nn} & \partial\sigma_{nn} / \partial\varepsilon_{ss} & 0 \\ \partial\sigma_{ss} / \partial\varepsilon_{nn} & \partial\sigma_{ss} / \partial\varepsilon_{ss} & 0 \\ 0 & 0 & (\sigma_{nn} - \sigma_{ss}) / 2(\varepsilon_{nn} - \varepsilon_{ss}) \end{bmatrix} \mathbf{T}(\phi) \tag{9}$$

3. Tension - shear model problem

To verify the elastic-softening formulations for the fixed and the rotating crack models, reference is made to the plane-stress model problem of biaxial tension and shear depicted in Figure 1 (Willam, Pramono and Sture, 1987 ; Rots 1988). It concerns an elastic-softening continuum of unit dimensions with a Young's modulus $E = 10000$ [N/mm^2], Poisson's ratio $\nu = 0.2$, tensile strength $f_{ct} = 1.0$ [N/mm^2] and a fracture energy $G_f = 0.15$ [Nmm/mm] over a unit crack band width $h = 1.0$ [mm]. Initially, the continuum is subjected to tensile straining in the x-direction accompanied by lateral Poisson contraction in the y-direction, i.e. $\Delta\varepsilon_{xx} : \Delta\varepsilon_{yy} : \Delta\gamma_{xy} = 1 : -\nu : 0$. Immediately after cracking, a switch is made to combined biaxial tension and shear according to $\Delta\varepsilon_{xx} : \Delta\varepsilon_{yy} : \Delta\gamma_{xy} = 0.5 : 0.75 : 1$, which causes the axes of principal strain to continuously rotate after cracking as is typical of fracture propagation in smeared finite element models. For both concepts an analysis has been made for both the fixed crack model as well as the coaxial rotating crack model. In the figures the following abbreviations have been used to indicate the different crack models in the two strain concepts:

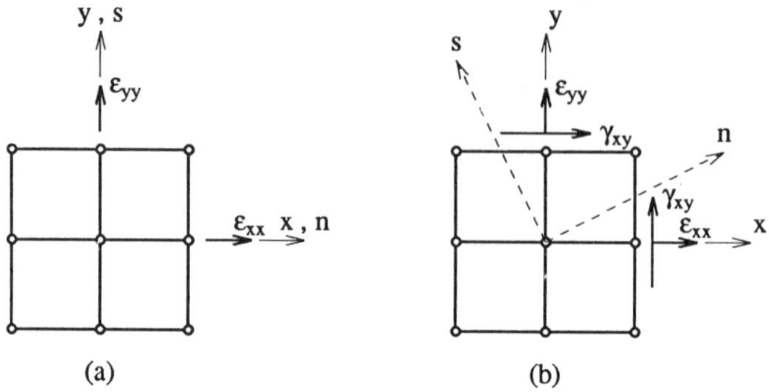

Figure 1 Lay-out tension-shear model problem.
(a) tension up to cracking;
(b) biaxial tension with shear beyond cracking

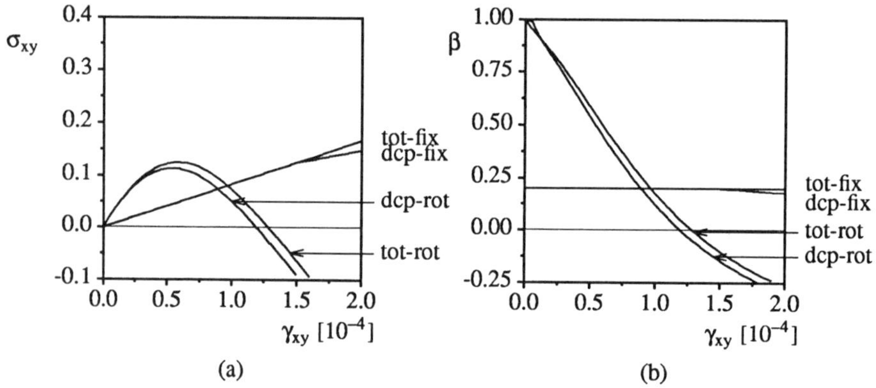

Figure 2 Behavior in shear for different strain concepts and crack models.
(a) shear stress-strain response in x, y-direction;
(b) shear retention factor $\beta = \sigma_{xy} / (G \gamma_{xy})$

decomposed strain concept with a fixed crack model: dcp-fix; decomposed rotating crack model: dcp-rot; total strain concept with a fixed crack model: tot-fix and total rotating crack model: tot-rot. The nominal $\sigma_{xy} - \gamma_{xy}$ shear response is shown in Figure 2(a). The fixed crack models both give a linear relation because of the constant shear-retention factor β. The equivalent shear-retention factor is given in Figure 2(b). This secant shear-retention factor is extracted from the stresses and strains in the global coordinate system. For the rotating crack models the results show that the shear-retention factor rapidly decreases and even becomes negative. The decomposed strain concept shows a small deviation compared with the total strain concept when a second crack arises. For the total formulation the shear stiffness is independent from the cracks in both directions. For the decomposed formulation the shear stiffness for each crack is dependent upon the shear-retention factor and this results in a decrease of the shear stiffness at the element level. The normal stress-strain response in the global coordinate system is depicted in Figure 3. The decomposed strain concept with a fixed crack model results in a

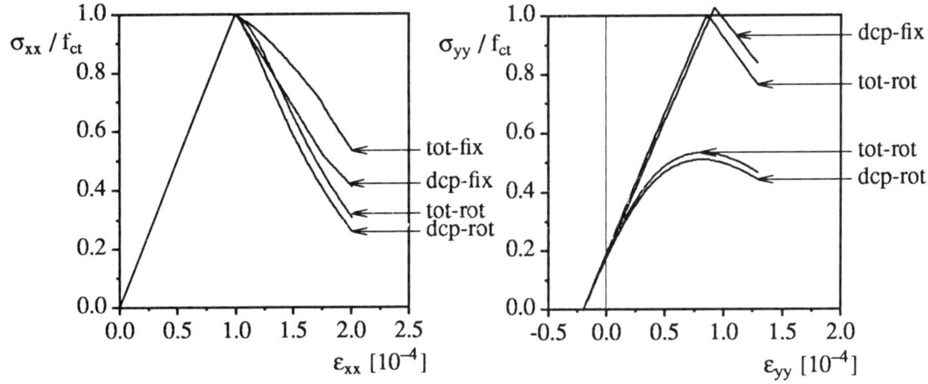

Figure 3 Normal stress-strain response in x-direction and y-direction, for different strain concepts and crack models.

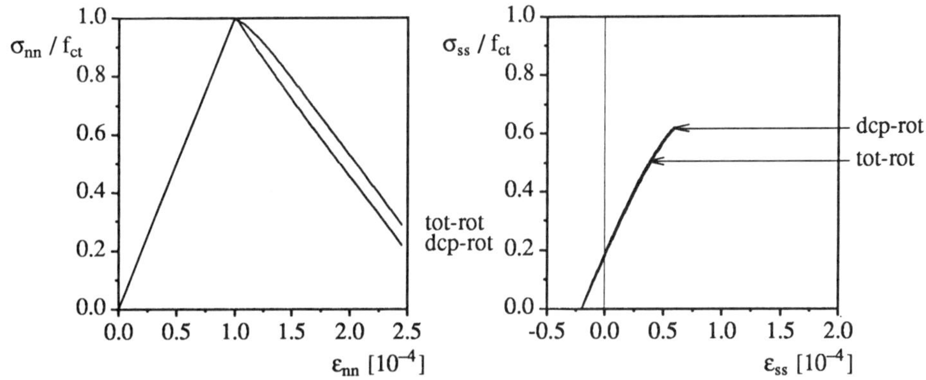

Figure 4 Principal tensile stress-strain response in n,s-direction for different strain concepts and a rotating crack model.

$\sigma_{xx} - \varepsilon_{xx}$ diagram which, until the second crack arises, is equivalent to the input softening diagram. The large deviation between the total and the decomposed strain concept for the fixed crack model is surprising, especially since, for the rotating crack model, both strain concepts give results that are in reasonable agreement. The $\sigma_{yy} - \varepsilon_{yy}$ response shows the the degradation of the strength in the lateral direction. The principal stress-strain response is depicted in Figure 4 for the rotating crack model. The principal stress-strain response for the fixed crack model has no physical meaning, since the directions of the principal strain and the principal stress are not equal. The $\sigma_{nn} - \varepsilon_{nn}$ response shows the linear softening diagram since the softening is explicitly monitored in the rotating, principal strain coordinate system. The deviation between the total and the decomposed strain concept is a result of the Poisson effect. The stress-strain diagram in the lateral direction $\sigma_{ss} - \varepsilon_{ss}$ shows that the lateral stresses remain small during the total loading process.

4. Constitutive model for reinforced concrete

The constitutive model for reinforced concrete assumed in this study is based on the assumption that reinforced concrete is a composite material. With this assumption it is possible to decompose the total stress in the stress components of the constituent materials, concrete and steel, and a stress component due to the interaction between concrete and steel. The reinforcing steel is modeled with an embedded formulation in which it is assumed that the strain for both materials is equal. The total stress is now given by:

$$\sigma_{x,y} = \sigma^{c}_{x,y} + \sigma^{s}_{x,y} + \sigma^{ia}_{x,y} \tag{10}$$

in which the superscript c denotes the concrete, s the reinforcing steel and ia the stresses due to the interaction between concrete and reinforcement. The constitutive relation for the concrete has already been derived (cf. eq.(3)):

$$\sigma^{c}_{x,y} = [\mathbf{T}^{T}(\phi)\, \mathbf{D}_{n,s}\, \mathbf{T}(\phi)\,]\, \varepsilon_{x,y}$$

The constitutive relation for the reinforcing steel looks very similar:

$$\sigma^{s}_{x,y} = [\, \mathbf{T}^{T}(\theta)\, \mathbf{D}_{s}\, \mathbf{T}(\theta)\,]\, \varepsilon_{x,y} \tag{11}$$

with θ the inclination angle between the reinforcing grid and the global x,y-coordinate system. The constitutive relation for the reinforcing grid is given by the matrix:

$$\mathbf{D}_{s} = \begin{bmatrix} \rho_p E^{ep}_{s} & 0 & 0 \\ 0 & \rho_q E^{ep}_{s} & 0 \\ 0 & 0 & 0 \end{bmatrix} \tag{12}$$

with the reinforcing ratios ρ_p, ρ_q in the p, q directions of the grid respectively and E^{ep}_{s} the secant elasto-plastic modulus.

The stress contribution owing to the interaction between concrete and reinforcement is given by:

$$\sigma^{ia}_{x,y} = \mathbf{T}^{T}(\phi)\, \sigma^{ia}_{n,s} \tag{13}$$

The stresses on the crack surface are in general functions of the angle between the reinforcement and the normal on the crack surface. For the first crack in the n-direction this transformation yields:

$$\begin{Bmatrix} \sigma_{nn} \\ \sigma_{ss} \\ \sigma_{ns} \end{Bmatrix}^{ia}_{1} = \begin{bmatrix} \cos\alpha & 0 & 0 \\ 0 & 0 & 0 \\ 0 & 0 & \cos\alpha \end{bmatrix} \begin{Bmatrix} \sigma_{pp} \\ \sigma_{qq} \\ \sigma_{pq} \end{Bmatrix}^{ia}_{1} + \begin{bmatrix} 0 & -\sin\alpha & 0 \\ 0 & 0 & 0 \\ 0 & 0 & \sin\alpha \end{bmatrix} \begin{Bmatrix} \sigma_{pp} \\ \sigma_{qq} \\ \sigma_{qp} \end{Bmatrix}^{ia}_{1} \tag{14}$$

with the angle α the angle between reinforcement and the normal of the crack. For the possible second crack in the s-direction the transformation of the stresses in the reinforcement direction to the crack coordinate system yields:

$$\begin{Bmatrix} \sigma_{nn} \\ \sigma_{ss} \\ \sigma_{ns} \end{Bmatrix}^{ia}_{2} = \begin{bmatrix} 0 & 0 & 0 \\ \sin\alpha & 0 & 0 \\ 0 & 0 & -\sin\alpha \end{bmatrix} \begin{Bmatrix} \sigma_{pp} \\ \sigma_{qq} \\ \sigma_{pq} \end{Bmatrix}^{ia}_{2} + \begin{bmatrix} 0 & 0 & 0 \\ 0 & \cos\alpha & 0 \\ 0 & 0 & \cos\alpha \end{bmatrix} \begin{Bmatrix} \sigma_{pp} \\ \sigma_{qq} \\ \sigma_{pq} \end{Bmatrix}^{ia}_{2} \tag{15}$$

The constitutive relation for the interaction stress in the p-direction of the reinforcement grid is given by:

$$\begin{Bmatrix} \sigma_{pp} \\ \sigma_{qq} \\ \sigma_{pq} \end{Bmatrix}_i = \begin{bmatrix} E_{bp,i} & 0 & 0 \\ 0 & 0 & 0 \\ 0 & 0 & E_{dp,i} \end{bmatrix} \begin{Bmatrix} \varepsilon_{pp} \\ \varepsilon_{qq} \\ \gamma_{pq} \end{Bmatrix} \tag{16}$$

and the stress in the q-direction of the grid:

$$\begin{Bmatrix} \sigma_{pp} \\ \sigma_{qq} \\ \sigma_{qp} \end{Bmatrix}_i = \begin{bmatrix} 0 & 0 & 0 \\ 0 & E_{bq,i} & 0 \\ 0 & 0 & E_{dq,i} \end{bmatrix} \begin{Bmatrix} \varepsilon_{pp} \\ \varepsilon_{qq} \\ \gamma_{pq} \end{Bmatrix} \tag{17}$$

The stiffness moduli E_{bp} and E_{bq} are the result of the bond stresses between the reinforcement and the concrete between the cracks. This contribution to the stiffness of the structure is commonly known as the tension-stiffening effect. The tension-stiffening moduli E_{bp} and E_{bq} are functions of the strain in the reinforcement and the strain in the crack coordinate system. This tension stiffening component is assumed to be a trilinear function, according to Cervenka, Pukl and Eligehausen (1990), see Figure 5(a). At ε_0 the crack arises and the ascending branch coincides with the softening of the concrete until the ultimate crack strain ε_u. The constant part is a fraction of the tensile strength of concrete, as a rough approximation 0.4 f_{ct}. The descending branch starts when the yielding of the reinforcement begins at a yield stress f_{sy} and a yield strain ε_{sy}.

The stiffness moduli E_{dp} and E_{dq} are the dowel stiffnesses of the reinforcement bars. These stiffness moduli are dependent upon the bar diameter, the crack width, the stress of the reinforcement and possibly on other parameters as well. In this study the dowel stiffness is assumed to be a fraction of the shear stiffness G_s of the reinforcement bar, $E_{dp} = E_{dq} = \delta\, \rho G_s$, $0 \le \delta \le 1$.

The tangential formulation of the stress components in reinforced concrete is essential for the solution of the nonlinear equations via a Newton-Raphson method. The tangential formulation is obtained by differentiation of the total stress-strain relation. For the concrete the tangential formulation has already been derived, eq.(5) and eq.(6). The tangential formulation for the reinforcement reads

$$\dot{\boldsymbol{\sigma}}^{s}_{x,y} = [\, \mathbf{T}^{T}(\theta)\,{}^{t}\mathbf{D}_s \mathbf{T}(\theta)\,]\, \dot{\boldsymbol{\varepsilon}}_{x,y} \tag{18}$$

in which the tangent operator for the grid is given by the matrix

$${}^{t}\mathbf{D}_s = \begin{bmatrix} \rho_p\,{}^{t}E^{ep}_s & 0 & 0 \\ 0 & \rho_q\,{}^{t}E^{ep}_s & 0 \\ 0 & 0 & 0 \end{bmatrix} \tag{19}$$

where ${}^{t}E^{ep}_s$ is the tangential elasto-plastic modulus: ${}^{t}E^{ep}_s = d\sigma_i \,/\, d\varepsilon_i$.

The tangent operator of the interaction stresses is more complicated because of the rotation of the crack direction if a rotating crack model is used. Using the formulation of the interaction stresses of eq.(13), the total derivative is given by

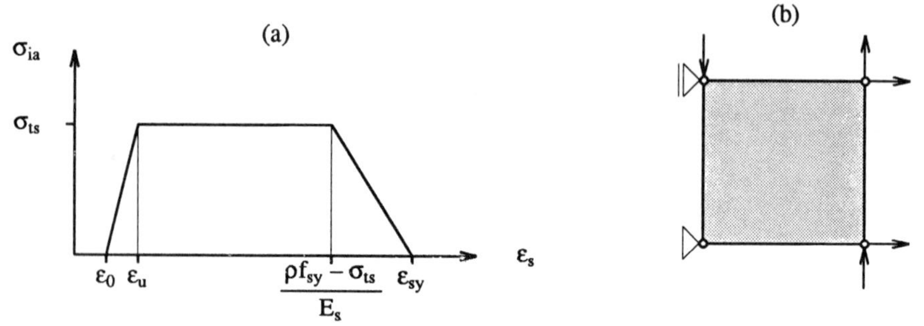

Figure 5 (a) Modeling of the tension-stiffening effect;
(b) Model panel Bhide and Collins (1987).

$$\dot{\sigma}^{ia}_{x,y} = \left[T^T(\phi)\frac{\partial \sigma^{ia}_{n,s}}{\partial \varepsilon_{x,y}} + \frac{\partial T^T(\phi)}{\partial \phi}\sigma^{ia}_{n,s}\frac{\partial \phi^T}{\partial \varepsilon_{x,y}} \right]\dot{\varepsilon}_{x,y} \tag{20}$$

in which the stress in the crack direction $\sigma^{ia}_{n,s}$ is given by eq.(14) for the first crack and eq.(15) for the second crack. The partial derivatives of these stress components with regard to the global strain vector result in a complicated formulation of the tangent stiffness operator for the interaction stress component of the total stress vector, Feenstra (1991).

5. Panel PB21 Bhide and Collins

Numerical experiments with the above described model for reinforced concrete were carried out for panel PB21 of Bhide and Collins (1987), with dimensions 870x870x70 [mm^3], a reinforcement percentage of 2.195 % in the element x-direction and a reinforcement percentage of 0.0 % in the element y-direction. The concrete has been modeled with a Young's modulus $E_c = 20000$ [N/mm^2], Poisson's ratio $\nu = 0.2$, tensile strength $f_{ct} = 2.0$ [N/mm^2] and a fracture energy $G_f = 0.89$ [Nmm/mm] over a crack band width of 890 [mm]. A crack spacing of 100 [mm] has been assumed which means that approximately 9 cracks are distributed over the element size. The total amount of energy consumed in the panel is then 8.9 times the fracture energy for one crack. The total strain concept and linear tension softening have been used. Both the rotating crack model and the fixed crack model have been utilized. For the latter model a shear retention factor $\beta = 0.2$ has been adopted. The reinforcing steel has been modeled with a linear-elastic plastic model with a Young's modulus $E_s = 210000$ [N/mm^2] and a yield stress $f_{sy} = 402$ [N/mm^2]. The deformed bars of the reinforcing grid can sustain much more bond stress than plain bars. These bond characteristics have been modeled with a maximum bond stress of 1.4 [N/mm^2]. The dowel action and the nonlinearity in compression have been neglected. The finite element configuration for the analysis is one four-noded element with four integration points for both the reinforcement and the concrete, see Figure 5(b). The analyses have been performed under arc-length control. The results of the analyses are given in Figure 6 which shows the nominal shear stress-principal tensile strain diagram.

6. Concluding remarks

The crack models embedded in a total strain concept and a decomposed strain concept show reasonable agreement when the models are applied to a tension-shear model problem. The

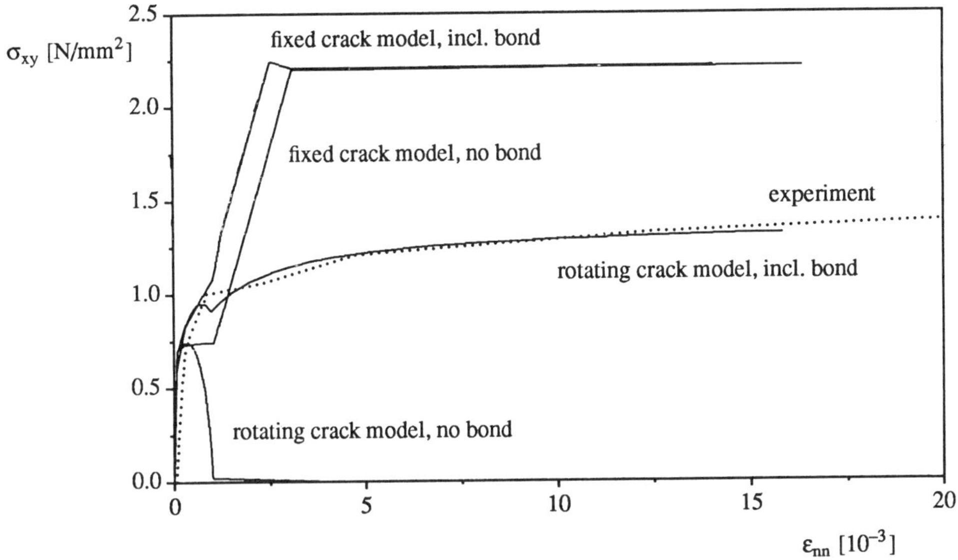

Figure 6 Shear stress - principal tensile strain response for panel PB21.

crack models embedded in a total strain concept are conceptually simple and less CPU consuming than the crack models embedded in a decomposed strain concept.

The results of the simulation of the reinforced panel demonstrate the capability of the presented model for simulating reinforced concrete behavior. Further research is necessary on the constitutive components of the model. The main advantage of the proposed model is the separate treatment of the softening behavior of the concrete and the tension-stiffening behavior of the reinforced concrete due to bond stresses between the cracks.

7. Acknowledgements

The calculations have been carried out with the DIANA finite element code of the Netherlands Institute for Applied Scientific Research, Building and Construction Research (TNO-BOUW). The research was supported financially by the Netherlands Technology Foundation (STW) under grant DCT-77.1405 and by committee A30 "Concrete Mechanics" of the Netherlands Organization for Research, Codes and Specifications (CUR A30).

8. References

Bažant, Z.P. (1983). Comment on orthotropic models for concrete and geomaterials. **J. Engrg. Mech.** , ASCE, 109(3), 849-865.

Bažant, Z.P. and Oh, B.H. (1983). Crack band theory for fracture of concrete. **Materials and Structures**, RILEM, 93(16), 155-177.

Bhide, S.B. and Collins, M.P. (1987). Reinforced concrete elements in shear and tension. Publication 87-02, Univ. of Toronto, Dept. of Civil Engrg., Canada.

de Borst, R. and Nauta, P. (1985). Non-orthogonal cracks in a smeared finite element model.

Eng. Comput. , (2), 35-46.

Cervenka, V.,Pukl, R. and Eligehausen, R. (1990). Computer simulation of anchoring technique in reinforced concrete beams, in **Proc. Int. Conf. Computer Aided Analysis and Design of Concrete Structures** , Part I, eds. N. Bićanić et al., Pineridge Press, Swansea, 1-21.

Crisfield, M.A. and Wills, J. (1989). Analysis of R/C panels using different concrete models. **J. Engrg. Mech.** , ASCE, 115(3), 578-597.

Feenstra, P.H. (1991). Modeling of reinforced concrete panels. Report 25.2.91-5-2, Delft University of Technology, Delft, Netherlands.

Rots, J.G., Nauta, P., Kusters, G.M.A. and Blaauwendraad, J. (1985). Smeared crack approach and fracture localization in concrete. **Heron** , 30(1), 3-48.

Rots, J.G. (1988). **Computational modeling of concrete fracture.** Dissertation, Delft University of Technology, Delft, Netherlands.

Willam, K.J., Pramono, E. and Sture, S. (1987). Fundamental issues of smeared crack models in **Proc. SEM/RILEM Int. Conf. Fracture of Concrete and Rock** , eds. S.P. Shah et al., Houston, 142-157.

56 CYCLIC MICROPLANE MODEL FOR CONCRETE

J. OŽBOLT
Stuttgart University, Germany
Z.P. BAŽANT
Department of Civil Engineering, Northwestern University, Evanston, IL, USA

Abstract
A recently proposed microplane model, which describes not only cracking but also general nonlinear triaxial response, is extended to cyclic loading and the rate effect, and is implemented in a three dimensional finite element code. To avoid spurious localization instabilities due to strain-softening and consequent mesh-sensitivity problems in finite element analysis, the concept of nonlocal continuum with local strain is adopted. The rate effect is introduced by combining the damage model with the Maxwell rheological model. As an illustration, the results of the finite element analysis for some basic cases on the material level as well as for plain concrete specimen loaded in three-point bending are demonstrated. The calculated responses yield hysteretic loops of an approximately correct area and for shear, the loops have well known pinched form.
Keywords: Microplane model, Cycling, Rate effect, Creep, Fracture, Damage.

1 Introduction

Currently, no material model for concrete is available which is able to predict concrete behavior, not only in the case of three dimensional monotonic loading, but also for general three dimensional cyclic, including strain-softening and rate effect.

The microplane model represents a constitutive model in which the material is characterized by a relation between the stress and strain components on planes of various orientations, which may be imagined to represent the damage planes or weak planes in the microstructure, such as contact layers between aggregate pieces in concrete. The latest, Bažant and Prat (1988) version of the microplane model, was shown capable to predict the behavior of concrete in monotonic loading for wide range of stress and strain conditions using only a few material parameters. Together with the nonlocal concept, this model was implemented in a finite element code. Its capabilities to realistically predict the structural response was demonstrated in a number of numerical examples.

Here, this microplane model is further improved and extended to cyclic loading, as well as the rate effect, which is important for such loading. A kinematic constraint is used, i.e. the total strain

Fracture Processes in Concrete, Rock and Ceramics Edited by J.G.M. van Mier, J.G. Rots and A. Bakker.
 Published by E. & F.N. Spon, 2–6 Boundary Row, London SE1 8HN. ISBN 0 419 15870 7

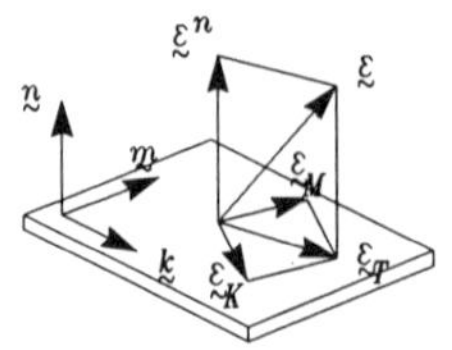

Fig. 1 Stress-strain components on each microplane

vector on each microplane is assumed to be the resolved component of the macroscopic strain tensor. The microplane strains are split into volumetric, deviatoric and shear components. The shear strain vector is further split into two orthogonal components. For each microplane strain component the main characteristics of the load history are stored during calculations. The numerical integration of the time step is based on the previously proposed exponential algorithm. The new model is implemented in a three dimensional finite element code.

2 Review of Microplane model

2.1 basic hypotheses and strain components

The basic hypotheses used in the present cyclic microplane model are as follows:

Hypothesis I

The strains on any microplane are the resolved components of the macroscopic strain tensor ε_{ij}. This hypothesis represents a kinematic constraint and yields the relations (Fig. 1):

$$(\boldsymbol{\varepsilon}_R)_i = n_i \varepsilon_{ij}; \qquad \varepsilon_N = n_i n_j \varepsilon_{ij}; \qquad \varepsilon_D = n_i n_j \varepsilon_{ij} - \varepsilon_V \tag{1}$$

$$\boldsymbol{\varepsilon}_T = \varepsilon_M \boldsymbol{m} + \varepsilon_K \boldsymbol{k}; \quad \varepsilon_M = \boldsymbol{m}\,\boldsymbol{\varepsilon}_T = m_i n_j \varepsilon_{ij}; \quad \varepsilon_K = \boldsymbol{k}\,\boldsymbol{\varepsilon}_T = k_i n_j \varepsilon_{ij} \tag{2}$$

where latin lower case subscript refer to Cartesian coordinates x_i (i= 1, 2, 3); $\boldsymbol{\varepsilon}_R$ = the strain vector on a microplane whose unit normal is n_i; ε_R= magnitude of $\boldsymbol{\varepsilon}_R$; $\boldsymbol{\varepsilon}_T$ = the tangential vector component of $\boldsymbol{\varepsilon}_R$; $\boldsymbol{m}$ and $\boldsymbol{k}$ are unit coordinate vectors defined in advance for each microplane (Fig. 1); ε_M and ε_K are the inplane components of vector $\boldsymbol{\varepsilon}_T$ in the direction of vectors $\boldsymbol{m}$ and $\boldsymbol{k}$; and ε_N is normal strain component of the microplane (Fig. 1). The normal strain vector is separated into volumetric strain $\varepsilon_V = \varepsilon_{kk}/3$ and deviatoric strain ε_D (i.e., $\varepsilon_N = \varepsilon_V + \varepsilon_D$). In contrast to the originally proposed model (Bažant and Prat, 1988), the shear strain vector $\boldsymbol{\varepsilon}_T$ is split into two components ($\boldsymbol{\varepsilon}_M$ and $\boldsymbol{\varepsilon}_K$). In Bažant and Prat's original version of the general microplane model, the shear component $\boldsymbol{\varepsilon}_T$ has been characterized by its magnitude ε_T (which is always non negative) and the corresponding tangential stress vector $\boldsymbol{\sigma}_T$ has been considered to be always parallel to $\boldsymbol{\varepsilon}_T$. This simplification seems adequate for the monotonic loading or nearly monotonic loading and in a crude manner perhaps also for the first unloading, but is obviously inadequate for cycling loading. The choice of coordinate vector $\boldsymbol{m}$ within the microplane is arbitrary, and the other coordinate vector is obtained as $\boldsymbol{k} = \boldsymbol{m} \times \boldsymbol{n}$ (Fig. 1). To minimize directional bias, directions of microplane shear strain components are chosen as follows: For the first microplane $m \perp z$, $m_3 = 0$; for the second $m \perp x$, $m_1 = 0$; for the third $m \perp y$; $m_2 = 0$ etc.

Hypothesis II

Each microplane resists only normal and shear components. As a consequence of the **Hypothesis I**, the shear stress σ_T does not have the same direction as shear strain ε_T.

Hypothesis III

The response on each microplane is assumed to depend on the mean lateral strain ε_L, which is equivalent assuming that it depends on the volumetric strain $\varepsilon_V = \varepsilon_{kk} / 3$ (and of course ε_N and ε_D). This feature was found to be necessary for modeling triaxial test data for very high confining pressures (but not other data).

Hypothesis IV

The stress-strain curves of each microplane are assumed to be path-independent as long as there is no unloading on this microplane. During each unloading and reloading, which is defined separately on each microplane, the curves of the stress and strain differences from the state at the start of unloading are also assumed to be path-independent. In this way, all the macroscopic path-dependence is produced by various combinations of loading and unloading on all the microplanes. It may be noted that some microplanes may get unloaded even for macroscopically monotonic or virgin loading, thus making the response path-dependent.

Hypothesis V

The volumetric, deviatoric and shear responses on each microplane are assumed to be mutually independent.

2.2 Microplane stress-strain relations

According to the foregoing hypotheses, in the case of virgin loading, the behavior on the microplane for each microplane strain component is described by path-independent total stress-strain relations of the form:

$$\sigma_V = F_V(\varepsilon_V), \quad \sigma_D = F_D(\varepsilon_D), \quad \sigma_M = F_M(\varepsilon_M), \quad \sigma_K = F_K(\varepsilon_K) \tag{3}$$

To represent unloading and application of the nonlocal damage concept, it is convenient to define the total stress-strain relation in the form of continuum damage mechanics:

$$\sigma_V = C_V \varepsilon_V, \quad \sigma_D = C_D \varepsilon_D, \quad \sigma_M = C_M \varepsilon_M, \quad \sigma_K = C_K \varepsilon_K \tag{4}$$

in which, except for volumetric compression,

$$C_V = C_V^o(1-\omega_V), \quad C_D = C_D^o(1-\omega_D), \quad C_M = C_M(1-\omega_M), \quad C_K = C_K(1-\omega_K) \tag{5}$$

where C_V, C_D, C_M, C_K represent the secant moduli; $C_V = F_V(\varepsilon_V)/\varepsilon_V$, $C_D = F_D(\varepsilon_D)/\varepsilon_D$, $C_M = F_M(\varepsilon_M)/\varepsilon_M$, $C_K = F_K(\varepsilon_K)/\varepsilon_K$; C_V^o, C_D^o, C_M^o, C_K^o are the initial values of C_V, C_D, C_K, C_M; and ω_V, ω_D, ω_K, ω_M are the volumetric damage, deviatoric damage and shear damage on the microplane level. The secant shear moduli C_M and C_K must be defined by the same functions of the corresponding shear strain components. The initial shear moduli for both shear components are equal for both directions, $C^0_M = C^0_K$. Best fits of selected typical test data for concrete have been obtained using the following approximation for virgin loading:

$$\text{for } \varepsilon_V \geq 0: \quad \omega_V = 1-\exp\left[-\left(\frac{\varepsilon_V}{e_1}\right)^m\right] \tag{6}$$

$$\text{for } \varepsilon_D \geq 0: \quad \omega_D = 1-\exp\left[-\left|\frac{\varepsilon_D}{e_1}\right|^m\right] \tag{7a}$$

$$\text{for } \varepsilon_D < 0: \quad \omega_D = 1-\exp\left[-\left|\frac{\varepsilon_D}{e_2}\right|^n\right] \tag{7b}$$

$$\omega_M = 1-\exp\left[-\left|\frac{\varepsilon_M}{e_5}\right|^k\right] \tag{8a}$$

$$\omega_K = 1-\exp\left[-\left|\frac{\varepsilon_K}{e_5}\right|^k\right] \tag{8b}$$

in which $a_5 = a_3$ if $\varepsilon_V \geq 0$, and $a_5 = a_3 - a_4\varepsilon_V$ if $\varepsilon_V < 0$, where e_1, e_2, e_3, e_4, e_5, m, n and k are empirical material constants. The dependence of a_5 on the volumetric strain ε_V reflects internal friction and represents an additional kinematic constraint of scalar type. In the volumetric behavior there is no damage ($\omega_V = 0$) and the response for virgin loading is described by:

$$\text{for } \varepsilon_V < 0 \quad C_V = C^0_V \left[(1+|\varepsilon_V/a|)^{-p} + |\varepsilon_V/b|^q\right] \tag{9}$$

where a, b, p, q are empirical constants.

3 Generalization to Unloading, Reloading and Cycling Loading

Since the previous work (Bažant and Prat, 1988) was focused on monotonic loading, the rule for unloading was very simple but it was possible to represent only the first unloading, and even that with considerable errors compared to experiments. To model unloading, reloading and cycling loading and do so even for general triaxial stress states, more complex rules on the microplane level are needed. After many trials, the following unloading-reloading rules, which are different for each microplane strain component, have been chosen and verified.

In contrast to virgin loading, stress-strain relation must be in the incremental form and is as follows:

$$d\sigma_V = C_V\, d\varepsilon_V, \quad d\sigma_D = C_D\, d\varepsilon_D, \quad d\sigma_M = C_M\, d\varepsilon_M, \quad d\sigma_K = C_K\, d\varepsilon_K \tag{10}$$

where, C_V, C_D, C_M, C_K represent unloading-reloading tangent moduli. Principally, the unloading and reloading moduli for each microplane component are defined as follows (See Fig. 3):

$$C = C^0\,\alpha + (1-\alpha)\,\sigma/(\varepsilon-\varepsilon_1)$$

$$\text{for } \varepsilon > \varepsilon_p;\ \varepsilon_1 = \varepsilon_p - \sigma_p/C^0 + \beta\,(\varepsilon-\varepsilon_p) \tag{11}$$
$$\text{for } \varepsilon \le \varepsilon_p;\ \varepsilon_1 = 0.$$

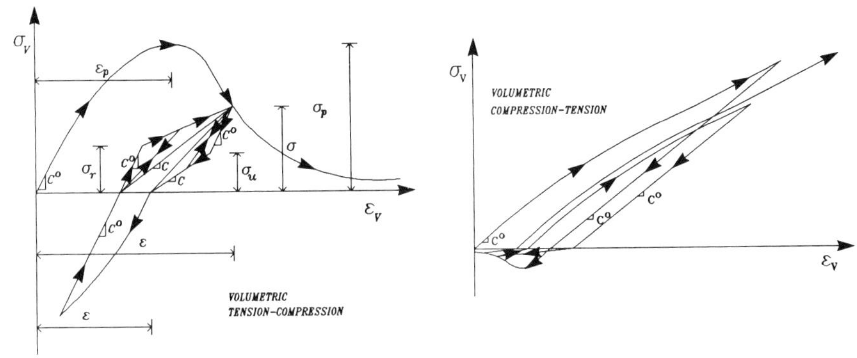

Fig. 2 Cyclic rule for volumetric tension and compression component

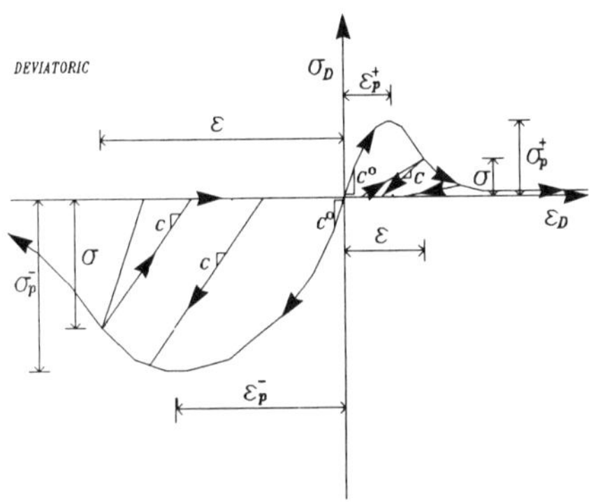

Fig. 3 Cyclic deviatoric rule

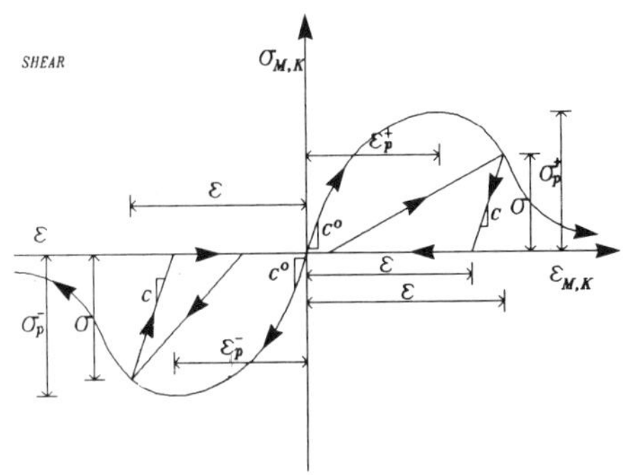

Fig. 4 Cyclic shear rule

where, σ_p and ε_p denote the positive or negative peak stress and the corresponding strain for each microplane component, taking values σ_p^+, ε_p^+ and σ_p^-, ε_p^- for the positive and negative peaks; α and β are empirically chosen constants between 0 and 1. The unloading-reloading rules for each microplane component, in tension (positive) and compression (negative), are defined generally in Figs. 2-4.

The virgin loading for each microplane strain component ε occurs if:

$$\varepsilon\,\Delta\varepsilon \ge 0 \text{ and } (\varepsilon-\varepsilon^{max})(\varepsilon-\varepsilon^{min}) \ge 0 \tag{12}$$

where ε^{max}, ε^{min} are the maximum and minimum values of ε that have occurred so far; otherwise unloading or reloading take place.

4 Incremental macroscopic stress-strain relations

As a consequence of the kinematic constraint, the microplane stresses cannot be exactly equal to the resolved components of the macrostresses. The micro-macro equivalence of the stresses can be enforced only approximately, which has been done by the means of the principle of virtual work. This provide an incremental macroscopic stress-strain relation of the form:

$$\Delta\sigma_{ij}=C_{ijrs}\Delta\varepsilon_{rs}-\Delta\sigma''_{ij} \tag{13}$$

in which

$$C_{ijrs}=\frac{3\Pi}{2}\int\Big[n_in_jn_rn_s\,C_D+\frac{1}{3}n_in_j\delta_{km}\,(C_V-C_D)$$
$$+\frac{1}{4}(m_im_j+m_jn_i)\,(m_rn_s+m_sn_r)\,C_M$$
$$+\frac{1}{4}(k_in_j+k_jn_i)\,(k_rn_s+k_sn_r)\,C_K\Big]\,F(\underset{\sim}{n})\,dS \tag{14}$$

$$\Delta\sigma''_{ij}=\frac{3\Pi}{2}\int_S\Big[n_in_j(\Delta\sigma''_V+\Delta\sigma''_D)+\frac{1}{2}(m_in_j+m_jn_i)\,\Delta\sigma''_M$$
$$+\frac{1}{2}(k_in_r+k_jn_r)\,\Delta\sigma''_K\Big]\,F(\underset{\sim}{n})\,dS \tag{15}$$

Here, C_{ijrs} is the macroscopic incremental material stiffness tensor, and $\Delta\sigma''_{ij}$ are the associated macroscopic inelastic stress increments. The integrals in (13), (14) and (15) are here evaluated numerically, employing an effective numerical integration scheme over the sphere using 21 plane directions (Stroud, 1971; Bažant and Oh, 1986).

The values of the material parameters used in the calculations are the same as described by Bažant and Prat (1988), except for two additional constants α and β for each microplane component. The optimal values of α and β can be obtained by fitting the set of cyclic test data. Present experience indicates that the α and β may be taken the same for all concrete types.

5 Rate effect

Experimental evidence indicates that concrete stiffness, strength and ductility are rather sensitive to the deformation rate. This is known as the rate effect. This effect is no doubt caused by creep in the bulk of test specimen as well as time dependent rupture of bonds in the fracture process zone, which both cause stress relaxation. The simplest model for relaxation is the Maxwell's spring dashpot model, which is here used for each of the microplane strain component ε_V, ε_D, ε_M and ε_K (Fig. 5). Dropping for the moment the subscripts V, D, M and K, the stress-strain relation for each microplane component may be written as

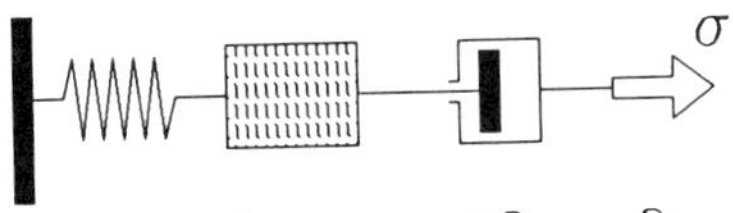

Fig. 5 Maxwell unit introduced for each microplane component

$$\dot{\sigma} + \frac{\sigma}{\rho} = C_t \dot{\varepsilon} \tag{16}$$

where ε is the total microplane strain including the flow (creep) strain, the superior dots denote the time derivatives, ρ is a material parameter representing the relaxation time, and C_t is the tangent material modulus. Since we use a microplane stress-strain relation in the secant form, it is convenient to rewrite (16) in the secant form:

$$\dot{\sigma} = C \dot{\varepsilon} - \frac{\sigma}{\lambda} \tag{17}$$

in which the following notations are used:

$$\frac{1}{\lambda} = \frac{1}{\rho} - \varepsilon \frac{\Delta C}{\Delta \varepsilon} \frac{\dot{\varepsilon}}{\sigma}, \quad C_t = (C + \varepsilon \frac{\Delta C}{\Delta \varepsilon}), \quad \dot{\sigma} = \frac{\Delta \sigma}{\Delta t}, \quad \dot{\varepsilon} = \frac{\Delta \varepsilon}{\Delta t}$$

Here, λ is an apparent relaxation time, C represents the secant modulus for the microplane strain component. Eq. (17) is of the same form as incremental secant formulation of Eq. (4) (Bažant and Ožbolt, 1990), for each microplane direction. Note that upon introducing $\rho = \infty$ (no rate effect), Eq. (17) becomes identical to the incremental formulation of Eq. (4). In the present model, the relaxation time ρ is a constant and the same for all microplane components and directions.

6 Numerical algorithm in each load step

In every iteration of the load or time step r, based on the kinematic constrain (Eq. 1), the known macrostrains $\varepsilon_{ij(r)}$ and their known increments $\Delta\varepsilon_{ij(r)}$ can be used to calculate the strains and strain increments on each microplane. Further, the known values of $\varepsilon_N = \varepsilon_V + \varepsilon_D$, ε_M, ε_K, $\Delta\varepsilon_N = \Delta\varepsilon_V + \Delta\varepsilon_D$, $\Delta\varepsilon_M$ and $\Delta\varepsilon_K$, are used to calculate the stresses on each microplane by solving the Eq. (17). Each of these equations could be solved by using a central difference approximation. However, such an approximation often appears unstable in the case of strain softening, and, even if the computations remain stable, a large error is usually accumulated, with the result that the stress is not reduced exactly to zero at very large strain.

These drawbacks can be eliminated by the use of so-called exponential algorithm, initially developed for aging creep of concrete (Bažant 1971) and later extended to creep with strain softening and applied by Bažant and Chern (1985) and Bažant and Ožbolt (1990).

Another possibility is to take advantage of the secant formulation of the microplane model and from known total microplane strains ($\varepsilon^{el} + \varepsilon_{soft}$) using Eq. (4) directly calculate the microplane stresses while nonelastic deformations due to rate effect can be calculated directly from Eq. (16) or more accurate using the exponential algorithm. In both cases the basic stress-strain curve for each microplane must be followed.

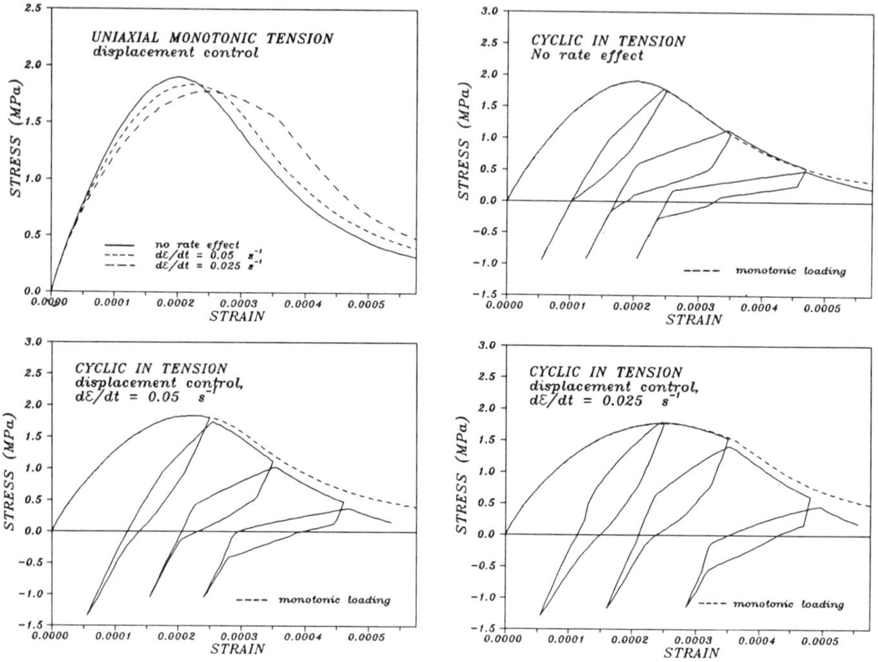

Fig. 6 Stress-strain curves for monotonic and cyclic tension

Implementation of the above numerical procedures as well as an numerical algorithm employed in each load step of the finite element analysis is described in more detail by Bažant and Ožbolt (1990). Furthermore, the nonlocal concept proposed by Pijaudier-Cabot and Bažant (1987) is also implemented into the present cycling model. More detailed description of the nonlocal implementation is given by Bažant and Ožbolt (1990).

7 Numerical examples

The capability of the present cyclic microplane model in predicting the cyclic behavior of concrete including the rate effect is demonstrated in numerical simulations using different stress-strain histories and different rates. The behavior of the model is demonstrated on the material level, using only one uniformly strained finite element loaded in tension, compression and shear. Cyclic behavior of the three-point-bend specimen in plane stress is also simulated. It should be noted that the plane stress state is a relatively complex state to simulate with the microplane model. Namely, the microplane model is full three-dimensional model and in plane-stress finite elements the lateral strains (out-of-plane

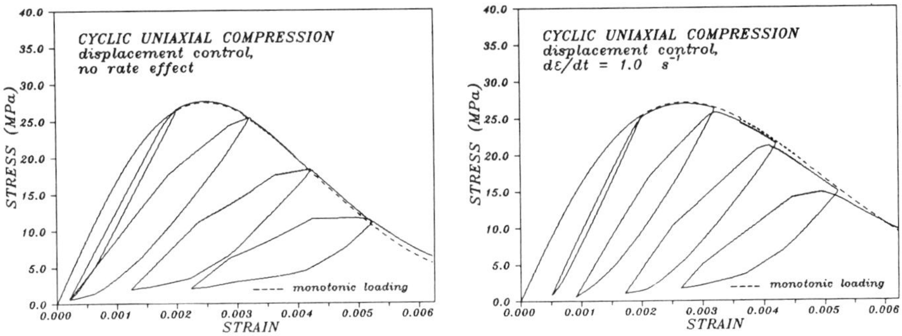

Fig. 7 Stress-strain curves for cyclic compression

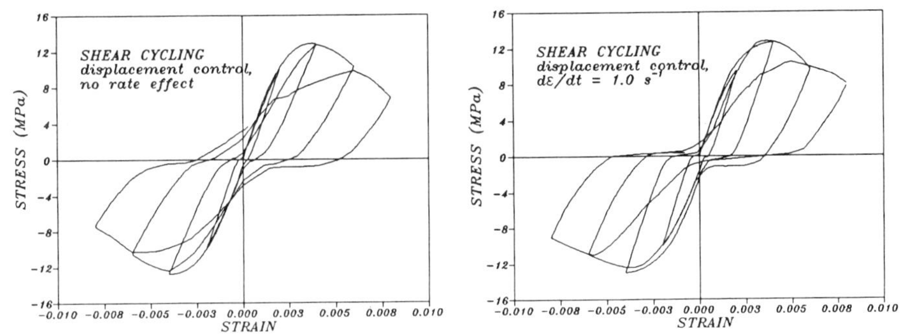

Fig. 8 Stress-strain curves for cyclic shear

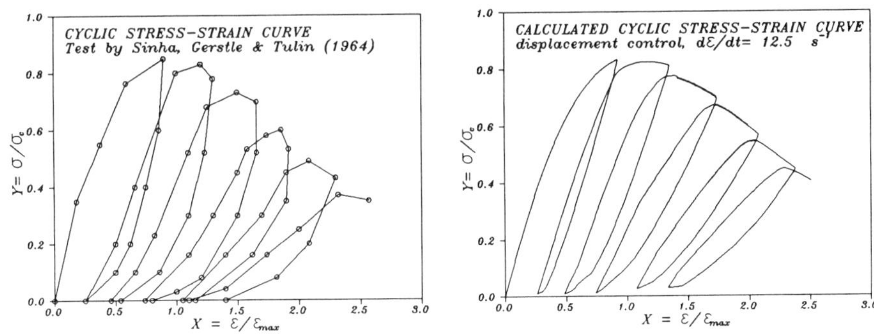

Fig. 9 Test-calculations comparison for uniaxial cyclic compression

strains) need to be calculated from the condition that the lateral stresses are zero. The basic material parameters used are: initial Young's modulus E^0= 20000 MPa, Poisson's ratio ν= 0.18 and relaxation time ρ= 0.01. The microplane material parameters are chosen as follows: a = 0.005, b = 0.043, p = 0.75, q = 2.00, e_1 = 0.00007, e_2 = 0.0020, e_3 = 0.0020, e_4=0., m = 0.85, n = 2.25, k = 2.25. Most of these values are the same as in Bazant and Prat (1988), except m, n and k (which have been adjusted such that descending stress-strain curves were more steep). The load was introduced by prescribing the displacement increases with different rates.

The stress-strain curves in the cases of monotonic and cyclic tension, with different loading rates, are shown in Fig. 6. Similar calculations, using again different rates of loading, are carried out in the case of cyclic compression (Fig. 7) and cyclic shear (Fig. 8). Fig. 9 shows comparison between the uniaxial cyclic test results of Sinha, Gerstle and Tulin (1964) and the present calculations.

The present results indicate that cyclic microplane model can realistically predict the cyclic behavior of concrete in different stress-strain conditions using the same material parameters (van Mier, 1984; Yankelevski and Reinhard, 1989). The results shown are only one among many possible responses that can be modeled. The microplane model parameters, the parameters that control loading and unloading on each microplane in each direction as well as relaxation times are here kept constant. Varying these parameters as material functions of the strain state and history it would be possible to obtain a number of different kinds of responses.

Furthermore, the results indicate that the rate effect has significant influence on the stress-strain response. If the rate of loading decreases, the peak stress also decreases while the postpeak descending stress-strain curve becomes less steep. The present model can predict the drop of stresses after repeated unloading-reloading cycles in post-peak strain-softening. As will be demonstrated later, this effect is significant when simulating the structural behavior.

Finally, to demonstrate cyclic response of the present model in

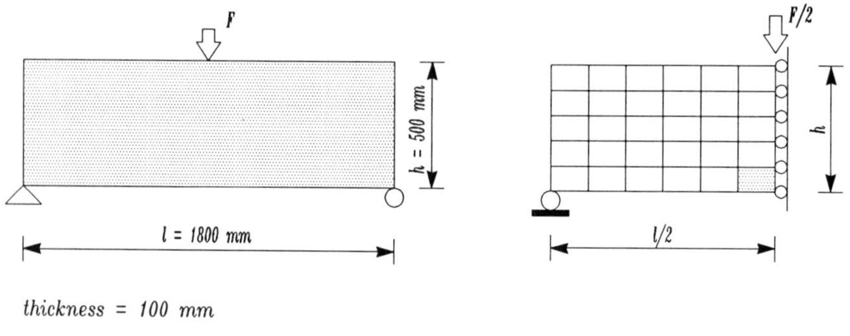

Fig. 10 Three-point-bend specimen, geometry and FE mesh

finite element applications, behavior of the three-point-bend specimen (Fig. 10) is demonstrated. In the analysis, four-point isoparametric quadrilateral plane stress finite elements with four integration points are used. Only one half of the specimen is analyzed, i.e. symmetry is forced. The material parameters were the same as in previous examples except in the case of the notched specimen (e_1= 0.00004 and m= 0.5).

The load-displacement curve for the the specimen loaded using displacement control at the upper node (see Fig. 10) is shown in Fig. 11. The displacement rate used in the analysis was du/dt= 62.5 mm/s. One cycle is performed prior to the peak load, and the next one late in the softening range. It is interesting that after the second cycle the material was damaged, and so after repeated loading a significant decrease of strength is observed.

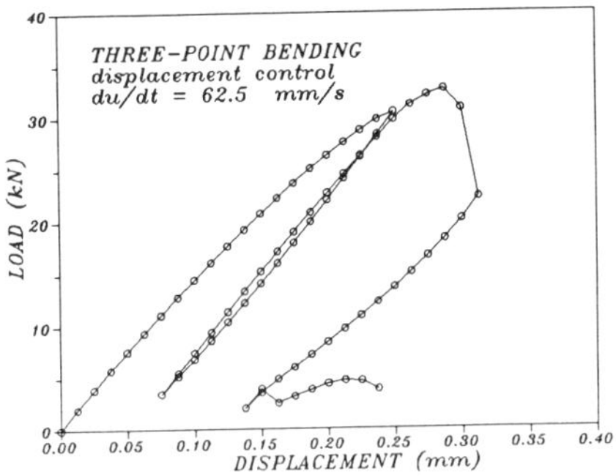

Fig. 11 Load-displacement curve, cyclic in softening range

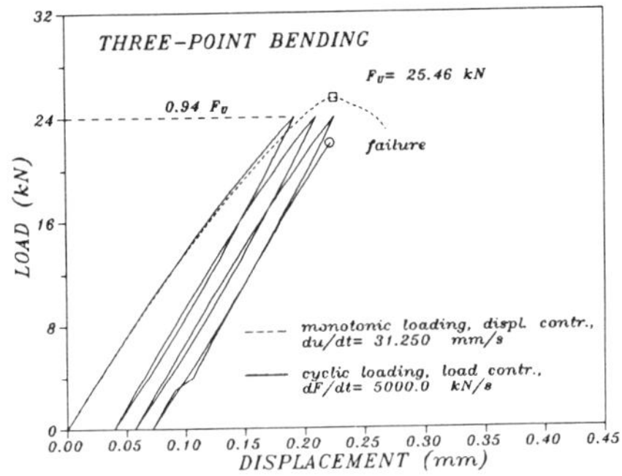

Fig. 12 Load-displacement curve, cyclic before peak load

In the subsequent calculations of the same specimen, one finite element at the bottom of the specimen was assumed to be weaker, having a tensile strength limit 10% lower than the other elements, (the shaded element in Fig. 10). The specimen has been loaded up to 94% of failure load (obtained by monotonic loading), followed by cyclic loading between 0 and 94% of peak load. The loading consisted of prescribing a force at the specimen top with the rate of increase dF/dt= 5000 kN/s. The results indicate (Fig. 12) a significant increase of the displacements due to material damage. As a consequence of the material damage due to cycling, failure is obtained after the third cycle, slightly before reaching the monotonic descending branch (See Fig. 12).

Results show that the calculated hysteretic loops are wider when the rate effect is taken into account and have approximately the correct area. The unloading begins by a steep slope, in agreement with laboratory observations. In the case of the shear cycling, the calculated hysteretic loops exhibit the characteristic pinched form with a nearly zero stiffness at the coming of the stress axis, also known from experiments.

8 Conclusions

1. The present cyclic microplane model is capable of realistically predicting the behavior of plain concrete under a wide range of strain states and histories using the same material parameters. The model is full three dimensional.
2. The rate effect can be implemented in the microplane model by combining damage with Maxwell rheological model. The model is able to give an approximately correct hysteretic loop area and a steep initial unloading slope.

3. The examples indicate that the cyclic microplane model, together with rate effect, predicts material damage due to repeated unloading-reloading cycles realistically.

9 References

Bažant, Z.P., and Prat, P.C., Microplane model for brittle-plastic material - parts I and II. J. of Eng. Mech., ASCE, 114(10), pp. 1672-1702, 1988.

Bažant, Z.P., and Ožbolt, J., Nonlocal microplane model for fracture, damage and size effect in structures. J. of Eng. Mech., ASCE, 116, (11), 1990, pp. 2485-2505.

Bažant, Z.P., "Numerically stable algorithm with increasing time steps for integral-type aging creep", Proc., First Int. Conf. on Struct. Mech. in Reactor Technology, West Berlin, T.A. Jaeger, ed., Vol. 4, Part H, 1971, pp. 119-126.

Bažant, Z.P., and Chern, J.-C., Strain softening with creep and exponential algorithm. J. of Eng. Mech., ASCE, 111(3), pp. 381-390, 1985.

Bažant, Z.P., and Oh, B.-H., Efficient numerical integration on the surface of a sphere. Zeitschrift für angewandte Mathematik und Mechanik ZAMM, 66(1), pp. 37-49, 1986.

van Mier, J.G.M., Strain-Softening of concrete under multiaxial loading conditions. Dissertation, Eindhoven Univ. of Techn., 1984.

Pijaudier-Cabot, G., and Bažant, Z.P., Nonlocal damage theory. J. of Eng. Mech., ASCE, 113, pp. 1512-1533, 1987.

Sinha, B.P., Gerstle, K.H. and Tulin, L.G., Stress-Starin Relations for Concrete Under Cyclic Loading, J. Am. Concr. Inst., 61(2), pp. 195-211, 1964.

Stroud, A.H., Approximate calculation of multiple integrals, Prentice Hall, Englewood Cliffs, N.J., 1971.

Yankelevski, D.Z. and Reinhard, H.W., Uniaxial Behavior of Concrete in Cyclic Tension. J. of Struc. Eng., ASCE, 115, (1), 1989, pp. 166-182.

57 COMPUTATIONAL ASPECTS OF LOCALIZED FAILURE SIMULATIONS IN PLAIN CONCRETE

G. ETSE and P. STEINMANN
Institute of Mechanics, University of Karlsruhe, Germany
K.J. WILLAM
CEAE Department, University of Colorado, Boulder, CO, USA

ABSTRACT
This paper focus on computations of plain concrete when different finite element formulations are utilized. At constitutive level, an elastoplastic concrete model is adopted which was developed recently on the basis of the EXTENDED LEON failure criterion. In the pre-peak regime the model is based on an isotropic work-hardening formulation. Decohesion and the strength degradation is modeled with a fracture energy-based work-softening mechanism which features an 'equivalent lenght' depending on the size and type of the element. The influence of the finite element formulation to capture sharp failure mechanisms is illustrated by means of the weak localization test. Numerical solutions are presented for the analysis of a plane strain problem with nonhomogeneous boundary conditions. A mesh refinement analysis was carried out which revealed that enriched element expansions render extraordinary accurate results even for coarse meshes. Furthermore, they allow to reproduce localized failure bands which are not aligned with favourite directions of the spatial discretization. This lack of directional bias and mesh size dependence are not inherent when standard elements are utilized.

1 Introduction

Accurate numerical predictions of progressive failure in concrete structures with excessive localization of deformations into concentrated cracks require not only a realistic model but also an appropiated element strategy to capture the correct failure mode towards the boundaries at collapse.

The weak localization test for discontinuous bifurcation recommends the use of enriched element strategies in progressive failure computations. However, if elastoplastic models based only on stress- strain constitutive equations are used together with such enriched element expansions, the width of the failure band depends strongly on the element size. The model considered here is based on the nonassociated theory of plasticity with isotropic hardening in the pre-peak regime. The plastic softening model for the post-peak regime is based on fracture energy concepts for tensile as well as for compressive strength degradation. This formulation introduces an equivalent length which reduces the mesh sensitivity of the

Fracture Processes in Concrete, Rock and Ceramics Edited by J.G.M. van Mier, J.G. Rots and A. Bakker.
 Published by E. & F.N. Spon, 2–6 Boundary Row, London SE1 8HN. ISBN 0 419 15870 7

post-peak regime.

The ability of different element formulations to reproduce the localized failure mode of a rectangular concrete panel under plane strain condition is analysed. Attention is focused on the mesh size dependence of the post-peak regime associated with each element strategy. The analysis includes a description of the model and of the weak localization condition.

2 The Extended Leon Model

The constitutive model of the numerical analysis reported in this study follows the original LEON MODEL developed by PRAMONO & WILLAM [1]. The failure envelope was modified in the deviatoric region to generate a smooth C^1 continuous surface based on the elliptic approximation by WILLAM & WARNKE [4]. This model was further extended to include isotropic work-hardening in the pre-peak regime and isotropic work-softening in the post-peak regime below the so called transition point of brittle-ductile failure, see WILLAM & ETSE [6]

The all encompassing equations of the loading surface in the hardening and softening regime for the EXTENDED LEON MODEL are:

$$F(p,\rho,\theta,k,c) = \left[(1-k)\left(\frac{p}{f_c'}+\frac{\rho g(\theta)}{\sqrt{6}f_c'}\right)^2+\frac{\rho}{f_c'}g(\theta)\sqrt{\frac{3}{2}}\right]^2 + \\ +\frac{k^2 m}{f_c'}\left(p+\frac{\rho g(\theta)}{\sqrt{6}}\right)-k^2 c = 0 \quad (1)$$

$$g(\theta) = \frac{4\,(1-e^2)\cos^2\theta+(2e-1)^2}{2\,(1-e^2)\cos\theta+(2e-1)\,[4\,(1-e^2)\cos^2\theta+5e^2-4e]^{1/2}} \quad (2)$$

where:

$$p = \frac{1}{3}(\boldsymbol{\sigma}:\mathbf{I}) \quad (3)$$

$$\rho = \|\mathbf{s}\| \quad \mathbf{s}=\boldsymbol{\sigma}-p\mathbf{I} \quad (4)$$

$$\cos 3\theta = \sqrt{54}\frac{\det(\mathbf{s})}{\rho^3} \quad (5)$$

$$e = \frac{\rho_{tens}}{\rho_{comp}} \quad (6)$$

Here θ denotes the similarity angle in the π-plane and f_c' the compressive strength. The different loading surfaces in the hardening regime are generated by the scalar parameter k which varies between 0.1 and 1.0, while c and m, the cohesive and frictional parameter remain constant. For $k = 1$ the modified failure envelope of LEON is regained. The parameter k depends on the work-hardening ductility measure κ_h, defined by the evolution law $\dot{\kappa}_h = H_h\dot{\varepsilon}_p$, where H_h is a function of the first invariant of the stress tensor and $\dot{\varepsilon}_p$ the equivalent plastic strain. Fig.1 illustrates the evolution of the loading surface in the hardening regime.

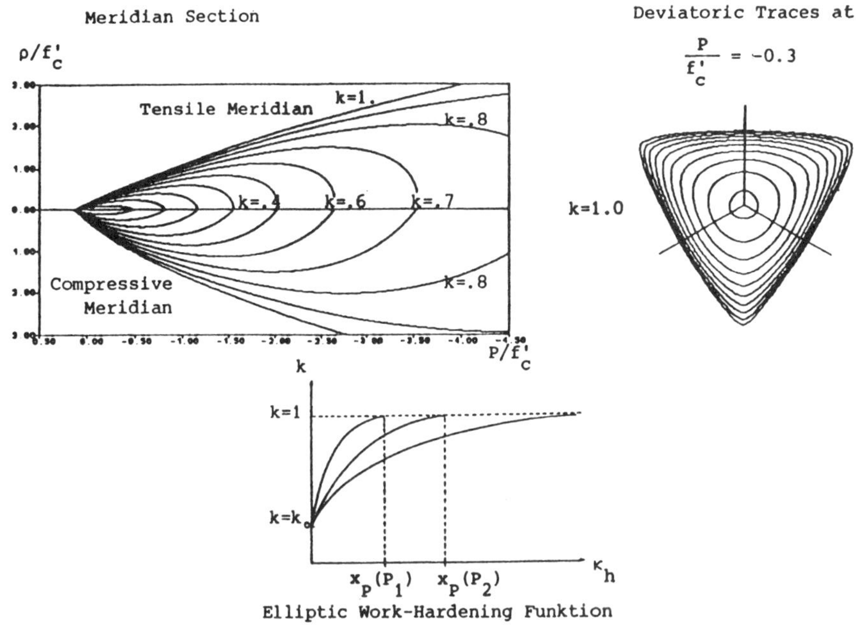

FIG.1: Isotropic Hardening of EXTENDED LEON MODEL.

The softening stages below the transition point of brittle-ductil fracture is defined by the decohesive c-parameter and by the hardening of the frictional m-parameter as a dependent parameter of c. The mechanism which controls the variation of c is the degradation of the tensile strength σ_t. A key assumption in this theory is that the degradations process of σ_t for any loading path can be described via an equivalent fracture energy release rate measure $\dot{\kappa}_s$ in analogy to a work-softening criterion, which is chosen according to the rate low:

$$\dot{\kappa}_s = \frac{G_f^I}{G_f^{II}} \dot{\epsilon}_s h_t \tag{7}$$

where $\dot{\epsilon}_s$ is the equivalent tensile fracture strain, and $\frac{G_f^I}{G_f^{II}}$ is the ratio between the fracture energy release rate in direct tension and that in shear, which is expressed as a function of the stress invariants. The geometric length h_t is introduced by the fracture energy argument developed earlier by WILLAM, HURLBUT & STURE [5] for tensile cracking and represents the crack spacing. Fig.2 describes the exponential relation between σ_t and κ_s and the evolution of the loading surface in the softening region for the EXTENDED LEON MODEL.

To reduce excessive dilatation a non associated flow rule was developed for the EXTENDED LEON MODEL which is based on a modification of the yield surface in the hydrostatic plane.

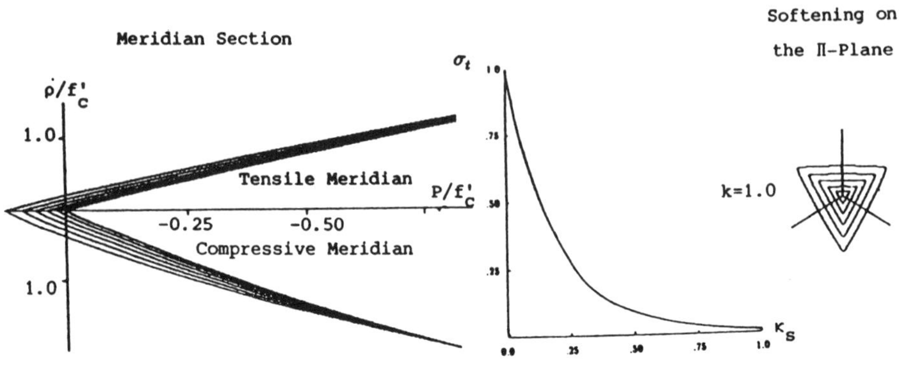

FIG.2: Isotropic Softening of EXTENDED LEON MODEL.

3 Single Element Weak Localization Test

Boltzmann continuum theory signals the condition for which the assumption of a smooth strainfield fails to hold, so that bifurcation at the constitutive level may develop. This localization condition describes the formation of a discontinuity in the field of displacement- or rather velocity gradients across a surface with orientation $\mathbf{N}$. Discontinuous bifurcation is signalled by a singularity of the acoustic tensor or localization tensor respectively, i.e. when $\det(\mathbf{Q}^{ep}) = \det(\mathbf{N} \cdot \mathbf{E}^{ep} \cdot \mathbf{N}) = 0$. For numerical computations, loss of ellipticity causes severe mesh dependence because of the underlying singularity of $\mathbf{Q}$. Thereby the post-peak behaviour depends crucially on the mesh design and the choice of the element formulation. If the fixed mesh approach is utilized, the finite element grid has to transmit the local singularity throughout the structural system without mesh adaption to capture the particular failure mode.

The discontinuity in the strainrate is defined by the tensor product of the unit vector $\mathbf{N}$ and the polarization direction $\mathbf{M}$

$$[[\dot{\epsilon}]] = [[\nabla^s \dot{\mathbf{u}}]] = \frac{1}{2}\gamma(\mathbf{M} \otimes \mathbf{N} + \mathbf{N} \otimes \mathbf{M}) \tag{8}$$

Let us consider localization under the assumption of uniform stress- and state variables within the element domain. When a singularity in the localization tensor is detected at the element centroid a discontinuity develops with the normal vector $\mathbf{N}$ which divides the element area A. The nodal degrees of freedom $\dot{\mathbf{d}}_{loc}$ which are equivalent to discontinuous bifurcation within the element domain are determined. However since discontinuous fields are usually not permitted within the element domain, the discontinuity is redistributed by the shape functions and the $\mathbf{B}$-operator respectively. This leads to the smeared form of localization which is energetically equivalent to the formation of the localized discontinuity.

$$[[\dot{\epsilon}]]_{dis} = \mathbf{B}\dot{\mathbf{d}}^{loc} \neq [\mathbf{N} \otimes \mathbf{M}]^s \tag{9}$$

Because of the minor symmetry of the material operator the jump condition can be restated in terms of the pointwise localization condition

$$\rho c^2 = \mathbf{M} \otimes \mathbf{N} : \mathbf{E}^{ep} : \mathbf{N} \otimes \mathbf{M} = [\mathbf{M} \otimes \mathbf{N}]^s : \mathbf{E}^{ep} : [\mathbf{N} \otimes \mathbf{M}]^s = [[\dot{\epsilon}]]^t : \mathbf{E}^{ep} : [[\dot{\epsilon}]] = 0 \tag{10}$$

In general the distributed approximation of the strainrate jump violates pointwise localization

$$[[\dot{\epsilon}]]^t_{dis} : \mathbf{E}^{ep} : [[\dot{\epsilon}]]_{dis} \neq 0 \tag{11}$$

However, $[[\dot{\epsilon}]]_{dis}$ satiesfies the localization condition in the integral sense if the element description captures discontinuous bifurcation. The weak form is obtained by integrating the localization condition over the element domain

$$\mathrm{d}^2 W_{loc} = \int_{V^e} \rho c^2 \mathrm{d}V = \int_{V^e} [[\dot{\epsilon}]]^t_{dis} \mathbf{E}^{ep} [[\dot{\epsilon}]]_{dis} \mathrm{d}V \doteq 0 \tag{12}$$

In other terms, the singularity $\det(\mathbf{Q}^{ep}) = 0$ of the acoustic tensor is reproduced at the element level, if the localization kinematics above is associated with zero energy. Analysis of the localization tensor determines the kinematics of discontinuous bifurcation and furnishes the nodal values $\dot{\mathbf{d}}_{loc}$. For each constitutive model, the localization kinematics $\dot{\mathbf{d}}_{loc}$ identifies those eigenmodes which are active and which participate in the formation of discontinuity surface during failure. In that case, all eigenvalues associated with those eigenmodes must vanish in order to reproduce discontinuous bifurcation at the element level without numerical dissipation.

An analysis of the weak localization test for several finite element formulations was performed by STEINMANN & WILLAM [3]. Here we merely adopt the basic results. The standard bilinear displacement formulation **Q4** is not able to capture localized failure. In other terms this element produces a positive energy dissipation during localization as long as the element edges are not exactly aligned with the discontinuity surface. On the other hand mixed strain-displacement formulations advocated recently e.g. by SIMO & RIFAI [2] proved superior to reflect the bifurcation at constitutive level within the element domain. To this end the compatible part of the strain field approximation is augmented by an enhanced strain field. In this study we refer to the **QE5** element proposed by SIMO & RIFAI [2] with 5 additional internal degrees of freedom to interpolate the enhanced strain field. The **QE5** element satiesfies the weak localization test for a whole range of constitutive models and is not restricted to certain alignments with respect to the discontinuity surface.

4 Nonhomogeneous Uniaxial Compression Problem

Most problems of practical interest are concerned with nonuniform rather than uniform deformation conditions. Therefore a plane strain compression problem with non homogeneous boundary conditions is considered and the connection of the element response to the weak localization test is highlighted. The rectangular solid concrete block of Fig.3 is discretized by 4-node quadrilateral elements and the constitutive behaviour is modelled by the **EXTENDED LEON MODEL**. The vertical displacement is prescribed at the upper surface while the horizontal displacements at the top and the bottom of the specimen are fixed. The material parameters are also specified in Fig.3.

The load-deflection curve for a discretization with (8 × 14) **Q4** elements is depicted in Fig.4-a. The deformed mesh magnified by a factor of 30 in Fig.4-b corresponding to the last increment of deformations shows a fairly diffuse failure mode. Clearly there is a one to one relation between the poor performance of the standard bilinear element expansion in the weak localization test and the distribution of the active failure zone in the BVP. The results of the localization analysis at every gauss point depicted in Fig.4-d and the variation of principal strains in Fig.4-c indicate that the current discretization smears and suppresses the progression of a failure band. The normals to the possible discontinuity surface at every gauss point are shown in Fig.4-d and serve as an indicator how accurate the spatial discretization resolves the discontinuity.

The results for (16 × 28) **Q4** elements are obtained for the upper right quadrant of the panel taking into account the double symmetrie of the problem. Clearly this excludes nonsymmetric failure modes. As alluded to above the mesh refinement results in a more localized deformation pattern. The post-peak slope of the load-deflection curve reflects the brittle behaviour of concrete under compression more accentuated than that obtained by the coarse mesh. The load-deflection curve, the deformed mesh, the principal strains and the visualization of the loacalization analysis are depicted in Fig.5-a to Fig.5-d. Obviously more mesh refinements are neccessary to resolve the final failure mode.

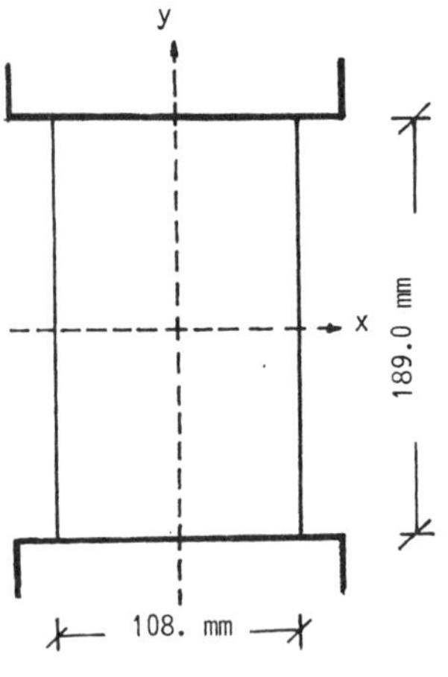

Elastic modulus $E_0 = 19,305 MPa$
Poisson's ratio $\nu_0 = 0.2$
Compressive strength $f_c' = 22.06 MPa$
Tensile strength $f_t' = 2.76 MPa$
Rupture displacement $u_r = 0.127 mm$
Crack spacing $h_t = 108 mm$

FIG.3: Dimensions of the Rectangular Panel and Material Parameters.

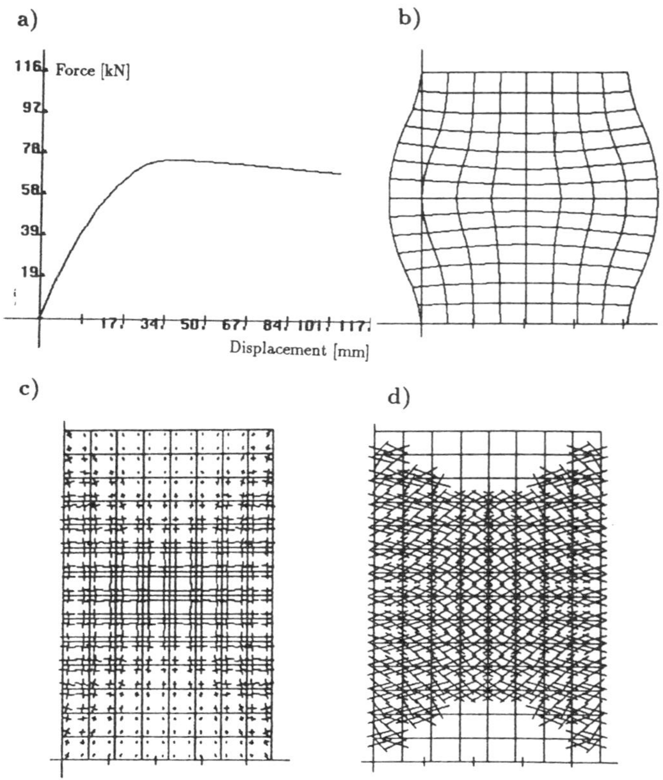

FIG.4: Mesh with (8×14) **Q4** Elements: force-displacement curve, deformed mesh, principal strains distribution and distribution of active localization planes.

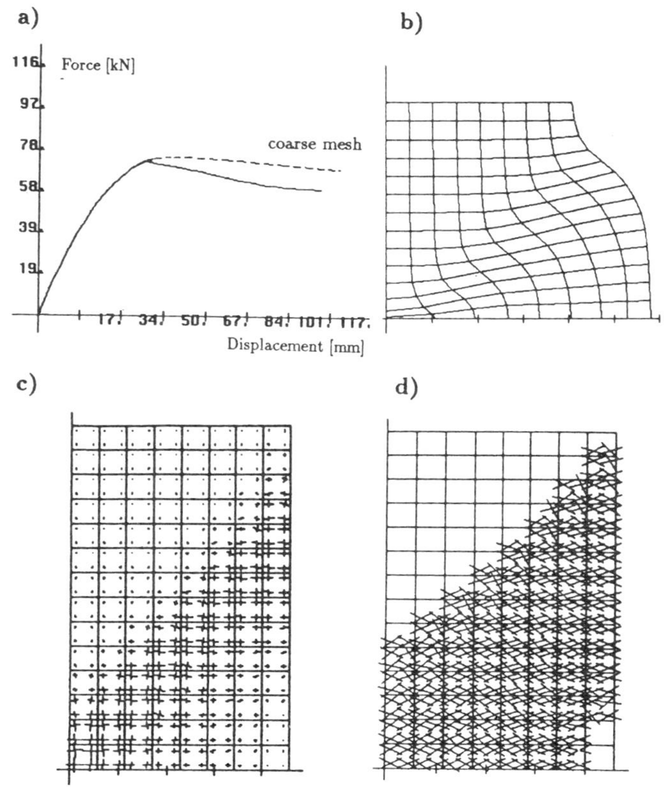

FIG.5: Mesh with (16×28) **Q4** Elements: force-displacement curve and deformed mesh, principal strains distribution and distribution of active localization planes for the upper right quadrand.

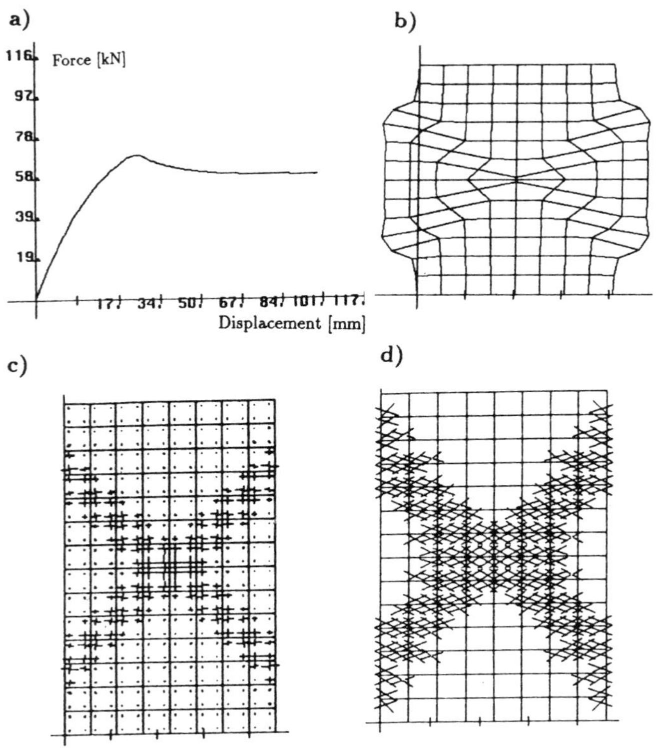

FIG.6: Mesh with (8×14) **QE5** Elements: force-displacement curve, deformed mesh, principal strains distribution and distribution of active localization planes.

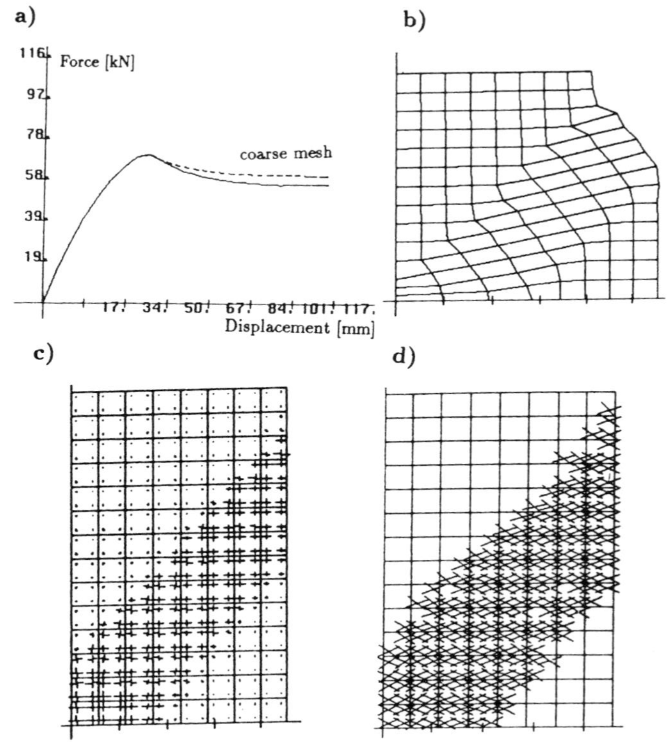

FIG.7: Mesh with (16×28) **QE5** Elements: force-displacement curve and deformed mesh, principal strains distribution and distribution of active localization planes for the upper right quadrant.

The weak localization test indicates that the enriched element strategies are superior in capturing localization phenomena. This assessment is confirmed by the results for the nonuniform compression problem obtained with the **QE5** element formulation. Even with the coarse (8×14) discretization the **QE5** renders pretty localized failure pattern, see Fig.6-a to Fig.6-d. The deformed mesh in Fig.6-b reveals that the whole deformation is confined to two symmetric narrow bands. It is of utmost importance to predict realistic dimensions of the localization zone. Severe mesh dependence of the solution, which is usually acompanied by the loss of ellipticity of the incremental BVP signalled by a singularity of the acoustic tensor, is remedied in terms of the 'equivalent length' inherent to the fracture energy-based softening concept.

The consequence of this internal lenght scale is revealed by the mesh refinement to (16×28) **QE5** elements. Again only a quarter of the panel is dicretized due to the double symmetry. As highlighted in Fig.7-a to Fig.7-d the width of the failure band is not reduced in comparison to the solution of the coarse mesh. The same convergence properties are observed for the overall load-deflection curve in Fig.7-a. The peak load and assoziated deflection concide with the result of the (8×14) mesh which is in contrast to the results predicted by the standard **Q4** element expansion.

5 Conclusion

Different 4-node element strategies were used to simulate the localized failure of concrete under nonuniform compression. A realistic pressure sensitive material model was considered. The model predictions encompass brittle failure modes in tension and compression with low confinement and ductil failure mode under high confinement. The post failure behavior is modeled in terms of the fracture energy-based softening concept.

From the comparison between the numerical results of a rectangular panel with a mesh discretized by 8×14 elements and with 16×28 elements it is concluded that the **Q4** element strategy leads to a stiff behavior with unconservative predictions of the limit load and the ductility for the coarse mesh. To reproduce the localized failure mode further mesh refinement would be neccessary. Contrary the **QE5** element formulation in conjunction with the fracture energy-based softening concept not only indicates mesh insensitivity for the limit load, ductility and post-peak regime, but also for the width of the failure band. This results coincide with the findings of the weak localization test which determines the element capability to reproduce discontinuos bifurcation.

References

[1] E. Pramono and K. Willam, [1989], "Fracture Energy-Based Plasticity Formulation of Plain Concrete", *ASCE J. Eng. Mech*, Vol. 115, No.6, pp. 1183-1203.

[2] J.C. SIMO & M.S. RIFAI, [1990], "A Class of Mixed Assumed Strain Methods and the Method of Incompatible Modes", submitted for publication in *Comp. Meth. Appl. Mech. Engr.*.

[3] P. STEINMANN & K. WILLAM, [1990], "Finite Elements Performance in Localized Failure Computations", submitted for publication in *Comp. Meth. Appl. Mech. Engr.*.

[4] K. WILLAM & E. WARNKE, [1974], "Constitutive Model for Triaxial Behaviour of Concrete", seminar on "Concrete Structures Subjected to Triaxial Stresses", ISMES Bergamo-Italy, IABSE Intern. Assoc. of Bridge and Struct. Eng. Proc.

[5] K. WILLAM, B. HURLBUT & S. STURE, [1986], "Experimental and Constitutive Aspects of Concrete Failure", *Proc. US-Japan Seminar on FE Analysis of RC Structures*, Tokyo, 1985, ASCE, Special Publication, New York, pp. 226-254.

[6] K. WILLAM & G. ETSE, [1990], "Assessment of the extended Leon model for plain concrete", *Proc. Int. Conf. Computer Aided Analysis and Design of Concrete Structures*, Zell am See, Austria, Vol II, pp. 851-870.

58 SOLUTION METHODS FOR LOCALIZATION IN FRACTURE DYNAMICS

L.J. SLUYS
Department of Civil Engineering, Delft University of Technology,
The Netherlands
R. de BORST
Department of Civil Engineering, Delft University of Technology,
and TNO Building and Construction Research, Rijswijk,
The Netherlands

Abstract
Rate dependence is shown to introduce an internal length scale into initial value problems. As a result thereof solutions of initial value problems that involve softening no longer exhibit excessive mesh dependence. A linear rate-dependent model and a model which has been proposed earlier by Needleman have been investigated. Numerical results have been obtained for one-dimensional localisation problems.
Keywords : Rate Dependent Models, Strain Softening, Dynamic Loading, Localisation, Finite Element Analysis.

1. Introduction

To carry out proper transient failure analyses of concrete structures it is important to examine cracking and strain localisation in cracked zones. In a cracked section the capacity to transmit tensile forces decreases. This phenomenon is commonly called softening and is attended by the formation of narrow bands of intense straining. Mapping of measured load-displacement data onto stress-strain relations leads to a negative slope in the stress-strain curve. A numerical description of strain localisation reveals a severe dependence on the fineness of the mesh.

Recent studies (Read and Hegemier 1984, Sandler 1984, Lasry and Belytschko 1988) have elucidated the mathematical background of the mesh-sensitivity problem. The field equations which describe the dynamic motion of the body lose hyperbolicity as soon as softening occurs. This means that the wave speeds become imaginary and no longer have physical significance. Solutions of initial value problems that involve softening (Bazant and Belytschko 1985) show localisation zones of zero thickness with infinite strains which develop in a time span that approaches zero. Numerical simulation of a softening solid exhibits the same features upon mesh refinement (Sluys 1989). Strain localisation occurs in a zone the width of which is entirely determined by the element size. Energy dissipation in the fracture zone tends to zero upon mesh refinement and the extent of wave reflection on cracks depends on the discretisation of a structure.

Three types of methods have been suggested in the literature to correct the above-mentioned deficiencies. The first class of methods consists of non-local models (Pijaudier-Cabot and Bazant 1987) and gradient models (Schreyer and Chen 1986, de Borst and Mühlhaus 1991, Mühlhaus and Aifantis 1991) in which higher-order displacement gradients are introduced in the material equations. A second approach is the micro-polar or Cosserat theory which allows mesh-objective analyses to be carried out under mode-II and mixed-mode conditions. This holds true for static (de Borst 1990, Mühlhaus, de Borst and Aifantis 1991) as well as dynamic loading conditions (Sluys and de Borst 1990, Sluys 1990). The solution strategy followed here

Fracture Processes in Concrete, Rock and Ceramics Edited by J.G.M. van Mier, J.G. Rots and A. Bakker.
 Published by E. & F.N. Spon, 2–6 Boundary Row, London SE1 8HN. ISBN 0 419 15870 7

is based on the inclusion of the strain rate in the constitutive equations (Needleman 1988, Wu and Freund 1984). Numerical analyses with rate-dependent models are included to assess the performance with respect to mesh dependence.

2. Rate-dependent crack models

Rate dependence is introduced to prevent the character of the set of equations that describes the dynamic motion of the softening solid from becoming elliptic. The dynamic equations should possess real characteristics in the *x-t* plane which is only possible when the initial value problem is not elliptic. Then well-posedness of the initial value problem is preserved and meaningful results can be obtained for localisation zones of intense straining. Because failure modes are usually attended by high strain rates the inclusion of the strain rate in the constitutive equations seems natural. It is essential for the solution of the mesh-sensitivity problem that rate dependence naturally introduces an internal length scale into the initial value problem, although the constitutive equations do not contain a parameter with the dimension of length. Two different models will be treated next in which the strain rate vector is incorporated in the material equations and for which it will be proved that the initial value problem preserves well-posedness. Both models have been implemented within the framework of the fixed smeared crack concept (de Borst and Nauta 1985, Rots 1988). In this concept a cracked zone is conceived to be a continuum which permits a description in terms of stress-strain relations.

2.1. Linear rate-dependent model

A crack model that exhibits both softening and strain rate dependence is chosen according to

$$\sigma = f(\varepsilon_{cr}) + M_1 \dot{\varepsilon}_{cr} \tag{1}$$

where σ and ε_{cr} denote stress and crack strain respectively, a superimposed dot denotes differentiation with respect to time, $f(\varepsilon_{cr})$ is a linear softening function and M_1 is the rate-sensitivity parameter. Note that rate dependence is chosen to be a function of the inelastic strain ε_{cr} and not of the total strain. When a linear softening model is utilised as in this study

$$f(\varepsilon_{cr}) = f_t + D_1 \varepsilon_{cr} \tag{2}$$

eq.(1) becomes

$$\sigma = f_t + D_1 \varepsilon_{cr} + M_1 \dot{\varepsilon}_{cr} \tag{3}$$

In eqs.(2) and (3) f_t is the initial tensile strength under static loading conditions and D_1 is a constant softening modulus. In the reference calculations the parameters have been chosen as : Young's modulus $E = 25000$ N/mm^2, $D_1 = -6000$ N/mm^2, $M_1 = 10$ Ns/mm^2 and $f_t = 3.0$ N/mm^2.

In Figure 1 computed stress-strain curves are shown for various imposed strain rates. The curves show an increase of the tensile strength under dynamic loading, which is also observed in several experiments (Körmeling 1986, Zielinski 1982). In these experiments the increase of tensile strength f_t is measured as a function of a constant strain rate. However, from these tests as well as from the numerical simulations of the subsequent section it appears that the strain rate not only varies in time but also has large gradients along the localisation zone. It is therefore difficult to derive the material rate sensitivity parameter M_1 from experimental data. For the increase of fracture energy per unit area and the ultimate strain there exists currently less experimental support, although Körmeling observed the same tendencies in softening behaviour under high strain rates. For unloading a secant model without rate-dependence is

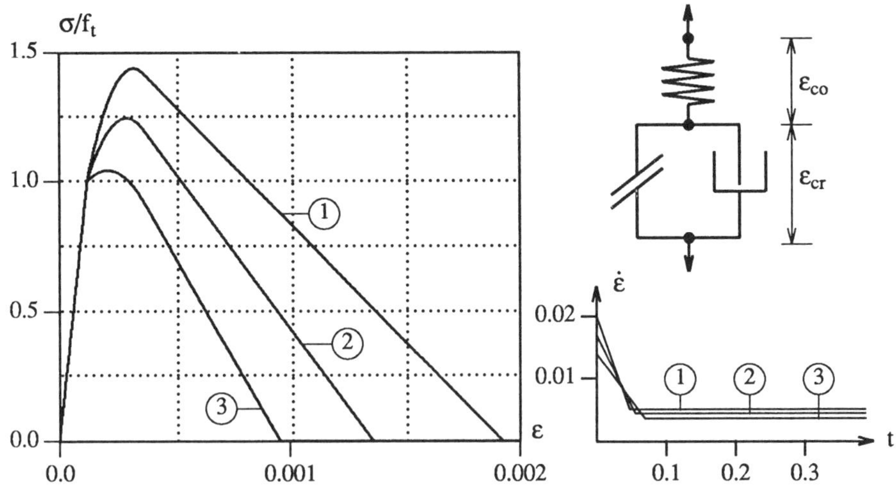

Figure 1. Linear rate-dependent crack model :
Left : Stress-strain curves for different imposed strain rates.
Right above : Schematic representation of the model.
Right bottom : Imposed strain rates.

used.

In a smeared crack approach we apply a decomposition of total strain into elastic strain $\boldsymbol{\varepsilon}_{co}$ and crack strain $\boldsymbol{\varepsilon}_{cr}$. Therefore the incremental stress-strain relation can be written as

$$\Delta\boldsymbol{\sigma} = \mathbf{D}_{co}\,(\Delta\boldsymbol{\varepsilon} - \Delta\boldsymbol{\varepsilon}_{cr}) \tag{4}$$

where the matrix $\mathbf{D}_{co}$ contains the instantaneous moduli of the concrete and $\Delta\boldsymbol{\varepsilon}_{cr}$ is the incremental crack strain. Generalising eq.(3) to two dimensions we obtain

$$\Delta\boldsymbol{\sigma} = \mathbf{D}_{cr}\Delta\boldsymbol{\varepsilon}_{cr} + \mathbf{M}\Delta\dot{\boldsymbol{\varepsilon}}_{cr} \tag{5}$$

in which $\mathbf{D}_{cr} = \mathrm{diag}[D_1, 0]$ and $\mathbf{M} = \mathrm{diag}[M_1, 0]$, so that only mode-I effects are considered. $\Delta\dot{\boldsymbol{\varepsilon}}_{cr}$ is the difference between the crack strain rates in the beginning and the end of the time step respectively. Combination of eqs.(4) and (5) yields,

$$\Delta\boldsymbol{\varepsilon}_{cr} = (\mathbf{D}_{cr} + \mathbf{D}_{co})^{-1}[\mathbf{D}_{co}\Delta\boldsymbol{\varepsilon} - \mathbf{M}\Delta\dot{\boldsymbol{\varepsilon}}_{cr}] \tag{6}$$

This equation can be substituted in eq.(4), which gives the global constitutive equation :

$$\Delta\boldsymbol{\sigma} = \mathbf{D}_{co}[\mathbf{I} - (\mathbf{D}_{cr} + \mathbf{D}_{co})^{-1}\mathbf{D}_{co}]\Delta\boldsymbol{\varepsilon} + \mathbf{D}_{co}[\mathbf{D}_{cr} + \mathbf{D}_{co}]^{-1}\mathbf{M}\Delta\dot{\boldsymbol{\varepsilon}}_{cr} \tag{7}$$

For the sake of simplicity only one crack has been considered in this derivation and the axes of the crack plane are taken aligned with the global element axes. For a general treatment of the linear strain rate dependent model is referred to Sluys (1991).

Next, we shall integrate eq.(7) over a finite time step Δt. In this derivation we define $\Delta\boldsymbol{\varepsilon}_{cr}^{t+\Delta t}$ as the incremental crack strain in time interval $t \le \tau \le t + \Delta t$. By utilising the crack strain rate at the beginning of the time step $\dot{\boldsymbol{\varepsilon}}_{cr}^{t}$ and that at the end of the time step $\dot{\boldsymbol{\varepsilon}}_{cr}^{t+\Delta t}$ the incremental crack strain is chosen according to

$$\Delta\boldsymbol{\varepsilon}_{cr}^{t+\Delta t} = ((1-\Theta)\dot{\boldsymbol{\varepsilon}}_{cr}^{t} + \Theta\dot{\boldsymbol{\varepsilon}}_{cr}^{t+\Delta t})\,\Delta t \tag{8}$$

where Θ is an interpolation parameter for which $0 \leq \Theta \leq 1$. For $\Theta = 0$ we obtain the fully explicit Euler scheme. On the other hand $\Theta = 1$ gives a scheme that is fully implicit and the case $\Theta = 1/2$ represents an implicit scheme according to the trapezoidal rule. From the definition of $\Delta\dot{\boldsymbol{\varepsilon}}_{cr}$ we have

$$\Delta\dot{\boldsymbol{\varepsilon}}_{cr}^{t+\Delta t} = \dot{\boldsymbol{\varepsilon}}_{cr}^{t+\Delta t} - \dot{\boldsymbol{\varepsilon}}_{cr}^{t} \tag{9}$$

which can be combined with eq.(8) to give

$$\Delta\dot{\boldsymbol{\varepsilon}}_{cr}^{t+\Delta t} = \frac{1}{\Theta}\left(\frac{\Delta\boldsymbol{\varepsilon}_{cr}^{t+\Delta t}}{\Delta t} - \dot{\boldsymbol{\varepsilon}}_{cr}^{t}\right) \tag{10}$$

We estimate $\dot{\boldsymbol{\varepsilon}}_{cr}^{t}$ by a central difference approximation :

$$\dot{\boldsymbol{\varepsilon}}_{cr}^{t} = \frac{1}{2}(\dot{\boldsymbol{\varepsilon}}_{cr}^{t-1/2\Delta t} + \dot{\boldsymbol{\varepsilon}}_{cr}^{t+1/2\Delta t}) = \frac{1}{2\Delta t}(\Delta\boldsymbol{\varepsilon}_{cr}^{t} + \Delta\boldsymbol{\varepsilon}_{cr}^{t+\Delta t}) \tag{11}$$

By substituting eq.(11) into eq.(10) we obtain an expression for the incremental strain rate vector,

$$\Delta\dot{\boldsymbol{\varepsilon}}_{cr}^{t+\Delta t} = \frac{1}{2\Theta\Delta t}(\Delta\boldsymbol{\varepsilon}_{cr}^{t+\Delta t} - \Delta\boldsymbol{\varepsilon}_{cr}^{t}) \tag{12}$$

Eq.(12) is now combined with eq.(7) to obtain the incremental constitutive relation

$$\Delta\boldsymbol{\sigma}^{t+\Delta t} = \mathbf{D}_{tan}\Delta\boldsymbol{\varepsilon}^{t+\Delta t} - \Delta\mathbf{q} \tag{13}$$

where

$$\mathbf{D}_{tan} = \mathbf{D}_{co}[\mathbf{I} - (\mathbf{D}_{cr} + \mathbf{D}_{co} + \frac{\mathbf{M}}{2\Theta\Delta t})^{-1}\mathbf{D}_{co}] \tag{14}$$

$$\Delta\mathbf{q} = \mathbf{D}_{co}[\mathbf{D}_{cr} + \mathbf{D}_{co} + \frac{\mathbf{M}}{2\Theta\Delta t}]^{-1}\frac{\mathbf{M}}{2\Theta\Delta t}\Delta\boldsymbol{\varepsilon}_{cr}^{t} \tag{15}$$

Matrix $\mathbf{D}_{tan}$ is not only determined by material parameters since the time integration parameters Δt and Θ also appear. $\Delta\mathbf{q}$ can be considered as a pseudo-load vector.

The material parameters can be grouped into a length scale parameter l via,

$$l = \frac{M_1\, c}{E} \tag{16}$$

where c is the longitudinal wave velocity $\sqrt{E/\rho}$ with ρ the density. If the length scale parameter vanishes due to a decreasing viscosity the rate-independent solid emerges as the limit of the rate-dependent model. In the preceding it has been said that the inclusion of rate dependence prevents the initial value problem from becoming elliptic. This can be proved by considering the quasi-linear system of equations,

$$\sigma = f(\varepsilon_{cr}) + M_1\frac{\partial v}{\partial x} - \frac{M_1}{E}\frac{\partial \sigma}{\partial t} \tag{17}$$

$$\frac{\partial \sigma}{\partial x} = \rho\frac{\partial v}{\partial t} \tag{18}$$

where the velocity v is the derivative of the displacement with respect to time. Combining eqs.(17) and (18) results in the second order differential equation,

$$\frac{\partial^2 v}{\partial x^2} - \frac{1}{c^2}\frac{\partial^2 v}{\partial t^2} - \frac{\rho}{M_1}\frac{\partial v}{\partial t} = -\frac{1}{M_1}\frac{\partial f(\varepsilon_{cr})}{\partial x} \tag{19}$$

The character of a partial differential equation is determined by the terms of the highest order. This means eq.(19) is a hyperbolic equation with a source term depending on the softening function. The characteristics of this system are real and have the physical meaning of a wave velocity. So, the problem remains well-posed and describes the localisation behaviour of the underlying physical material as will be shown in the next part.

2.2 Rate-dependent model of Needleman

For the rate-dependent model proposed by Needleman and Asaro (Needleman 1988, Asaro 1983) the strain rate is chosen according to a power law function,

$$\dot{\varepsilon}_{cr} = a\,(\frac{\sigma}{g})^{1/m} \tag{20}$$

where a is a material parameter of dimension 1/s, g is a hardness parameter which depends on the inelastic strain. The parameter m characterises the strain rate sensitivity of the material. The distinction between this model and the linear rate-dependent model is that the strain rate is a function of the stress and the current state and does not depend on incremental quantities. Needleman has used the power law formulation in a plasticity model for the investigation of shear band formation in structural metals. Here we use it within the framework of a principal stress criterion model for the description of fracture zones in brittle materials. For the parameter g a bilinear diagram is used with an evolution law according to

$$\dot{g} = \begin{cases} H_1\dot{\varepsilon}_{cr} & \text{for} \quad \varepsilon_{cr} \le \varepsilon_{crm} \\ H_2\dot{\varepsilon}_{cr} & \text{for} \quad \varepsilon_{cr} > \varepsilon_{crm} \end{cases} \tag{21}$$

with $g(0) = f_t$ and ε_{crm} the transition strain between hardening ($\varepsilon_{cr} < \varepsilon_{crm}$) and softening ($\varepsilon_{cr} > \varepsilon_{crm}$). In Figure 2 computed stress-strain curves are plotted for different strain rates. Additional to the parameters of the previous model we have chosen : $a = 0.08$ 1/s, $m = 0.02$, $\varepsilon_{crm} = \frac{1}{2}f_t/E$, $H_1 = 3000$ N/mm^2 and $H_2 = -500$ N/mm^2. Under strain reversal the material unloads elastically. It is difficult to give a physical explanation for the choice of parameters because, so far, application of the model has only been done for structural metals.

Again we use the incremental stress-strain equation (4) and the linear interpolation formula (8) for the crack strain rate. In this formulae the strain rate $\dot{\varepsilon}^t_{cr}$ at time t follows directly from the current state of stress σ^t and hardness g^t. The strain rate in time interval $t \le \tau \le t + \Delta t$ can be estimated via an Euler forward scheme,

$$\frac{{}^*\Delta\varepsilon_{cr}}{\Delta t} = a\,(\frac{\sigma^t}{g^t})^{1/m} \tag{22}$$

We can use this strain rate as a first approximation for calculation of the stress ${}^*\sigma^{t+\Delta t}$ and the hardness ${}^*g^{t+\Delta t}$. Now, the strain rate at time t+Δt can be determined, which completes the data set required for the incremental constitutive equation,

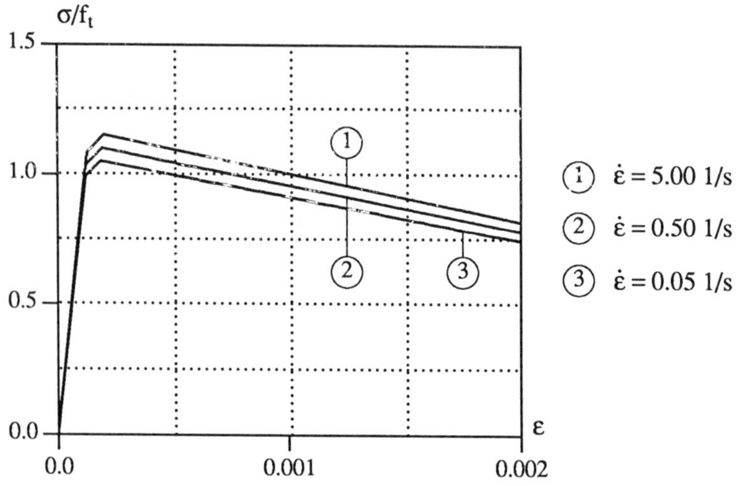

Figure 2. Rate-dependent model of Needleman for different strain rates.

$$\Delta\sigma = E\,\Delta\varepsilon - \Delta q \tag{23}$$

with

$$\Delta q = a\,\Delta t\,E\,[\,(1-\Theta)(\frac{\sigma^t}{g^t})^{1/m} + \Theta(\frac{{}^*\sigma^{t+\Delta t}}{{}^*g^{t+\Delta t}})^{1/m}\,] \tag{24}$$

When choosing $\Theta = 1/2$ this procedure corresponds to an implicit Heun integration scheme. Since the elastic stiffness matrix is used the total inelastic response is transferred via the pseudo-load Δq. The length scale parameter l is defined by

$$l = \frac{c}{a} \tag{25}$$

which can be considered as the distance that an elastic wave travels in a characteristic time $T = 1/a$. Also this model results in initial value problems that are well-posed and have real characteristics under strain softening conditions (Needleman 1988).

3. Wave propagation in a rate-dependent solid

To investigate the performance of the rate-dependent crack models numerical simulations have been carried out for a bar in uniaxial tension. The problem of mesh-sensitivity and the solution thereof can be demonstrated clearly by means of this one-dimensional problem. The attention is focused on the influence of mesh spacing on the strain localisation in the bar and the wave reflection on cracked zones. The problem is sketched in Figure 3 : the magnitude of the impact load is taken 75 % of the maximum tensile load in the static case, the loading rate is determined by time span t_d. The bar has a length of 5 m and is divided in 10, 20, 40 and 80 elements respectively. Use has been made of eight-noded elements with a nine-point Gauss integration scheme. The longitudinal wave velocity c = 1000 m/s, which is a material property in this non-dispersive continuum. If not specifically stated otherwise the material parameters are

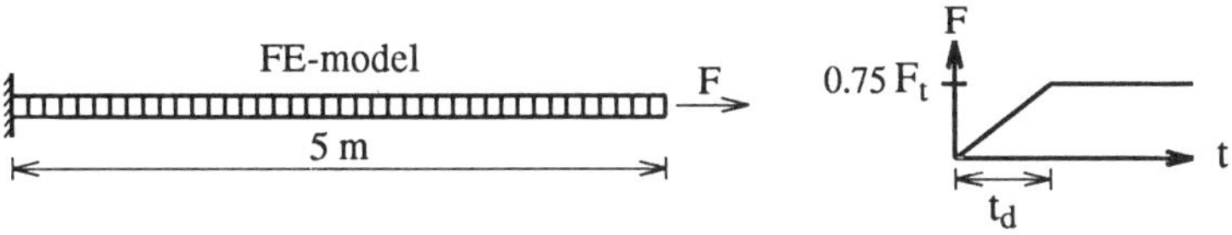

Figure 3. Finite element model (40 elements) and time-load diagram.

as given in Figure 1 for the linear rate-dependent model or as given in Figure 2 for the rate-dependent model of Needleman. The response of the bar is linearly elastic until the loading wave reaches the left boundary where reflection of the wave causes crack initiation. The material enters the softening regime and a localisation zone of intense straining occurs. The time integration of the field equations has been done with a Newmark scheme ($\beta = 1/4$, $\gamma = 1/2$).

3.1. Rate-independent crack model

The analyses with a rate-independent solid have been carried out to demonstrate the mesh-dependence of strain localisation and wave reflection. The comparison with the linear rate-dependent model is difficult because we cannot employ exactly the same softening diagram. A linear model is used with an ultimate strain equal to the peak strain which occurs in the localisation band of the rate-dependent analysis to be discussed next. This results in a softening modulus $D_1 = -1000\,N/mm^2$. We consider a shock wave with a vertical stress front which corresponds to $t_d = 0$. The results for the different meshes are plotted in Figure 4. Mesh sensitivity is obvious from the strain plots: strain localisation occurs in only one integration point which is the smallest possible zone. The results for the discretisations with 40 and 80 elements have not been plotted because at t = 0.009 s the bar has already failed. Instead, the results for a 5 element mesh are given. The stress plots show that the amount of wave reflection depends on the mesh: the more elements are present, the larger is the reduction in stress of the reflected wave. Furthermore, the energy dissipation decreases under mesh refinement and approaches the limiting case of failure without energy dissipation.

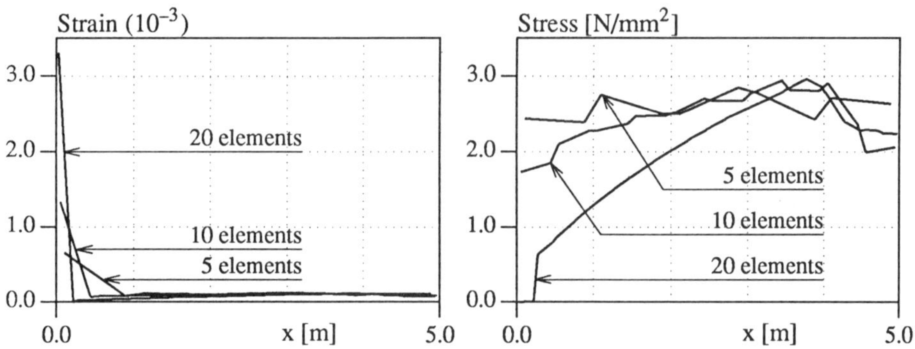

Figure 4. Strain localisation (left) and stress distribution (right) along the bar at t = 0.009 s for three different meshes in a rate-independent solid.

3.2. Linear rate-dependent crack model

Rate-dependence is applied according to the constitutive equations (13)-(15). It appears that the choice $\Theta = ½$ for the interpolation parameter gives stable and accurate results, but that the time step Δt must be taken smaller than h/5c, where h is the element size. This is a strong requirement that is caused by the sensitivity of the edge between the localisation zone and the elastically unloading remaining part of the bar. If the time step is taken larger than h/5c the strain profile is disturbed in these edge zones. Use of the reference material parameter set of par. 2.1 results in a length scale parameter $l = 400$ mm. The same loading pulse with $t_d = 0$ is applied. For the rate-dependent solid a localisation zone emerges that converges to a finite, constant band width upon mesh refinement. In Figure 5 the strain localisation is plotted for different meshes to demonstrate the uniqueness of the solution. The coarse meshes (10 and 20 elements) still deviate somewhat but the fine meshes (40 and 80 elements) give identical results, not only in the sense that the band width is constant but also that the energy dissipation is constant and the wave reflection pattern is mesh objective.

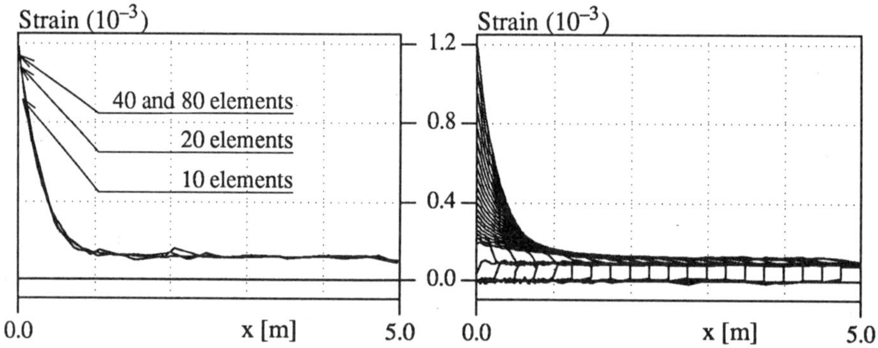

Figure 5. Linear rate-dependent model with $t_d = 0.$ s :
Left : Strain profiles for different discretisations at t = 0.01 s.
Right : Development of the localisation band (80 elements).

A second analysis has been carried out for a different loading pulse. In the time-load diagram the time span t_d is chosen as $2.5 \cdot 10^{-3}$ s. So, firstly the loading pulse increases linearly in time before it becomes constant. Again the effect of the inclusion of the length scale can be observed from the strain localisation plots in Figure 6. Convergence to a unique solution with a finite localisation zone characterises the mesh independence. It is noted that the strain distribution in the localisation zone has a different shape for this loading case. In the previous analysis a sharp peak in the strain occurs at the left boundary, whereas in this analysis the strain profile is more uniformly divided and has a lower peak value. This is due to the strain rate profiles in the bar at the moment of cracking. In the previous analysis cracking is initiated in one point at the left boundary. From there the localisation zone extends in the direction of the reflected wave with a rapidly decreasing strain rate. In this analysis the static tensile strength is exceeded over a zone with a certain length (exact length = 833 mm). In this zone much lower and almost constant strain rates occur compared to the peak strain rate in the previous analysis.

Finally, the localisation band width has been analysed. Firstly the influence of the length scale parameter on the observed localisation width was investigated in an analysis with $t_d = 0$ by using three different values for l, namely 200, 400 and 600 mm. From the left plot of Figure

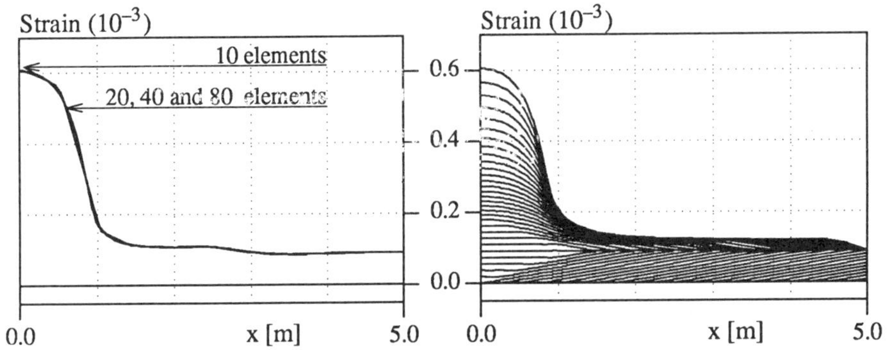

Figure 6. Linear rate-dependent model with $t_d = 2.5 \cdot 10^{-3}$ s :
Left : Strain profiles for different discretisations at t = 0.0125 s.
Right : Development of the localisation band (80 elements).

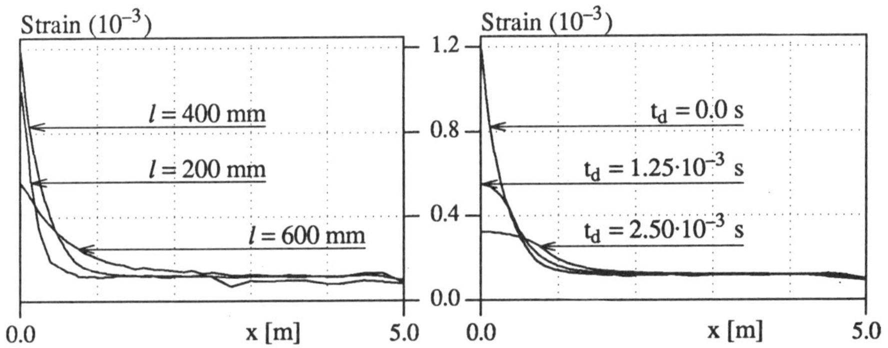

Figure 7. Variation of the length scale parameter (left $l = 200$ mm : t = 0.0075 s and $l = 400$ mm and 600 mm : t = 0.010 s) and variation of the loading rate (right at t = 0.010 s) for a 40 element analysis.

7 it appears that the width of the localised zone is proportional to the length scale parameter. These results agree with the fact that the localisation zone vanishes when the length scale parameter approaches zero. From the earlier analyses it already became clear that the shape of the loading wave influences the strain rate distribution in the localisation zone and therefore also the localisation band width. This effect is shown in the right plot of Figure 7 by taking three different values for the time span t_d in which the load is increased from zero to its maximum value.

3.3. Rate-dependent model of Needleman

We use the bar problem with a time span t_d equal to $2.5 \cdot 10^{-3}$ s for a mesh-sensitivity analysis of the Needleman model. Mesh independence has been obtained only for an analysis with a rather small negative slope ($H_2 = -500$ N/mm^2, see par.2.2). In Figure 8 it is shown that the localisation zone is more or less constant, independent of the discretisation of the bar. It

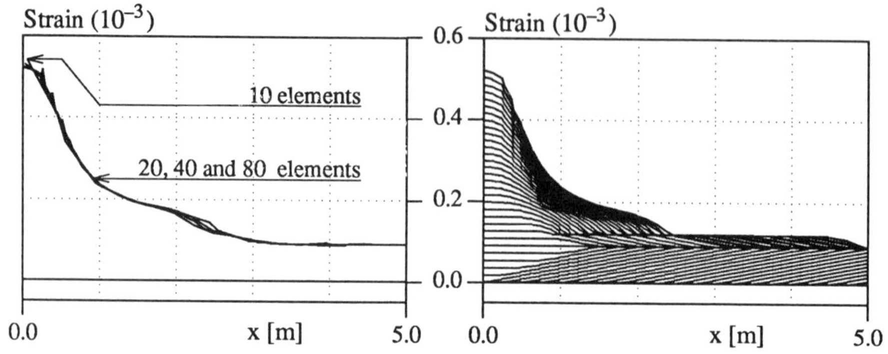

Figure 8. Rate-dependent model of Needleman with $t_d = 2.5 \cdot 10^{-3}$ s :
Left : Strain profiles for different discretisations at t = 0.0125 s.
Right : Development of the localisation band (40 elements).

appeared, however, that at at higher strain level localisation again becomes mesh-dependent. This type of behaviour becomes worse when a strain softening model with a steeper branch is used. In brittle materials we deal with softening models that do show a rapid stress drop after cracking which suggests that the Needleman model of viscous regularisation is less appropriate for modelling brittle materials.

From the right plot of Figure 8 we observe that the localisation zone increases in thickness in the course of time. In a brittle material as concrete the band width is almost constant which apparently is not predicted by this rate-dependent model. A third point of caution is the choice of the parameters of the power law model for à material as concrete.

4. Concluding remarks

It has been demonstrated that numerical solutions for localisation problems are mesh-sensitive, show a spurious strain localisation behaviour and a disturbed wave reflection pattern when use is made of a rate-independent strain-softening model. Therefore a model has been proposed that includes a dependence on the strain rate. This prevents the initial value problem from becoming elliptic. A linear rate dependence is used in which the strain rate vector is integrated with an implicit algorithm. The model incorporates a length scale parameter which is necessary to overcome the mesh-sensitivity problem. Numerical analyses have proved the mesh objectivity of the model with respect to the width of the localisation zone, the dissipation of energy and the wave reflection on the cracked zone. It appears that the width of the localisation zone is proportional to the length scale parameter. A second rate-dependent model proposed by Needleman also gives mesh-independent results for ductile softening models. These results, however, cannot be obtained for more brittle materials as concrete.

5. Acknowledgements

The calculations described in this paper have been carried out using the DIANA finite element package of TNO Building and Construction Research. This project is partially funded by the Prins Maurits Laboratory of TNO.

6. References

Asaro, R.J. (1983). Mechanics of crystals and polycristals. **Adv. Appl. Mech.** , 23, 2-115.

Bazant, Z.P. and Belytschko, T.B. (1985). Wave propagation in a strain softening bar - exact solution. **ASCE J. Engrg. Mech.** , 111(3), 381-389.

de Borst, R. and Nauta, P. (1985). Non-orthogonal cracks in a smeared finite element model. **Eng. Comput.** , 2(1), 35-46.

de Borst, R. (1990). Simulation of localisation using Cosserat theory, in **Proc. 2nd Int. Conf. Computer Aided Analysis and Design of Concrete Structures,** (eds N. Bićanić and H.A. Mang), Pineridge Press, Swansea, 931-944.

de Borst, R. and Mühlhaus, H.-B. (1991). Continuum models for discontinuous media. This volume.

Körmeling, H.A. (1986). **Strain rate and temperature behaviour of steel fibre concrete in tension** Dissertation, Delft University of Technology, Delft, Netherlands.

Lasry, D. and Belytschko, T. (1988). Localization limiters in transient problems. **Int. J. Solids Structures**, 24, 581-597.

Mühlhaus, H.-B., de Borst, R. and Aifantis E.C. (1991). Constitutive models and numerical analyses for inelastic materials with microstructure, in **Proc. Seventh Conf. Int. Assoc. Comp. Methods and Advances in Geomechanics,** (eds G. Beer and J.R. Booker), Balkema, Rotterdam and Boston.

Mühlhaus, H.-B. and Aifantis, E.C. (1991). A variational principle for gradient plasticity. **Int. J. Solids Structures**, in press.

Needleman, A. (1988). Material rate dependence and mesh sensitivity on localisation problems. **Comp. Meth. Appl. Mech. Eng.** , 67, 69-86.

Pijaudier-Cabot, G. and Bazant, Z.P. (1987). Nonlocal damage theory. **ASCE J. Eng. Mech.** , 113, 1512-1533.

Read, H.E. and Hegemier, G.A. (1984). Strain softening of rock, soil and concrete - A review article. **Mech. of Mat.** , 3, 271-294.

Rots, J.G. (1988). **Computational modeling of concrete fracture.** Dissertation, Delft University of Technology, Delft, Netherlands.

Sandler, I.S. (1984). Strain softening for static and dynamic problems. **Constitutive Equations - Macro and Computational Aspects - ASME**, 217-231.

Schreyer, H.L. and Chen, Z. (1986). One-dimensional softening with localization. **J. Appl. Mech.** 53, 791-979.

Sluys, L.J. (1989). Strain softening under static and dynamic loading conditions. TU-Delft report, nr.25.2-89-5-09.

Sluys, L.J. and de Borst, R. (1990). A numerical study of concrete fracture under impact loading, in **Proc. Conf. Micromechanics of Failure of Quasi-Brittle Materials,** (eds S.P. Shah, S.E. Swartz and M.L. Wang), Elsevier Applied Science, London and New York, 524-535.

Sluys, L.J. (1990). Localisation in a Cosserat continuum under dynamic loading conditions. TU-Delft report, nr.25.2-90-5-15.

Sluys, L.J. (1991). Rate-dependent modelling for dynamic localisation. TU-Delft report, nr.25.2-91-5-03.

Wu, F.H. and Freund, L.B. (1984). Deformation trapping due to thermoplastic instability in one-dimensional wave propagation. **J. of Mech. Phys. Solids**, Vol.32 no.2, 119-132.

Zielinski, A.J. (1982). **Fracture of concrete and mortar under uniaxial impact tensile loading.** Dissertation, Delft University of Technology, Delft, Netherlands.

59 HIGH ORDER CRITERION FOR FAILURE PREDICTION OF THIN BERYLLIUM SHEETS

P.P. PAPADOS and P.N. ROSCHKE
Texas A&M University, College Station, TX, USA

Abstract
A macroscopic failure criterion that describes behavior of anisotropic materials is described. The new failure criterion is derived in terms of a six-dimensional stress space. Invariant requirements for coordinate transformation, material symmetries, multiaxial stress cases, and differences in strengths are taken into account. In addition to reviewing existing failure criteria and outlining development of a higher order failure theory, application is made to thin cross-rolled beryllium sheets that are used in spacecraft structures.
Keywords: Beryllium, Closure, Failure, Orthotropic, Strength Tensor, Stress

1 Introduction

Certain materials, such as cross-rolled beryllium sheets, are known to fail in a sudden, brittle manner, when deformed by out-of-plane loadings. The majority of research dedicated to predicting failure of such materials uses a fracture mechanics, micro-oriented, point of view. A relative paucity of continuum mechanics, stress analysis studies for materials in thin structural shapes have been reported. Primary objective in the current effort is to approach failure prediction from a macroscopic point of view.

Failure criteria available today frequently incorporate two-dimensional methodology for prediction of failure stresses and/or strains. Primary motivation for development of a new multi-dimensional failure criterion is the lack of a mathematical model that accounts for behavior of a brittle material with anisotropic properties. Here the term brittle refers to material failure which is preceded with either negligible or very small inelastic deformations. Moreover, development of at least one separation surface within the body is required. Such a criterion should be able to predict a reliable maximum load estimate. By supplementing and modifying failure criteria suggested by previous investigators, a multi-dimensional failure surface is proposed that describes failure of thin structures that includes though-thickness stress parameters.

In general, criteria for failure prediction of a brittle anisotropic material are required to satisfy the following (Priddy 1974):

(a) Stability conditions and a smooth, continuous, convex, non-singular surface are required to satisfy uniqueness.
(b) The criterion should be invariant with respect to coordinate axis transformations.
(c) Because of assumed symmetry of strength tensors it should be possible to relate components of the tensors with other well established properties of anisotropic material, such as the elastic compliances.
(d) Strength interaction coefficients should be used that depend on mechanical properties of the material at different ultimate strengths.
(e) Applicability of the criterion for multiaxial and complex states of stress is necessary.
(f) Only a finite number of tests can be required to evaluate strength coefficients.

Fracture Processes in Concrete, Rock and Ceramics Edited by J.G.M. van Mier, J.G. Rots and A. Bakker.
 Published by E. & F.N. Spon, 2–6 Boundary Row, London SE1 8HN. ISBN 0 419 15870 7

(g) Each complex state of stress should be described by a combination of strength parameters, not only by one component of the strength tensor.

2 Review of existing failure criteria

In order to formulate an accurate criterion for failure prediction of brittle anisotropic materials, variation of tensile and compressive strengths needs to be taken into account. Von Mises' criterion is generalized by Hill (1950) in one of the first attempts to account for these variations. For orthotropic polycrystalline metals Hill proposes the following equation:

$$A\,(\sigma_x - \sigma_y)^2 + B\,(\sigma_x - \sigma_z)^2 + C\,(\sigma_y - \sigma_z)^2 + 2D\,\sigma_{xy}^2 + 2E\,\sigma_{yz}^2 + 2F\,\sigma_{xz}^2 = 1 \qquad (1)$$

where A, B, C, D, E, and F are material constants. Since it is assumed that hydrostatic pressure does not influence failure, Hill uses only the difference of the normal components of stress. One of the disadvantages of this criterion is that it implies the same tensile and compressive strengths for all possible specimen orientations and, therefore, does not take into consideration the Bauschinger effect. Moreover, quadratic terms in which any one shear stress occurs linearly are not included.

A second approach is proposed by Hoffman (1967), who includes terms that are odd functions of the material strength, so that Hill's equation becomes:

$$C_1\,(\sigma_x - \sigma_z)^2 + C_2\,(\sigma_z - \sigma_x)^2 + C_3\,(\sigma_x - \sigma_y)^2 + C_4\,\sigma_x + C_5\,\sigma_y + C_6\,\sigma_z$$
$$+ C_7\,\sigma_{yz}^2 + C_8\,\sigma_{zx}^2 + C_9\,\sigma_{xy}^2 = 1 \qquad (2)$$

where C_1 through C_9 are independent coefficients that are to be determined from nine independent, basic experiments. Incorporation of linear terms in Hoffman's criterion provides a first formulation for failure prediction of brittle anisotropic material that takes into account different tensile and compressive strengths. Moreover, first order tension and compression terms partly account for the Bauschinger phenomenon that is completely ignored by the Hill and Von Mises criteria. Hoffman's criterion contains symmetry and is consistent with other well-established isotropic and anisotropic failure conditions. It provides a smooth and adoptable formula of interpolating between basic strength data.

A more general approach is presented by Tsai and Wu (1971), who propose the following tensor formulation:

$$f(\sigma_k) = F_i\sigma_i + F_{ij}\sigma_i\sigma_j = 1 \qquad (3)$$

where, $\sigma_1 = \sigma_{11}$, $\sigma_2 = \sigma_{22}$, $\sigma_3 = \sigma_{33}$, $\sigma_4 = \sigma_{13}$, $\sigma_5 = \sigma_{23}$, $\sigma_6 = \sigma_{12}$, and $i, j, k = 1, ..., 6$; F_i and F_{ij} are contracted equivalents of the second and fourth order strength tensors, respectively. The main advantage of the Tsai-Wu criterion compared to earlier failure theories is that it can account for multi-dimensional stress space as well as different material symmetries. Only first and second order contracted strength tensors are incorporated in order to achieve mathematical simplicity and to maintain a determinate number of linear equations that yield strength coefficients. After taking symmetry conditions of the strength tensors into consideration, twenty-seven coefficients describe the behavior of anisotropic materials. This approach avoids incorporation of even higher order tensors due to mathematical complexity in evaluating strength interaction coefficients and indeterminacy of the linear system of equations that arises from such an inclusion.

Although it is widely used, the Tsai-Wu envelope, which yields an ellipse for an isotropic material, does not yield accurate correlations with experimental data for tension-tension and compression-compression quadrants of the failure surface. Furthermore, a shortcoming in this criterion is the fact that tension-tension and compression-compression interactions may not be treated independently.

In an attempt to obtain more generality, Priddy (1974) includes products of stress components of order greater than two in a failure criterion. He proposes the following general expression:

$$f(\sigma_k) = F_i\,\sigma_i + F_{ij}\sigma_i\sigma_j + F_{ijk}\sigma_i\sigma_j\sigma_k = 1 \tag{4}$$

where F_{ijk} is the contracted equivalent of its sixth order tensor. Due to complexity of the contracted sixth order tensor, Priddy uses only a limited number of mathematically independent cubic terms. For the special case of orthotropic material having material principal coordinate axes coinciding with orthotropic axes, Eqn. (4) becomes:

$$\begin{aligned} &a_1\,\sigma_6^2 + a_2\,\sigma_5^2 + a_3\,\sigma_4^2 + b_1\,\sigma_1^2 + b_2\,\sigma_2^2 + b_3\,\sigma_3^2 + c_1\sigma_1\sigma_2 \\ &+ c_2\sigma_2\sigma_3 + c_3\sigma_1\sigma_3 = 1 + d_1\sigma_1 + d_2\sigma_2 + d_3\sigma_3 + (e_1\sigma_1 + e_2\sigma_2 + e_3\sigma_3) \\ &(\sigma_1\sigma_2 + \sigma_2\sigma_3 + \sigma_1\sigma_3 - \sigma_6^2 - \sigma_5^2 - \sigma_4^2) + f(\sigma_1\sigma_2\sigma_3 + 2\sigma_4\,\sigma_5\,\sigma_6 - \sigma_1\sigma_5^2 \\ &-\sigma_2\sigma_4^2 - \sigma_3\sigma_6^2) \end{aligned} \tag{5}$$

in which $a_i = (F_{ij})^{-2}$, $b_i = (F_{ti}\,F_{ci})^{-1}$, and $d_i = -(F_{ci} - F_{ti})\,(F_{ti}\,F_{ci})^{-1}$ are strength coefficients that are to be obtained from uniaxial strength tests, and F_{ti} and F_{ci} are tensile and compressive strengths, respectively, of the material. In order to reduce the number of coefficients required to describe the failure surface, approximations for biaxial and triaxial strengths are used. For example, the triaxial tensile strength is considered to be linearly related to the biaxial strength. These coefficients are represented in Eqn. (5) by c_i, e_i, and f. Furthermore, a stability condition is introduced such that the failure surface is forced to be open for the case of hydrostatic pressure.

Other higher order criteria include those of Tennyson and Elliott (1983), and Tennyson and Jiang (1989). The former contribution is similar to that of Priddy in that independent biaxial tests are incorporated for calculation of interaction coefficients. Although the latter model only considers specially orthotropic material under in-plane loading, it serves as a fundamental reference for the proposed criterion.

In summary, of currently available theories for failure prediction of brittle materials, Priddy's goes to the greatest lengths to include attributes that describe brittle behavior in multi-dimensional space.

3 New proposed failure criterion

Criteria reviewed in the previous discussion are approximations of the sixth order tensor and, consequently, overall failure predictions are not expected to be extremely accurate for all possible states of stress. Moreover, they do not guarantee closure of the cubic surface in multi-dimensional space. Thus, situations where infinite strengths are predicted can occur.

In what follows a new criterion is proposed to overcome these limitations. A cubic form of the tensor polynomial surface is forced to satisfy a number of constraints that are associated with the image of this failure surface function that is projected onto the three orthogonal Cartesian planes $(\sigma_1,\ \sigma_2)$, $(\sigma_2,\ \sigma_3)$, and $(\sigma_1,\ \sigma_3)$. For simplicity, only orthotropic materials are considered.

3.1 General state of stress

A basic assumption is that a change in sign of the shear stress does not affect failure strength (Tsai and Wu 1971). Redundancy of the F_{iii} terms, as indicated by Wu and Scheublein (1974), is also considered. Expansion of terms in Eqn. (4) and application of assumptions concerning material behavior lead to the following scalar equation:

$$\begin{aligned} &F_1\,\sigma_1 + F_2\,\sigma_2 + F_3\,\sigma_3 + F_{11}\,\sigma_1^2 + F_{22}\,\sigma_2^2 + F_{33}\,\sigma_3^2 + F_{44}\,\sigma_4^2 + F_{55}\,\sigma_5^2 \\ &+ F_{66}\,\sigma_6^2 + 2F_{12}\,\sigma_1\,\sigma_2 + 2F_{13}\,\sigma_1\,\sigma_3 + 2F_{23}\,\sigma_2\,\sigma_3 + 3F_{112}\,\sigma_1^2\,\sigma_2 \end{aligned}$$

$$+ 3F_{113}\sigma_1^2\sigma_3 + 3F_{122}\sigma_1\sigma_2^2 + 3F_{223}\sigma_2^2\sigma_3 + 3F_{133}\sigma_1\sigma_3^2 + 3F_{233}\sigma_2\sigma_3^2$$
$$+ 3F_{144}\sigma_1\sigma_4^2 + 3F_{244}\sigma_2\sigma_4^2 + 3F_{344}\sigma_3\sigma_4^2 + 3F_{155}\sigma_1\sigma_5^2 + 3F_{255}\sigma_2\sigma_5^2$$
$$+ 3F_{355}\sigma_3\sigma_5^2 + 3F_{166}\sigma_1\sigma_6^2 + 3F_{266}\sigma_2\sigma_6^2 + 3F_{366}\sigma_3\sigma_6^2 = 1 \quad (6)$$

The principal tensor components, F_i and F_{ii} (no summation on i), are identical to those derived by Tsai-Wu (1971) (for $i = 1, 2, 3$). The remaining coefficients are to be evaluated so that closure of the cubic failure surface is accomplished. Rearrangement of Eqn. (6) leads to the following form:

$$(3F_{144}\sigma_1 + 3F_{244}\sigma_2 + 3F_{344}\sigma_3 + F_{44})\sigma_4^2 + (3F_{155}\sigma_1 + 3F_{255}\sigma_2$$
$$+ 3F_{355}\sigma_3 + F_{55})\sigma_5^2 + (3F_{166}\sigma_1 + 3F_{266}\sigma_2 + 3F_{366}\sigma_3 + F_{66})\sigma_6^2$$
$$= -(F_1\sigma_1 + F_2\sigma_2 + F_3\sigma_3 + F_{11}\sigma_1^2 + F_{22}\sigma_2^2 + F_{33}\sigma_3^2 + 2F_{12}\sigma_1\sigma_2$$
$$+ 2F_{13}\sigma_1\sigma_3 + 2F_{23}\sigma_2\sigma_3 + 3F_{112}\sigma_1^2\sigma_2 + 3F_{113}\sigma_1^2\sigma_3 + 3F_{122}\sigma_1\sigma_2^2$$
$$+ 3F_{223}\sigma_2^2\sigma_3 + 3F_{133}\sigma_1\sigma_3^2 + 3F_{233}\sigma_2\sigma_3^2 - 1) \quad (7)$$

Closure is ensured if the following two conditions are met by the failure surface (Tennyson and Jiang 1989):

(a) Images of the cubic curve projected onto the (σ_1, σ_2), (σ_1, σ_3), and (σ_2, σ_3) planes are closed.
(b) Real values of σ_4, σ_5, and σ_6 exist for given values of σ_1, σ_2, and σ_3.

3.2 Investigation of necessary conditions
To examine failure surface images on the three orthogonal Cartesian axes and to satisfy the set of requirements for condition (a), crossing of the failure surface on the three projection planes must be examined when $\sigma_4 = \sigma_5 = \sigma_6 = 0$. For this special case Eqn. (6) becomes:

$$F_1\sigma_1 + F_2\sigma_2 + F_3\sigma_3 + F_{11}\sigma_1^2 + F_{22}\sigma_2^2 + F_{33}\sigma_3^2 + 2F_{12}\sigma_1\sigma_2 + 2F_{13}\sigma_1\sigma_3$$
$$+ 2F_{23}\sigma_2\sigma_3 + 3F_{112}\sigma_1^2\sigma_2 + 3F_{113}\sigma_1^2\sigma_3 + 3F_{122}\sigma_1\sigma_2^2 + 3F_{223}\sigma_2^2\sigma_3$$
$$+ 3F_{133}\sigma_1\sigma_3^2 + 3F_{233}\sigma_2\sigma_3^2 - 1 = 0 \quad (8)$$

In addition, criteria need to be specified that are satisfied by interaction terms $F_{12}, F_{13}, F_{23}, F_{112}, F_{113}, F_{122}, F_{223}, F_{133}$, and F_{233}. This can be accomplished by investigating the constraining planes of Eqn. (8). Asymptotes are obtained by collecting quadratic terms of σ_1, σ_2, and σ_3, and setting the coefficients of each term to zero.

3.2.1 Evaluation of F_{ij} and F_{iij} for $i \neq j$
In the event that there are data from n1, n2, and n3 sets of biaxial load tests that correspond to $(\sigma_{1i}, \sigma_{2i})$ $(i = 1, 2, ..., n1)$, $(\sigma_{1i}, \sigma_{3i})$ $(i = 1, 2, ..., n2)$, and $(\sigma_{2i}, \sigma_{3i})$ $(i = 1, 2, ..., n3)$, respectively, it is possible to evaluate the interaction coefficients by a least-square fit of the cubic Eqn. (8). However, this approach may not be sufficient to produce closed curves in the projected (σ_1, σ_2), (σ_1, σ_3), and (σ_2, σ_3) subspaces. In the event that closure is not accomplished, one or more of the constraints that intersect the failure surface are shifted in space (i.e. their coefficients are modified) in such a manner that all constraints are satisfied and the surface is closed.

As an illustration, suppose that the asymptotic plane,

$$F_{22} + 3F_{122}\sigma_1 + 3F_{223}\sigma_3 = 0 \quad (9)$$

which is obtained by rewriting Eqn. (8) as a quadratic of σ_2, is parallel with the σ_2 axis and intersects the open-ended failure surface as shown in Fig. 1a. Adjustment in the F_{122} and F_{223} terms orients this plane with respect to the σ_2 axis. Closure is accomplished by requiring that the plane of

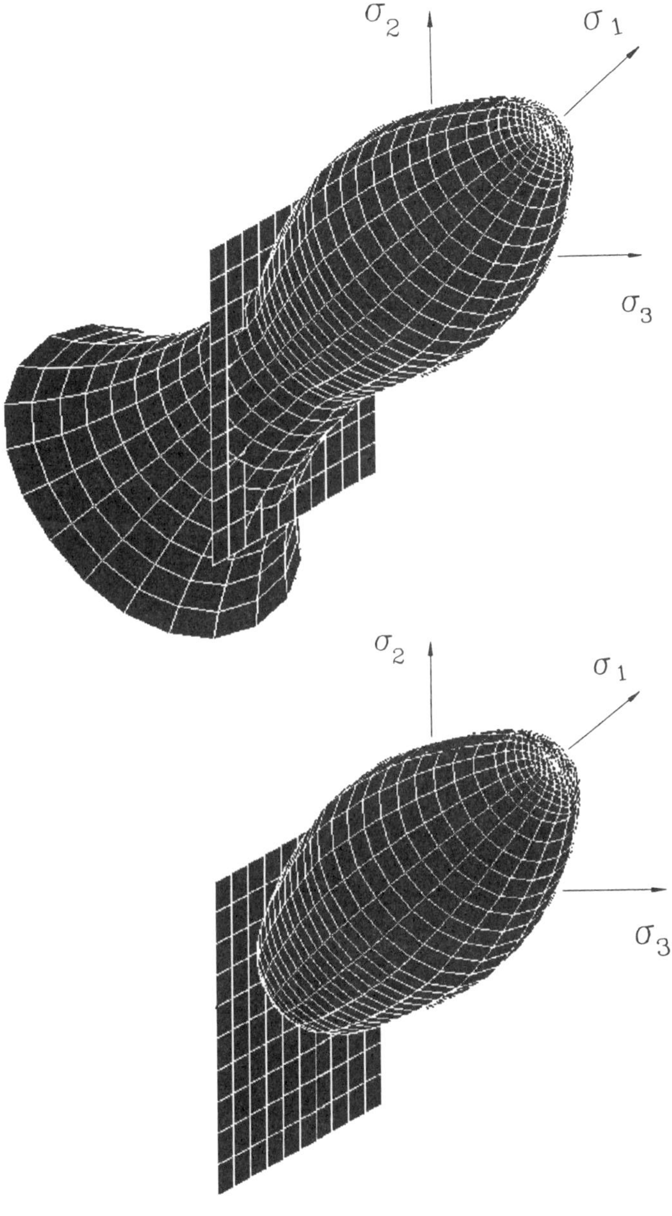

Fig. 1. Failure surface: (a) original with constraining plane; (b) closed with asymptotic plane

Eqn. (9) pass through the line (-X', σ_2, -Z'), where -X' and -Z' are uniaxial compressive strengths along the σ_1 and σ_3 axes, respectively. Thus, Eqn. (9) becomes an asymptotic plane for the failure surface (Fig. 1b). $\sigma_1 = -X'$ and $\sigma_3 = -Z'$ may be substituted into Eqn. (8) to obtain the following relation:

$$(-3F_{122}X' - 3F_{223}Z' + F_{22})\,\sigma_2^2 + (3F_{112}X'^2 + 3F_{233}Z'^2 - 2F_{12}X' - 2F_{23}Z' + F_2)\,\sigma_2$$
$$+ (-3F_{133}X'Z'^2 - 3F_{113}X'^2Z' + 2F_{13}X'Z' + F_{11}X'^2 + F_{33}Z'^2 - F_1X' - F_3Z' - 1) = 0 \quad (10)$$

Subsequently, for an infinity of σ_2 roots to exist it follows that:

$$-3F_{122}X' - 3F_{223}Z' + F_{22} = 0 \quad (11)$$

$$3F_{112}X'^2 + 3F_{233}Z'^2 - 2F_{12}X' - 2F_{23}Z' + F_2 = 0 \quad (12)$$

$$-3F_{133}X'Z'^2 - 3F_{113}X'^2Z' + 2F_{13}X'Z' + F_{11}X'^2 + F_{33}Z'^2 - F_1X' - F_3Z' - 1 = 0 \quad (13)$$

Eqn. (13) has the same form that would result if $\sigma_1 = -X'$ and $\sigma_3 = -Z'$ were substituted into the cubic criterion of Eqn. (6).

Isolation of σ_1 and σ_3 quadratic terms in Eqn. (8) leads to constraint plane equations that are similar in form to Eqn. (9) and parallel to the σ_1 and σ_3 axes, respectively. Following the procedure outlined above leads to four additional equations (Roschke, Papados, and Mascorro 1990):

$$-3F_{112}Y' - 3F_{113}Z' + F_{11} = 0 \quad (14)$$

$$3F_{122}Y'^2 + 3F_{133}Z'^2 - 2F_{12}Y' - 2F_{13}Z' + F_1 = 0 \quad (15)$$

$$-3F_{133}X' - 3F_{233}Y' + F_{33} = 0 \quad (16)$$

$$3F_{113}X'^2 + 3F_{223}Y'^2 - 2F_{13}X' - 2F_{23}Y' + F_3 = 0 \quad (17)$$

Using Lagrange multipliers to incorporate Eqns. (11), (12), and (14)-(17) as constraint conditions the following functional is obtained for calculation of the interaction parameters:

$$\Phi = \sum_{i=1}^{n1}(F_1\sigma_{1i} + F_2\sigma_{2i} + F_{11}\sigma_{1i}^2 + F_{22}\sigma_{2i}^2 + 2F_{12}\sigma_{1i}\sigma_{2i} + 3F_{112}\sigma_{1i}^2\sigma_{2i}$$
$$+ 3F_{122}\sigma_{1i}\,\sigma_{2i}^2 - 1)^2 + \sum_{i=1}^{n2}(F_1\sigma_{1i} + F_3\sigma_{3i} + F_{11}\sigma_{1i}^2 + F_{33}\sigma_{3i}^2 + 2F_{13}\sigma_{1i}\sigma_{3i}$$
$$+ 3F_{113}\sigma_{1i}^2\sigma_{3i} + 3F_{133}\sigma_{1i}\sigma_{3i}^2 - 1)^2 + \sum_{i=1}^{n3}(F_2\sigma_{2i} + F_3\sigma_{3i} + F_{22}\sigma_{2i}^2 + F_{33}\sigma_{3i}^2$$
$$+ 2F_{23}\sigma_{2i}\sigma_{3i} + 3F_{223}\sigma_{2i}^2\sigma_{3i} + 3F_{233}\sigma_{2i}\sigma_{3i}^2 - 1)^2 + \lambda_1(-3F_{122}X' - 3F_{223}Z' + F_{22})$$
$$+ \lambda_2(3F_{112}X'^2 + 3F_{233}Z'^2 - 2F_{12}X' - 2F_{23}Z' + F_2) + \lambda_3(-3F_{112}Y' - 3F_{113}Z' + F_{11})$$
$$+ \lambda_4(3F_{122}Y'^2 + 3F_{133}Z'^2 - 2F_{12}Y' - 2F_{13}Z' + F_1) + \lambda_5(-3F_{133}X' - 3F_{233}Y' + F_{33})$$
$$+ \lambda_6(3F_{113}X'^2 + 3F_{223}Y'^2 - 2F_{13}X' - 2F_{23}Y' + F_3) \quad (18)$$

where λ_i for $i = 1, 2, ..., 6$ represent Lagrange multipliers, and summations range over the number of tests carried out.

Interaction coefficients are determined by minimizing the functional Φ. Fifteen equations are obtained with an identical number of unknowns. Coefficients F_i and F_{ii} ($i = 1, 2, 3$) need not be obtained from this expression since they are identical to the Tsai-Wu coefficients stated earlier. The system of equations is determinate and calculation of the coefficients is feasible. Thus, it is possible to construct a closed cubic surface for the particular quadrant (compression-compression) under consideration. The same approach can be applied for each of the other seven quadrants of the failure surface.

3.2.2 Evaluation of F_{iij} for $i = 1, 2, 3$, and $j = 4$ or 5 or 6

For the condition (b) to be met the following asymptotic equations, that do not intersect the curves of condition (a), must exist:

$$3F_{144}\,\sigma_1 + 3F_{244}\,\sigma_2 + 3F_{344}\,\sigma_3 + F_{44} = 0 \tag{19}$$

$$3F_{155}\,\sigma_1 + 3F_{255}\,\sigma_2 + 3F_{355}\,\sigma_3 + F_{55} = 0 \tag{20}$$

$$3F_{166}\,\sigma_1 + 3F_{266}\,\sigma_2 + 3F_{366}\,\sigma_3 + F_{66} = 0 \tag{21}$$

For this condition to be satisfied Eqns. (19)-(21) should not intersect the failure surface defined by Eqn. (8). This condition occurs only if the planes described by these three equations are, at most, tangent to the cubic surface. For simplicity, only one of these conditions is explicitly considered in the present discussion. Later, the concept is generalized to incorporate equations for the other two planes.

Eqn. (21) may be rewritten as follows:

$$\sigma_3 = -(3F_{166}\sigma_1 + 3F_{266}\sigma_2 + F_{66})/3F_{366} \tag{22}$$

The following convenient form is obtained by substitution of σ_3 from Eqn. (22) into Eqn. (7) and rearranging the result:

$$(A_{16}\sigma_1^3 + A_{26}\sigma_1^2 + A_{36}\sigma_1 + A_{46}) + (B_{16}\sigma_2^3 + B_{26}\sigma_2^2 + B_{36}\sigma_2 + B_{46})$$
$$+ (C_{16}\sigma_1^2\sigma_2 + C_{26}\sigma_1\sigma_2^2 + C_{36}\sigma_1\sigma_2) = 0 \tag{23}$$

where, $A_{16}, A_{26}, A_{36}, A_{46}, B_{16}, B_{26}, B_{36}, B_{46}, C_{16}, C_{26}$, and C_{36} are constants which are functions of F_i, F_{ij}, and F_{i66} for $i = 1, 2, 3$ (Roschke, Papados, and Mascorro 1990).

A functional, Ψ, is defined from the closed-form solution of the bicubic Eqn. (23), that yields repeating σ_1 and σ_2 roots at points of tangency, and the side conditions $C_{16} = C_{26} = C_{36} = 0$. Assuming that there are n1 sets of failure data, $(\sigma_{1i}, \sigma_{2i}, \sigma_{6i})$ for $i = 1, 2, ..., n1$, Ψ takes the following form:

$$\Psi = \sum_{i=1}^{n1}(F_1\,\sigma_{1i} + F_2\,\sigma_{2i} + F_{11}\,\sigma_{1i}^2 + F_{22}\,\sigma_{2i}^2 + F_{66}\,\sigma_{6i}^2 + 2F_{12}\,\sigma_{1i}\,\sigma_{2i}$$
$$+ 3F_{112}\,\sigma_{1i}^2\,\sigma_{2i} + 3F_{122}\,\sigma_{1i}\,\sigma_{2i}^2 + 3F_{166}\,\sigma_{1i}\,\sigma_6^2 + 3F_{266}\,\sigma_{2i}\,\sigma_6^2$$
$$+ 3F_{366}\,\sigma_3\,\sigma_6^2 - 1)^2 + \mu_1(27A_{16}^2A_{46}^2 - 18A_{16}A_{26}A_{36}A_{46} + 4A_{16}A_{36}^3$$
$$- A_{26}^2A_{36}^2 + A_{26}^3A_{46}) + \mu_2(27B_{16}^2B_{46}^2 - 18B_{16}B_{26}B_{36}B_{46} + 4B_{16}B_{36}^3$$
$$- B_{26}^2B_{36}^2 + B_{26}^3B_{46}) + \mu_3C_{16} + \mu_4C_{26} + \mu_5C_{36} \tag{24}$$

where μ_1 through μ_5 are Lagrange multipliers.

The functional, Ψ, reaches an extreme value (maximum or minimum) when:

$$\partial\Psi/\partial F_{166} = 0, \quad \partial\Psi/\partial F_{266} = 0, \quad \partial\Psi/\partial F_{366} = 0, \quad \partial\Psi/\partial\mu_1 = 0,$$
$$\partial\Psi/\partial\mu_2 = 0, \quad \partial\Psi/\partial\mu_3 = 0, \quad \partial\Psi/\partial\mu_4 = 0, \quad \partial\Psi/\partial\mu_5 = 0 \tag{25}$$

A set of eight nonlinear simultaneous equations is obtained that yields coefficients F_{166}, F_{266}, and F_{366} and numerical values for the five Lagrange multipliers.

Similarly, Eqns. (19) and (20) may be rearranged as follows:

$$\sigma_1 = -(3F_{255}\,\sigma_2 + 3F_{355}\,\sigma_3 + F_{55})/3F_{155} \tag{26}$$

$$\sigma_2 = -(3F_{144}\,\sigma_1 + 3F_{344}\,\sigma_3 + F_{44})/3F_{244} \tag{27}$$

Relationships similar to Eqn. (24) can be derived from Eqns. (26) and (27) in the event that biaxial test data, (σ_{1i}, σ_{3i}, σ_{4i}) for $i = 1, 2, ..., n2$ and σ_{2i}, σ_{3i}, σ_{5i} for $i = 1, 2, ..., n3$, respectively, exist. By an analogous analysis the interaction coefficients F_{144}, F_{244}, F_{344}, F_{155}, F_{255}, and F_{355} may be obtained, thus, yielding all interaction coefficients necessary for determining the failure surface.

4 Experimental evaluation of strength coefficients

A number of experimentally determined failure coefficients are required for estimating the principal (F_i, F_{ii}, F_{iii}) and interaction (F_{ij}, F_{iij} for $i \neq j$) strength parameters (see Table 1). Geometrical limitations on their determination arise, however, due to the fact that cross-rolled beryllium sheets are only available in plate form. Thus, special experiments are required to simulate the necessary states of stress. A total of fifteen tests are necessary to evaluate failure coefficients of an orthotropic material: three tensile, three compressive, three pure shear, three pure biaxial (σ_i, σ_j for $i \neq j$ and $i, j = 1, 2, 3$), and three biaxial tests with shear ((σ_1, σ_2, σ_6), (σ_1, σ_3, σ_4), and (σ_2, σ_3, σ_5)). When extreme accuracy is desired or the material under consideration is highly anisotropic, the number of experiments should be increased to include positive and negative shear tests and a large number of biaxial tests combined with shear.

Numerical modeling is an essential design aid for these experiments. Here, PATRAN-II, a graphical pre- and post-processor, and ABAQUS, a non-linear finite element code are used. As an example, numerical simulation of an in-plane shear test is presented. Determination of the principal coefficient F_{66}, as well as interaction coefficients F_{166} and F_{266}, require knowledge of in-plane shear strength, σ_6. A two-dimensional numerical model is used to simulate the in-plane shear test prior to actual physical testing (Fig. 2). Material and geometrical nonlinearities are incorporated in the finite element simulation. The actual specimen is a 114.3-mm x 25.4-mm x 2.54-mm (4.5 in. x 1.0 in. x 0.1 in.) coupon with two 45° slits located antisymmetrically with respect to the longitudinal and transverse planes of symmetry.

Table 1. Experiments required for evaluation of coefficients

Experiment	Coefficients
Uniaxial:	
Tension *and* Compression	
X-axis	F_1, F_{11}
Y-axis	F_2, F_{22}
Z-axis	F_3, F_{33}
Pure Shear	
X-Y Plane	F_{66}
X-Z Plane	F_{44}
Y-Z Plane	F_{55}
Biaxial:	
Tension *and/or* Compression (X-Y plane)	F_{12}, F_{112}, F_{122}
Tension *and/or* Compression (X-Z plane)	F_{13}, F_{113}, F_{133}
Tension *and/or* Compression(Y-Z plane)	F_{23}, F_{223}, F_{223}
Tension *and/or* Compression and Shear (X-Y plane)	$F_{166}, F_{266}, F_{366}$
Tension *and/or* Compression and Shear (X-Z plane)	$F_{144}, F_{244}, F_{344}$
Tension *and/or* Compression and Shear (Y-Z plane)	$F_{155}, F_{255}, F_{355}$

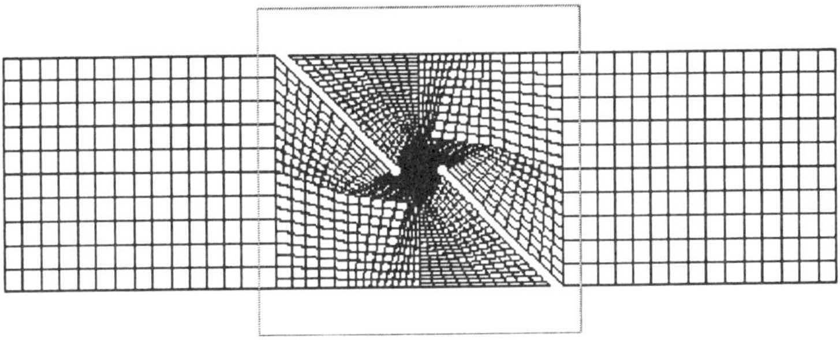

Fig. 2. Finite element mesh for two-dimensional in-plane shear specimen.

The two-dimensional model is comprised of two-thousand, eight-noded plate elements, which is equivalent to approximately 40,000 degrees-of-freedom. Five through-thickness points of integration are provided for each of the nine integration points within a given plate element. Numerical output is requested at each node for the top and bottom surfaces of the elements. Fig. 3 shows numerical results for shearing stresses at the top surface with the experimental failure load of 3,890 N (875 lbf).

Fig. 4 illustrates contours of in-plane failure envelopes for cross-rolled beryllium sheet for different levels of shearing stress. At present, failure for beryllium sheets loaded in-plane can be adequately predicted with the coefficients available. For out-of-plane loadings, however, not all failure conditions can be predicted due to the fact that some of the interaction coefficients are yet to be determined.

5 Conclusion

Three-dimensional failure envelopes of brittle anisotropic material can be derived using the new proposed criterion. The criterion is mathematically simple, although it includes formulation of the sixth order strength tensor. Moreover, it complies with all requirements, such as invariance, convexity, and non-singularity, that are necessary for establishing a failure surface. It should be noted that the proposed multi-dimensional criterion can be reduced to that of Tennyson and Jiang (1989), if out-of-plane stresses are omitted.

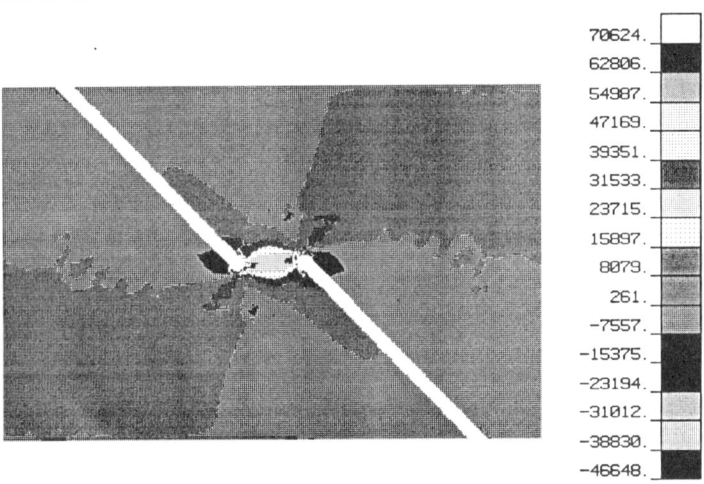

Fig. 3. Shear stress distribution (psi) from two-dimensional finite element analysis.

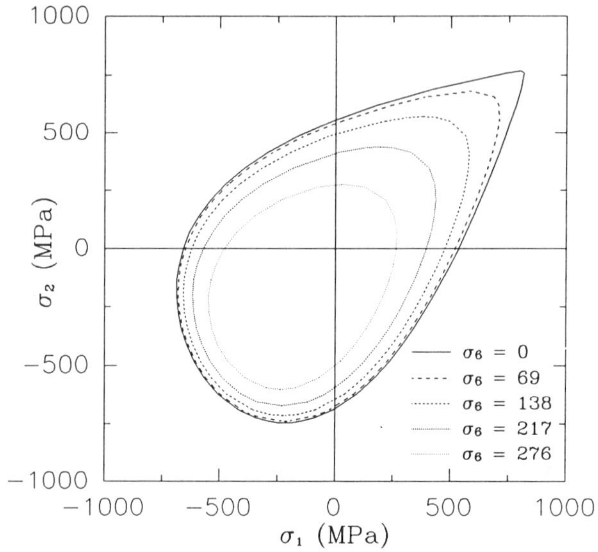

Fig. 4. Contours of in-plane failure surfaces for beryllium sheet.

The new criterion takes advantage of independent biaxial tests to accurately establish coefficients of the sixth order strength tensor. In the event that limited data are available, the failure surface can be bounded by utilizing asymptotic equations. Experimental efforts are underway to determine the remaining coefficients and thus, establish a complete model for cross-rolled beryllium sheet material.

6 Acknowledgement

This study was supported in part by grant NAG 9-280 from the National Aeronautics and Space Administration, Johnson Space Center, Houston, Texas, U.S.A.

7 References

Hill, R. (1950) **The Mathematical Theory of Plasticity.** Oxford University Press, London.

Hoffman, O. (1967) The brittle strength of material. **J. Compos. Mater.**, 1, 200-206.

Priddy, T.G. (1974) A fracture theory for brittle anisotropic materials. **J. Engng. Mater. and Techn.**, 91-96.

Roschke, P.N., Papados, P.P. and Mascorro, E. (1990) Failure prediction of thin beryllium sheets used in spacecraft structures. **NASA:** NAG 9-280, Annual report.

Tennyson, R.C. and Elliott, W.G. (1983) Failure analysis of composite laminates including biaxial compression. **NASA** CR 72192.

Tennyson, R.C. and Jiang, Z. (1989) Closure of the cubic tensor polynomial failure surface. **J. Compos. Mater.**, 23, 208-231.

Tsai, S.W. and Wu, E.M. (1971) A general theory of strength for anisotropic material. **J. Compos. Mater.**, 5, 58-80.

Wu, E.M. (1974) Optical experimental measurements of anisotropic failure tensors. **J. Compos. Mater.**: Testing and Design (Third Conference), ASTM STP 546.

Wu, E.M. and Scheublein, J.K. (1974) Laminate strength - a direct procedure. **J. Compos. Mater.**: Testing and Design (Third Conference), ASTM STP 546.

PART TEN

MIXED-MODE FRACTURE

60 DETERMINATION OF MIXED MODE PROPERTIES OF CONCRETE

M. HASSANZADEH
Lund Institute of Technology, Sweden

Abstract
A testing arrangement has been developed in order to determine the properties of a fracture process zone when it is exposed to both normal and shear displacements. The properties are expressed through a square stiffness matrix which relates the displacements within the fracture zone to the stresses. By means of the test results, functions have been derived in order to evaluate the elements of the stiffness matrix.
Keywords: Concrete, Fracture Mechanics, Mixed Mode

1 Introduction

Crack propagation in concrete structures is preceded by formation of a thin, but extended, fracture process zone, also called damage zone. Material inside the fracture process zone softens and acquires properties which differ from the other parts of the structure. Using models which take into account the effects of the fracture process zone have demonstrated the very important role which softening of concrete plays in governing the failure process of concrete structures of medium size. It is now broadly agreed that an appropriate fracture mechanics model should take into account the effects of the softening behaviour of concrete. The Fictitious Crack Model, Hillerborg (1989a), is one of the models in which the effects of the fracture process zone are considered.

In the Fictitious Crack Model, FCM, crack growth is described in the following manner. Since concrete is a homogeneous and isotropic material with much lower tensile strength than compressive strength, when the first principal tensile stress reaches the tensile strength of the material, a fracture process zone starts to develop which is oriented perpendicular to the first principal tensile stress direction. The stress state inside the fracture process zone at the instant of initiation is pure tensile, mode I. This is irrespective of the mode of loading. However, in the subsequent

Fracture Processes in Concrete, Rock and Ceramics Edited by J.G.M. van Mier, J.G. Rots and A. Bakker.
 Published by E. & F.N. Spon, 2–6 Boundary Row, London SE1 8HN. ISBN 0 419 15870 7

stages, the stress state inside the fracture process zone may be altered if the direction of the first principal tensile stress is changed. In this case, shear stresses may be developed inside the fracture process zone, i.e. the stress state becomes mixed mode. Thus, according to the FCM, a fracture process zone is formed due to mode I conditions. Furthermore, mixed mode conditions only occur inside the fracture process zone due to alterations of the first principal stress direction. This definition of the mixed mode crack propagation differs from that given by Linear Elastic Fracture Mechanics, LEFM. The differences are that in LEFM the effect of the fracture process zone is disregarded, and the stress state in the tip of a crack is mixed mode, i.e. both K_I and K_{II} are non-zero, whereas, according to the FCM, the stress state at the tip of a crack prior to the formation of the fracture process zone is mode I.

As mentioned above, according to the FCM, the mixed mode stress state occurs only inside the fracture process zone. Thus, mixed mode behaviour of the fracture process zone should be determined in order to carry on mixed mode analyses by means of FCM. The mixed mode behaviour of the fracture process zone is generally expressed by means of the following relation:

$$\begin{bmatrix} d\sigma \\ d\tau \end{bmatrix} = \begin{bmatrix} K_{11} & K_{12} \\ K_{21} & K_{22} \end{bmatrix} \begin{bmatrix} dw \\ d\delta \end{bmatrix} \tag{1}$$

$d\sigma$ and $d\tau$ are the normal- and shear-stress increments. dw and $d\delta$ are normal- and shear-displacement increments. The square matrix is the crack stiffness matrix, and K_{11}, .., K_{22} are crack stiffness coefficients which depend on normal, and shear displacements. Furthermore they depend on the history of the displacement, i.e. the displacement path on the plane of the normal and shear displacement.

A test method has been developed in order to determine the above-mentioned stiffness coefficients. The test method is described in the previous publications to which the reader is referred (refs. 1,2,4).

2 Mixed mode tests

Figure 1. shows test equipment which, together with a closed-loop testing machine, constitutes an arrangement for mixed-mode tests. The tests are displacement-controlled and are run in a semi-automatic manner, which means that the closed-loop machine continuously runs a programmed normal displacement scheme and that the shear displacement is manually imposed by rotating the crank indicated in Figure 1. By means of continuous adjustment of the shear displacement, it is possible to perform mixed-mode tests according to any arbitrary displacement path, or any relation between normal and shear displacements. Figure 2 shows the geometry of the specimens used in this investigation.

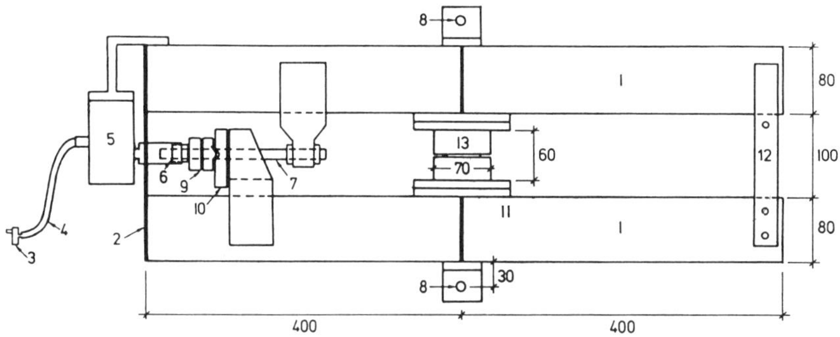

Fig. 1. Test equipment.
1=box beams, 2=steel sheet, 3=crank, 4=flexible axle, 5=gear box, 6=nut, 7=load-cell, 8=connection to the testing machine, 9=roller, 10=slip support, 11=steel plates, 12=lock, 13=specimen.

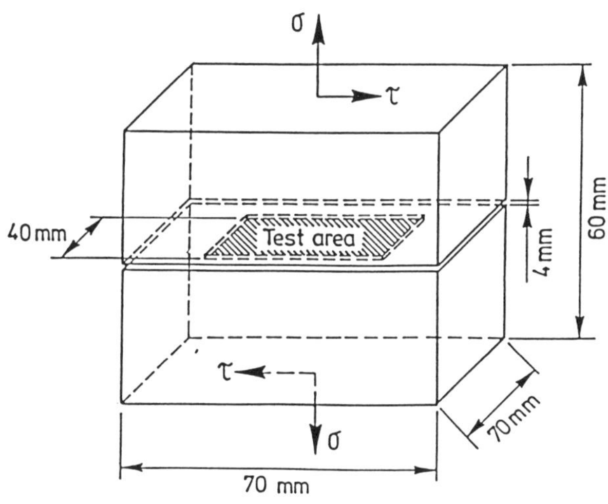

Fig. 2. Specimen used for the mixed mode tests.

Mixed-mode tests have been conducted in such a way that they satisfy the FCM's definition of the mixed-mode crack propagation, i.e. normal displacement is first imposed so that a tensile fracture zone starts to form. Shortly after the normal stress starts to decrease, shear displacement is also imposed. In the subsequent stages, the test is continued in such a way that the normal and shear displacements follow a predefined displacement path. Since the displacement path is not a material property and there is no knowledge about the displacement paths occurring in practice, the displacement paths used in this investigation have been chosen arbitrarily. Two different sets of displacement paths have been tested, namely linear paths and parabolic paths. The functions are shown below:

$$w = (\tan\alpha)\,\delta \qquad (2)$$

$$w = \beta\sqrt{\delta} \qquad (3)$$

In the tests, values of the α and β have been varied in a wide range. However, much part of the investigation has dealt with the parabolic paths, because this is one of the most realistic ways of describing the shear failure of concrete. In this paper the results of the $\alpha=60^\circ$ and $\beta=0.4\ mm^{1/2}$, $0.6\ mm^{1/2}$ are presented. Besides the paths mentioned another displacement path is tested, i.e. a combination of both linear path and parabolic path, denoted mixed path, paths a-b-c-d indicated in Figure 3.

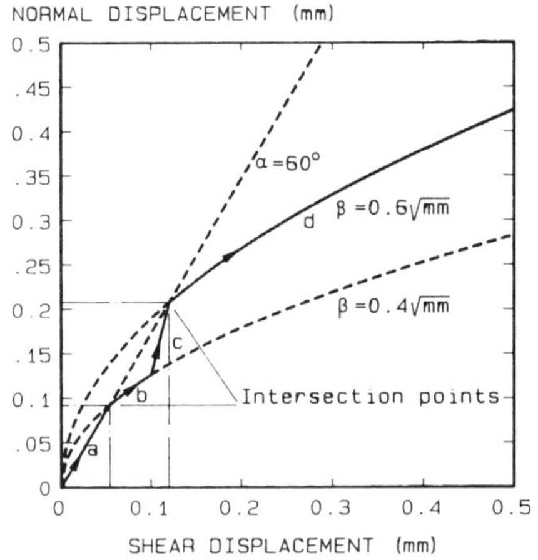

Fig. 3. Displacement paths.

All the tests have been performed on ordinary concrete with 50 MPa compressive strength and 8 mm maximum aggregate size at the age of at least 28 days. None of the specimens have been exposed to drying.

Results of the parabolic paths are the mean value of three tests while the results of the other paths represent a single test. Although, more than one test has been performed with the $\alpha=60^{\circ}$, the mean curve is not calculated. Good reproducibility has been observed in all the displacement paths.

3 Comparison of the displacement paths

Figure 4 shows the results of the parabolic, linear and mixed paths. It is interesting to compare the shape of the curves. It can be observed that shear stress - shear displacement curves pertaining to the parabolic paths increase continuously, within the range of the test, while the same curves for the linear and mixed paths show a clear maximum. The same trends can be observed in the normal stress - normal displacement curves but in the opposite direction. Furthermore, it can be observed that the ultimate compressive stresses developed in the case of linear and mixed path are very low and may be negligible.

The above-mentioned observations may be explained by means of the theoretical calculations. Walraven (1979) has derived a theoretical relation between the stresses and the contact surface between the matrix and the aggregates. The relation is shown below:

$$\frac{\sigma}{\sigma_{pu}} = A_x - \mu A_y \tag{4}$$

$$\frac{\tau}{\sigma_{pu}} = A_y + \mu A_x \tag{5}$$

In this case σ is compressive stress, σ_{pu} is the matrix compressive yield stress and μ, approximately 0.4, is the coefficient of friction. A (mm^2/mm^2 crack surface) is the total contact surface between aggregates and the matrix, see Figure 5. The indices denote projections of the A where, A_x is the projected contact surface which is perpendicular to the direction of the normal stress and A_y is the projected contact surface perpendicular to the direction of the shear stress. A_x and A_y are functions of crack displacements and the intersection between crack plane and aggregates, i.e they are also functions of the

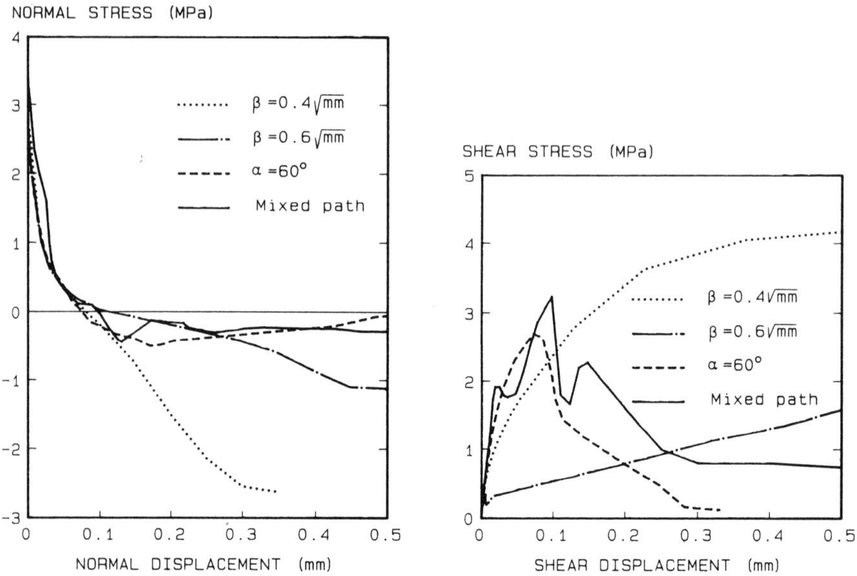

Fig. 4. Test results.

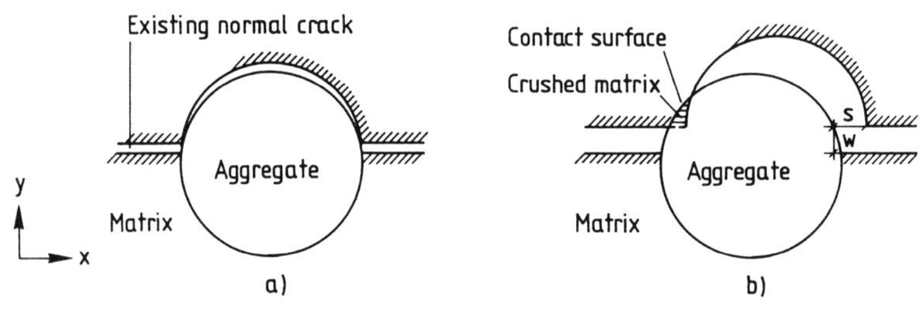

Fig. 5. Crack plane before (a), and after (b) applied normal and shear displacement.

particle size distribution, maximum aggregate size and the relative volume of the aggregates. The contact surfaces are not functions of the displacement paths, i.e. the parabolic path and the linear path give the same stress at their intersection points. Walraven has derived equations for calculation of the contact surfaces for spherical particles. Equations for A_x and A_y have been derived on the assumption that a normal crack exists before the shear and normal displacements are applied, see Figure 5a. Furthermore, it has been assumed that the matrix is a rigid-plastic material. The equations are not given here, but some results are demonstrated in Figure 6 for different displacement paths for concrete with 8 mm maximum aggregate size and 75 % aggregates by volume.

Figure 6.a shows A_y as a function of shear displacement and the displacement paths. It can be observed that in the case of the linear paths the A_y is high from the beginning and descends with increased shear displacement, while in the case of parabolic paths A_y is low and increases with increased shear displacement. It can be observed that the shape of the calculated curves pertaining to the parabolic paths is similar to the experimental curves shown in Figure 4. Furthermore, the calculations also verify the sudden change of the stress-displacement relationship when ß increases from 0.4 $mm^{1/2}$ to 0.6 $mm^{1/2}$.

Figure 6.b shows dimensionless stresses (σ/σ_{pu}, τ/σ_{pu}) as functions of crack displacements. It should be noted that the solid curves demonstrate shear stress - shear displacement relationships while dashed curves demonstrate normal stress - normal displacement relationships. Furthermore, it should be noted that the shear stress - shear displacement curves pertaining to the linear path are not zero at origin. This is due to the fact that the matrix has been assumed to be a rigid-plastic material. It can be observed that in the case of the linear paths, shear stress is high in the beginning and decreases with increased shear displacement. Furthermore the developed normal stress is low compared to the parabolic paths.

4 Mathematical expression of the test results

As was mentioned earlier, much of the investigation has dealt with the parabolic paths. The parabola constant ß has been varied between 0.4 $mm^{1/2}$ and 0.9 $mm^{1/2}$ with 0.1 $mm^{1/2}$ at each step. The lower limit of the ß has been chosen with regard to the capability of the testing arrangement. At $ß \leq 0.30$ $mm^{1/2}$ the final shear stresses and compressive stresses are so high that specimens failed due to diagonal cracks. The upper limit has been set at

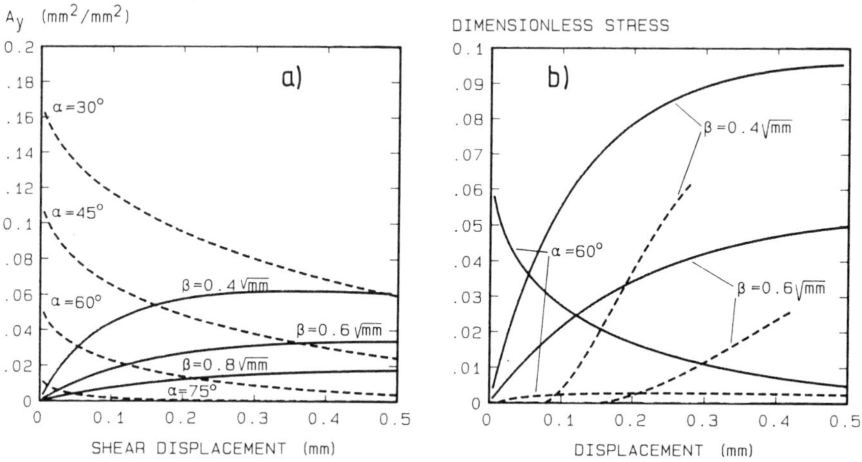

Fig. 6. Theoretical results based on equations derived by Walraven.

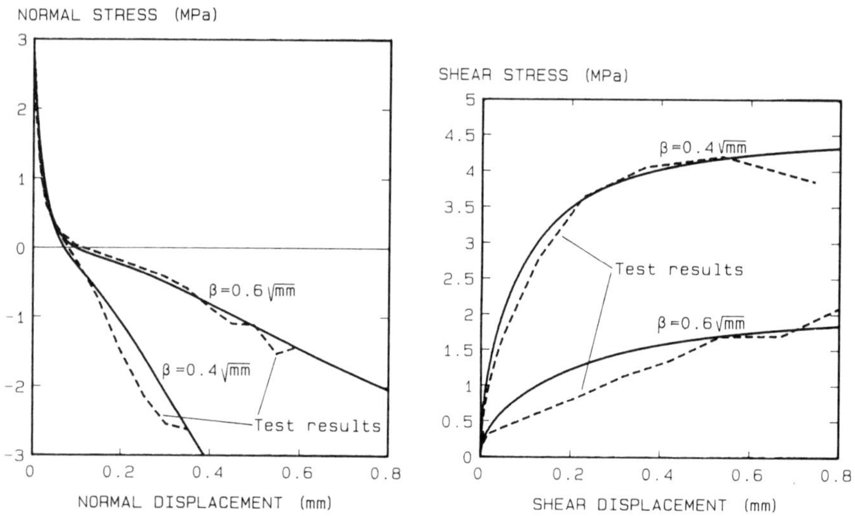

Fig. 7. Fit of mathematical expressions to the test results.

0.9 $mm^{1/2}$ because at this level no considerable shear stresses or compressive stresses have been observed. Results of the tests have published in Hassanzadeh (1989). The functions which fit the test results are given below:

$$\sigma = 2.8(e^{-40w} - 2.5e^{-5\beta^3}\arctan(\frac{w}{\beta})^2) \tag{6}$$

$$\tau = 10e^{-3.6\beta^{1.4}}(\arctan\frac{\delta}{\beta^2})^{0.5} \tag{7}$$

$$\beta = \frac{w}{\sqrt{\delta}}$$

It has been difficult to find simple functions which fit the test results. However, the given expressions are best fitted to the path ß=0.4 $mm^{1/2}$ and worst fitted to the path ß=0.6 $mm^{1/2}$, see Figure 7.

The advantages of the functions are that they are easy to differentiate, and for the cases of constant parabolic paths, or paths which can be approximated to a certain parabola, ß is constant which leads to a symmetric stiffness matrix, equation 1, with zero off-diagonal terms.

5 Conclusions

The test method which is described here is suitable for determination of the mixed-mode properties of concrete. However, since the test method requires predefined displacement paths, further research is needed to better understand the relations between normal and shear displacement during the process of fracture. Such research should combine both laboratory and theoretical experiments.

References

1. Hassanzadeh, M., Hillerborg, A., Zhou, F.P. (1987) Tests of material properties in mixed mode I and II, in Fracture of Concrete and Rock, RILEM-SEM Int. Conf., Huston (ed. S.P. Shah and S.E. Swartz), pp 353-358.

2. Hassanzadeh, M. (1989) Determination of fracture zone properties in mode I and II, proc. of the 1989 SEM Spring Conference on Experimental Mechanics, pp 521-527. Cambridge, MA, U.S.A.

3. Hillerborg, A. (1989-a), Discrete crack approach, in Fracture mecahnics of concrete structures, RILEM report, (ed. L. Elfgren), pp 128-137.

4. Hillerborg, A. (1989-b) Mixed mode fracture in concrete, Seventh International Conference on Fracture, Houston, Texas, March 20-24.

5. Walraven, J.C. (1979), Experiments on shear transfer in cracks in concrete, part II: analysis of results. Stevin report 5-79-10, Delft University of Technology.

61 INFLUENCE OF FRICTION ON COHESIVE CRACK PROPAGATION

S. VALENTE
Department of Structural Engineering, Politecnico di Torino, Italy

Abstract
In materials such as concrete, rock, brick, fibre-reinforced composites, the prevailing form of non-linear behaviour is strain-softening, which means that a reduction in the material's load-carrying capacity takes place due to inelastic deformation processes. If an initial imperfection is present in the structure, this results in the localization of a non-linear strain within a very narrow band whilst the behaviour of the material outside this band remains linear. This happens independently of the boundary conditions. The cohesive model, originally proposed by Barenblatt and Dugdale, describes the narrow band by means of a crack extension, called the fictitious crack, carrying a stress which decreases with increasing crack opening displacement. In this paper the cohesive crack model is extended to take into account the effects of friction across the crack. This makes it possible to explain the experimental results of pull-out tests. In such tests, in fact, experimental results show a much higher energy dissipation with respect to the energy strictly necessary to produce concrete separation (Mode I).
Keywords: Fracture, Cohesive Crack, Friction, Pull-out, Concrete.

1 Introduction

The ultimate behaviour of several materials, such as concrete, rock, bricks and fibre-reinforced composites, is characterized by the localization of a non-linear zone within a very narrow band, whilst the rest of the material retains its linear behaviour. The cohesive model describes this band as a fictitious crack (or process zone) where the material, albeit damaged, can still transfer stresses. The stresses involved are described as decreasing (strain-softening) functions of the relative displacements of the opposing crack surfaces (displacement discontinuity). The process zone originates perpendicularly to the principal tensile stress when the latter reaches

Fracture Processes in Concrete, Rock and Ceramics Edited by J.G.M. van Mier, J.G. Rots and A. Bakker.
 Published by E. & F.N. Spon, 2–6 Boundary Row, London SE1 8HN. ISBN 0 419 15870 7

its ultimate value, σ_u. The point at which this condition occurs is called fictitious crack tip: it represents the boundary between the integral and the damaged material. According to the foregoing assumptions, there are no stress singularities in this model.

The cohesive model was first proposed by Barenblatt (1959) and, based on independent studies, by Dugdale (1960) as a tool for the analysis of collinear propagation (Mode I) problems. Later on, Hillerborg et al. (1976) applied it to concrete for the interpretation of three-point bending tests. In the same field, Carpinteri (1985, 1989) has explained the phenomenon of ductile-brittle transition, by defining the brittleness number, s_E, which depends on the properties of the material and structural dimensions (size-effect). Recently (in Carpinteri and Valente, 1988; Carpinteri et al., 1989; Bocca et al., 1989, 1990), the cohesive model has been employed to cope with problems in which the crack path is unknown a priori (Mixed Mode). These applications, based on the finite elements method, concern the simulation of the four-point shear test and the pull-out test on concrete specimens (axial-symmetrical case, Fig. 1). During the latter test, which is the subject-matter of this investigation, it has been assumed that the front of the fictitious crack retains its circular form and is not ovalized as it gets larger. This assumption is substantiated by experimental evidence: the portions of material pulled out together with the metal bolt, in fact, are approximately axial-symmetrical.

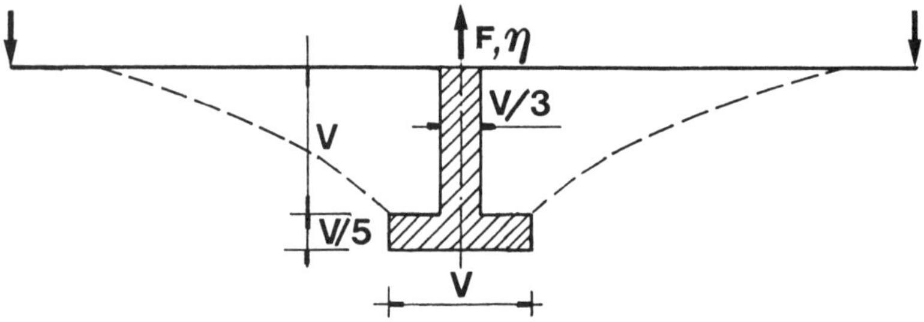

Fig. 1 Axial-symmetrical pull-out testing apparatus

2 The constitutive law of the fictitious crack

The cohesive stresses (σ_c, τ_c) transferred through the process zone (fictitious crack) are decreasing functions of the displacement discontinuities (w_n, w_t). As already observed in earlier testing programs (Bocca at al, 1989), the amount of fracture energy dissipated during pull-out tests is greater than that strictly necessary to cause the spread of a crack in Mode I conditions. Hence, the crack has been characterized by a constitutive law designed to take τ_c, and hence friction dissipation, into account. This law is based on the following observations:

1. Numerical simulations have shown that the displacement discontinuity evolves following, with good approximation, the following parabolic law:

$$w_n = \beta w_t^{1/2} \tag{1}$$

where w_n is a monotone increasing function during the irreversible process of crack growth ($\dot{w}_n > 0$, $\dot{w}_t > 0$, where the dot denotes derivation with respect to time).

2. Using a testing set-up designed to impose a pre-determined crack opening path, Hillerborg (1989) and Hassanzadeh (1990) have observed that a small departure from eq.(1) produces a negligible variation in the cohesive stresses. These Authors have also published the test results obtained by imposing strict compliance with the opening path denoted by eq.(1). The curves shown in Fig. 2 have been obtained through piecewise linear interpolation of said results. Stresses σ_c and τ_c relating to values of β not explicitly shown in Fig. 2 are determined by linear interpolation.

As illustrated in greater detail by Bocca et al. (to appear b), from the curves shown in Fig. 2 it is possible to derive the following incremental law:

$$\begin{Bmatrix} d\sigma_c \\ d\tau_c \end{Bmatrix} = \begin{bmatrix} \dfrac{\partial \sigma_c}{\partial w_n} & \dfrac{\partial \sigma_c}{\partial w_t} \\ \dfrac{\partial \tau_c}{\partial w_n} & \dfrac{\partial \tau_c}{\partial w_t} \end{bmatrix} \begin{Bmatrix} dw_n \\ dw_t \end{Bmatrix} \tag{2}$$

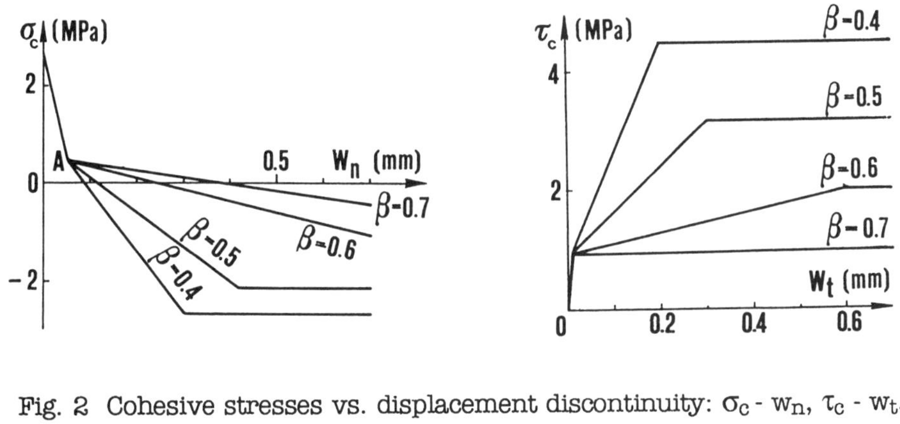

Fig. 2 Cohesive stresses vs. displacement discontinuity: σ_c - w_n, τ_c - w_t.

Assuming $\dot{w}_n > 0$, $\dot{w}_t > 0$ to have been verified, the constitutive law can be expressed in holonomic terms:

$$\mathbf{P}_c^{(i+1)} = \mathbf{P}_0^{(i)} + \mathbf{L}^{(i)} \mathbf{w}^{(i+1)}, \qquad (3)$$

where $\mathbf{L}$ is the 2x2 matrix contained in eq.(2) and

$$\mathbf{P}_c = [\sigma_c , \tau_c]^T , \qquad \mathbf{w} = [w_n , w_t]^T , \qquad \mathbf{P}_0^{(i)} = \mathbf{P}_c^{(i)} - \mathbf{L}^{(i)}\mathbf{w}^{(i)} .$$

This being a non linear evolutionary phenomenon, the index (i) denotes the step the magnitudes refer to.

3 Solution for a single crack growth step

By subdividing the domain into a finite number of elements, arranged so that the (real and fictitious) crack is at the interface between two elements, considering the constitutive law (3) and making use of the Principal of Virtual Works (see Carpinteri and Valente, 1988; Carpinteri et al., 1989; Bocca et al., to appear a), we get the following system of linear equations:

$$\left(\mathbf{K} - \mathbf{C}^{(i)}\right) \mathbf{u}^{(i+1)} = \mathbf{F}_0^{(i)} + \lambda \mathbf{F}_1 , \qquad (4)$$

where:

$\mathbf{K}$ = stiffness matrix (n x n)
$\mathbf{C}^{(i)}$ = non-symmetrical strain-softening matrix, assembled from $\mathbf{L}$
$\mathbf{u}^{(i+1)}$ = vector of the n unknown nodal displacements
$\mathbf{F}_0^{(i)}$ = load vector, assembled from $\mathbf{P}_0^{(i)}$
$\mathbf{F}_1$ = external load vector
λ = external load multiplier

If the matrix $(\mathbf{K} - \mathbf{C}^{(i)})$ is non-singular, it is useful to solve the linear system of eqs.(4) separately for the two right hand side vectors:

$$\left(\mathbf{K} - \mathbf{C}^{(i)}\right) \mathbf{u}_0^{(i+1)} = \mathbf{F}_0^{(i)} , \qquad \left(\mathbf{K} - \mathbf{C}^{(i)}\right) \mathbf{u}_1^{(i+1)} = \mathbf{F}_1 . \qquad (5)$$

Denoting by $\boldsymbol{\sigma}_0^{(i+1)}$ and $\boldsymbol{\sigma}_1^{(i+1)}$ the stresses at the tip of the fictitious crack (centroid of the dashed element in Fig. 3), relating to $\mathbf{u}_0^{(i+1)}$ and $\mathbf{u}_1^{(i+1)}$, respectively, we can write (axial-symmetrical problem):

$$\boldsymbol{\sigma} = [\sigma_x , \sigma_y , \sigma_z , \tau] = \boldsymbol{\sigma}_0^{(i+1)} + \lambda \boldsymbol{\sigma}_1^{(i+1)} . \qquad (6)$$

The load multiplier λ may therefore be obtained by imposing that

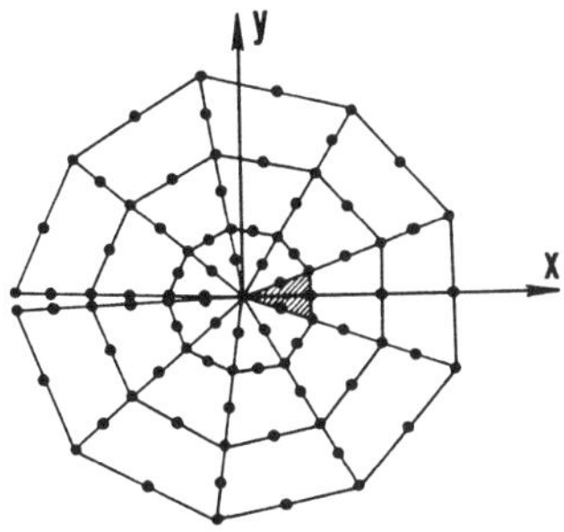

Fig. 3. Finite element rosette at the fictitious crack tip.

the main tensile stress, at the tip of the fictitious crack, should reach the concrete tensile strength, σ_u. From Mohr's circle we therefore get:

$$\sigma_u^3 - (\sigma_x + \sigma_y + \sigma_z)\,\sigma_u^2 + \left(\sigma_x\,\sigma_y + \sigma_x\,\sigma_z + \sigma_y\,\sigma_z - \tau^2\right)\sigma_u + \\ + \sigma_x\,\sigma_y\,\sigma_z - \sigma_z\,\tau^2 = 0\,. \tag{7}$$

By substituting eq.(6) into eq.(7) we obtain a cubic equation in λ and we can determine its value. The vector of nodal displacements then turns out to be:

$$\mathbf{u}^{(i+1)} = \mathbf{u}_0^{(i+1)} + \lambda\mathbf{u}_1^{(i+1)} \tag{8}$$

Since the coefficients and known terms of the linear system of equations (4) depend on $\mathbf{u}$, it proves necessary to assemble matrix $\mathbf{C}^{(i+1)}$ and vector $\mathbf{F}_0^{(i+1)}$, to increment index (i) by a unit and to solve the system of equations (4) and (7) again. This sequence of equilibrium iterations comes to an end when the Euclidean norm of the vector $(\mathbf{u}^{(i+1)} - \mathbf{u}^{(i)})$ drops below a predetermined tolerance value. However, before the solution found in this manner can be deemed acceptable, it is necessary to verify that $\dot{w}_n > 0$, $\dot{w}_t > 0$, and that, at all points of the domain, the main tensile stress does not exceed σ_u.

If the solution is found to be acceptable, by substituting λ into eq.(6) the state of stress at the fictitious crack tip is determined. If we denote the corresponding principal stresses by $\sigma_3 \le \sigma_2 \le \sigma_1$, and take the tensile stresses to be positive, either of the following situations may occur:

$$\sigma_3 \le \sigma_2 < \sigma_1 = \sigma_u\,, \tag{9.a}$$
$$\sigma_3 \le \sigma_2 = \sigma_1 = \sigma_u\,. \tag{9.b}$$

In the case of (9.a), the crack is assumed to propagate orthogonally to stress σ_1. In the case of (9.b), one of Mohr's three circles has degenerated into a point and hence the above mentioned propagation criterion cannot be applied.

4 Controlling the numerical procedure through the length of the fictitious crack

Based on the considerations made in the previous paragraph, it can be seen that the external load, represented by the multiplier λ, is not the control variable, but rather an unknown which is determined, at each crack growth step, by substituting eq.(6) into eq.(7).

Similarly, the displacements **u** are not control variables, as they have to be determined by substituting the current value of λ into eq.(8).

According to a method called "Fictitious Crack Length Control Scheme" (see Carpinteri and Valente, 1988; Carpinteri et al., 1989; Bocca et al., to appear a), the length of the fictitious crack serves as the control variable, since it is a surely monotone increasing function during the irreversible cracking process. This does not apply to the load since, once it has reached its maximum value, it decreases. Likewise, it may happen that the load point displacement decreases with increasing crack opening. The latter phenomenon is known as catastrophic collapse or snap back.

In the initial stage of the cracking process the matrix (**K** - **C**) is positive definite and hence all its eigenvalues are positive. This property derives from the fact that the matrix **K**, representing elastic stiffness, is positive definite, whereas matrix **C** initially is negligible. Under these conditions, the system of linear eqs.(5) has one and only one solution. Since the cubic equation in λ which is obtained by substituting eq.(6) into eq.(7) offers only one physically acceptable solution, it can be concluded that system(4) admits one and only one solution.

It is important to note that this uniqueness of the solution holds under the assumptions illustrated in the foregoing sections. If the assumption of the crack propagating orthogonally to the main tensile stress (eq. (7)), for instance, is replaced with the assumption of the crack propagating according to Mohr-Coulomb's failure criterion, we obtain a different crack path and a different maximum load. Numerical tests, however, have yielded negligible differences. Similarly, if we carry out a three-dimensional analysis and assume that the front of the fictitious crack is ovalized, we shall probably obtain slightly different solutions in respect of the ones illustrated in Figs. 4 and 5.

With increasing fictitious crack length, the smallest eigenvalue of (**K** - **C**) decreases, till it vanishes. Before this last condition is verified, in all the geometrical cases subjected to numerical analysis, condition (9.b) has been seen to occur. This condition corresponds to the degeneration of one of Mohr's three circles and hence entails the impossibility of determining the direction of next crack growth step. A qualitative illustration of the phenomenon from the mechanical point of view is provided by Bocca et al. (to appear b).

In order to go on with the numerical simulation, from this point on, the crack path has been imposed a priori, by adopting a curve based on the experimental results (see Bocca et al., to appear b).

Proceeding with the analysis in this manner, the smallest eigenvalue of the (**K** - **C**) matrix is found to decrease monotonically until it vanishes. Now, making use of the singular value decomposition, as illustrated by Valente (1990), it is possible to demonstrate that this is not a bifurcation point.

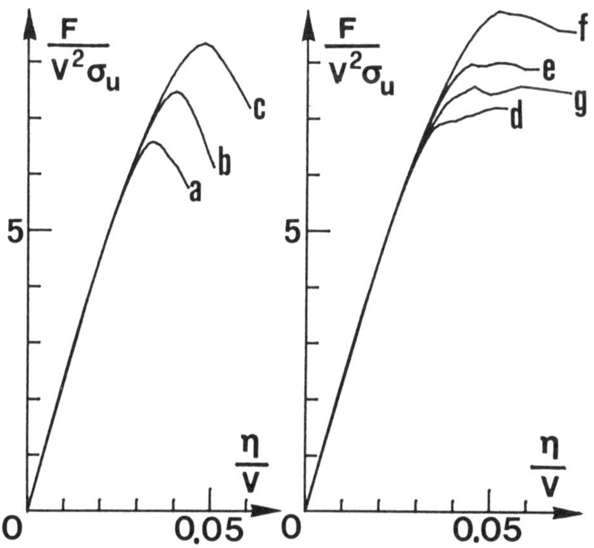

Curves	$s_E \cdot 10^3$	τ_c
a	1	no
b	1.44	no
c	2	no
d	1	yes
e	1.44	yes
f	2	yes
g	1.44	

Fig. 4 Dimensionless diagram of load vs. loading point displacement. Cohesive crack model (a - f) and experimental results (g) from Bocca et al. (to appear b).

The analysis is carried on with matrix (**K** - **C**) no longer positive definite until the finite elements rosette, which has the fictitious crack tip as its centre, becomes tangential to the surface of the concrete slab.

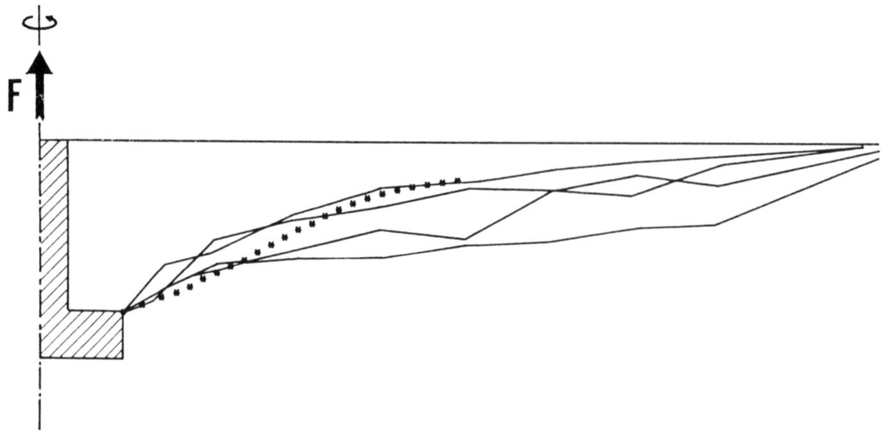

Fig. 5. Crack trajectories for $s_E = 1.44 \cdot 10^{-3}$. Dotted line: cohesive crack model; solid line: experimental results from Bocca et al. (to appear b).

5 Numerical results

Numerical simulations were performed for three different embedment depths (v) while keeping all geometric relationships and properties of the materials ($\nu = 0.1$, $\varepsilon u = 9.71 \cdot 10^{-5}$, $E_a/E_c = 7$) unaltered. The results have been plotted in Fig. 4, as a function of the brittleness number introduced by Carpinteri (1985), $s_E = \mathcal{G}_F^I/(\sigma_u v)$. Mode I fracture energy, $\mathcal{G}_F^I$, is assumed, as a first approximation, as the area lying under the positive portion of the σ_c - w_n curve, relating to $\beta = 0.7$. Fig. 6 shows the finite element rosette and the displacement field at step 26.

The irregular form of the post-critical branch, relating to the case of $\tau_c \neq 0$, is due to the assumption of a constitutive law of an experimental character.

6 Conclusions

1. By varying the brittleness number, s_E, we get marked variations in maximum load, while the variations in the slope of the post-critical branch are seen to be negligible. This observation applies for $\tau_c = 0$ as well as for $\tau_c \neq 0$.
2. The taking into account of τ_c eliminates the negative slope of the post-critical branch.
3. A comparison with the experimental data (published by Bocca et

al., to appear b) shows that the assumption of $\tau_c \neq 0$ is more realistic.

4. The taking into account of τ_c results in a slight increase in maximum load.
5. The taking into account of τ_c and the variation of s_E have a negligible effect on the non-dimensional crack trajectory. This is why in Fig. 5 only the trajectory relating to $s_E = 1.44 \cdot 10^{-3}$ is given.
6. The trajectories obtained numerically are in good agreement with the ones obtained experimentally by Bocca et al. (to appear b) for $s_E = 1.44 \cdot 10^{-3}$.

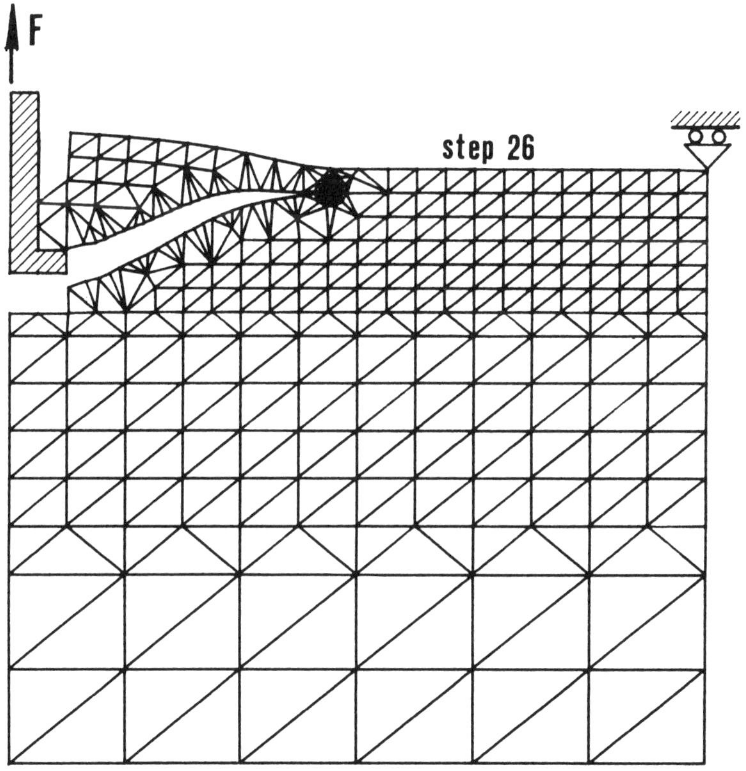

Fig. 6. Finite element mesh and displacement field (enlarged 80 times) at step 26.

7 Acknowledgements

This study was carried out with the financial support of the Ministry of University and Research (M.U.R.S.T.) and the National Research Council (C.N.R.).

8 References

Barenblatt, G.I. (1959) The Formation of Equilibrium Cracks during Brittle Fracture: General Ideas and Hypotheses. Axially-Symmetric Cracks. **J. Appl. Math. Mech.**, 23, 622-636.

Bocca, P., Carpinteri, A., Valente, S. (1989) Evaluation of Concrete Fracture Energy Through a Pull-Out Testing Procedure. in **Proc. Int. Conf. on Recent Developments on the Fracture of Concrete and Rock** (eds. S.P. Shah, S.E. Swartz, B. Barr), Elsevier Applied Science, pp. 347-356.

Bocca, P., Carpinteri, A., Valente, S. (1990) Size Effects in the Mixed Mode Crack Propagation: Softening and Snap-Back Analysis. **Eng. Fracture Mech.**, 35, 159-170.

Bocca, P., Carpinteri, A., Valente, S. (to appear a) Mixed Mode Fracture of Concrete. **Int. J. Solids and Structures**.

Bocca, P., Carpinteri, A., Valente, S. (to appear b) Fracture Mechanics Evaluation of Anchorage Bearing Capacity in Concrete, in **Proc. Int. Workshop on the Application of Fracture Mechanics to Reinforced Concrete** (ed. A. Carpinteri), Elsevier Applied Science.

Carpinteri, A. (1985) Interpretation of the Griffith Instability as a Bifurcation of the Global Equilibrium, in **Proceedings of NATO Advanced Research Workshop on Application of Fracture Mechanics to Cementitious Composites** (ed. S.P. Shah), Martinus Nijhoff, pp. 284-316.

Carpinteri A. (1989) Softening and Snap-Back Instability in Cohesive Solids. **Int. J. Numer. Methods Eng.** , 28, 1521-1537.

Carpinteri, A. and Valente S. (1988) Size-Scale Transition from Ductile to Brittle Failure: a Dimensional Analysis Approach, in **Proc. CNRS-NSF Workshop on Strain Localization and Size Effects Due to Cracking and Damage** (eds. J. Mazars & Z.P. Bazant), Elsevier Applied Science, pp. 477-490.

Carpinteri, A., Valente, S., Bocca, P. (1989) Mixed Mode Cohesive Crack Propagation, in **Proc. 7th Int. Conf. on Fracture (ICF-7)** (eds. K. Salama, K. Ravi-Chandar, D.M.R. Taplin, P. Rama Rao), Pergamon Press, pp. 2243-2257.

Dugdale, D.S. (1960) Yielding of Steel Sheets Containing Slits. **J. Mech. Phys. Solids**, 8, 100-104.

Hassanzadeh, M. (1990) Determination of Fracture Zone Properties in Mixed Mode I and II. **Eng. Fracture Mech.** ,36, 845-853.

Hillerborg, A., Modeer, M., Petersson, P.E. (1976) Analysis of Crack Formation and Crack Growth in Concrete by means of Fracture Mechanics and Finite Elements. **Cement and Concrete Research**, 6, 773-782.

Hillerborg, A. (1989) Mixed Mode Fracture in Concrete, in **Proc. 7th Int. Conf. on Fracture (ICF-7)** (eds. K. Salama, K. Ravi-Chandar, D.M.R. Taplin, P. Rama Rao), Pergamon Press, pp. 2259-2268.

Valente, S. (1990) Bifurcation phenomena in the Growth of Cracks Subjected to Cohesive Softening Forces (in Italian), in **Proc. 10th Nat. Conf. A.I.M.E.T.A.**, ETS Editrice, Pisa (Italy), pp. 111-116.

62 BOUNDARY EFFECTS IN MIXED MODE I AND II FRACTURE OF CONCRETE

E. SCHLANGEN and J.G.M. van MIER
Delft University of Technology, Stevin Laboratory,
The Netherlands

Abstract
In this paper results of experiments on single edge notched beams are presented. The beams are loaded in four point shear. It seems that, even when the loading situation is mainly mode II, the beams fail due to exceeding of the maximum tensile (mode I) stress. There is no sliding over the crack plane. A numerical procedure has been developed for simulating tensile as well as mixed mode I and II experiments. This procedure is based on a lattice of brittle breaking beam elements with tensile strengths which are randomly chosen out of a normal distribution.
Keywords: Mixed Mode Fracture, Fracture Mechanics, Concrete, Numerical Modelling, Experimental Research.

1 Introduction

Cracking is one of the most critical considerations in the behaviour of concrete structures, and much research has been done to determine the fracture response of concrete loaded in uniaxial tension. At present a great deal of research is being carried out to determine crack initiation and growth under different loading types, in particular mixed mode combinations, Carpinteri & Swartz (1991).

Mixed mode I and II loading is a combination of tensile and in plane shear loading. In the present mixed mode study a single edge notched beam specimen subjected to four point shear is used according to a proposal for a Round Robin organized by RILEM TC89-FMT, Carpinteri & Swartz (1991). The beams are loaded under displacement control using a servo-controlled machine. Variables in the investigation are specimen size and concrete composition, as well as the boundary conditions. Main goal of the present study is to come to a better understanding of mixed mode I and II cracking in concrete and to develop a numerical tool which can be used in combination with existing continuum based FE approaches.

To compare these results with mode I experiments, deformation controlled uniaxial tensile tests are carried out in addition to the beam tests. From both the mode I and mixed mode tests the fracture energy G_f is calculated. The crack

Fracture Processes in Concrete, Rock and Ceramics Edited by J.G.M. van Mier, J.G. Rots and A. Bakker.
 Published by E. & F.N. Spon, 2–6 Boundary Row, London SE1 8HN. ISBN 0 419 15870 7

opening in the beam specimen is measured and studied with a long distance microscope during the tests.

Since cracks in concrete follow the path with the lowest resistance they are not straight, but rather tortuous. Numerical models based on continuum mechanics cannot produce crack patterns as found in experiments on heterogeneous materials like concrete. Lately Van Mier (1990) showed that cracks in concrete are not continuous: there are overlaps behind the crack tip. A numerical model is required which can simulate tortuous crack patterns as well as crack bridging for concrete under different loading conditions.

2 Experiments

2.1 Test rig and measurement procedure

The experiments are carried out in the test rig shown in Figure 1. The test setup differs from the loading scheme proposed by the RILEM committee TC89-FMT. The load is introduced into the specimen via double hinged tensile bars in contrast to the proposed loading system where the beams are loaded in compression. The setup is changed to minimize the influence of friction in the supports. For more information about the loading system the reader is referred to Schlangen & Van Mier (1990).

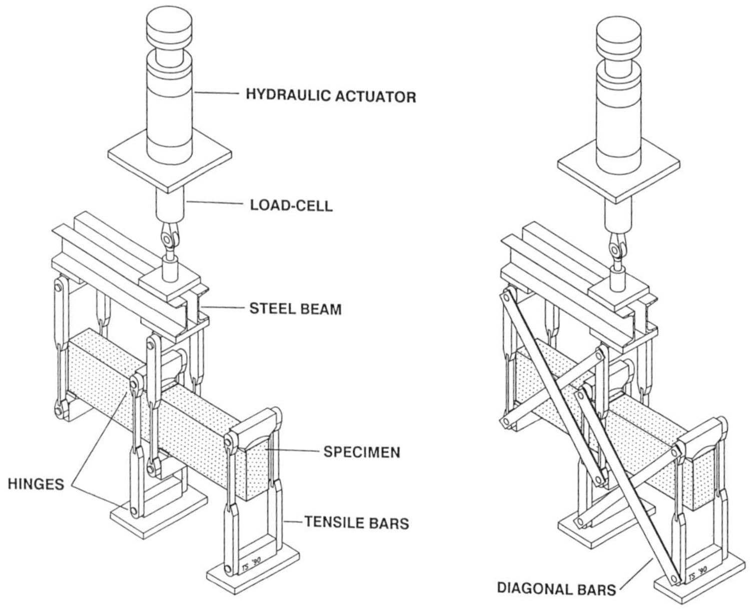

Fig.1. Loading configuration; free supports (loading type I and III) at the left; fixed supports (loading type II) at the right.

The beams are loaded under displacement control. The average signal of two LVDTs (Sangamo type A6GH, linear stroke ± 1 mm), measuring the deformation δ_2 (Fig. 2a) at front and rear of the specimen, was used as feed back signal in a closed-loop servo-control. Further deformations measured are δ_1 (Fig. 2a), CMOD, CMSD (Fig. 7a), all with LVDTs at front and rear. In addition to this, local deformations over the crack zone (Fig. 7a) were measured with extensometers (clip gauges with strain gauges in full-bridge configuration, accuracy 0.5 μm). It is important to know the exact load distribution as this influences crack paths and load-displacement curves, Schlangen & Van Mier (1990). In the present tests the loads and reactions are directly measured with strain gauges (full bridge configuration) glued onto the tensile bars.

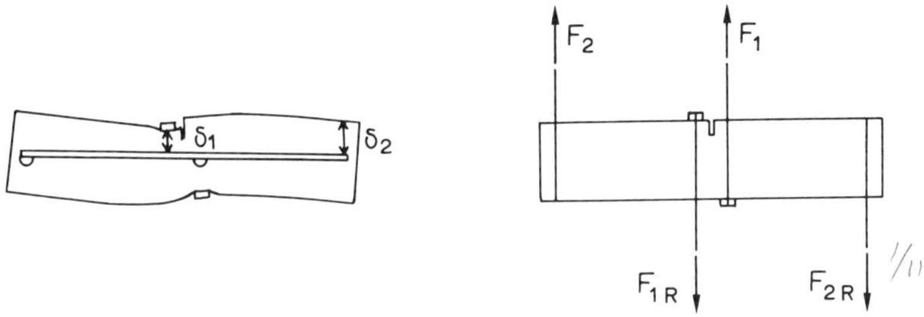

Fig.2. Deformations δ_1 and δ_2 (a); loads F_1 and F_2 and reactions F_{1R} and F_{2R} (b).

2.2 Test series

Three different loading types are used to fracture the specimens. Loading type I has free supports and the notch is situated at the top of the beam. The supports are fixed in loading type II. In order to study the influence of friction in the supports in loading type II, diagonal bars are mounted on the loading frame (Fig. 1). Loading type III again has free supports, but the notch is now situated at the bottom of the beam in order to check the loading frame for possible eccentricities.

Further variables in the investigation are the specimen size and the concrete mix. Two different concretes, with maximum aggregate size $d_{max}=8$ and 16 mm respectively, were selected. The concrete mix proportions are given in the appendix. Each batch consists of two beams of size 840∗260∗100 mm^3 and two smaller beams of size 440∗160∗100 mm^3. After two days of casting the specimens were demoulded and cured under water. After 14 days the beams were sawn to the required size: 840∗200∗100 mm^3 and 440∗100∗100 mm^3 with a notch depth of 40 and 20 mm respectively. The width of the notch is 5 mm. After sawing the specimens were again placed in a fresh water basin. After 28 days of casting the specimens were taken out of the water and left in the laboratory to dry until testing. The age at testing varied between 45 and 65 days.

In total 12 loading combinations are possible: loading configuration(3) ∗ beam size(2) ∗ concrete mix(2). For each combination at least three tests have been carried out.

In addition to the beams, plates (300*300*50 mm^3) were casted. The plates were sawn to specimens of size 150*60*50 mm^3 for deformation controlled uniaxial tensile tests. The tensile specimens are double notched (5 mm deep) at half height. The net cross area where the crack will appear is 50*50 mm^2. For each batch three tensile tests were carried out. Also included in each batch were six cubes for determining the compressive and tensile splitting strength of the material. The respective values are given in the appendix.

3 Experimental results

3.1 Global specimen response

A significant difference is observed between beams tested with free supports (type I and III) on the one hand and fixed supports (type II) on the other hand. Figures 3a and 3b show plots of the total load P versus the deformation δ_2 for small and large beams tested in the three different loading configurations. The crack patterns are shown in Figure 4. In the situation with fixed supports, type II, the descending branch is significantly higher compared to the other two configurations. At the end of the descending branch the load increases and in the small beams a flexural crack develops in the specimen (Fig. 4). Uncontrollable failure occurs now because δ_2 decreases. There is hardly any difference between the load displacement curves of loading types I and III.

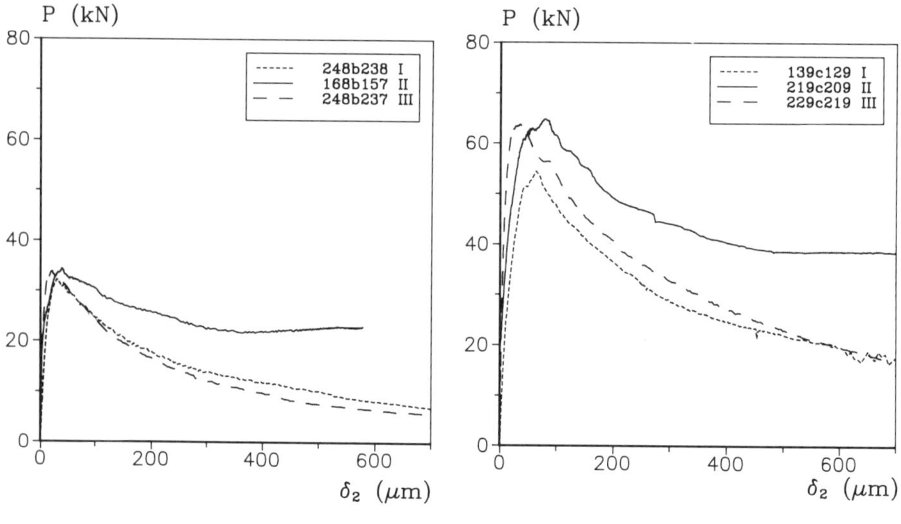

Fig.3. Total load P versus δ_2 for tests of small beams (a) and large beams(b) in the three loading configurations.

Friction in the supports and eccentricities will influence the load distribution. In Figures 5a and 5b the ratio of the loads F_1 and F_2 and the ratio of the reaction loads F_{1R} and F_{2R} (Fig. 2b) are plotted versus the deformation δ_2. Without friction and eccentricities these ratios should be equal to 10. For loading type I and II, the ratio F_{1R}/F_{2R} increases and F_1/F_2 decreases if δ_2 increases. For loading type III, with the notch at the bottom of the beam, the ratio F_1/F_2 increases and F_{1R}/F_{2R} decreases if δ_2 increases. If the beam is fractured into two pieces separated by the curved crack, in all three loading situations the load which acts on the smaller beam half (F_{2R} in loading type I and II; F_2 in loading type III) will decrease with respect to the three loads which act on the other beam half. In loading type II with fixed supports, the ratios deviate more from 10 as compared to loading types I and III. The load F_{2R} on the smaller specimen half approaches 0 kN. In the larger beam half the loading situation resembles three point bending, which leads to flexural failure.

Fig.4. Crack patterns.

From the above observations it can be concluded that the load distribution in the specimen changes during crack propagation.

The difference between the two concretes tested is small. The beams with the 16 mm concrete have a peak load which is only a few percent higher than the beams with the 8 mm concrete (Table 1).

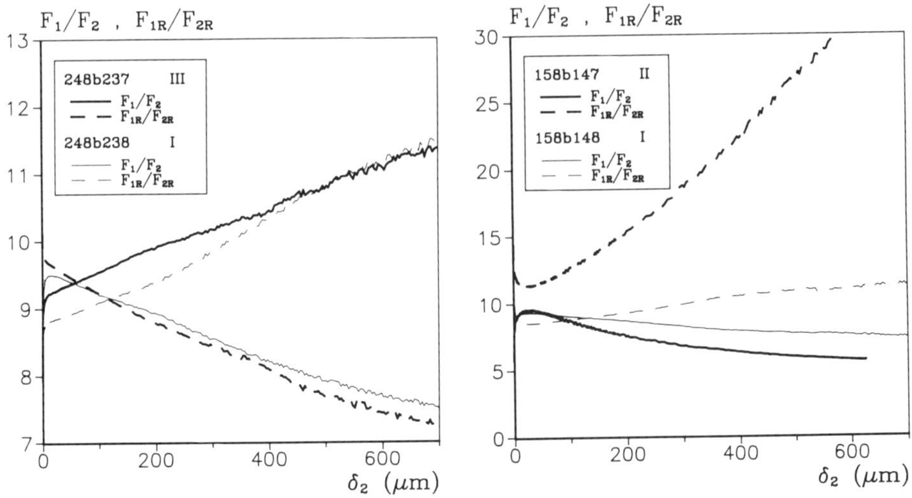

Fig.5. Ratios F_1/F_2 and F_{1R}/F_{2R} versus δ_2; loading type I and III (a); loading type I and II (b).

d_{max} (mm)	size	type	P (kN)	F_2,F_{2R} (kN)	$G_{f,i}$ (J/m^2)		$G_{f,ii}$ (J/m^2)		G_f^I (J/m^2)
		I	26.4	2.74	98.4		116.4		
		I	38.0	3.70	119.8	average	126.9	average	
		I	32.2	3.26	118.4	112.1	133.8	125.6	
	small	III	27.0	2.59	120.5	st.dev.	137.1	st.dev.	
		III	27.8	2.81	102.8	8.2	126.2	7.4	
		III	35.7	3.39	116.3		116.3		average
		III	33.8	3.31	108.6		122.3		115.1
8		I	58.2	5.59	84.9		83.5		st.dev.
		I	56.2	5.05	111.1	average	127.2	average	8.3
		I	56.6	5.17	120.1	114.0	145.4	132.1	
	large	III	48.2	4.75	143.6	st.dev.	189.5	st.dev.	
		III	48.1	4.61	104.2	16.5	115.5	29.6	
		III	52.0	5.04	113.9		132.8		
		III	55.7	5.39	120.5		131.1		
		I	29.8	2.89	143.7		164.3		
		I	27.2	2.73	121.5	average	136.1	average	
		I	30.4	2.95	87.1	114.6	91.8	127.8	
	small	I	35.9	3.34	108.9	st.dev.	122.3	st.dev.	
		I	36.8	3.33	118.5	20.1	127.7	22.1	
		III	44.4	4.31	136.1		144.7		average
		III	33.7	3.30	87.6		107.7		125.4
16		I	54.4	5.45	128.9		168.2		st.dev.
		I	48.9	4.38	94.3	average	115.6	average	11.5
	large	I	60.4	5.67	130.4	121.7	149.0	144.6	
		III	56.9	5.49	104.9	st.dev.	123.8	st.dev.	
		III	62.9	6.04	134.8	16.2	154.0	18.7	
		III	63.7	6.14	137.2		157.0		

Table 1. Peak load and fracture energy.

3.2 Fracture energy

In mode I experiments the area under the stress deformation curve represents the fracture energy G_f of concrete (Fig. 6). How this fracture energy should be determined and by which parameters G_f is influenced is not clear. Is pre-peak cracking, area A_1 in Figure 6a, important? What is the influence of the bulk dissipation, Guinea et al. (1990)? To what extent should the tail of the descending branch, area A_3 in Figure 6a, be taken into account? All these confusing components lead to G_f-values which differ sometimes more than 100%, see for example the results of the round robin test on mode I beams, Hillerborg (1985).

G_f-values for both four point shear tests and uniaxial tensile tests are determined, see Table 1.

For the beam experiments the fracture energy is calculated by adding the area under the F_{1R}-δ_1 and F_{2R}-δ_2 curves (for loading type III the area under the F_1-δ_1 and F_2-δ_2 curves) and dividing the total by the area of the curved crack. If the loads F_1 and F_2 are taken equal to 10/11P and 1/11P respectively, as done by Carpinteri et al. (1990), the G_f-values are overestimated. In this paper the measured loads and reactions are used. The area under the F_{1R}-δ_1 curve is almost zero. For the area under the F_{2R}-δ_2 curve the question arises how to extend the curve. The deformation δ_2 is measured up to 700 μm, at which moment a considerable amount of load is still carried in the beams. Two methods are used. In the first method, $G_{f,i}$, a straight line is drawn from the end of the measured curvé ($\delta_2 = 700\,\mu m$) to the point 1000 μm on the x-axis. $G_{f,i}$ is equal to area A_5 shown in Figure 6c. In the second method, $G_{f,ii}$, the tangent to the F_{2R}-δ_2 curve is extended till it crosses the x-axis. $G_{f,ii}$ is equal to areas A_5 plus A_6. Pre-peak cracking as well as elastic deformations are included in $G_{f,i}$ and $G_{f,ii}$. $G_{f,i}$ and $G_{f,ii}$ are compared with G_f^I derived from the deformation controlled uniaxial tensile tests. For G_f^I also the total area under the load displacement (35 mm measuring length) curve is taken up to a displacement of 150 μm where the curve is cut off. G_f^I is equal to area A_4 in Figure 6b. In Table 1 G_f^I, $G_{f,i}$ and $G_{f,ii}$ (for loading type I and III) are shown. Also the maximum total load P and the corresponding F_{2R} for loading type I and F_2 for loading type III are listed.

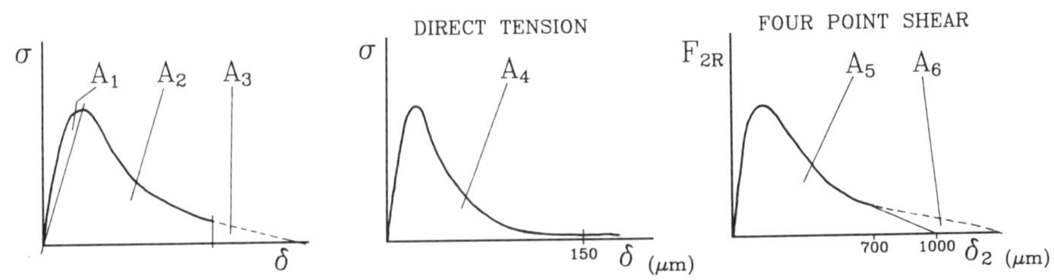

Fig.6. Areas for determining the fracture energy G_f..

The difference between G_f^I on the one hand and $G_{f,i}$ and $G_{f,ii}$ on the other hand is small. If mode II has an influence on G_f, this influence will be small. What can be concluded is that G_f for the 16 mm concrete is larger than for the 8 mm concrete. The fracture energy G_f seems size dependent, which is probably caused by crack interface bridging, Van Mier (1990). The crack interface bridges will probably be larger not only when larger aggregates are used but also when the absolute length of the crack increases. More energy is needed to break larger bridges. Of great importance is the observation that for a good comparison between G_f-values found with different test procedures and G_f-values found by different researchers, the above effects are important sources for errors and should be eliminated.

3.3 Crack patterns

Local deformations (Fig. 7a) on front and rear of the specimen are measured in order to determine the crack width at different places on the specimen surface during the experiments. Figure 8 shows these deformations plotted versus the total load for a large beam specimen (249c239). In Figure 7b the local deformations and also CMOD and CMSD are plotted at the stage in this test where the load is 20 kN in the descending branch.

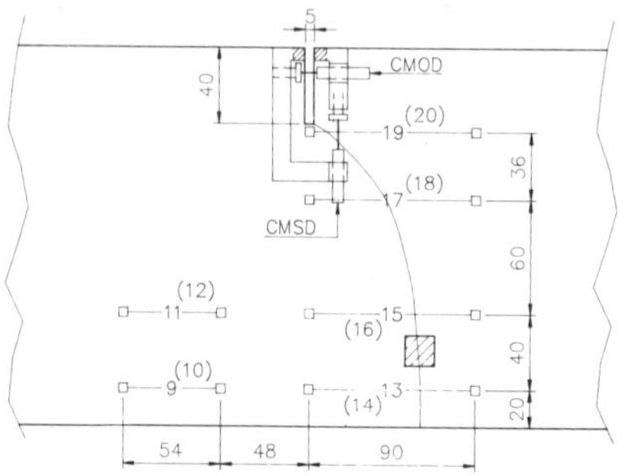

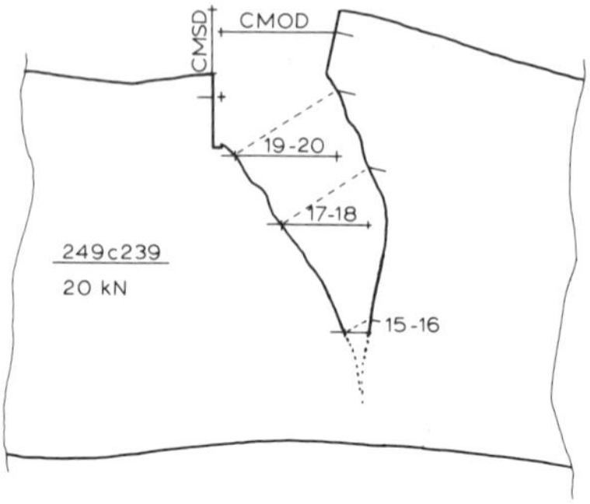

Fig.7. Local deformations measured (a); open crack with measured deformations in test 249c239 at 20 kN in the descending branch (b).

The left beam half with the final crack pattern for this particular test is drawn. At the places where the extensometers are fixed to the specimen a horizontal line is plotted with a length corresponding to the measured deformation starting on the left crack surface. For CMSD a vertical line is plotted. The right beam half with the same crack curvature is drawn starting at the top of the beam, because this point is defined by CMOD and CMSD. The horizontal distance between the points on the right crack surface at the places where local deformations are measured and the corresponding points on the left crack surface has to be equal to the lengths of the horizontal lines plotted. The results indicate that the crack opens perpendicular to the crack surface, except maybe for the region near the crack tip.

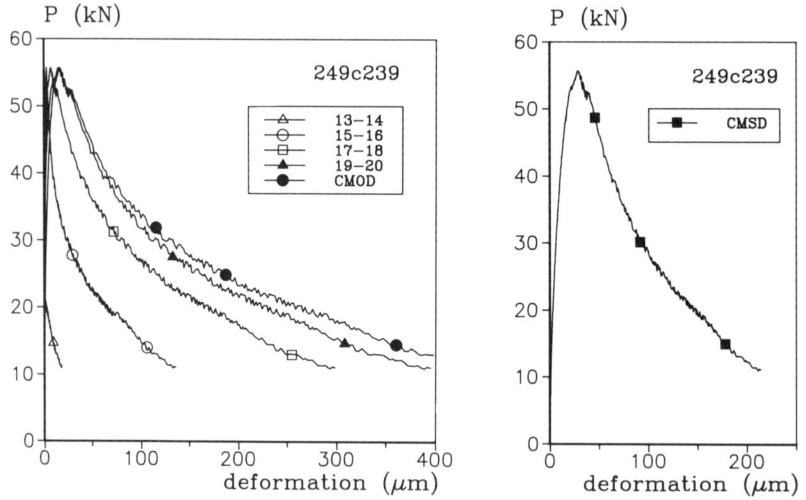

Fig.8. Total load P versus local deformations in test 249c239.

The crack opening is studied by means of a long distance microscope (Questar). The procedure is described in Van Mier (1991). Figure 9 shows the crack growth in the specimen 249c239 at different stages in the descending branch. The pictures are taken in the middle of the shaded square in Figure 7a. From these results we also see that there is only opening of the two separating crack sides and no sliding. The crack is not a single continuous crack, but there are overlaps or so-called crack interface bridges as described earlier, Van Mier (1990).

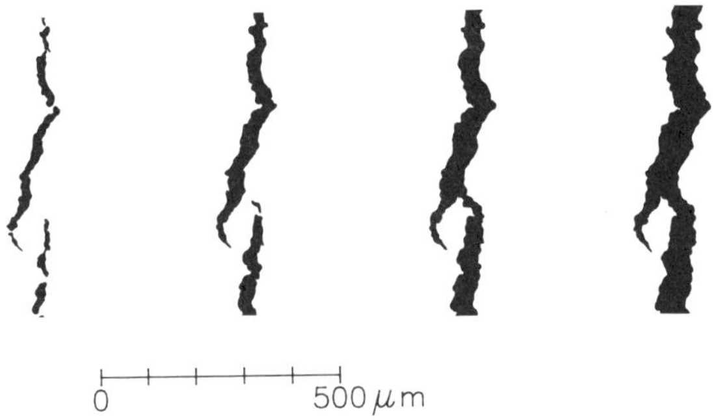

Fig.9. View of cracks studied with a microscope in test 249c239 at stages where the load was 26.7 kN, 22.9 kN, 19.4 kN and 11.1 kN in the descending branch.

From the above observation it can be concluded that there is no sliding over the crack surfaces, but only opening of the crack. CMSD which is measured arises due to opening of the curved crack. If there is an influence of mode II, it is at the crack initiation and maybe near the crack tip during propagation.

4 Numerical model

The above results indicate that the determination of global fracture parameters, such as fracture energy G_f, is hampered by experimental deficiencies. Numerical simulation of the growth process "Fracture" may be a method to overcome the experimental problems.

A numerical model is required which predicts crack patterns and load displacement curves for all types of loading. The model should be able to determine overlapping cracks in good agreement with experimental results. Lattice models as described by Herrmann (1988) seem a promising tool.

In this investigation such a model is combined with a finite element program. In the area where cracks are expected to appear the mesh contains bars with two nodes in a two dimensional triangular lattice (Fig. 10). In the nodes there is a fixed connection between the bars. The bars have tensile strengths f_t which are randomly chosen out of a normal distribution with an average value of 3 N/mm^2 and a standard deviation of 1 N/mm^2. The system is loaded via nodal displacements or nodal forces.

Fig.10. Triangular lattice of bars with fixed nodes.

The normal forces F in the bars and the moments M_i and M_j in the nodes (i and j) of the bars are computed linearly elastic. From this, the stress σ_t for each bar is derived as follows:

$$\sigma_t = F/A + \max(|M_i|,|M_j|)/W$$

$A = b*h =$ cross area of bar
$W = 1/6*b*h^2$

The bar with the lowest ratio f_t/σ_t is removed from the mesh, after which the procedure is repeated until the lattice is separated into two pieces.

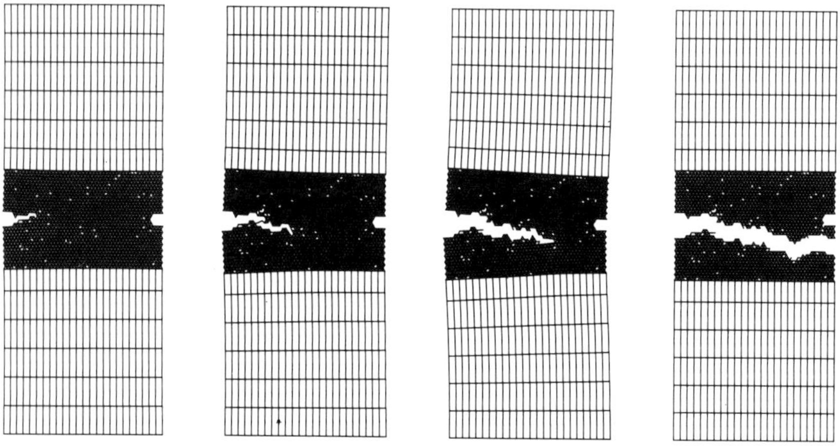

Fig.11. Crack propagations in simulated uniaxial tensile test.

Variables in this model are the cross area and the moment of inertia of the bars, as well as the distribution of tensile strengths. Simulations of uniaxial tensile tests with the same specimen dimensions and boundary conditions as in the experiments are used to tune the various variables, after which simulations on the four point shear beams are carried out. Some preliminary results (crack patterns) are shown in Figures 11 and 12. In the simulations of the four point shear beams the crack starts at the notch and propagates to the opposite support. It is a curved crack as found in the experiments. Numerical simulations with finite element programs based on continuum mechanics are generally not capable of simulating curved cracks, Schlangen & Van Mier (1990), Rots (1988).

Further research is required to optimize the present model.

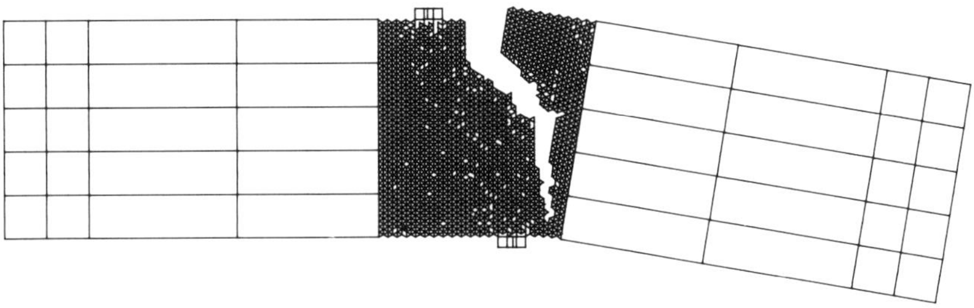

Fig. 12. Deformed mesh of simulated four point shear test.

5 Conclusions

The behaviour of the four point shear beams is significantly influenced by boundary condition effects. The load distribution, the crack patterns and the load displacement curves, depend to a large extent on the amount of friction in the supports.

It is not clear how the fracture energy G_f should be determined. A comparison between G_f values determined in different tests and found by different researchers is not possible, because different procedures are used for analysing the results. G_f is size dependent. This is probably caused by crack interface bridging, which may be more prominent in larger specimens. The influence of mode II on G_f is negligible.

Although the loading situation in the four point shear beams is mixed mode I and II the fracture growth is in mode I. Fracture nucleation is likely to be more affected by the mixed mode loading.

Lattice models are considered an excellent tool for studying crack growth in heterogeneous materials. Realistic crack patterns for mode I and mixed mode I and II loading were obtained in the present study. Further research is needed for determining the input parameters for the model.

6 Acknowledgement

Financial support from the Dutch Technology Foundation (STW) is gratefully acknowledged. The authors are greatly indebted to Mr. A.S. Elgersma for his substantial assistance in carrying out the experiments.

7 References

Carpinteri, A. Ferrara, G. Melchiorri, G. and Valente S. (1990) The four point shear test on single notched specimens: an experimental and numerical analysis, in *Fracture Behaviour and Design of Materials and Structures* (ed. D. Firrao), EMAS, UK, pp. 667-675.

Carpinteri, A. Swartz, S.E. (1991) Mixed mode crack propagation in concrete, in *Fracture Mechanics of Concrete - Test Method* (ed. S.P. Shah), Chapman and Hall, London/New York.

Guinea, G.V. Planas, J. and Elices M. (1990) On the influence of the bulk dissipation on the average specific fracture energy of concrete, in *Fracture Behaviour and Design of Materials and Structures* (ed. D. Firrao), EMAS, UK, pp. 715-720.

Herrmann, H.J. (1988) Introduction to modern ideas on fracture patterns, in *Random Fluctuations and Pattern Growth: Experiments and Models* (eds. H.E. Stanley and N. Ostrowsky), Kluwer Academic Publishers, Dordrecht, pp. 149-160.

Hillerborg, A. (1985) Results of the comparative test series for determining the fracture energy G_f of concrete. *Materials and Structures*, vol. 18, no. 107, 33-39.

Rots, J.G. (1988) *Computational Modeling of Concrete Fracture*. PhD. thesis, Delft University of Technology.

Schlangen, E. and Van Mier J.G.M. (1990) A fe-supported investigation of mixed mode fracture in concrete, in *Proceedings of the 9th International Conference on Experimental Mechanics*, Technical University of Denmark, Copenhagen, Denmark, pp. 1403-1412.

Van Mier, J.G.M. (1990) Internal crack detection in single edge notched concrete plates subjected to uniform boundary displacements, in *Micromechanics of Failure of Quasi-Brittle Materials* (eds. S.P. Shah, S.E. Swartz and M.C. Wang), Elsevier Applied Science Publishers, London/New York, pp. 33-42.

Van Mier, J.G.M. (1991) Crack face bridging in normal, high strength and lytag concrete, *this volume*.

8 Appendix

CONCRETE MIX PROPORTIONS:

Portland cement Type B, 375 kg/m^3; w/c-ratio = 0.5;

<u>8 mm concrete:</u> 127 kg/m^3, 0.125-0.25 mm; 234 kg/m^3, 0.25-0.5 mm; 272 kg/m^3, 0.5-1 mm; 272 kg/m^3, 1-2 mm; 363 kg/m^3, 2-4 mm; 540 kg/m^3, 4-8 mm;

<u>16 mm concrete:</u> 128 kg/m^3, 0.125-0.25 mm; 165 kg/m^3, 0.25-0.5 mm; 202 kg/m^3, 0.5-1 mm; 256 kg/m^3, 1-2 mm; 256 kg/m^3, 2-4 mm; 365 kg/m^3, 4-8 mm; 457 kg/m^3, 8-16 mm.

STRENGTH RESULTS:

The average 28 days compressive strength $f_{cc} = 48.89$ N/mm^2 (st. dev. 1.95 N/mm^2) for the 8 mm concrete and 50.10 N/mm^2 (st. dev. 1.98 N/mm^2) for the 16 mm concrete. The average 28 days splitting tensile strength $f_{c,spl} = 3.30$ N/mm^2 (st. dev. 0.17 N/mm^2) for the 8 mm concrete and 3.38 N/mm^2 (st. dev. 0.11 N/mm^2) for the 16 mm concrete.

63 NUMERICAL AND EXPERIMENTAL STUDY OF DEVELOPMENT OF FRACTURE PATH UNDER MIXED MODE LOADING

J. DAVIES
The Polytechnic of Wales, Pontypridd, UK

Abstract
The study of the distribution of elastic stresses between notches represents a special interest in the understanding of the mechanism of fracture. The Finite Element Analysis was used to determine the influence of specimen geometry and loading on the stress distribution between notches of varying configurations. The experiments indicated that in some cases where the shear stress concentration was very high the fracture path followed the direction of maximum shear planes rather than the expected direction perpendicular to the maximum tension. Using a photographic technique it was possible to demonstrate that a number of almost vertical discontinuous fissures originated from the notches and in the direction which was in a good agreement with the direction of maximum in-plane shear stresses. These findings imply that in some situations failure mechanism other than tensile may develop and this distinctly different mechanism can be classed as shear fracture.
Keywords: Fracture, Concrete, Mortar, Mixed-Mode, Shear Fracture, Notches, Finite Element Analysis, Fracture Path, Shear Stress,Cracks,Tensile Fracture, Microcracks, Mode I Mode II, Failure Mechanism.

1 Introduction

The fracture process in mortar or concrete takes place over a relatively large fracture process zone and the tip of the visible crack is often blunted by a zone of microcracks which is relative to the size of material homogeneity. The fracture in these materials is initiated when stresses over a fracture process zone reach a critical value, rather than their value at some point as it is in the

Fracture Processes in Concrete, Rock and Ceramics Edited by J.G.M. van Mier, J.G. Rots and A. Bakker.
 Published by E. & F.N. Spon, 2–6 Boundary Row, London SE1 8HN. ISBN 0 419 15870 7

case of homogeneous and elastic materials. Cracks are generally assumed to propagate in the direction normal to the maximum principal stress, which represents the tensile opening mode (Mode I). The tensile fracture mechanism alone, however, is not sufficient to cause fracture of specimens subjected to compression or shear loading and clearly some other failure mechanism becomes activated at some stage of the failure process.

The Linear Elastic Finite Element Analysis was used to study a stress distribution in specimens of varying geometries and notch configurations subjected to mixed-mode loading. The element size in the ligament zone was generally greater than aggregate size, therefore the stress singularities were not reflected in the analysis. The analyses carried out on standard beam and cube specimens enables a study of stress contours and directions of internal stresses and consequently models describing most probable crack were established. This was achieved by superimposing the directions perpendicular to the maximum principal tensile stresses and the fracture path obtained from the experiments. The fracture in most cases was very fast and this fact made it difficult to capture an early stages of the fracture process. The monitoring of the fracture zone formation was carried out with the aid of a video camera with the shutter speed 500 - 1000 fames/sec. It was shown that in some cases discontinuous fissures appeared in almost vertical direction,and this direction was in a good agreement with the direction of planes of the maximum in-plane shear. The mortar specimens tested in this program were 250X100X100mm beams and 100 mm cubes with the following notch configurations:

Two opposite single notches

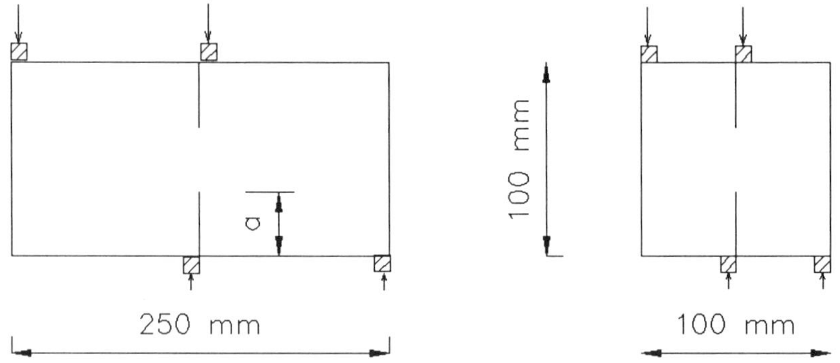

a) double-notched shear beam

b) double-notched shear cube

Fig.1.

Two pairs of opposite notches of the different length

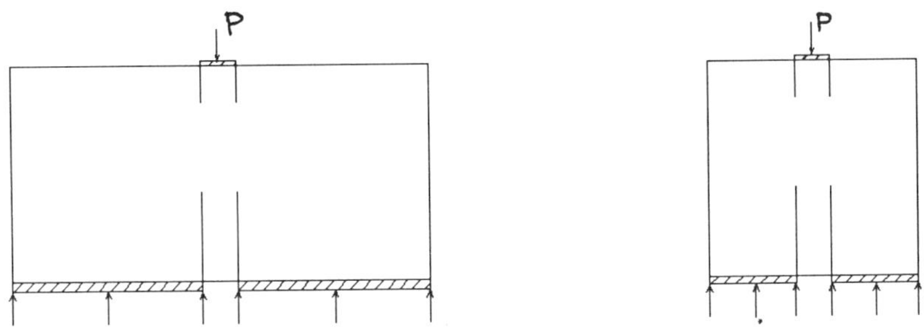

a) punch-through shear beam

b) punch-through shear cube

Fig.2.

Two pairs of opposite notches of the same length

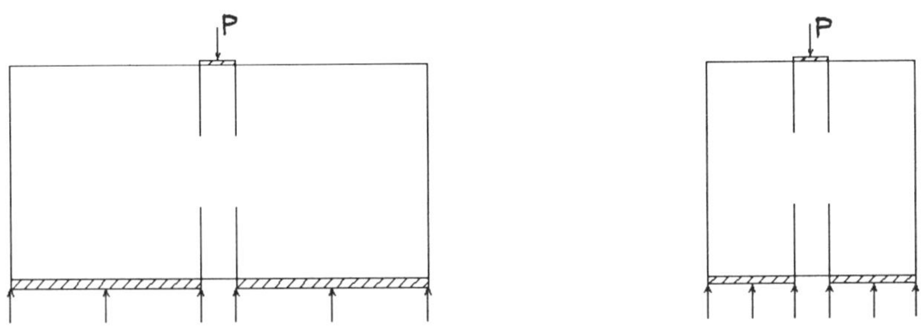

a) double-punch shear beam

b) double-punch shear cube

Fig.3.

2 Results

2.1 Double-notched shear beam

The geometry of these specimens is shown in Fig.1a. The distribution of the principal stresses in the ligament containing the notch as obtained from the numerical analysis is shown in Fig.4. and the directions perpendicular to the maximum principal tensile stresses are shown in Fig.5. Inclination of these stress directions were generally varying between 37° to 41° from the vertical direction.

It was observed during the experiments that very faintly

defined inclined cracks running 30° to 40° from the both notches and the almost vertical crack, inclined 0° to 12° appeared simultaneously, Fig.6. With the increasing loading additional vertical cracks appeared, both emanating from two crush zones at the end of each notch and producing a final failure. The inclined crack, however, did not increase in length or width during the additional loading, Fig.7.

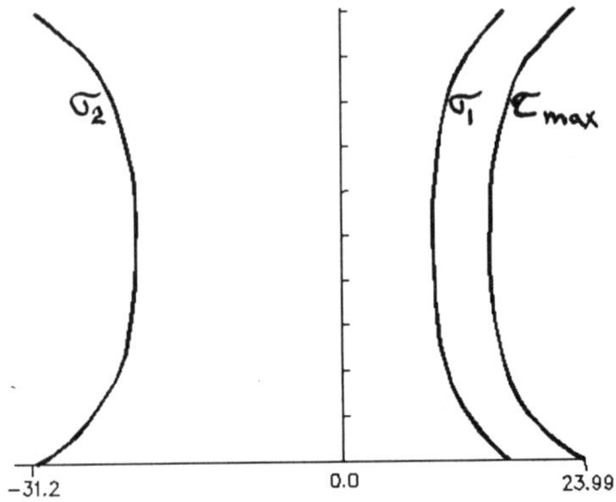

Variation of principal stresses

Fig.4.

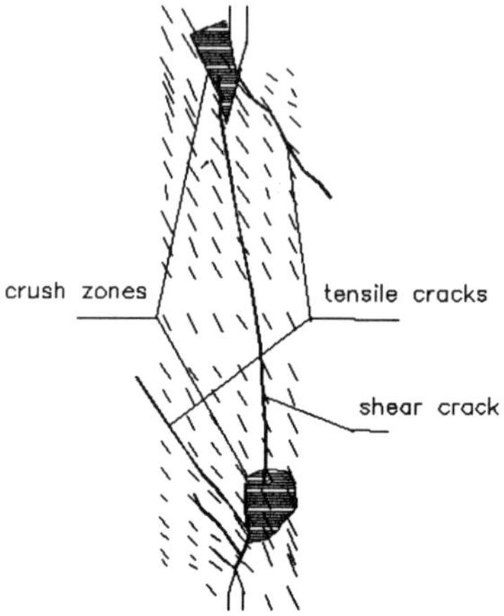

Predicted and actual fracture paths

Fig.5.

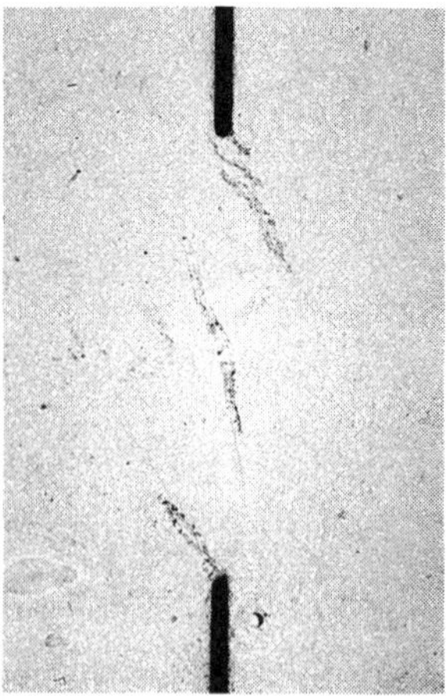

Initial fracture

Fig.6.

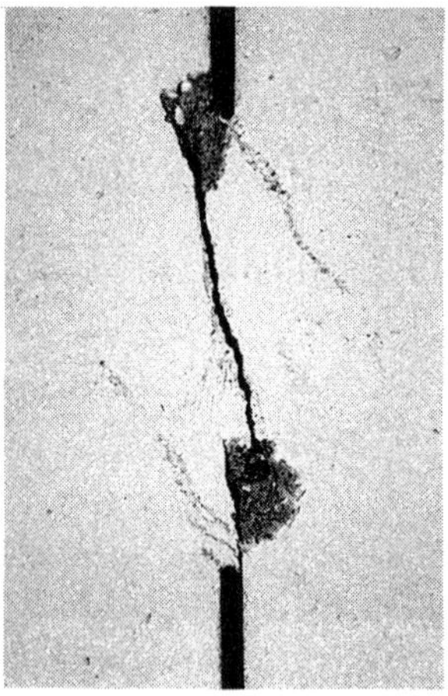

Final failure

Fig.7.

Fig.5. also shows a superposition of the actual crack path

with the predicted directions of the tensile fracture. As can be seen the inclined cracks follow very well the predicted direction of the tensile fracture, but not propagating any further. The main vertical crack,however, follows very closely direction of maximum in-plane shear predicted from a numerical analysis which was in order of 4° to 8° .

It would appear that the subcritical tensile cracks formed at the notch tips caused a local weakening of the material and that some different failure mechanism than tensile was activated during the loading process. The presence of the vertical cracks and crushing zones at the notch tips also suggests that the final failure is unlikely due to tension but must be shear compression failure.

2.2 Double-notched shear cube

The geometry of the specimen is shown in Fig.1b.The variation of principal stresses in a cube specimen is shown in Fig.8. and Fig.9 represents the actual crack path superimposed with the predicted directions of the tensile fracture. As can be seen the experimentally obtained crack path does not follow the strongly inclined direction of tensile fracture,implying that the driving force during fracture process is unlikely tension.The fracture mechanism in this specimen is most likely activated by the action of highly concentrated shear stresses within the ligament, producing almost vertical crack.

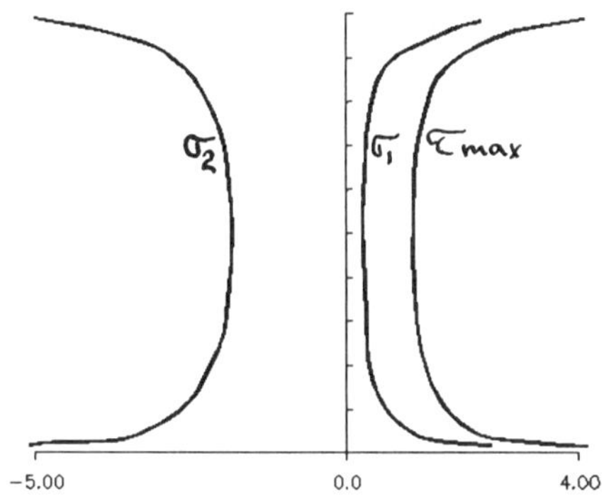

Variation of principal stresses

Fig.8.

Predicted and actual fracture paths

Fig.9.

Fig.10a,b show the initial and the final stages of the fracture, in this case no visible traces of inclined subcritical cracks were detected during the experiments. It is interesting to note that the shorter specimen behaved in a considerably more brittle manner than the beam specimen and the fracture path in these specimens was formed by a series of discontinuous almost vertical fissures.

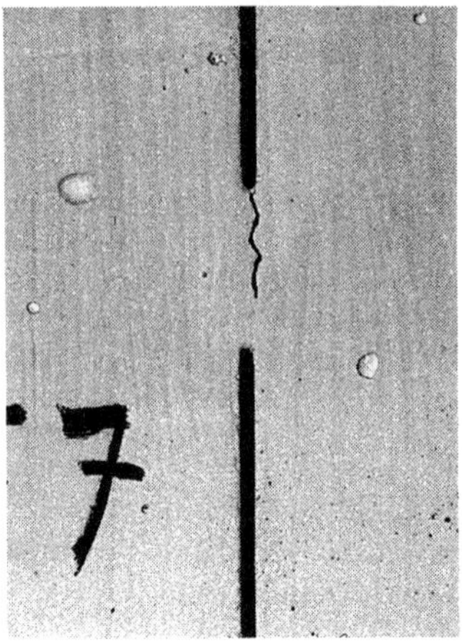

Initial Fracture

Fig.10a.

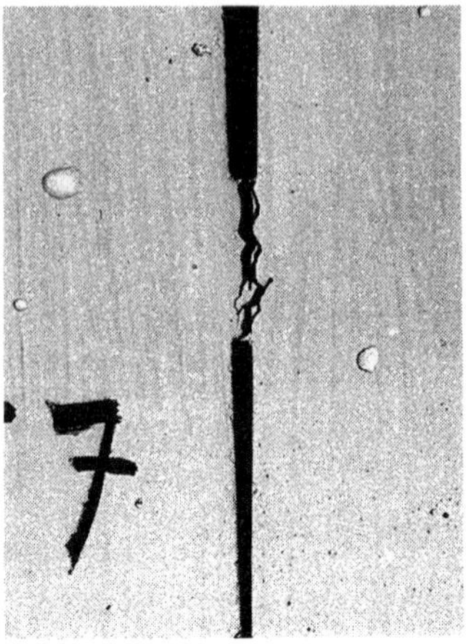

Final Fracture

Fig.10b.

2.3 Punch-through shear beam

Beam specimens were prepared from the standard beams by cutting them in half and inserting series of notches 25mm apart, Fig.2a. Fig.11 shows the variation of principal stresses in the specimen and Fig. 12 represents a superposition of the actual crack path with the predicted direction of the tensile fracture.Due to symmetry only one half of the specimen was considered in the numerical analysis.

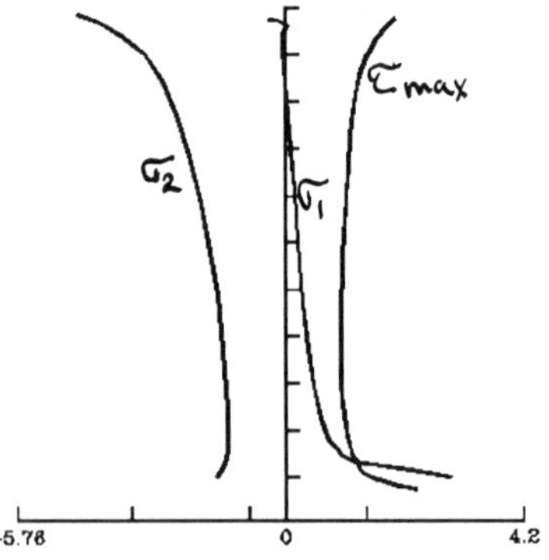

Variation of principal stresses

Fig.11.

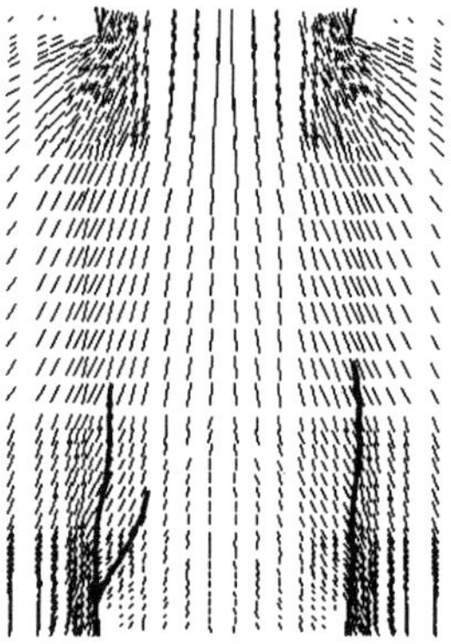

Predicted and actual fracture paths

Fig.12.

The fracture process zone in these specimens consisted of a number of discontinuous cracks with the initial crack inclined by about 29° from the vertical direction, Fig.13. The fracture in this specimen was very fast. Fig.14. represents 1/25th second time interval between Fig.13 and Fig.14 the shearing action being clearly demonstrated. It is also interesting to note that the final fracture path follows the direction of discontinuous fissures and not the initial inclined direction of the tensile crack. Many beams in this test series failed by a single crack .

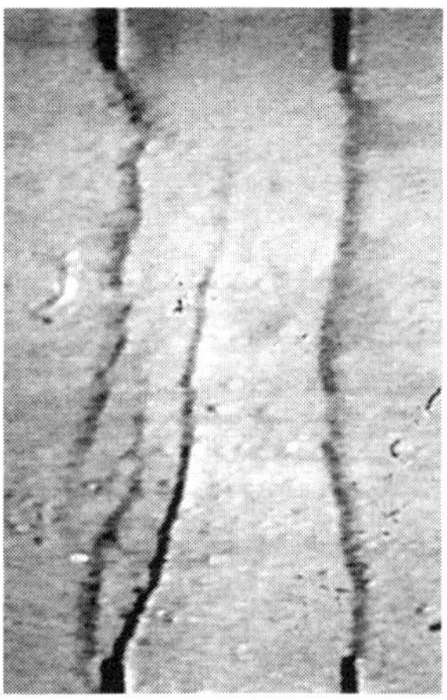

Formation of a fracture zone

Fig.13

Final fracture

Fig.14.

2.4 Punch-through shear cube.

The notch configuration in cube specimens, Fig.2b, was the same as that in the punch-through shear beam. The variation of the principal stresses in the ligament is shown in Fig.15. Strongly inclined directions of predicted tensile fracture is shown in Fig.16. where a superposition with the directions of initial fissures is also indicated. The development of the fracture zone can be seen in Fig.17 and Fig.18., indicating a remarkable similarity with the fracture zone recorded for a beam. In this case, however, the inclination of initial fissures measured from the vertical varied from 0 to 16 the direction which is in an agreement with the direction of the maximum in-plane shear determined numerically.

It was observed that in the cube specimens the failure between the notches was succesfully achieved in almost all specimens whereas in beam specimens considerable amount of bending taking place produced mostly single crack failure. This observation suggests that shorter specimens may be more suitable for the study of shear fracture.

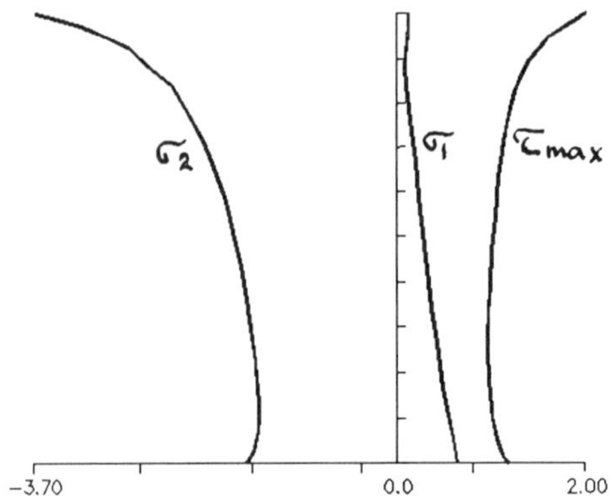

Variation of principal stresses
Fig.15.

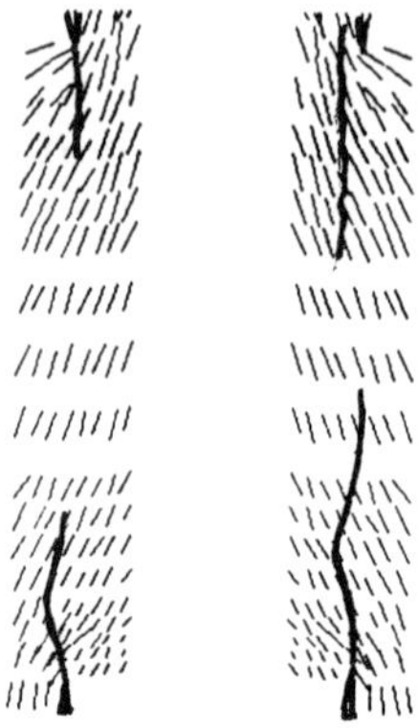

Predicted and actual fracture
Fig.16.

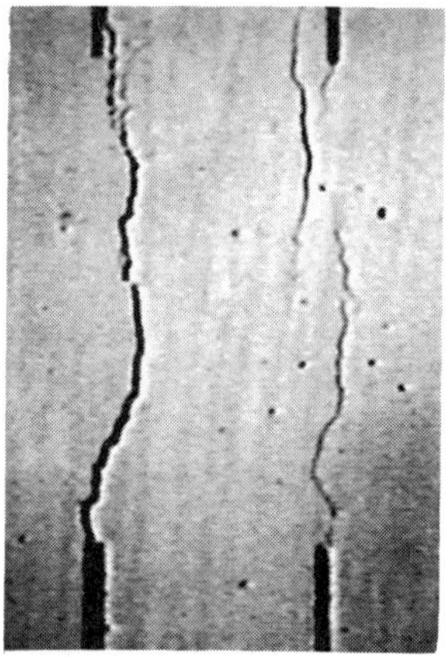

Initial fracture
Fig.17.

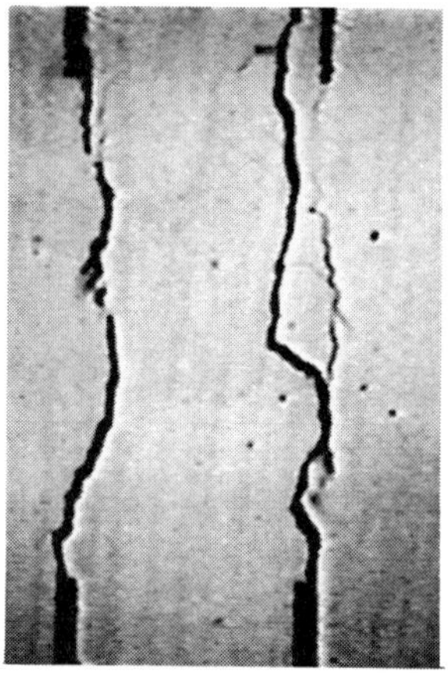

Final fracture
Fig.18.

2.5 Double-punch shear beam

The geometry of this specimen is shown in Fig.3a. An insertion of two identical parallel notches was made in an attempt to furter increase a shear stress concentration and therefore to encourage a shear failure. The stress analysis indicated that there is no significant increase in a shear stress concentration at the notch tips and that the tensile stresses are significant. Fig.19. indicates a superposition of predicted and actual initial tensile crack. The final fracture is shown in Fig.20, indicating clearly that the mixed mode fracture has taken place.

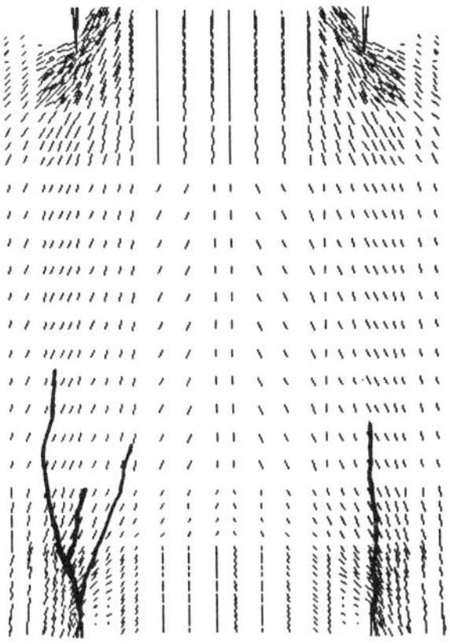

Predicted and actual fracture

Fig.19.

Final fracture

Fig.20.

2.6 Double-punch shear cube.
Double-punch shear cube specimen, Fig.3b. was more succesful in producing a predominantly shear failure but again a faint inclined crack appeared from one of the bottom notches followed by a narrow band of discontinuous fissures, Fig.21 and Fig.22.

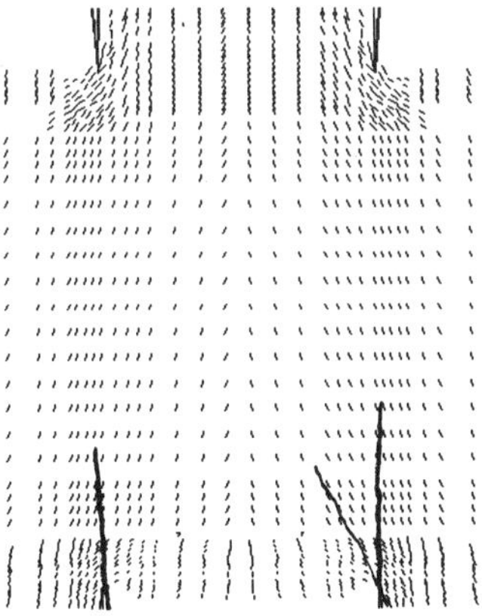

Predicted and actual fracture path

Fig.21.

Final fracture

Fig.22.

3 Conclusion

It would appear that in specimens, where the shear stress concentration was very high,the fracture path followed the direction of maximum shear stresses rather than the direction perpendicular to maximum tension. This phenomenon was observed in double notched and punch-through shear cube specimens. In beams,the maximum principal tensile stresses generated by the shear force acting in the section would encourage a formation of a faint initial inclined crack followed by a series of discontinuous fissures in almost vertical direction finally producing a narrow crush zone.

The experimental monitoring of the fracture zone formation was not easy because the majority of specimens failed suddenly without any signs of initial cracking.A video camera with a high shutter speed was used to monitor experiments and then viewed with the aid of a MERLYN - X vision mixer with effects.Using this technique it was possible to split a fracture process into frames 1/25 seconds apart, which after freezing were photographed from the monitor screen using a PENTAX P30N still camera.

Using this advanced vision technique it was possible to demonstrate that the failure zones in cube specimens with high shear stress concentration between the notches consisted of a number of discontinuous vertical microfissures that originated from the crack tips.The direction of these microfissures was in a good agreement with the direction of maximum shear stresses obtained from the numerical analysis.

4 Acknowledgment

The author wishes to express her gratitude to the members of the Media Resources Department, namely Viv Cole, Chris Thomas, Malcom Coundley and John Thomas for their infinite patience during the production of still photographs.

5 References

Bazant, Z. and Pfeiffer, P.A. (1986) Shear fracture tests of concrete, Mat. Const., 19, 111-21

Carpinteri,A and Valente, S. (1990) Numerical methods in mixed mode cohesive crack simulation, Numeta 90, (eds. G.N.Pande & J.Middleton), Elsevier Applied Science,pp, 804-811.

Davies, J. (1988) Numerical study of punch-through shear specimen in Mode II testing for cementitious materials, Int. J. Cem. Comp. and Light. Concrete, 10, 3 -14.

Davies, J. (1989), Numerical study of specimens subjected to shear compression stress field, Civil. Comp 89, (ed. B.H.V. Topping), Civil-Comp Press, pp, 351-360.

Rots, J.G. and deBost, R. (1987) Analysis of mixed-mode fracture in concrete. J.Appl. Mech., 53, 573-79.

64 MEASURING TOUGHNESS INDICES ON BRITTLE MATERIALS

M.P. LUONG
CNRS-LMS, École Polytechnique, Palaiseau, France

Abstract
An experimental study is presented using new testing devices in order to point out the salient features of fracture processes in brittle disordered materials such as path dependence (mode I tension fracture, mode II or mode III shear fracture) and shear transfer due to aggregate interlock (crack dilatancy), cross effect (frictional contact slip). Experimental observations show that the shear displacement or shear slip is affected not only by the shear stress but also by the normal displacement owing to rough asperities of aggregate or surface roughness and variable rock strength. The normal stresses across the interface in mode II shear tests and mode III torsion tests are measured and applied in different ways, which results in the different numbers of Belleville springs ensuring different shear stiffnesses. These factors must be taken into account when measuring toughness indices of the brittle disordered materials.
Keywords: Fracture testing, Dilatancy effects, Toughness indices.

1 Introduction

Fracture processes in brittle disordered materials are substantially influenced by the stiffness and strength of their constituent particles. An important aspect of crack propagation in concrete and rock is to examine the energy dissipation or toughening mechanisms involved during severe loadings. Measuring toughness indices by means of fracture mechanics test-methods should provide a more comprehensive knowledge of their mechanical behaviour and their failure mechanisms including strength and the evolution of failure. These data should lead to a more realistic numerical modelling for the design of earthquake resistant structures where are needed measures of maximum displacement, energy absorption and loading levels. Usually the overall ductility considers a particular pattern of displacement corresponding to the preferred mode of deformation of the structure in a response condition in which inelastic energy is absorbed as generally as it is possible to develop such deformation throughout the structure.

To survive a severe earthquake, the structural members based on brittle disordered materials must present a sufficient toughness or energy-absorbing ability by undergoing substantial inelastic deformations at nearly constant load, without suffering excessive damage or loss of strength in face

Fracture Processes in Concrete, Rock and Ceramics Edited by J.G.M. van Mier, J.G. Rots and A. Bakker.
 Published by E. & F.N. Spon, 2–6 Boundary Row, London SE1 8HN. ISBN 0 419 15870 7

of subsequent load applications. This requires a designer to realistically assess the possible levels of strength and to ensure adequate toughness with little decrease in strength. The available toughness is mainly dependent on the stress path, stress level and development of dilatancy during fracture processes.

In addition, strength and deformation characteristics of brittle disordered materials such as cementious and rock materials has focussed an increasing interest in recent years because they are directly applicable to the design of large and challenging engineering structures. Practical loading situations in structures often result in complex stress states within individual members, particularly near connections. For engineers, macroscopic fracture mechanics provides an aid in the design against fracture in engineering structures.

Brittle failure is characterized by the appearance of discrete discontinuities which can range from a single crack to any degree of fracturing. Generally after reaching the failure condition, a brittle material presents : (a) a discontinuous structure, (b) the ability nevertheless to sustain shear stresses, (c) a non-elastic bulking wherein shear strain is accompanied by large volumetric strains, (d) the dissipation of energy through inelastic work [Luong, 1987] and (e) a failure which is strain history dependent.

The strength of engineering brittle materials represents the maximum resistance that they are capable of developing under the particular system of loading. The strength does not therefore represent the stress with which the structure begins to be damaged by loading. Ideally, a parameter for material toughness should be a characteristic, independent of the test specimen geometry, dimensions and loading conditions. For economic reasons, it should be easily obtained from simple tests and be expressed in terms of a single parameter for ease of comparison and quality control. However in practice, the methods, often used to assess toughness, do not meet these requirements.

2 Toughness indices

The use of conventional fracture mechanics to characterise cementious material toughness has been discussed by [Hillerborg, 1980] who has shown that the fracture process can be characterised by the stress-crack separation curve which could be obtained from a direct tensile test. The area under the curve corresponds to the fracture energy of the material. However the use of a curve rather a single parameter makes it difficult to compare the toughness among different materials, and the determination of such curve requires complicated test arrangements to ensure a stable loading condition and uniform strain distribution across the specimen.

In theoretical analysis, complete accurate tests are preferred which may give insights into the material characteristics. For practical purposes, a simple method to evaluate the toughness is clearly needed. Many quantities have been proposed which are derived from areas under test curves obtained from simple tests such as flexural and compact tension tests. They are often expressed in terms of nondimensional quantities such as toughness indices.

The toughness index is a measure of the energy absorbed during fracture in the non linear portion of the load-deflection curve. It gives the numerical value to the toughness and is independent of the units used. The index is particularly useful in analysing the various toughening mechanisms. It readily discriminates between high and low energy absorbing materials when subjected to a given loading path.

A simple method of evaluating the post-cracking performance of materials is based on a toughness index that ideally should be a material characteristic, independent of test specimen geometry, dimensions and loading conditions. For ease of comparison and quality control, it should be readily obtained from a load-deflection curve of simple tests. Some toughness indices are reported in literature [Barr et al, 1982 ; Wang & Backer, 1989].

3 Mode I toughness index

Tensile strength of brittle materials is one of its most fundamental and important properties, particularly because these materials are very much weaker in tension than in compression.

$$R_I = F/\pi(r^2_e - r^2_i)$$

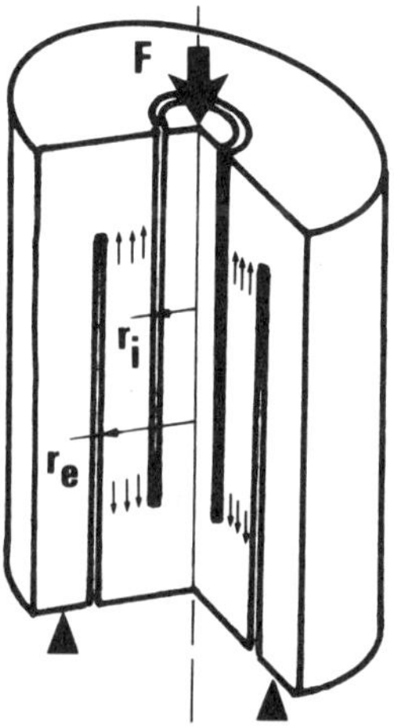

Fig. 1 - Geometry of the tensile test specimen.

Most concrete and rock materials are more or less brittle. When unconfined, the test samples cannot yield plastically to relieve the stress concentrations that are produced at the localized points around the specimens, where these are gripped to be pulled apart by the testing machine. Consequently premature failure is generated from these points. Difficulties in ensuring truly axial loading also exist, so that the specimen is liable to be twisted or bent when gripped and pulled from either end. It has been reported that different fracture energy values were obtained if different specimen geometries were used [Jenq & Shah, 1989]. Does this difference result from different testing set-ups involved : different degree of eccentricity, alignment and support conditions, etc..? The proposed test specimen [Luong, 1989] is a cylindrical tube (Fig. 1) easily prepared, with two parallel flat ends and two inversed tubular coaxial borings. The external surface requires no particular preparation.

The proposed configuration converts the applied load on the specimen into a tension, so that the usual compressive test machine can be used. The cylindrical symmetry permits self alignment of loads parallel to the axis of the specimen. The test requires no special device for specimen gripping and compressive loads for example can be applied without any precaution. There will be no tendency to cause bending so that abnormal stress concentrations are avoided.

The main advantage of the proposed direct tensile test is evidenced by the uniformity of uniaxial tension stress throughout the test volume. Thanks to its cylindrical symmetry, there is no bending or torsional stresses, no stress concentrations arising from geometrical irregularities of the sample and no end restraint effects perturbing the stress field : most of observed failures occur in the central part of the specimen. Mode I toughness index is readily derived from the load-displacement curve (Fig. 2).

Barr et al Toughness Index = 0.288

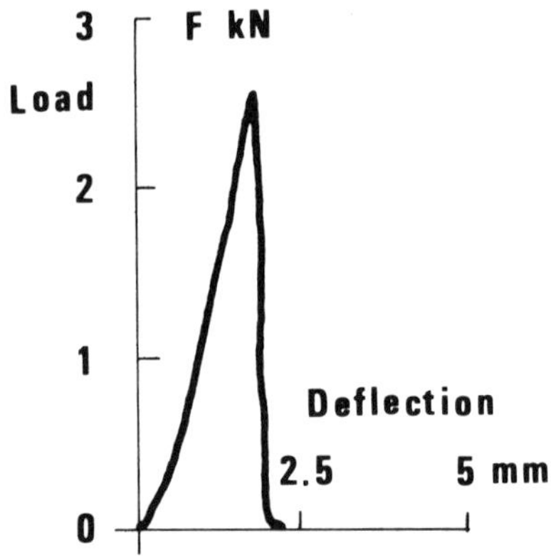

Fig. 2 - Experimental load-displacement curve.

4 Shear mechanisms in brittle disordered materials

The interest in the constitutive relations of cracks in concrete is considerably enhanced during the development of analytical studies on cracked concrete introducing the dilatancy ratio and the frictional coefficient in order to reflect in a realistic manner the typical characteristics such as shear transfer due to aggregate interlock, cross effect such as crack dilatancy and frictional contact slip, and path dependence.

In earthquake engineering, attention has been focussed on the variations of the ratio of travel times for shear waves and compressional waves prior to seismic events because the volumetric dilation is believed to occur in the focal zone prior to fault development. It has been thought that the resulting dilatancy hardening has a temporary stabilizing effect on the potential failure. In others respects, measurements of frictional strength of rocks are of fundamental importance in problems such as the state of stress and the mechanism of earthquakes along crustal faults, slope stability and mechanical behaviour of jointed and faulted rocks [Voropinov, 1967]. Patton [1966] has investigated the mechanism of shear failure in rock by studying over 300 rock slopes in the Rocky Mountains, making laboratory sliding friction tests on rock samples and direct shear tests on simulated rock surfaces. He concluded that the curved Mohr's failure envelope reflects the multiple modes of shear failure.

Concrete and rocks are geomaterials composed of particles, mineral grains, that are bonded together by a combination of their intermolecular forces and mineral cements and that are physically interlocked to varying degrees. Systems of cracks, such as joints and bedding planes, usually divide the material mass into discrete blocks that may be regular in shape and size, or extremely irregular with non planar separations and random sizes. Cracks can thus transmit significant amounts of shear by means of aggregate interlock also termed interface shear. A shear displacement or slip roughly parallel to the plane of the crack is essential to mobilize shear transfer by aggregate interlock. This displacement is accompanied by a displacement in the normal direction or dilation owing to rough asperities of aggregate and mortar tips projecting across the sliding path. Shear transfer by aggregate interlock is of a frictional nature with the normal compressive forces being provided by the embedded reinforcement.

Thus the resistance against shear loading originates in three components : cohesion, shear friction and dilatancy. As the resistance given by friction and dilatancy increases with increasing normal stress σ, so also does the shear strength at failure τ_f . Therefore the strength can be represented in the Mohr diagram by an inclined line which rises with the increasing σ. The normal stresses across the interface in direct shear tests can be applied in three different modes, which result in three different boundary conditions for this problem : (1) it can be kept constant, which corresponds to zero normal stiffness across the interface ; (2) it can be varied such that there is no crack dilation, which corresponds to infinite normal stiffness ; or (3) it can be allowed to vary as the crack slides and dilates, which corresponds to variable finite normal stiffness. The latter type corresponds to the most practical situation as there is always a finite normal stiffness provided by the embedded reinforcement. This also means that in most practical situations, the shear and normal stresses (τ, σ) as well as the displacements (δ_t , δ_n) are coupled and vary according to some stress or displacement history.

4.1 Mode II toughness index

Shear strength is often used to cover several different concepts [Everling, 1964] such as (a) strength against pure shear, (b) shear stress required for failure without normal stress, (c) shear diagram on solid interface depending upon normal stress and (d) Mohr's stress envelope. Several methods have been devised for measuring the shear strength of rock : torsion test, single shear test, double shear test, punch test, shear loaded beam with starter notches according to mode II of fracture mechanics. Attempts to realize mode II crack propagation on a laboratory scale usually fail because mode I growth become predominant. On the other hand, traces of mode II growth are often detected at or after earthquake slipping. The following part of the present paper uses a direct shear test [Luong, 1989] that maximizes mode II conditions of crack propagation in concrete and rock materials. The axial load F is applied on the central cylinder of the specimen which is retained by the four concentric tubular parts. The notches define the sheared zone S = 4s. Several series of tests have been performed on specimens (Fig. 3) of brittle materials subject to different values of normal stress in order to analyse the dilatancy effect. The nominal mode II shear strength is given by $R_{II} = F/4s = \tau$ in relation with the normal stress $\sigma = F_n /s$ where F_n denotes the lateral force directly applied on the failure surface using a variable number of coned discs or Belleville springs for the control of normal loading stiffness.

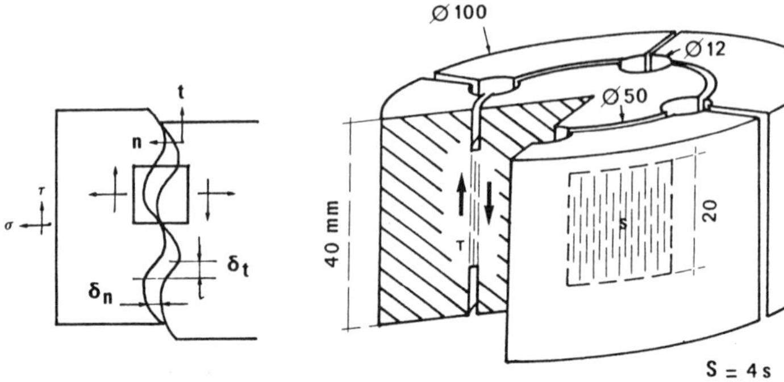

Fig. 3 - Shear test specimen and dilatancy parameters.

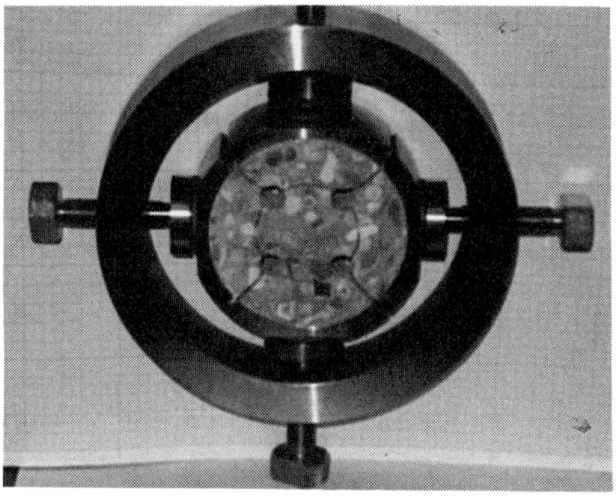

Fig. 4 - Mode II shear testing set-up allowing the dilatancy effect evaluation.

It is commonly considered that the strength of concrete or rock will become constant if sufficient relative movement is developed [Barton, 1976]. Some authors have suggested that the displacement necessary to develop the residual state increases with rock strength [Krsmanovic et al, 1966]. Various rocks behave differently with regard to relations of the peak and the residual strengths. The main parameter is the stiffness of the normal loading, able to partially prevent the aggregate interlock breakdown, which softens the mechanical response of the materials. When the tangential displacement or shear slip δ_t occurs along the crack interface, the shear stress τ working parallel to the crack is induced and is accompanied by the compressive stress σ and the normal displacement δ_n or crack width normal to the crack plane. These four parameters define the deformational characteristics of the cracked interface [Yoshikawa et al, 1989].

To monitor the dilatancy effect, the specimen subjected to direct shearing, defined as mode II of fracture mechanics, is loaded as shown in (Fig. 4).

Typical responses of brittle materials under direct shear loading illustrate the effect of dilatancy phenomenon as a function of the number of Belleville springs which decreases the stiffness of the normal stress loading (Fig. 5).

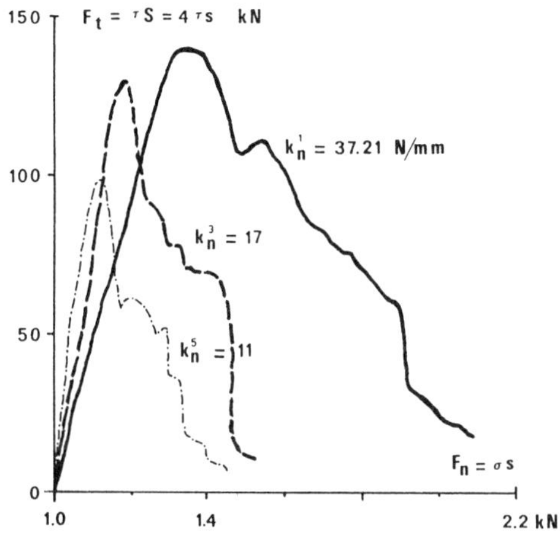

Fig. 5 - Normal loading stiffness influencing dilatancy behaviour.

4.2 Mode III toughness index

When studying the torsional response of members, it is convenient to identify two different modes of torsional resistance : (a) circulatory torsion and (2) warping torsion [Krpan & Collins, 1981]. In the former case, the torsion shear stress remains constant in magnitude from point to point around the cross section and there is no change in magnitude of the longitudinal stresses. In the later case, torsion shear stresses are not uniform and account must be taken of the variable effects of dilatancy in regard to the difference of the normal stiffnesses.

Conventional techniques of determining shear strength of rock materials have used indifferently tests which promote predominantly mode II or mode III of fracture mechanics.

For this study, a tubular cylindrical specimen for direct torsion testing [Luong, 1989] is used for the determination of mode III fracture strength. The specimen (Fig. 6) resists the applied torque by circulatory torsion. The shear flow lies some small distance inside the outside surface of the notch tip. The two ends are prepared as square cross section prisms. A thin cut is turned in the central part of the specimen. It is assumed that, in spite of the low tensile strength, fracture will occur from shearing stresses in mode III, since the screw-shaped surface of tensile fracture would be much larger than the thin round neck under torsion. When it is subjected to torsion T by holding its ends in chucks, maximum mode III shear stress develops at the outermost fibre. With an external radius r and an internal radius mr of the same order of magnitude, the nominal mode III strength can be given by $R_{III} = T / \pi r^3 (1 - m^3)$.

Fig. 6 - Direct torsion specimen.

The fracture surfaces appear complex, consisting of inclined facets, known as "factory roof" fracture surfaces by metallurgist workers [Hourlier et al, 1985], formed by the intersection of a large number of independently initiated mode I cracks. Thanks to the thin round neck, mode III or torsion fracture has been preferred as shown by the fracture surfaces, mode I fracture being controlled by the stiffness of normal stress loading or dilatancy effect.

Fig. 7 - Direct torsion testing under compression.

The normal stress $\sigma = F / \pi r^2 (1 - m^2)$ is applied on the specimen by means of a long screw. The stiffness of normal stress loading on the failure surface is controlled by one or several Belleville springs just as described by (Fig. 7).

4.3 Dilatancy effects on toughness indices

Using the toughness indices defined in literature based on the load-deflection curves, (Table 1) shows the effect of dilatancy for instance in the case of mode II shear tests, as a function of the normal stiffness.

Table 1. Toughness indices from direct shear tests under an initial normal stress σ_0 = 3.68 MPa.

Normal stiffness ($N.mm^{-1}$)	37	17	13
Toughness indices			
ACI	2.55	2.50	2.90
Johnston (n = 3)	2.91	3.88	4.16
Barr et al	0.80	0.81	0.83
Wang (n = 20)	4.47	4.51	4.80
Toughness ($kJ.m^{-2}$)			
Wang	25.3	14.7	11.5

5 Concluding remarks

The versatility of these new types of experimental set-up allows a ready evaluation of the behaviour of brittle disordered materials during mode I, mode II direct shear and mode III direct torsion loadings.

The proposed testing arrangements for both direct shear and torsion shear strength measurements simplify the loading equipment by the use of a uniaxial test machine with the combined compression, shear and torsion loading frames described in this paper. The test set-up allows the normal stresses to be monitored and measured during the experiment under varying shear stresses. The resulting responses are shear stress versus slip and shear stress versus normal stress, which are of course very useful in the interpretation of conventional pull-out tests. They are practical and reliable. They also facilitate the testing procedure when the materials are subjected to severe and hostile environmental conditions. And finally they can all be used as routine tests more readily achieved than on the conventional triaxial tests in the cases of low mean stress level.

The experimental results show a very strong influence of dilatancy on the shear strength of quasi brittle materials as shown by the variation of toughness indices, especially in cases of low mean stress loadings. They indicate that shear stress versus slip response at constant normal stresses exhibits a peak and a residual value according to the monitored stiffness of normal loading. This elucidates the mechanism of softening occurring in the ultimate behaviour of brittle materials under loading. In addition, they demonstrate the importance of fracture modes on the mechanical behaviour of brittle disordered materials.

6 References

Barr, B.I.G., Liu, K. and Dowers, R.C. (1982) A toughness index to measure the energy absorption of fibre reinforced concrete, Int. J. of Cement Composites and Lightweight Concrete, 4(4), November 1982, pp. 221-227.

Barton, N. (1976) Rock Mechanics Review, The shear strength of rock and rock joints, Int. J. Rock Mech. Min. Sci. & Geomech. Abstr., 13, pp. 255-279.

Everling, G. (1964) Comments upon the definition of shear strength, Int. J. Rock Mech. Mining Sc., 1, pp. 145-154.

Hillerborg, A. (1980) Analysis of fracture by means of the fictitious crack model, particularly for fibre reinforced concrete, The International Journal of Cement Composites, Vol. 2, No. 4, pp.177-184.

Hourlier, F., d'Hondt, H., Truchon, M. and Pineau, A. (1985) Fatigue crack path behavior under polymodal fatigue, Multiaxial Fatigue, ASTM STP 853, K.J. Miller and M.W. Brown, Eds., ASTM, Philadelphia, pp. 228-248.

Jenq, Y.S. and Shah, S.P. (1989) Geometrical effects on mode I fracture parameters, Report to RILEM Committee 89-FMT, January 1989.

Krsmanovic, D., Tufo, M. and Langof, Z. (1966) Shear strength of rock masses and possibilities of its reproduction on models, Proc. 1st Cong. Int. Soc. Rock Mech., Lisbon, 1, pp. 537-542.

Krpan, P. and Collins, M.P. (1981) Predicting torsional response of thin-walled open RC members, J. Struct. Div., ASCE 107(6), pp. 1107-1127.

Luong, M.P. (1987) Infrared thermography of fracture of concrete and rock, Fracture of Concrete and Rock, Shah & Swartz ed., Springer-Verlag, pp. 343-353.

Luong, M.P. (1989) Fracture behaviour of concrete and rock under mode II and mode III shear loading, Fracture of Concrete and Rock - Recent Developments, edited by S.P. Shah, S.E. Swartz, B. Barr, Elsevier Applied Science, pp. 18-26.

Patton, F.D. (1966) Multiple modes of shear failure in rock, Proc. 1st Congr. Int. Soc. Rock Mech., Lisbon, 1, pp. 509-513.

Roberts, A. (1977) Geotechnology - An introductory text for students and engineers, Pergamon Press.

Voropinov, J. (1967) Rheological research on rocks in shear, Proc. Geotechnical Conf. Shear Strength Properties of Natural Soils and Rocks, Oslo, 1, pp. 305-309.

Wang, Y. and Backer, S. (1989) Toughness determination for fibre reinforced concrete, Int. J. of Cement Composites and Lightweight Concrete, 11(1), Feb. 1989, pp. 11-19.

Yoshikawa, H., Wu, Z. and Tanabe, T. (1989) Analytical model for shear slip of cracked concrete, J. Struct. Div., ASCE, 115(4), pp. 771-788.

65 MIXED MODE FRACTURE – INCLUDING TORSION IN A NEW COMPACT TEST SPECIMEN GEOMETRY

A. ARSLAN, T.G. HUGHES and B.I.G. BARR
School of Engineering, University of Wales, Cardiff, UK

Abstract
A new compact fracture test geometry, which can be used to investigate the full mixed-mode (I-II-III) behaviour of cementious materials, has been developed. A range of variations of this geometry have been studied numerically using 3-D finite element linear elastic modelling. The main variations in the parameters used in the analysis being in the main notch depth and the presence and depth of the secondary vertical notches.

Experimental studies have been carried out by means of displacement controlled tests on the selected geometry. The fracture performances of plain and fibre reinforced concrete have been investigated with this geometry using both steel and polypropylene fibre containing a range of fibre contents.
Keywords: Mixed-mode Fracture, Fibre Reinforced Concrete, Steel Fibres, Polypropylene Fibres, Shear Strength.

1 Introduction

The primary aim of the work was the development of a new mixed-mode fracture specimen geometry capable of being used to investigate the behaviour of materials in a full mixed-mode condition. There is a shortage of research on the three-dimensional fracture behaviour of cementious materials, most of the research to date having concentrated on two-dimensional models. For this reason experiments and analysis have previously concentrated on only one, or combinations of two, modes of failure.

The application of fracture mechanics theory to engineering materials initially concentrated on Mode-I failure. Mode-II fracture of concrete occurs almost as frequently and it has received recent attention(1,2). A review of some of the shear test geometries in recent years has been given by Barr(3). However, interpretation of the test results is rather controversial. Numerous final failure mechanisms have been observed by different investigators, especially in the behaviour of the four-

Fracture Processes in Concrete, Rock and Ceramics Edited by J.G.M. van Mier, J.G. Rots and A. Bakker.
 Published by E. & F.N. Spon, 2–6 Boundary Row, London SE1 8HN. ISBN 0 419 15870 7

point shear specimen. These include tensile splitting failure in the middle region of the proposed failure plane[4], shear type failure between the roots[1,5] and flexural failure on the supports[6]. Similar controversy with various interpretations has also occured with the other mode-II based specimen geometries.

The third mode of failure, mode-III, has received little attention. A limited number of investigations have concentrated on this area[7,8]. The material behaviour is far from well understood, particularly in the full-mixed mode condition.

2 Numerical Analysis

The geometry and loading conditions of the mixed-mode specimen are illustrated in Fig.1. The original specimen is a standard 150 mm quality control cube. The initial numerical work concentrated on the effects of variations in the depth of the main (h) and secondary (a) notches. As summarized in Table.1, nine geometries have been investigated in order to quantify the effect of notch

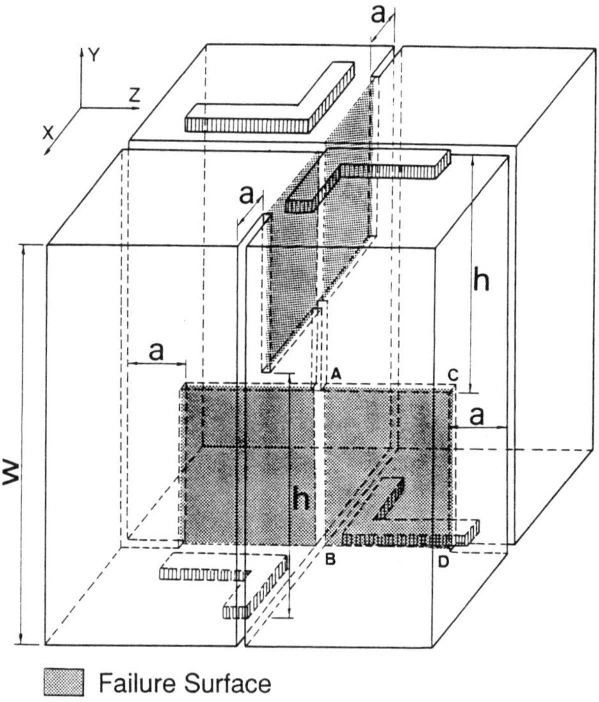

Fig.1. General geometry of the full mixed-mode fracture specimen

Table 1. Geometric variations in the notch depths used in the numerical analysis of the mixed mode specimen

Specimen No	h (mm)	a (mm)	Geometric ratio of shear plane, l_x/l_y
1	75	0	0.973
2	75	15	0.773
3	75	30	0.573
4	85	0	1.123
5	85	15	0.892
6	85	30	0.662
7	95	0	1.327
8	95	15	1.055
9	95	30	0.782

depths on the stress distribution on the proposed failure surfaces, detailed in Figure 1. Three different main notch depths with h=75, 85 and 95 mm and secondary vertical notch depths with a=0, 15 and 30 mm have been investigated. The secondary vertical notches were introduced to concentrate the shear surface to the central axis regions of the specimen and to also enable investigation of the effect of the torsional stresses on the overall behaviour of the specimen.

Three-Dimensional finite element analysis has been used to allow comparison of the nine compact specimens detailed. It is necessary to model the entire specimen because of the non-symmetry in the loading and supporting conditions. Standard 8-noded brick elements have been used throughout. Models generally consisted of approximately 5000 isoparametric elements. Larger elements were used at the midplane of the specimen with a fine discretization in the vicinity of the proposed plane of failure. The load was applied as line loads at the nodal points.

A typical 3-D plot of the stress blocks are given in Fig.2, for the model which has no secondary notches (a=0 mm) and h=85mm. The normal stress σ_{zz}, perpendicular to the proposed shear plane is illustrated together with the shear stress in the y-direction τ_{yz}, and in the x-direction τ_{zx}. The nearest corner to the front A represents the inner corner of the proposed failure plane, see Fig.2a. The left hand side, line AB, represent the height (along y-axis) with the width AC (along x-axis) to the right. Load was applied to each type of geometry proportional to the shear area of proposed failure plane to facilitate comparisons.

Considering initially the behaviour in the y-direction. From Fig.2a. It is apparent that the specimen generates tensile stresses near to the loading line, this was present to various degrees in all the specimens. This rapidly changes to compressive stresses at the points

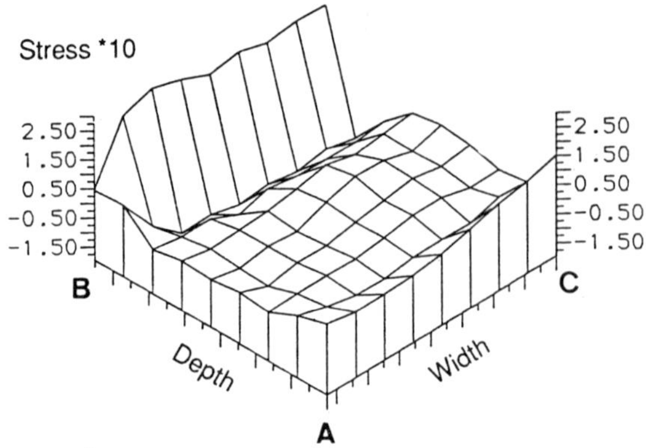

a) Normal stress surface, σ_{zz}

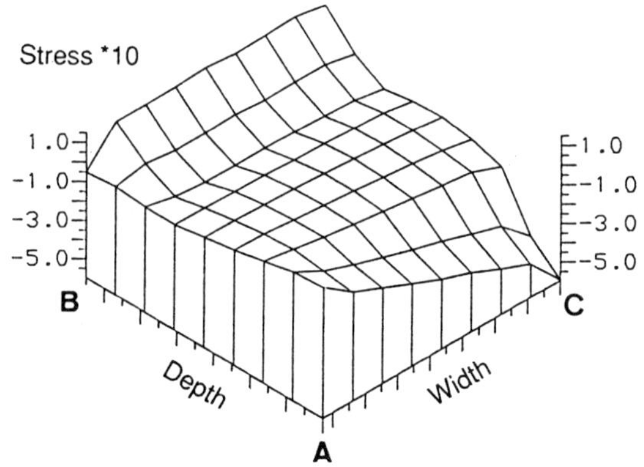

b) Shear stress surface, τ_{yz}

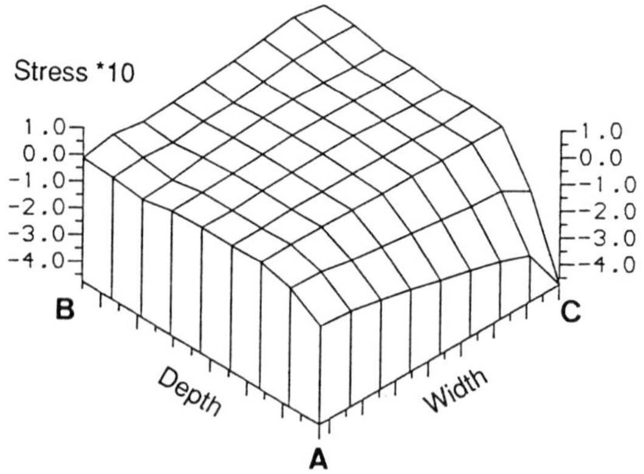

c) Shear stress surface, τ_{zx}

Fig.2. Stress distributions on the proposed failure surface

3-4 mm away. There is then an almost linear reduction in the magnitude of the compressive stresses with distance to the halfway point. The remaining compressive stresses are almost uniformly distributed through the remaining portion of failure surface but reduce to tensile stresses along AC. Distribution of the stresses in the x-direction indicates little variation except along the edges AC and BD.

At the outer edge of the proposed surface CD, the shear stresses in the y-direction, τ_{yz} is positive in magnitude at D and then immediately reduces being almost constant over the central region. This reduces suddenly again as C is approached. The overall stress block indicates an almost screw type of surface. (see Fig.2b). The difference in the magnitude of the stresses between the inner region and the outer region, towards the sides, indicates the degree of the torsional effect on the specimen behaviour.

Transverse shear stresses τ_{zx} in the x-direction, are small in magnitude over the region of the proposed failure surface. Near to the inner region of the shear surface CD the magnitude of the stress increases suddenly up to the main notch tip C and its direction turns outward, towards the x-axis. The distribution of the stresses on the plane shows almost the same screw type character as in the case of τ_{yz}, but to not such a great extend.

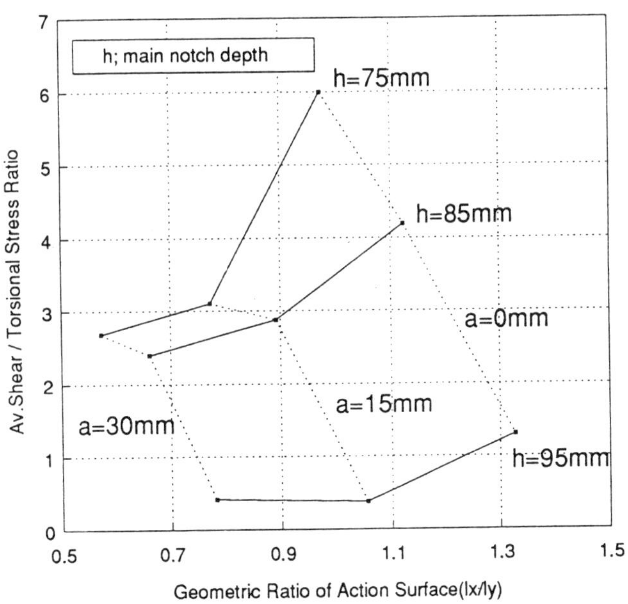

Fig.3. The variation of the average shear and torsional stress ratio with the geometric ratio of the proposed failure surface.

The stress distribution across the proposed failure plane varied with the inclusion of the secondary vertical notches. Average shear and torsional stress ratio are given as an aggregate of the geometric ratio of the proposed action surface (l_x/l_y) in Fig.3. Where l_x and l_y are the dimensions of the proposed failure surface and are given by:

$$l_x = \left(\frac{150-4}{2}\right) - a \quad \text{and} \quad l_y = 150 - h \quad \text{in mm.}$$

In the case of no vertical notches there is a decrease in the torsional stresses, here the shear stress dominates. The addition of the vertical notches increased the torsional stresses, particularly in the specimen with the deeper secondary notches (a=30 mm). Full details of the mesh discretisation used and the resulting stress distributions for all geometries are detailed elsewhere(9).

3 Experimental Study

The numerical investigations have been accompanied by experimental studies on a selected geometry. The geometry has a main notch depth of h=85 mm and no secondary vertical notches. This specimen geometry was selected because of its numerical generation of an average torsional effect within the specimens. The specimen is also the most easily produced test geometry because it has only two main notches.

The experimental programme included the testing of fibre reinforced concrete as well as plain concrete. The fracture performance of plain and fibre reinforced concrete has been investigated with this geometry using both steel and polypropylene fibre. Three fibre contents have been used, 1, 2 and 3% by weight in the case of steel fibre, and 0.1, 0.2 and 0.3% by weight in the case of polypropylene fibre. The straight steel fibres used were of 30 mm length and with a 1/100 aspect ratio. Polypropylene fibres were of the fibrillated type and have a 50 mm staple length and a linear density of 700 m/kg (1200 denier).

The standard concrete mix adopted the following mix proportions by weight, cement:sand:coarse aggregate 1:1.8:2.8 with a maximum aggregate size of 10mm. The Water/Cement ratio was kept constant at 0.45 thoughout the study.

The nominal strength of the concrete was relatively high at the test date. The tests were carried out at 3 months and gave an average compressive strength of 79 MPa.

Experimental studies were undertaken by means of displacement controlled tests. The shearing load was applied by means of a loading ram on a closed loop system

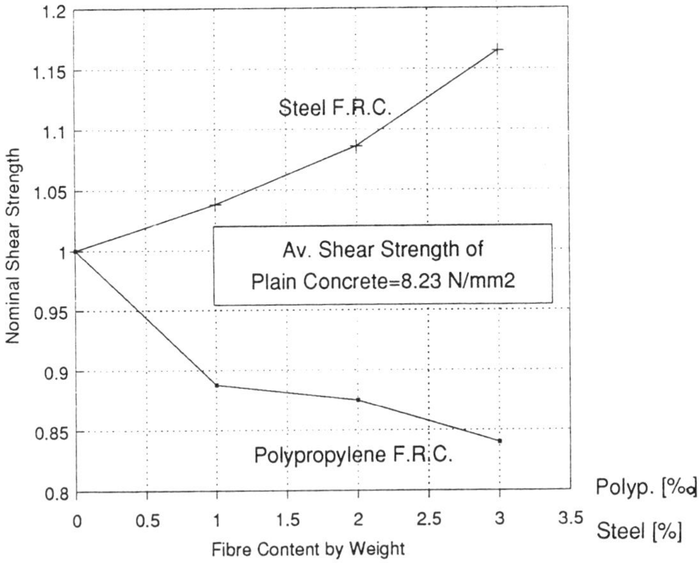

Fig.4. Variation of shear strength with fibre density

and was measured by a calibrated load cell beneath the lower platen. A pair of specially produced loading plates were used at both ends of the specimen in order to generate line loads on the specimen with the L shaped loading blocks detailed in Fig.1. Throughout the test programme the rate of loading was kept constant at 0.0015 mm/sec. The corresponding load-displacement responses were automatically recorded.

The non dimensional strength of fibre reinforced concrete and the variation of the effect of fibre types and volume on the shear strength of the material is illustrated in Fig.4. In the case of steel fibre R.C., with the higher fibre content, a substantial increase in the strength was exhibited. This enhancement was approximately 17% for 3% steel fibre.

For plain concrete the crack was seen to start at both notches approximately simultaneously and developed along asymmetric separate paths.(see Fig.5a). As the crack approached the centre portion of the throat one of the path dominated resulting in a complete failure. This behaviour indicates that at the final stages of the failure process, the tension mode of failure dominates.

With FRC, at higher percentage fibres, the fracture development is significantly different. At loads approaching the ultimate load (e.g. 90%) a band of micro cracks was detected, at higher increments of the load the failure occured by the joining of these micro cracks as a shear type failure.(see Fig.5b-c). This failure mechanism was particularly apparent in the behaviour of the

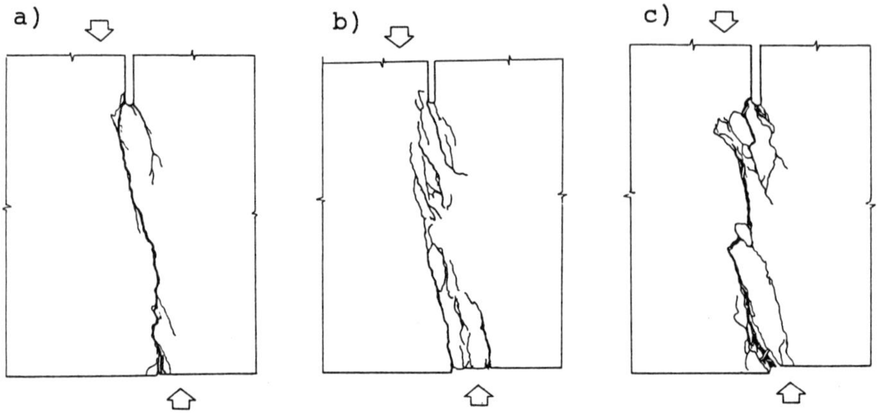

Fig.5. Typical crack paths on the specimen face, obtained by tracing
a) Plain Concrete
b) Polypropylene F.R.C. (0.10%)
c) Steel F.R.C. (2%)

polypropylene reinforced concrete. The maximum crack widths, at failure, were relatively small because of the generation of many more crack paths in this type of concrete.

In the case of steel fibre reinforced concrete (SFRC), the number of micro cracks is less than in the case of polypropylene F.R.C. The direction of the failure surface development is interrupted by the presence of local high densities of steel fibres. The other noticeable effect was the significance of the orientation of the fibres to the failure surface causing some kinks in the failure surface. The postcracking resistance of SFRC occurs mainly due to dowel action of intersected fibres along the failure plane.

After careful examination on the crack surfaces, the existence of all modes of fracture has been detected. An example of the crack surface after failure, for plain concrete, is given in Fig.6. The notched surface is at the top and the failure surface at the bottom as detailed in Fig.1. Shear load was applied to this block as indicated. Small microcracks near the bottom right corner D, and shear scoring due to aggregate interlock mechanism can be seen in the region indicated. The shape of the scratched surface parallel to the load direction clearly indicates the shear failure effect on the overall behaviour. It can be seen that the screw type of failure surface indicates the effect of the torsional stresses. These observations confirmed the results obtained from the three-dimensional FEM analysis.

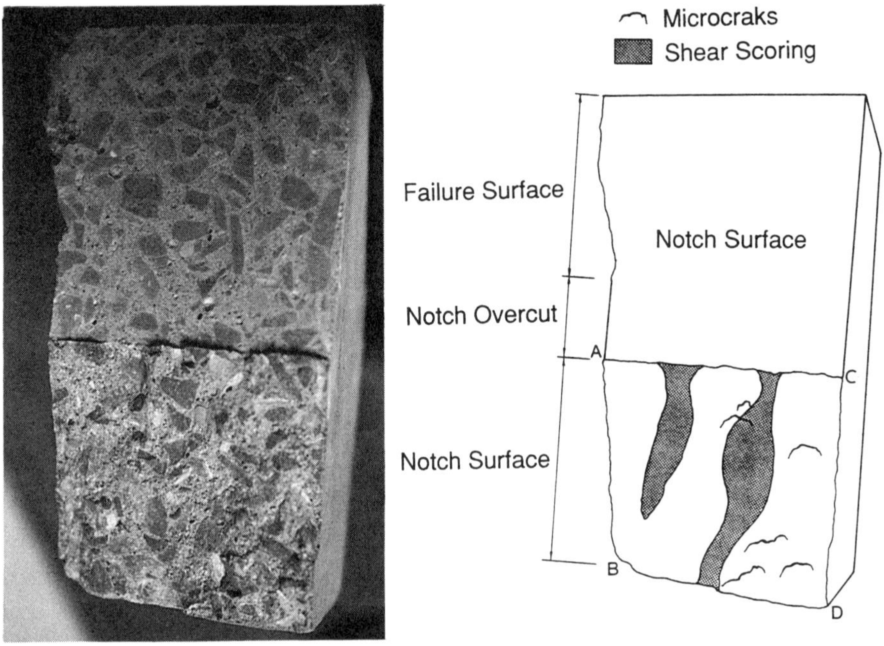

Fig.6. A typical failure surface for a plain concrete specimen.

4 Conclusions

The specimen without secondary notches has the advantage that it allows the largest failure area for the volume of concrete tested. By introducing four multiple failure surfaces in different regions of the samples the specimen allows for the inhomogeneous nature of the concrete specimen, i.e. poor compaction. Special care must be exhibited in the loading arrangement to make sure that a uniformly distributed load is generated. Early experiments resulted in multiple failure surfaces with individual surface failures occuring at different loading increments.

Use of this specimen, with properly designed full displacement instrumentation, allows the investigation of a variety of ratios of the fundamental failure modes. By varying the depths of the notches different combinations of failure modes can be simulated. It is concluded that this specimen geometry is capable of investigating the full-mixed mode behaviour of fibre reinforced as well as plain concrete.

The dowel action of fibres in SFRC under shear loading is an important issue in explaining the complete failure behaviour for FRC. A detailed investigation on this is currently being undertaken both experimentally and theoretically.

5. Acknowledgement

Mr. A. Arslan is a postgraduate research student sponsored by Firat University, Turkey, whose financial assistance is gratefully acknowledged.

References

1 Bazant, Z.P. and Pfeiffer, P.A.
Shear fracture tests on concrete, **RILEM Materials and Structures**, 1986, 19(110), 111-121.

2 Arslan, A. Hughes, T.G. and Barr, B.I.G.
Mixed-mode fracture in compact shear specimens, **ECF 8 Fracture Behaviour and Design of Materials and Structures**, Ed. D. Firrao, EMAS Publ., 1990, 676-681.

3 Barr, B.
The fracture charasteristics of FRC materials in shear **Fibre Reinforced Concrete Properties**, ACI Publication, SP-105, Detroit, 1987, 27-53.

4 Ingraffea, A.R. and Panthaki, M.J.
Analysis of shear fracture tests of concrete beams, **Seminar on F.E.A. of reinforced Concrete Structures**, May 21-24, 1985, Tokyo, 71-91.

5 Barr, B.I.G. and Thomas, W.F.
A study of the fracture charasteristics of glass reinforced cement, **Third Int. Symp. on Developments in Fibre Reinforced Cement And Concrete**, Eds. Swamy, Wagstaffe and Oakley, Sheffield, 1986.

6 Carpinteri, A.
Interaction between tensile strength failure and mixed-mode crack propagation in concrete, **Report to RILEM Commitee 89-FMT**, March 1987

7 Tokatly, Z.Y. Barr, B. and Norris, P.
Mode-III fracture - A tentative test geometry **Fracture of Concrete And Rock, Recent Developments**, Eds. S.P. Shah, S.E. Schwartz and B. Barr, Elsevier Applied Science Publ., 1989, 596-604.

8 Xu, D. and Reinhardt, H.W.
Softening of concrete under torsional loading **Fracture of Concrete And Rock, Recent Developments**, Eds. S.P. Shah, S.E. Schwartz and B. Barr, Elsevier Applied Science Publ., 1989, 39-50.

9 Arslan, A.
Stress developments in a new mixed mode specimen, **School of Engineering Report**, University of Wales, Cardiff, 1990.

PART ELEVEN

FRACTURE TESTING

66 EVALUATION OF THE COMPLETE COMPRESSIVE STRESS–STRAIN CURVE FOR HIGH STRENGTH CONCRETE

M. GLAVIND and H. STANG
Department of Structural Engineering, Technical University of Denmark

Abstract
This paper deals with an investigation and discussion of compressive stress-strain curves for high-strength concrete, and the applicability of the curves in a structural element calculation. The stress-strain curves are obtained in a closed loop testing machine, where different combinations of axial and circumferential deformation are used as the feed-back signal.
Keywords: Compressive stress-strain curves, High strength concrete, Feed-back signal, Gauge lengths, Snap-back phenomenon, Elastic recovery.

1. Introduction

The descending part of stress-strain curves in compression for high-strength concrete can be extremely difficult to establish, and many different attempts have been made in designing testing machines/systems, which can control compression failure of such a brittle material. Roughly, there are two ways to go when designing such a testing system: 1) using a stiff frame and 2) using a closed-loop feed-back system. In a stiff testing machine failure can be controlled if the stiffness of the machine is greater than the absolute value of the incremental, post-failure stiffness of the specimen. In a closed-loop servo controlled system, it is possible to control failure even if the absolute incremental stiffness of the specimen is greater than the stiffness of the machine.

One way of obtaining a stiff testing machine is to load the specimen in parallel with a steel device, so that the total load increases even when the specimen is unloading. This method was used by Wang, Shah and Naaman, (1978) who loaded the specimen in parallel with a steel tube. At the Department of Structural Engineering, Technical University of Denmark a similar device has been used. Here the specimen was loaded in parallel with three steel columns, Olsen (1990).

The basic principle of a closed-loop servo controlled system is to compare a desired program signal with a signal from an experimental condition. The difference between these two signals are called the error signal, and this signal is used to control the hydraulic system of the testing machine.

The success of a deformation controlled compressive test on high strength concrete specimen depends on the possibility of establishing a signal from the experiment which is increasing monotonically in time. The axial deformation signal measured either between the loading plates or directly on the specimen is often not monotonically increasing in time.

Fracture Processes in Concrete, Rock and Ceramics Edited by J.G.M. van Mier, J.G. Rots and A. Bakker.
 Published by E. & F.N. Spon, 2–6 Boundary Row, London SE1 8HN. ISBN 0 419 15870 7

Shah, Gokoz and An Sari (1981) have obtained stress-strain curves of high strength concrete by using the circumferential strain as a feed-back signal. Dahl and Brincker (1989a) have obtained stress-strain curves of high-strength concrete by using a combination of cross-head displacement, axial and circumferential deformation of the specimen as the feed-back signal.

At the Department of Structural Engineering a similar test arrangement has been developed. Here the test is controlled by a feed-back signal composed of axial deformation and circumferential deformation.

2. Description of experimental set-up

2.1 Testspecimens

The tests were performed on cylindrical specimens with a diameter of 100 mm and a length of 200 mm.

The materials used were: Rapid hardening cement ASTM Type III, Silica fume, aggregates from Danish marine deposits and superplastizers.

Three different mix proportions were used. Mix 1 was made with 267 kg/m^3 cement and had a water-cement ratio of 0.60. Mix 2 was made with 310 kg/m^3 cement and 31 kg/m^3 silica fume and had a water-cement ratio of 0.30 (silica fume is included with an activity factor of 2). Mix 3 was made with 390 kg/m^3 and 39 kg/m^3 silica fume and had a water-cement ratio of 0.24 (silica fume is included with an activity factor of 2).

The cylinders were cast in plastic moulds and consolidated using a vibrator table. The moulds were removed approximately 24 hours after casting. The cylinders were then cured in a water bassin until testing.

2.2 Measuring devices

The axial deformation used in the feed-back signal is measured between the loading plates. The signal from two LVDTs is averaged. The sensivity of the LVDTs is 0.7325 V/mm and the base is 2.5 mm.

The axial deformation was also measured with 88 mm strain gauges mounted on the cylinder and with transducers which were capable of measurering deformation over a length of 25 mm, 50 mm and 100. The extensometers were fixed on the specimens using knife edges and rubber bands.

A device to measure circumferential measurement was constructed with a pianowire wrapped around the cylinder, which was connected to a transducer. This device was suggested by Shah, Gokoz and Ansari (1981), and was also used by Dahl and Brincker (1989a). The transducer used here was a HP 1000 LVDT with a 25 mm base and a sensitivity of 0.2 V/mm.

No intermediary layers were used between the concrete cylinders and the steel loading plates.

2.3 Hardware

The servovalve is a Moog high-response type E 760-533 with a capacity of 38.5 l/min and a frequency of 200 hz. The maximum hydraulic pressure is 300 barr which corresponds to high pressure use.

The servo controller is a MTS with a 418.91 Profiler and a 458.20 Console which transmits the reference signal.

Programmable mV transmitters makes it possible to adjust the gain and the offset of each signal used to compose the feed-back signal. An analog calculator adds the two signals used as feed-back signal. The data are recorded via a datalogger to a computer.

2.4 Loading capacity

The test equipment was optimized in order to obtain controlled failure with the highest possible compressive strength. Success of a test is believed to be dependent on the following parameters:

- the response time of the servo controller
- the response time and the capacity of the servo valve
- the gain
- the combination of the feed-back signal
- the response time and the sensivity of the signals used to compose the feed-back signal
- the stiffness of the load frame
- the response time of the transmitters and the analog calculator

The servocontroller compares the program signal and the feed-back signal approximately 10.000 times a second. The servovalve is a high-response type as described in part 2.3.

The optimal gain was found to be about 9.5 of a scale from 0 to 10. This value of the gain was established by using a rectangular program signal, and recording both the program signal and the feed-back signal on an oscilloscope. The procedure was executed both with a specimen installed and without a specimen installed. The situation with a specimen installed corresponds to the ascending part of the stress-strain curve, and the situation without a specimen installed corresponds to the post-peak behavior.

The combination of the feed-back signal was found to be satisfactory at 0.225 V/mm for the circumferential signal and 1.65 V/mm for the axial signal, i.e. a combination of 1 :7.33. It was not decisive for the success of a test, that this combination was used, but it gave a convenient proportion between the time to reach the peak and the time at the descending part of the stress-strain curve. If a test is conducted in pure axial deformation control only concrete with a strength up to 50 MPa can be tested successfully.

The response time of the LVDTs is theoretically zero, but it seems reasonable that friction could delay the signal. Especially friction in the core of the HP 1000 used for circumferential measurement could be critical.

Furthermore, the sensitivity of the circumferential deformation measuring set up proved to be critical. It was not possible to obtain any reliable signal on the ascending part of the stress-strain curve. Thus it was not possible to conduct a test in pure circumferential deformation control. However, the registered circumferential deformation is sufficient to control a test, when there is also a contribution from the axial deformation to the feed-back signal.

An advantage of using the piano wire device to measure lateral strains, is that it measures deformation all around the specimen whereas strain gauge measurements are affected by the non-uniform cracking of the concrete.

The load frame is not especially stiff. The frame is build from hollow steel tubes which gives a rather soft load frame.

With the test equipment described above, it was possible to control a compression failure with concrete cylinders, which had a compressive strength of maximum 85 MPa.

The response time of the original transmitters turned out to be critical. However it was possible to modify the transmitters reducing the response time. With this change it was possible to control failure of specimens with a compressive strength up to 90 MPa. The optimization was stopped at that time, but it is believed that the weakest links now are the stiffness of the system, the sensivity of the circumferential measuring device and the transmitters and how they are configured.

3. Experimental investigations

3.1 Effect of composition on feed-back signal

A test series was made in order to investigate, if there is any difference in the shape of the stress-strain curve related to the combination of the feed-back signal. Three different combinations were chosen:

a) Circumferential strain to axial strain = 1 : 1.52
b) Circumferential strain to axial strain = 1 : 7.33
c) Circumferential strain to axial strain = 1 : 15.20

In order to avoid unsuccessful tests the test series was carried out on specimens made with mix 1 which obtained a compressive strength around 40 MPa. On the ascending part of the stress-strain curves there is hardly any response from the circumferential signal, so the test is controlled as in pure axial strain control for all combinations of the feed-back signal. Only the time to reach the peak is differing with the different feed-back combinations. The difference in loading rate is approximately 0.4 MPa/s for combination a) and 1.7 MPa/s for combination c. Fig. 1 show the stress-strain curves from the test.

In order to get information about the shape of the stress-strain curves, a previously proposed mathematical expression, Sargin (1971) and Wang, Shah and Naaman (1978) is fitted to the curves. The mathematical expression is shown in equation 1 for the ascending part of the stress-strain curve and in equation 2 for the descending part of the stress-strain curve, where x and y is dimensionless strain and stress, respectively.

$$y=\frac{A_1x+B_1x^2}{1+C_1x+D_1x^2} \quad (1) \qquad y=\frac{A_2x+B_2x^2}{1+C_2x+D_2x^2} \quad (2)$$

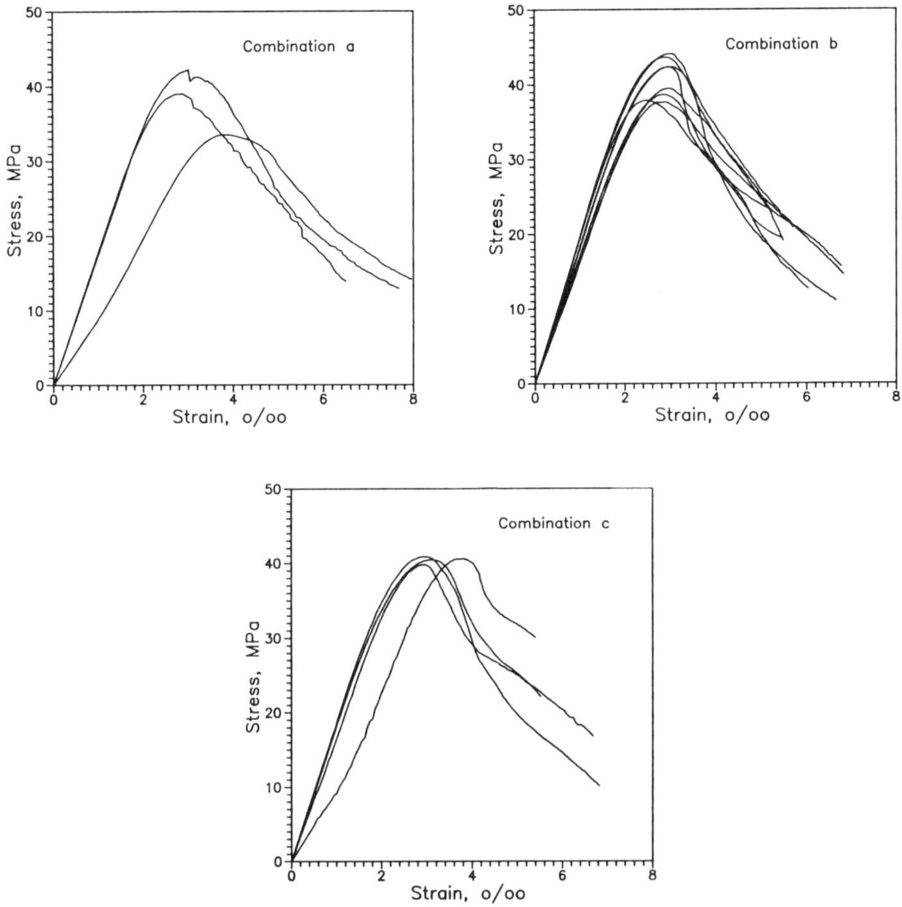

Fig. 1 Compressive stress-strain curves for the different combinations of the feed-back signal

Table 1 shows the peak stress, the strain corresponding to the peak stress (calculated on basis of deformation measurements over the whole length of the cylinder) and the constants in the mathematical expression.

As can be seen from fig. 1 there is apparently no difference between the shape of the stress-strain curves in the three test series. Statistical analysis also shows that the hypothesis that the peak stress, the strain corresponding to the peak stress and the constants in the mathematical expression are the same for the three combinations of the feed-back signal, can be accepted at all significance intervals less than 5 %. It is not surprising that the constants corresponding to the ascending part of the stress-strain curve for the three combinations are very close, because the circumferential deformation at the ascending part of the stress-strain curve is negligibly.

Table 1 Peak stress, peak strain (strain corresponding to the peak stress), constants in mathematical expression for combination a, b and c. The numbers in parenthesis is the standard deviation.

	a)	b)	c)
Num. of specim.	3	8	4
Peak stress, MPa	40.6 (2.3)	40.7 (2.6)	40.4 (0.5)
Peak strain, ‰	3.2 (0.5)	2.8 (0.2)	3.2 (0.4)
Constants			
A_1	1.21 (0.14)	1.28 (0.04)	1.24 (0.17)
B_1	-0.93 (0.09)	-0.89 (0.04)	-0.89 (0.12)
C_1	-0.79 (0.14)	-0.72 (0.04)	-0.76 (0.17)
D_1	0.07 (0.09)	0.11 (0.04)	0.11 (0.12)
A_2	0.75 (0.20)	0.30 (0.53)	-0.25 (0.50)
B_2	-0.18 (0.14)	0.07 (0.25)	0.40 (0.36)
C_2	-1.25 (0.20)	-1.70 (0.53)	-2.25 (0.50)
D_2	0.82 (0.14)	1.07 (0.25)	1.40 (0.36)

Thus, it is concluded that the combination of the axial- and circumferential deformation signal used as a feed-back signal do not have any significant influence on the shape of the stress-strain curve envelope. The only difference is, as can be seen from fig. 1 that the descending part of the stress-strain curve is more jagged the more the circumferential signal is weighed. This phenomenon gives useful information about crack development.

Furthermore, in fig. 1, a tendency to snap-back behaviour in the stress-strain curve is seen right after the peak. This phenomenon is discussed in section 3.2.

3.2 Snap-back phenomenon

On some of the stress-strain curves a tendency to snap-back on the descending part of the stress-strain curve just after peak can be seen.

Fig. 2 shows stress-strain curves and stress-circumferential strain curves for concrete with different strengths. These curves are obtained for the same composition of the feed-back signal and for specimens made with mix 3 tested at different ages.

Fig. 2 shows that the higher strength of the concrete, the more pronounced is the snap-back. Also, when the strength gets higher more than one snap-back can be observed. The snap-back always occurs just after peak. It can also be observed that the decrease in axial strain is accompanied by a drop in the stress and an increase of the circumferential strain. The concrete which has a compressive strength of 44 MPa has no snapback.

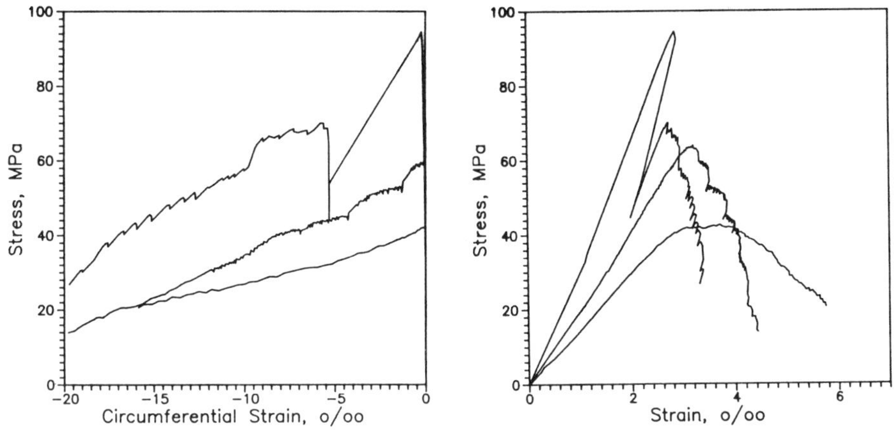

Fig. 2 Compressive stress-strain curves and stress-circumferential strain curves for concrete with different strengths.

It is suggested that the above observed phenomenon can be explained by four interacting parameters. First, the strength of the concrete seems to have an influence. Secondly, the composition of the feed-back signal has an influence (as mentioned in section 3.1). Thirdly, the frictional restraint at the end zones has an influence, and finally a strain localization has an influence on the stress-strain curves, where the strain is calculated on basis of the deformation measurements over the whole length of the cylinder.

The influence from the strength is explained in the following way from a fracture mechanics point of view. In normal strength concrete, the cracks get arrested by the aggregate. In high strength concrete, more strain energy will be assembled, (the stress intensity factor is bigger), and because the cement paste and the aggregate has the same strength and stiffness, the crack will extend through the aggregate. The cracking process in normal strength concrete is thus taking place in small steps, whereas in high strength concrete it is taking place in big steps. So when a crack develops in high strength concrete, there is a sudden increase in the circumferential strain, causing unloading to take place in the axial direction in order to regulate the error signal. Afterward loading is seen again, while the circumferential strain is constant, see fig. 2.

The second influential parameter is the composition of the feed-back signal. According to the above explanation it seems logical that the more the circumferential signal is weighed, the more pronounced is the snap-back. In section 3.1 it was also observed that the more the circumferential signal is weighed, the more jagged is the stress-strain curve. If a test is conducted in pure axial strain control, it would not be possible to obtain these curves as mentioned in section 2.4.

The influence from the frictional restraint and strain localization was investigated by changing gauge length as explained in the next section.

3.3 Effect of gauge length

Two test series were made - one with high strength concrete (mix 2) and one with normal strength concrete (mix 1) - in order to be able to derive information about the fracture zone length and development, and the influence of the frictional restraint of the end zones. Each series contained 16 specimens. The axial strain was measured with strain gauges in both test series and with transducers with variable gauge lengths as described in section 2.2 for the high strength concrete.

Fig. 3 and 4 show representative stress-strain relationships for normal strength concrete and high strength concrete respectively, where the strain is measured over different lengths.

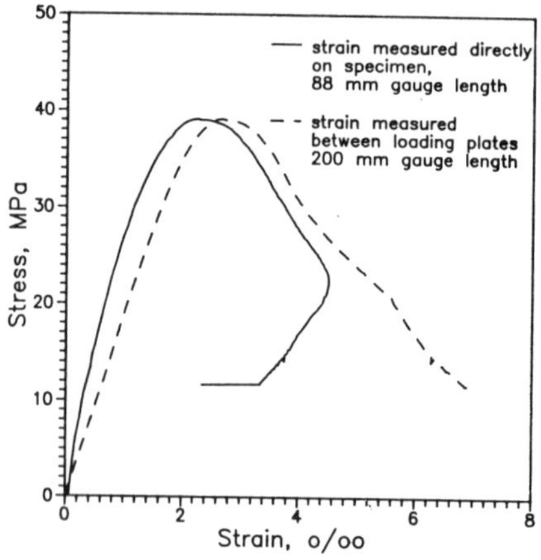

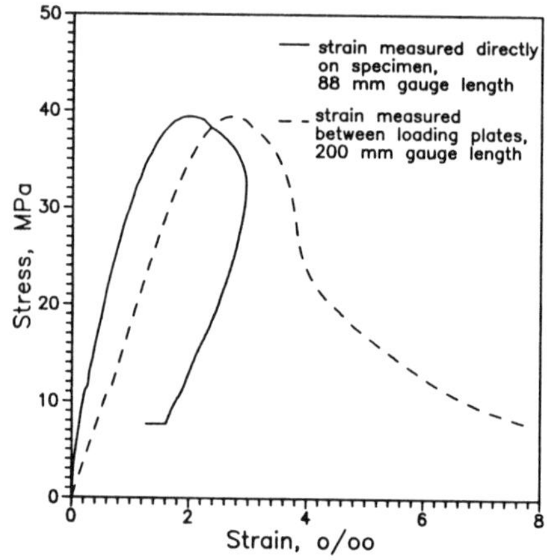

Fig. 3 Stress-strain curves for normal strength concrete

As seen in fig. 3 and fig. 4 there is a significant difference between the ascending part of the stress-strain curves when the strain is measured between the loading plates and the ascending part of the stress-strain curves when the strain is measured over the central part of the cylinder. This difference is also observed in tests performed by Van Mier, reported in Hordijk (1989b) and is probably due to end effects and non-perfect contact. There is a weak layer of cementpaste at one end of the cylinder - the top end at casting. Furthermore, the ends of the cylinders are not 100 % plane, and therefore only part of the cross-sectional area is loaded. This is supposed to result in the difference in the strain corresponding to the peak stress and in the initial modulus of elasticity.

It can be seen from fig. 4 that the ascending part of the stress-strain curve for high strength concrete measured with the strain gauges and the transducers are similar and independent of the length of the gauges. Statistical analysis show that the strain corresponding to the peak stress and the initial modulus of elasticity measured over a length of 25 mm, 50 mm, 88 mm and 100 mm are similar. The values for the strain corresponding to the peak stress measured at the central part of the cylinder are compatible with reported experimental information by Kotsovos (1983).

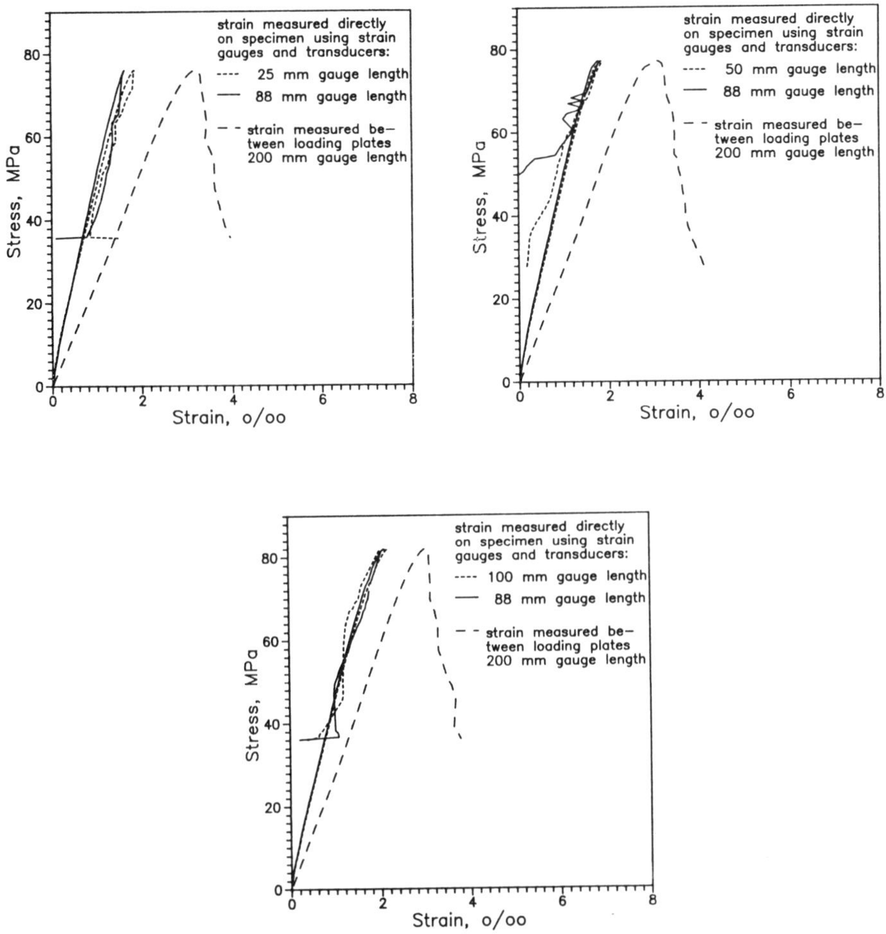

Fig. 4 Stress-strain curves for high strength concrete

As the strain corresponding to the peak stress and the initial modulus of elasticity are similar for stress-strain curves with 25 mm, 50 mm, 88 mm and 100 mm gauges, it can be concluded that effects resulting from the weak and non-uniform loaded end zones cannot be noticed over at least 100 mm at the middle of the specimen. Also, there does not seem to be any strain localization before peak.

At the descending part of the stress-strain curves an elastic recovery at peak is seen for the stress-strain curves for high strength concrete, when the strain is measured at the central part of the cylinder, see fig. 4. This phenomenon is also seen on the curves for the normal strength concrete, but the recovery starts after peak, see fig. 3. This recovery is considered to be due to longitudinal cracks separating the cylinder. In this process an outer layer is created which does not carry any load. Thus, the inner core carries the total load. This is also observed by Kotsovos, (1983). The difference in behaviour between cylinders with different strengths is considered to reflect the difference in the fracture process, as described in section 3.2. The zone in which the longitudinal cracks extend, is bigger the higher the strength. This explanation is also confirmed visually at the end of the test, where the lower strength cylinders can still be coherent. The same observations was made by Kotsovos, (1983).

For normal strength concretes no localization takes place for strains up to about 3 ‰, according to fig. 3 where stress-strain curves measured between the loading plates are compared to stress-strain curves obtained using a 88 mm gauge length.

4. Conclusion

It is shown that the composition of the feed-back signal used in a closed loop compression test set up does not have any influence on the shape of the stress-strain curve envelope. When the feed-back signal is composed of both axial- and circumferential deformation, more information about the fracture process is achieved. The axial stress-strain curve becomes more jagged and each drop in the curve corresponds to an extension of a crack.

A snapback of the stress-strain curves is not a material characteristic, but a phenomenon dependent on the frictional restraint of the test specimen, the strength of the concrete, and localization of the fracture. The phenomenon can only be observed, when the test is deformation controlled with a feed-back signal which includes the circumferential deformation. Furthermore, the snap-back is more pronounced the more the circumferential signal is weighed.

The tests indicate that stress-strain curves obtained from strain calculated on basis of deformation measurements between the loading plates are seriously affected by end zone effects. Stress-strain curves obtained from strain measurements over the central part of the surface of the cylinders made of high strength concrete show recovery due to separation of the cylinder. Thus none of the curves can be used for non-linear load carrying capacity calculations. When the specimen is separated and an inner core carries the load the stress is calculated on basis of a wrong cross-sectional area. The stress should have higher values, which again means that the descending part of the stress-strain curve in reality is less steep.

It is essential to measure deformation at the central zone of the specimen. In this way useful information is extracted about the coherency of the concrete, and thereby of how much of the stress-strain curve to take into account in load carrying capacity calculations. If no recovery is seen at these curves, (the concrete is coherent and the descending part is useful) measurements over different lengths of the cylinder can give information about the size of the zone of the specimen which is unaffected by end effect and about the fracture process and possible localization.

Techniques to study crack development are important in relation to compressive stress-strain tests. It is shown in this report that the circumferential deformation gives useful information about crack development.

In future investigations test series with normal strength concrete where the strain is measured over different lengths of the cylinder will be made in order to get more information about the fracture process. Measurement of lateral deformation near the end zones of the cylinder would make it possible to get more information about the longitudinal crack development.

Furthermore, the effect of the frictional restraint at the end zones on the splitting of high strength concrete will be investigated using different types of intermediary layers between the concrete cylinders and the loading plates.

5. References

Dahl, Henrik and Brincker, Rune (1989a) Fracture Energies of High-Strength Concrete in Compression. Proceedings from the International Conference on Recent Developments in the **Fracture of Concrete and Rock**, Cardiff, UK, 20-22 september.

Hordijk, Dirk A., Van Mier, Jan G.M. and Reinhardt, Hans W (1989b) Material Properties in **Fracture Mechanics of Concrete Structures From Theory to Applications** (edited by L. Elfgren), Report on the Technical Committee 90-FMA Fracture Mechanics to Concrete Applications RILEM, Chapman and Hall.

Kotsovos, M.D. (1983) Effect of Testing Techniques on the Postultimate Behaviour of Concrete in Compression. **Materials and Structures**, RILEM, Vol. 16, No. 91, January-February.

Olsen, N.H. (1990) Uniaxial Stress-Strain Curves of High Strength Concrete. **Ph.D. Thesis** at the Department of Structural Engineering, Technical University of Denmark, 2800 Lyngby, march.

Sargin, M. (1971) Stress-Strain Relationship for Concrete and the Analysis of Structural Concrete Sections, **Study** No.4, Solid Mechanics Division, University of Waterloo, Ontario, 1971.

Shah, S.P., Gokoz, Ulker and Ansari, Farhad (1981) An Experimental Technique for Obtaining Complete Stress-Strain Curves for High Strength Concrete. **Cement, Concrete and Aggregates**, Vol. 3, No. 1, pp. 21-27.

Wang, P.T., Shah, S.P. and Naaman, A.E. (1978) Stress-Strain Curves of Normal and Lightweight Concrete in Compression. **ACI Journal**, November 1978.

67 YIELD ENVELOPE INVESTIGATION OF FELSER SANDSTONE WITH TRUE TRIAXIAL COMPRESSION TESTS INCLUDING POST FAILURE BEHAVIOUR

W. KAMP and M.J. COCKRAM
Department of Mining and Petroleum Engineering, Delft University of Technology, The Netherlands

Abstract

In the Laboratory for Roçk Mechanics a true triaxial compression machine is available allowing for advanced stress and/or strain controlled tests with accurate displacement and acoustic P-wave velocity measurements in three perpendicular directions. A description of this machine is given including the force control system and the set-up of the various measurements.

A series of 17 tests is done on 105 mm cubical samples of Felser sandstone using a complex automatic loading sequence. A strain controlled loading is done along the Π-lane into the post failure region until approximately 5% linear strain in one of the directions is reached. With this procedure the yield surface is found, softening occurs and a residual strength is reached. Before and after failure the orthorhombic elastic parameters are measured. During the whole test, compressional wave velocities are measured in three directions.

The compressional wave velocities during the deviatoric loading indicate that the failure process starts long before the yield envelope has been reached and show that the microfractures are perpendicular to the lowest stress.

Keywords: Yield surface, Post failure behaviour, True triaxial testing, Compressional wave velocities.

1 Introduction

This research is carried out in the framework of the project "Cataclastic-Plasto-Elastic equilibria in brittle rock under triaxial differential loading conditions" of the E.C. R&D programme Raw Materials and Recycling, contract MA1M-0018-NL.

Good knowledge of material properties is essential for numerical calculations which include post failure behaviour. Usually parameter fitting is done based on a few test results. That way the material behaviour is predicted under all possible loading conditions. However, this approach introduces errors when extrapolating the material behaviour to loading conditions other than during the tests. This problem can only be resolved by a good understanding of the mechanisms that occur during non-linear deformation. The availability of the true triaxial compression machine with its sophisticated testing pos-

Fracture Processes in Concrete, Rock and Ceramics Edited by J.G.M. van Mier, J.G. Rots and A. Bakker.
 Published by E. & F.N. Spon, 2–6 Boundary Row, London SE1 8HN. ISBN 0 419 15870 7

sibilities makes it possible to carry out an extensive series of tests to investigate a large part of the yield envelope of a homogeneous sandstone including the post failure behaviour.

2 Description of true triaxial compression machine

The machine consists of three uniaxial compression systems assembled in perpendicular directions so that cubic samples can be compressed with sizes from 74 to 300 mm and up to a force of 3.5 MN. Each of the uniaxial systems consists of a piston with a pressure platen at one side and a pressure platen at the other side of the sample, connected with four tension bars. Two horizontal uniaxial systems can slide relative to each other in the horizontal plane. Vertical movements of the horizontal compression systems are possible as they are mounted on vertical springs. In this way shear forces on the sample surfaces are minimized.

On five of the six sides (the bottom side is fixed) of a cubical sample spherical seats are mounted in order to allow for small rotations of the pressure platens. This way proper loading conditions are realized, even in the case that the sample surfaces are not perfectly parallel or if the uniaxial compression systems are not positioned perfectly.

The main oil pressure acts on the piston surface of approximately 0.1 m^2 and a bearing pressure of a 100 bar acts against the main oil pressure on a ring shaped surface of approximately 0.01 m^2 (figure 1). The bearing pressure is used to retract the pistons and for the double hydrostatic bearings which are between piston and cylinder wall, Blok (1976).

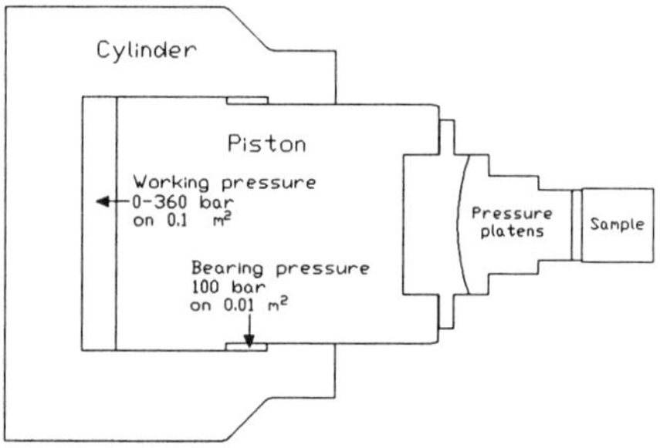

Fig. 1. Cylinder and piston (schematic).

The samples have to be made a few millimetres larger than the pressure platens, in order to avoid touching of the different platens during the large deformations of the samples that occur. A disadvantage of this is the introduction small stress free edges along the ribs of the cubes that may influence the fracture pattern.

The forces are controlled by an IBM compatible AT personal computer equipped with AD-converters which are connected to pressure transducers and DA-converters operating the hydraulic servo valves to control the main pressures in the three cylinders. The force control computer sends the actual forces to another AT personal computer and receives the desired forces (setpoints) for each compression system from this computer which serves as a function generator and data logger.

3 Measurements

The actual forces based on oil pressure measurements in the cylinders are received from the force control computer.

The displacements (length changes) of the sample are measured in three perpendicular directions with two LVDT's each with a range of 5 mm. They are mounted to a system of bars which are connected on the pressure platens near the sample in order to exclude deformations of the steel blocks or rotations of the spherical seats from the measurement (figure 2).

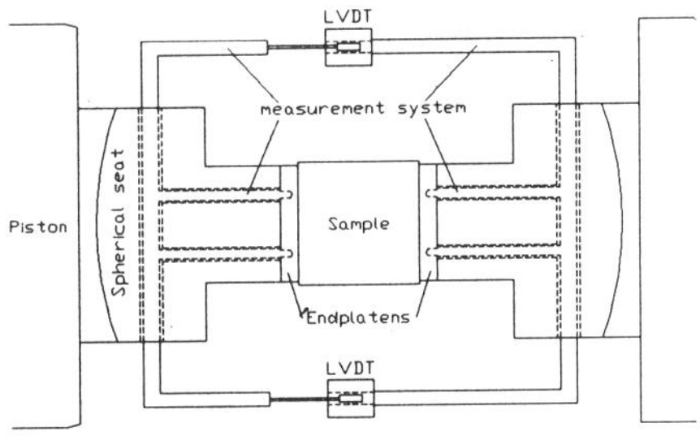

Fig. 2. Displacement measurement system.

The displacement measurements are made using an AD-converter in the test control computer and can be used for displacement controlled tests.

A schematic overview of the experimental set-up is shown in figure 3. In this set-up measurements of the compressional wave velocities are carried out by a stand alone system of which the "PC acoustic measurements" has a synchronization line with the "PC test control" for later correlation of the recorded data files.

The completely automatic compressional wave velocity measurements are carried out as follows. Using a computer controlled switch box two piezo-transducers on opposite sides of the cube are selected as transmitter and receiver of acoustic waves. The transmitter transducer then receives a pulse from the pulse generator which is first amplified to 300 V, at the same time the transient recorder is triggered and starts measuring the received signal. The received signal is amplified in two steps before it is sent to the transient recorder. A programmable gain amplifier makes it possible to have different amplifications in different directions of the sample. The amplification is automatically adjusted during a test based on previous measurements in a certain direction. The recorded signal is transferred to the "PC acoustic measurements" and a picking program determines the first arrival of the acoustic signal resulting in the travel time. The velocities are then obtained by dividing the sample length by the travel time. The sample length is corrected for deformations a result of applied stresses.

4 Sample description

Felser sandstone was selected a as material to carry out this test series. It is a homogeneous, isotropic, poorly sorted, yellowish sandstone. It consists mainly of quartz with some clay minerals between the quartz grains and it has a porosity of about 20%. The homogeneity of this material makes it possible to obtain many samples with the same properties. This material shows strong plastic defor-

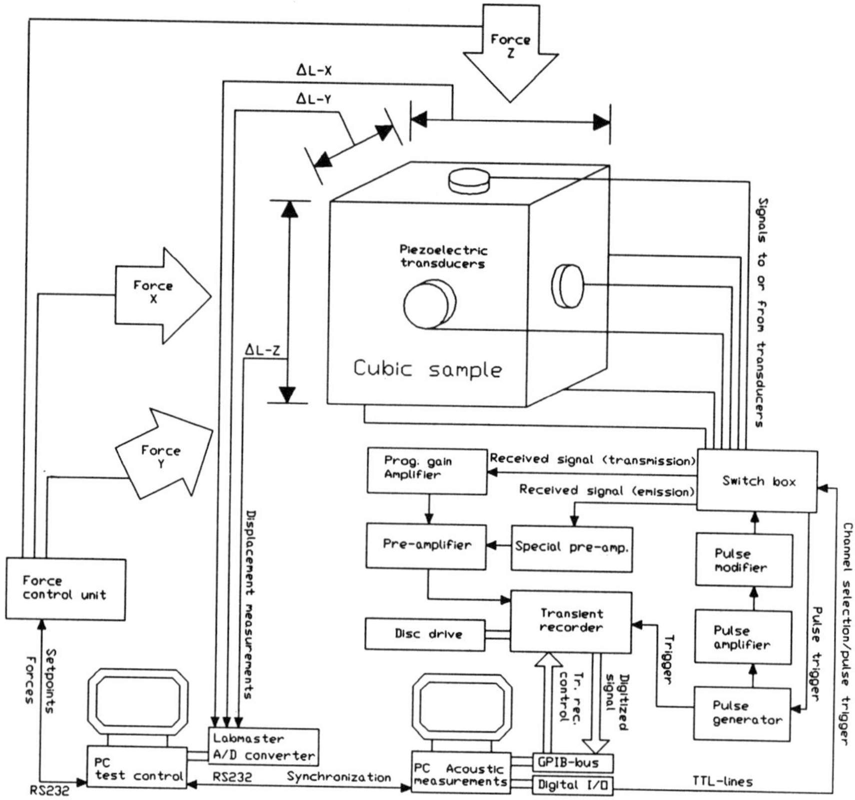

Fig. 3. Integrated equipment for fully automatic true triaxial testing.

mations with some strain softening while it is not extremely brittle, making it a suitable material for this type of testing.

Cubes are sawn and polished to 105 mm, with an accuracy of 0.2 mm but with larger accuracy of parallelism and right angles.
The cubes of the series are made from one large block and their orientation is recorded. All cubes are tested in the same position although the material appears isotropic before testing.

5 The test series

The loading path used for the test series (17 samples) consisted of the following steps:

- Hydrostatic loading to a certain stress (σ_h).
- Stress loops for determination of the elastic parameters.
- A displacement controlled Π-plane loading with a certain orienta-

tion (θ-value) in the Π-plane up to about 5% linear strain in one of the directions.

- Return to hydrostatic loading (σ_h).
- Stress loops to observe the change in elastic parameters.
- Unloading of the sample.

The stress loops consist of an alternating change in stress (25 MPa) in three directions (figure 4). From the resulting slopes of the displacement curves three Youngs moduli and three Poison ratios can be calculated. In order to measure only elastic strains the stress loops should be kept as small as possible. However, this reduces the accuracy of the measurements. Therefore, the stress loop is executed six times for averaging purposes.

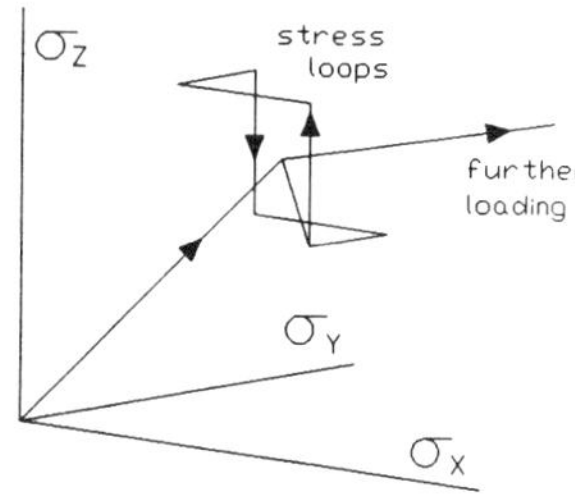

Fig. 4. Stress loops to determine anisotropic elastic properties.

During the Π-plane loading stresses are generated in such a manner that the loading path follows a line perpendicular to the hydrostatic axis resulting in a constant mean normal stress. By changing the mean normal stress and the orientation (θ: angle to the compressive meridian) between tests the different parts of the yield envelope can be investigated. For this type of loading the control of the compression machine is done in the following manner:

X-axis (intermediate): $$\sigma_x = \sigma_h - \frac{\cos(60-\theta)}{\cos(\theta)} * (\sigma_z - \sigma_h)$$

Y-axis (low): $$\sigma_y = \sigma_h - \frac{\sin(30-\theta)}{\cos(\theta)} * (\sigma_z - \sigma_h)$$

Z-axis (high): $\sigma_z =$ Displacement controlled with feedback of the displacement of the Z-axis at a rate of 2 μm/sec.

In table 1 a resume is given of the loading paths of all the tests. As an example of the results of a test, various graphs of the measurements of test no. 6 are given in figures 5, 6, 7 and 8.

5.1 The elastic part of the strains

A comparison of the elastic parameters from the stress loops and the compressional wave velocities can be made. More detailed information on this topic can be found in Kamp et al. (1990a). However, the actual values of the acoustic velocities are always higher than can be predicted from the elastic parameters determined from the stress loops.

The stress loops that are carried out at the beginning and end of each test can be analyzed resulting in the parameters of orthorhombic anisotropic elasticity. The material appears nearly isotropic, even at the end of the tests. Before failure the Youngs modulus is ≈30 GPa and after failure ≈15 GPa. The Poisson ratios are ≈0.22. Nevertheless, it can be clearly observed from the acoustic velocities that the material is significantly anisotropic during deviatoric loading and failure. When the stresses are brought back to hydrostatic loading after failure the velocities become approximately equal again in all directions.

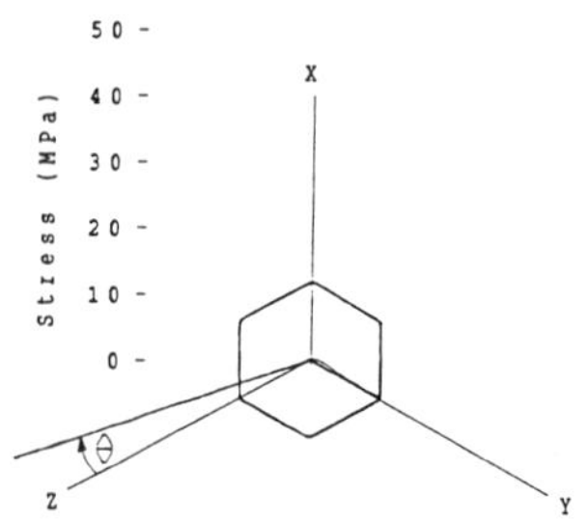

Fig. 5. Projection of stresses of test nr. 6 on the Π-plane, showing stress loops (hexagonal) and Π-plane loading with θ=10°.

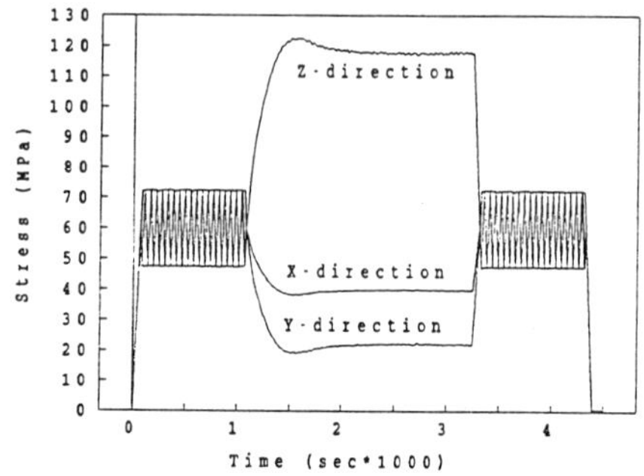

Fig. 6. Stress vs. time for test nr. 6 with stress loops, Π-plane loading and again stress loops.

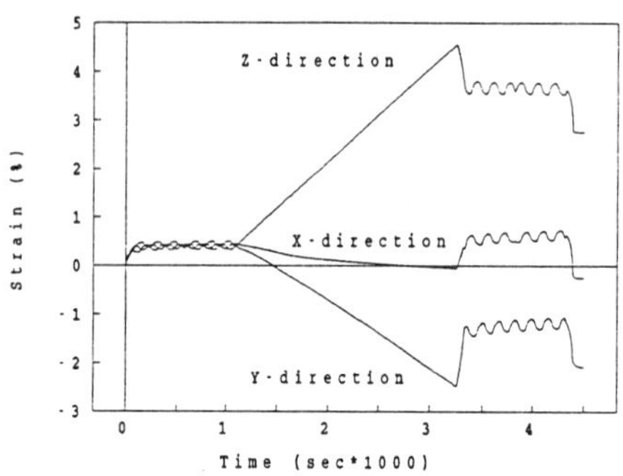

Fig. 7. Strain vs. time for test nr. 6 as a result of the loading sequence in fig. 6.

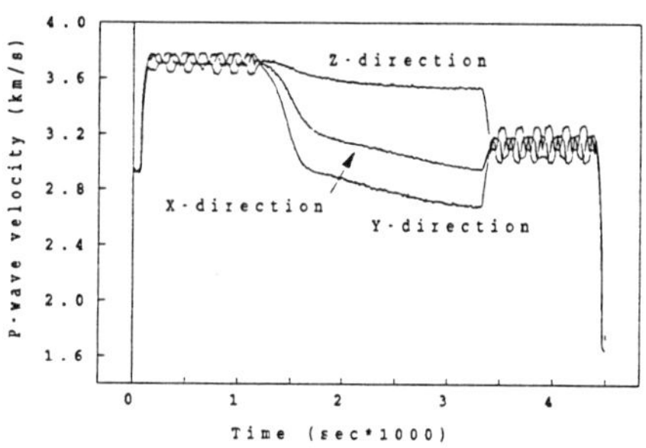

Fig. 8. P-wave velocity vs. time for test nr. 6 as a result of the stresses in fig. 6.

5.2 The yield points

From the measurements of the three stresses and three strains various plots can be made using the following notations:
Notation:

Mean normal stress: $$\sigma_{oct} = \frac{1}{3}(\sigma_1+\sigma_2+\sigma_3)$$

Equivalent shear stress: $$\tau_{oct} = \frac{1}{3}\sqrt{(\sigma_1-\sigma_2)^2+(\sigma_1-\sigma_3)^2+(\sigma_2-\sigma_3)^2}$$

Equivalent shear strain: $\gamma_{oct} = \frac{1}{3}\sqrt{(\epsilon_1-\epsilon_2)^2+(\epsilon_1-\epsilon_3)^2+(\epsilon_2-\epsilon_3)^2}$

The $\tau_{oct}-\gamma_{oct}$ plots seem to be the most appropriate to determine the yield points. In fact yielding (with work hardening) occurs already at very low stresses, trying to find the first yielding is not relevant for this material. In most of the tests a maximum bearing capacity (maximum τ_{oct}) can be found but tests at high σ_{oct} values show continuous work hardening.

The $\tau_{oct}-\gamma_{oct}$ plots of the tests with θ=30° (σ_2=constant) are shown in figure 9. The maximum occurring τ_{oct} points of all tests are plotted in the stress space in figure 10 (values can be found in table 2).

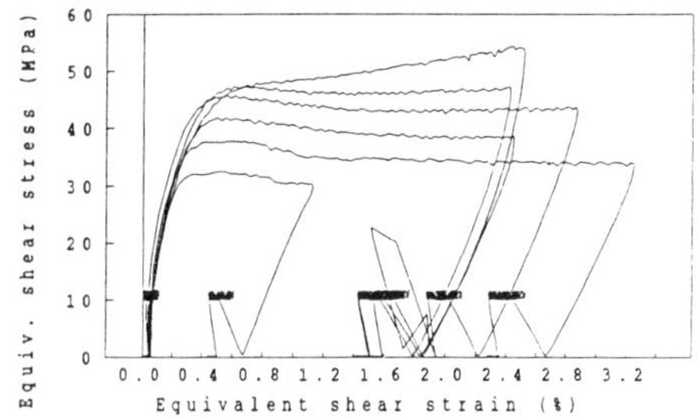

Fig. 9. Equivalent shear stresses vs. equivalent shear strains for the tests with θ=30°, showing work hardening, strain softening and residual strength.

The yield points of the tests with σ_{oct}=60 MPa are projected on the Π-plane and shown in figure 11. From this figure the smooth shape of the yield surface can be seen. The failure criterion is between the Mohr-Coulomb and Drucker-Prager criterion and can probably be described with the Willam-Warnke formulation as is given in Chen and Saleeb (1982). Parameter studies will be made in a later stage.

5.3 Non-associated flow

This material shows two types of non-associated flow.

The volume behaviour shows volume decrease until the ultimate strength is reached, slight volume increase during the strain softening and no significant volume change during further deformation at residual strength.

In the projection on the Π-plane in the strain space another type of non-associated flow is observed. The strain paths of the tests with σ_{oct}=60 MPa are projected on the Π-plane and shown in figure 12. It is seen that for θ=0° (σ_2=σ_3) or θ=60° (σ_1=σ_2) the strain paths are straight lines in the corresponding directions. For intermediate θ-angles the directions of the strains change in the sense that the intermediate principle strain ϵ_2 becomes very low.

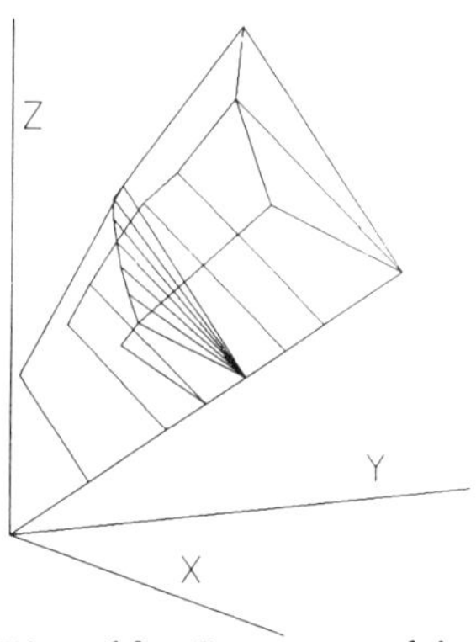

Fig. 10. Stereographic projection of the yield points in the stress space.

5.4 Fracture patterns

The series of tests include different Π-planes and different θ-angles resulting in information about the entire yield surface. For the post failure behaviour the Mohr-Coulomb mechanism predicts the development of shear zones. For θ=0° (σ_2=σ_3) or θ=60° (σ_1=σ_2) multiple orientations of the shear zones are found. For intermediate θ-values only one set of conjugate shear zones is found, resulting in $\epsilon_{2,p}\approx 0$. However, in the tests with high σ_{oct}-values fewer fractures are found with complicated patterns.

The observed plastic flow behaviour can be explained by studying the fracture patterns that develop in the samples during strain softening. The Mohr-Coulomb mechanism predicts the following pattern, which is also found from the actual fractures as they are found in the samples after the tests.

When two stresses are equal (θ=0° or θ=60°) multiple sets of conjugate shear zones are found.

For intermediate θ-values only one set of conjugate shear zones develops, visible in sections perpendicular to σ_2. In the tests with high σ_{oct}-values fewer fractures are found with complicated patterns.

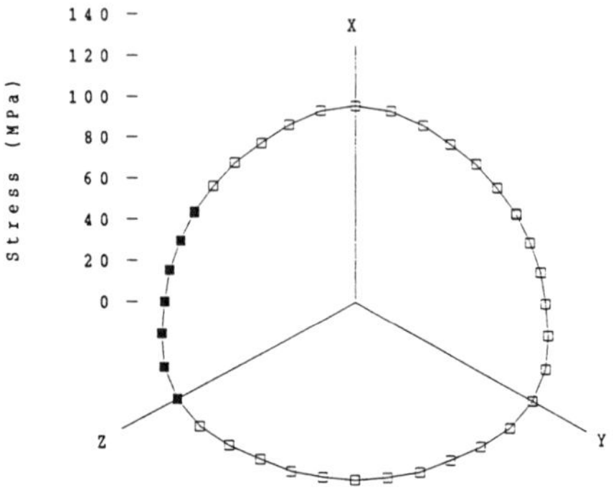

Fig. 11. Projection of yield points (black squares) on the octahedral plane of the tests with a mean normal stress of 60 MPa.

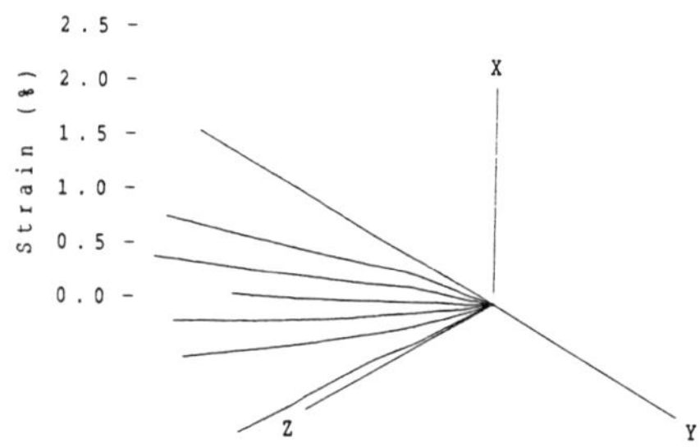

Fig. 12. Projection of the strains on the octahedral plane of seven tests with a mean normal stress of 60 MPa and θ-angles from 0° to 60° showing change of strain direction during strain softening.

To visualize the fracture patterns some of the samples are impregnated with coloured glue and sawn into slices. In figure 13 an example is shown of a section perpendicular to σ_2 of test nr.8. A set of conjugate shear zones is seen, starting from the corners of the cube, this is in agreement with the Mohr-Coulomb failure mechanism. Other fractures under another angle are also observed, these probably develop in a later stage and are most likely caused by friction on the sample surfaces resulting in a redistribution of the stresses.

5.5 Strain softening

The tests are continued far into the range of plastic deformations. In most of the tests some strain softening and a residual strength is

found.
The slope of the softening curves (down going σ–ϵ–curves) appears to be dependent of the actual value of σ_3 rather than the σ_{oct}–value.

6 Conclusions

Interesting material testing possibilities exist with the described equipment from which much can be learned about the failure behaviour of rock.

The fractures found in the Felser sandstone samples after a test and the results from the displacement measurements are consistent with the Mohr–Coulomb failure mechanism. However, the cross section of the yield envelope is rounded, which is not predicted by the Mohr–Coulomb failure mechanism.

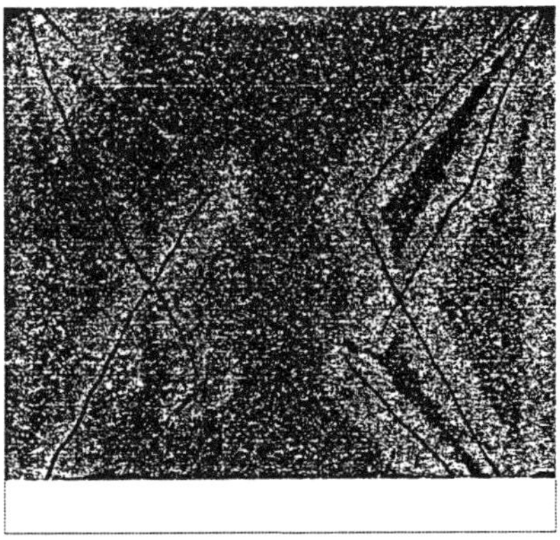

Fig. 13. Fracture pattern as a result of test no. 8. Fractures have been emphasized by hand. Change in color is caused by inhomogeneous impregnation.

The possibility to control and measure three independent stresses and strains, as described here, can result in a better understanding of the failure mechanisms of rock for complex stress conditions. Also more detailed input parameters for finite element or finite difference calculations can be obtained (eg. softening and flow behaviour) from such tests.

7 Acknowledgements

This work has been supported by the Commission of the European Community as a part of the programme "Primary raw Materials"

8 References

Blok P. (1976) Das Hydrostatischer Keilspaltlager: Berechnung und Anwendung bei Hydrozylindern. Delft.

Chen W.F., Saleeb A.F. (1982) Constitutive equations for engineering materials.

Kamp W., Cockram M.J., Roest J.P.A. (1990a) Cataclastic–Plasto–Elastic equilibria in brittle rock under triaxial differential loading conditions. EEC Contract number: MA1M–0018–NL(GDF). Second progress report.

Kamp W., Cockram M.J., Stam J.L., Roest J.P.A. (1990b) Cataclastic–Plasto–Elastic equilibria in brittle rock under triaxial differential loading conditions. EEC Contract number: MA1M–0018–NL(GDF). Third progress report.

Kamp W., Cockram M.J. (1990c) Possibilities of true triaxial experiments in the laboratory. I.A.E.G. Congress, Amsterdam.
Kamp W., Roest J.P.A. (1988) Overview of rock mechanics research for deep mining. Delft Progress report. Vol. 13, no. 1/2.
Mier J.G.M. van (1984) Strain softening of concrete under multiaxial loading conditions. Eindhoven.
Sayers C.M., Munster J.G. van, King M.S. (1990) Stress induced ultrasonic anisotropy in Berea Sandstone. Int. J. of Rock Mech. Min. Sci. & Geomech abstr., Vol. 27. No. 5.

9 Appendix

Table 1. Overview of the Π-plane loadings for the test series (test number is in the table).

θ \ σ_h (MPa)	20	30	40	50	60	70	80	100
0°	2				3			4
10°					6			
20°					7			
30°		9	10	11	8	12	13	5
40°					14			
50°					15			
60°				18	16			17

Table 2. Yield points during test series of Π-plane loadings.

Test Nr.	σ_h (MPa)	θ (Deg.)	σ_x (MPa)	σ_y (MPa)	σ_z (MPa)
2	20	0	2.27	2.29	55.83
3	60	0	28.24	28.18	124.01
4	100	0	58.91	58.62	181.98
5	100	30	100.04	33.47	166.48
6	60	10	38.25	19.06	122.01
7	60	20	49.38	13.11	117.43
8	60	30	59.92	8.75	111.45
9	30	30	29.93	0.39	72.81
10	40	30	40.01	0.35	80.01
11	50	30	49.89	3.68	95.58
12	70	30	69.87	13.92	126.21
13	80	30	79.88	21.94	138.11
14	60	40	70.16	4.84	105.05
15	60	50	79.61	2.95	97.37
16	60	60	88.71	1.98	88.95
17	100	60	135.68	28.44	135.79
18	50	60	79.85	0.04	80.03

68 ON THE VALIDITY OF INDIRECT MEASUREMENT OF THE LPD FOR SEN CONCRETE BEAMS

S.E. SWARTZ and Y.-C. KAN
Kansas State University, Manhattan, KS, USA

Abstract
The load point displacement (LPD) for single-edge-notched (SEN) beams in three-point-bending (TPB) may be measured directly as the movement of the load point or indirectly as the movement of a beam surface or other reference on the beam. As presented here, comparison of these measurement techniques shows that beam bending deflections vary depending on which surface is measured and are lower than the direct load point deflection. Reasons for this include support "punching", friction and support flexibility. It is suggested that even though errors of significance in the elastic energy may occur, the total fracture energy measurement, calculated from the entire load-displacement plot, may not be significantly affected by the method of displacement measurement.
Keywords: Fracture Energy, Beam Displacements, Errors, Concrete Beams, Load-Point-Displacement.

1 Introduction

The LPD measurement for TPB-SEN concrete beams is needed to determined fracture energy G_F by the method proposed by RILEM Committee TC50-FMC (1985). This measurement is also needed for other energy methods, Refai and Swartz (1988), and for proposed fracture toughness test methods for concrete, Karihaloo and Nallathambi (1990) and rock, Ouchterlony (1990).

Direct (δ_1) and indirect (δ_2, δ_3) measurements of the LPD are shown in Figure 1. It has been suggested by Carpinteri and Ballatore (1990) that "punching" at load and support points may severely affect the δ_1 measurement. Swartz and Kan (1990) expressed the error in δ_1 to be e_1 + 1/2 (e_2+e_3) + e_4 where e_1 - e_3 are generally given by

Fracture Processes in Concrete, Rock and Ceramics Edited by J.G.M. van Mier, J.G. Rots and A. Bakker.
 Published by E. & F.N. Spon, 2–6 Boundary Row, London SE1 8HN. ISBN 0 419 15870 7

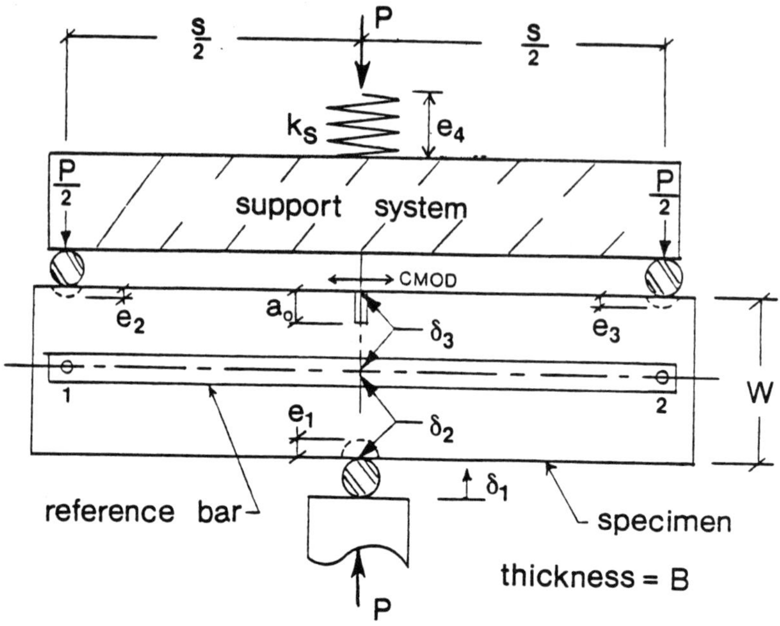

Fig. 1 Schematic of test setups

$$e = \frac{1}{2R}\left[\frac{F}{Bf_p}\right]^2 \tag{1}$$

based on plasticity concepts. In this R is the radius of the support roller, F is the applied force at the support, B is the beam width and f_p is taken to be the compressive strength of concrete. For beams with span/depth of 3.75, the error $e_1 + 1/2\ (e_2+e_3)$ ranged from 1.1% to 5.8%. The error, e_4 was not evaluated.

A device used to obtain δ_1 is described by Gillespie, et al. (1984) and devices to measure δ_2 or δ_3 are described by Gopalaratnum, et al. (1989) and Karihaloo and Nallathambi (1990) generally following recommendations of ASTM C1018 (1985). Gopalaratnum, et al. (1989) suggest that at peak loads, $(\delta_1-\delta_3)/\delta_3$ is about 3-4 with the difference due to "extraneous deformations" present in measuring δ_1.

2 Indirect displacement measurement

The indirect measurements discussed here assume a reference bar is attached to the beam specimen at its mid-depth--approximately the location of the neutral axis--as shown in Fig. 1. It is seen that the punching errors e_2, e_3 will not affect the measurement but e_1 will affect δ_2. Thus, it is expected that $\delta_2 > \delta_3$.

Other possible errors include friction effects at support points 1 and 2. These effects are clarified in Fig. 2 where the worst--and probably unrealistic--

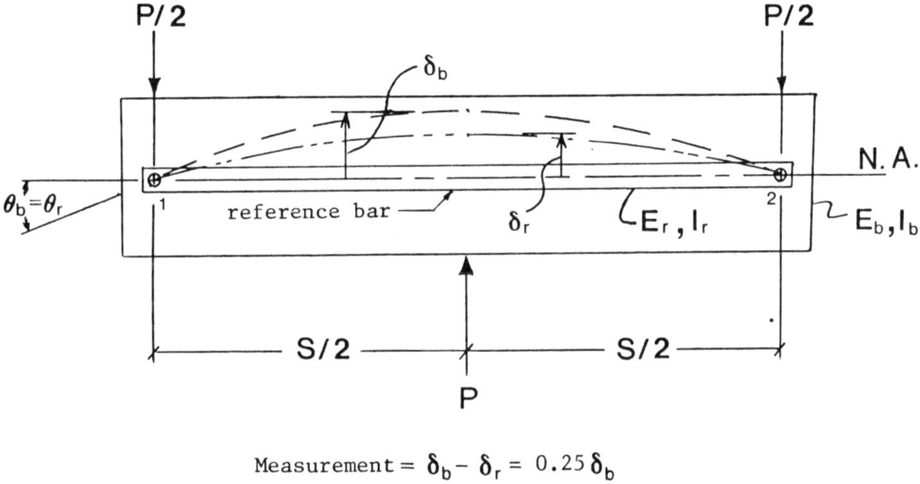

Fig. 2 Effect of fixing reference bar on displacement measurement

situation is envisaged, fixity at supports 1 and 2. By considering consistent bending deformations, it may be seen that $\delta_b = (S/3)\ \theta_b$, $\delta_r = (S/4)\ \theta_r$ and, if $\theta_b = \theta_r$, $4\delta_r = 3\delta_b$. The measurement is $\delta_m = \delta_b - \delta_r$. Thus

$$0.25\ \delta_b < \delta_m < \delta_b \tag{2}$$

For complete fixity of the reference bar at points 1 and 2, if $e_1 = 0$ and direct compression is neglected, $(\delta_1 - \delta_3)/\delta_3 = 3$.

3 Indirect measurement test

In order to make direct comparisons between δ_1, δ_2 and δ_3, a reference bar system was developed for simultaneous use with the direct measurement setup already available. The system is shown in Fig. 3 and consists of two reference bars. Each bar has a "clip" gage attached with each gage using two strain gages (A1, A2, B1, B2) as shown. By

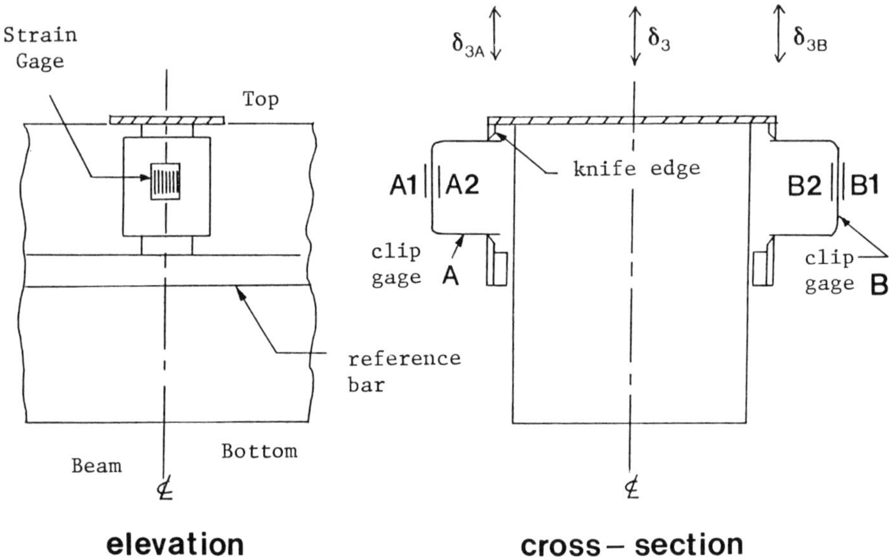

a. Arrangement of transducers

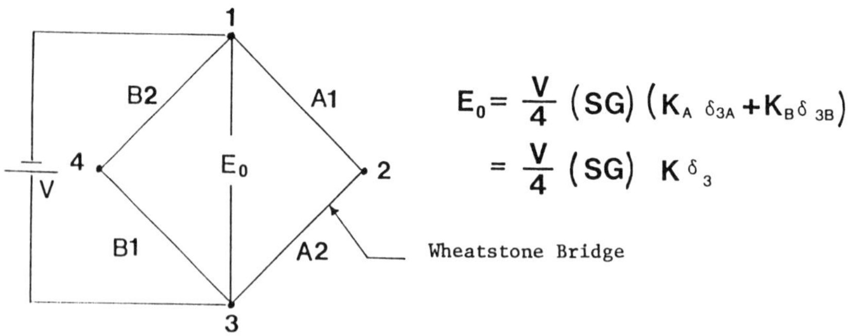

b. Full bridge connection for one-channel output

Fig. 3 Indirect LPD measurement

using a full bridge arrangement it is possible to obtain δ_2 or δ_3 as direct output to only one channel with beam twisting/rocking perturbations eliminated.

The calibration device using a micrometer with a direct least reading of 1 x 10^{-4} in. (2.54 x 10^{-6} m) is shown in Fig. 4. The calibration was performed using the strain conditioner in a closed-loop, testing system (MTS) and was δ = 2.14 x 10^{-4} m/V (r = 0.99953 using 75 data values) which gives a plotted sensitivity of 2.14 x 10^{-5} m per centimeter on the graph paper or a direct least reading of two microns.

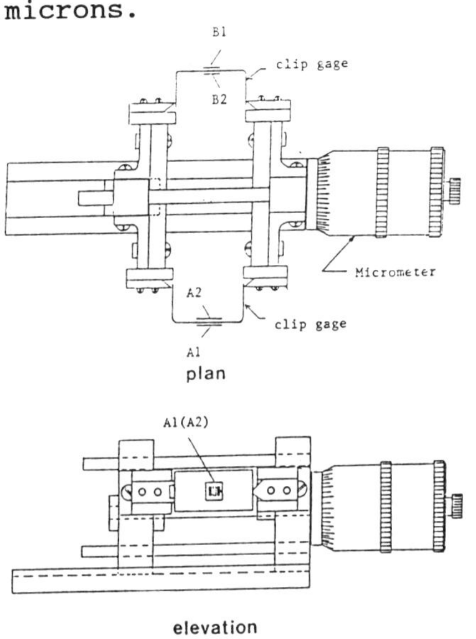

Fig. 4 Calibration of transducer

4 Beam tests

To compare directly δ_1, δ_2, δ_3 simultaneously was not possible since our system has only two strain conditioners. To enable a reasonable comparison with two tests, a wooden beam with S = 414 mm, W = B = 95 mm was tested. In both tests, δ_1 was measured and results for both were virtually identical. The average load-displacement slopes were: S_1 = 3.81 x 10^6 N/m, S_2 = 14.77 x 10^6 N/m and S_3 = 18.40 x 10^6 N/m. Thus δ_2/δ_1 = 0.258, δ_3/δ_1 = 0.207 and δ_3/δ_2 = 0.803. It is seen that the indirect displacements are about one-fourth the direct displacement for the elastic range of the material. Also δ_3 is about 80% of δ_2 indicating a local effect at the load point.

Using a notched concrete beam of 800 x 200 x 100 mm, plots of δ_1 and δ_3 were obtained as shown in Fig. 5. The material properties were $f_c' = 51.6$ MPa and $E_c = 32.4$ GPa (average of three cylinder tests). Again, it is seen that $\delta_3 > \delta_1$ in the elastic range. Using δ_1 and the formulas for E given by Karihaloo and Nallathambi (1990) for a notch depth of 0.2 W, the calculated value of E was

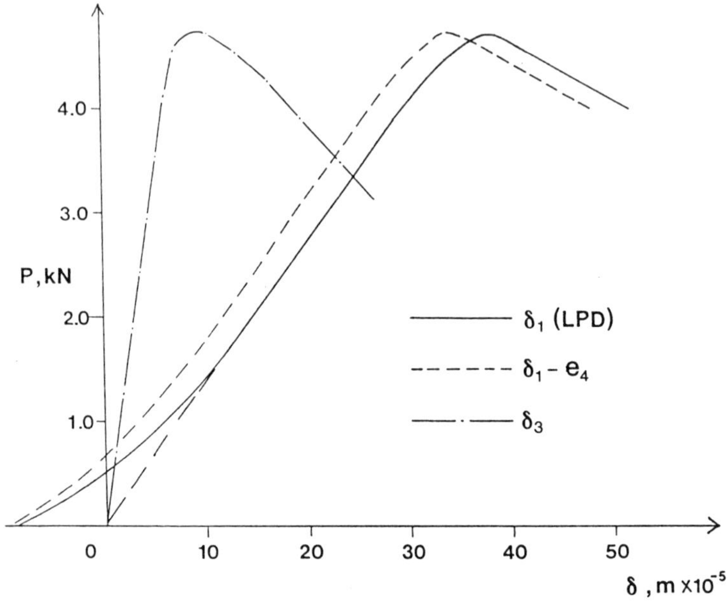

Fig. 5 Load-displacement for a concrete beam

6.03 GPA. When δ_3 was used, the calculated value of E was 31.5 GPa -- virtually the same as obtained from the cylinder compression tests. Note, however, that the total areas under the displacement diagrams to P = 0 may be quite similar.

The effect of e_4 was investigated as well and it was determined that k_s (c.f. Fig. 1) was 103.7 x 10^6 N/m. The maximum error--at maximum load--was about 14% of δ_1. A corrected curve of $\delta_1 - e_4$ is also shown in Fig. 5.

The effect that error e_4 has on fracture energy calculations based on δ_1 may be estimated by considering Fig. 6.

If OAC and OBC are approximated by straight lines, then the areas and hence G_F are the same regardless of e_4. If only the softening portion of each curve is considered, then the error is approximately in the ratio $e_4/(\delta_0 - \delta_1)$.

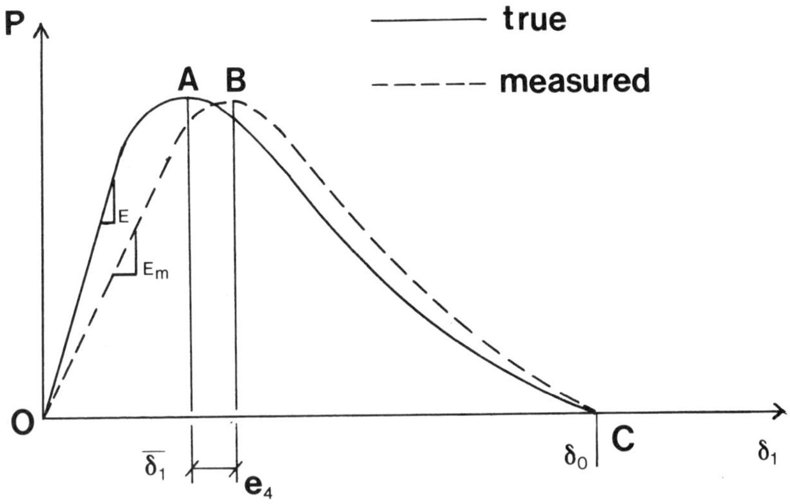

Fig. 6 Influence of support stiffness on W_0

5 Conclusions

Based on the work presented here the following may be concluded.

1. In the elastic range, the indirect method of measuring the beam displacement (δ_2 or δ_3) gives much smaller values than the direct LPD (δ_1) measurement. Further, with our data the indirect method gives a reasonable agreement between cylinder and beam values for the modulus of elasticity.
2. The support stiffness error can be measured and, for our data, gives a maximum error of 14% in δ_1. It was shown earlier (Swartz and Kan, 1990) that the maximum punching error in δ_1 is about 6%. Even if these are combined, the reduced value of δ_1 is still far greater than δ_2 or δ_3.
3. The error in δ_2 or δ_3 due to friction could be as high as 75% of δ_1. In our tests this effect was felt to be minimal.
4. In spite of these differences, which are primarily associated with the elastic response regime, the measured fracture energy may be relatively unaffected.

6 Acknowledgements

The work presented here was supported in part by the National Science Foundation through grants MSM-8317136 and MSM-8919449.

References

ASTM C1018. (1985) Standard test method for flexural toughness and first-crack strength of fiber-reinforced concrete (using beam with third-point loading).

Carpinteri, A. and Ballatore, E. (1990) Mixed mode fracture energy of concrete. **Engg. Fr. Mech.**, 35, 145-157.

Gillespie, R.L., Go, C.G., and Swartz, S.E. (1984) Vertical-displacement gage fixture for J_{IC} testing. **Exper. Tech.**, 8, 32-33.

Gjørv, O.E., Sorensen, S.E., and Arnessen, A. (1977) Notch sensitivity and fracture toughness of concrete. **Cem. and Conc. Res.**, 7, 333-344.

Gopalaratnum, V.S., Shah, S.P., Batson, G., Criswell, M., Ramakrishnan, V. and Wecharatana, M. (1989) Fracture toughness of fiber reinforced concrete. Report of an inter-university study sponsored by the ACI-CMRC and NSF, University of Missouri, Columbia, MO 65211, USA.

Karihallo, B.L. and Nallathambi, P. (1990) Notched beam tests: mode-I fracture toughness. In **Fracture Mechanics of Concrete-Test Method** (ed A. Carpinteri), Chapman and Hall, London (in press).

Malvar, L.J. and Warren, G.E. (1987) Fracture energy for three point bend tests on single edge notched beams: proposed evaluation. **Mat. and Str.**, 20, 440-447.

Ouchterony, F. (1990) Fracture toughness testing of rock with core based specimens. **Engg. Fr. Mech.**, 35, 351-366.

Refai, T.M.E. and Swartz, S.E. (1988) Mode I fracture-energy methods for concrete. **Exp. Mech.**, 28, 395-401.

RILEM TC50-FMC (1985) Determination of the fracture energy of mortar and concrete by means of three-point bend tests on notched beams. **Mat. and Str.**, 18, 285-290.

Swartz, S.E. and Kan, Y.-C. (1990) Effect of support conditions on fracture energy measurements for concrete beams. In **Fracture Behavior and Design of Materials and Structures** (ed. D. Firrao), EMAS, London, 660-666.

69 FRACTURE MECHANICS MATERIAL PARAMETERS OF MASS CONCRETE BASED ON DRILLING CORE TESTS – REVIEW AND DISCUSSION

H.N. LINSBAUER
University of Technology, Vienna, Austria

Abstract
Based on fracture-energy test results of dam concrete carried out on drilling core samples with 150 mm and 200 mm diameters the problematical nature of the testing procedure is discussed.
The influence of the aggregate attributes (size, mineral composition, bedding or foliation plane, etc.) as well as the sampling locality (construction joint area, homogeneous area) on the fracture behavior of dam concrete is shown up.
Keywords: Fracture Mechanics of Mass Concrete, Testing Procedure, Drilling Cores

1 Introduction

The design and construction of dams for creating water reservoirs to meet different demands as e.g. generation of electricity, irrigation, flood control, water supply etc., brings up a lot of problems which cannot be handled by usual engineering methods. Both on theoretical side, materials-development and construction technique problem-oriented procedures are necessary to guarantee safety and reliability in operation of the dam-structure. Despite careful recording of the local (in situ) conditions as topography, geology, rock quality on one hand and use of high technology in analysis and material processing on the other hand, crack formation cannot totally be excluded (Linsbauer et al.,1989).

The conventional rules and practices of design, construction and failure assessment with respect to cracking in mass concrete structures are restricted to the concepts of structural engineering (DIN 19702,1966) within which is not possible to fulfill continuum mechanics requirements. Fracture mechanics on the other hand offers possibilities to overcome these shortcomings.

The progress of linear and nonlinear fracture mechanics for cementitious materials may be seen such that the

Fracture Processes in Concrete, Rock and Ceramics Edited by J.G.M. van Mier, J.G. Rots and A. Bakker.
 Published by E. & F.N. Spon, 2–6 Boundary Row, London SE1 8HN. ISBN 0 419 15870 7

physical process of cracking has been comprehended and mathematically simulated. However, the main practical problem consists in the fact, that the formulation of adequate fracture criteria is based on **appropriate material parameters**. This requirement is particular difficult to fulfill for the analysis of cracking in mass concrete structures (dams), where the size of the aggregates in the concrete can be up to 150 mm across.

Charakteristic dimensions of concrete test samples for LEFM-conditions in fracture mechanics materials testing should be, by rule of thumb within the range of ten to twenty times the maximum aggregate size (Hilsdorf,1979). Another rule based on material properties - modulus of elasticity (E), fracture energy (G_f), and tensile strength (f_t) - calls for approximately ten times the "characteristic length" ($lch = EG_f/f_t^2$, Hillerborg,1976).

Using the above mentioned requirements for the appropriate material parameters for dam concrete, experimental test samples must have dimensions of 3 meters and between 8 and 15 meters respectively. Such sample sizes are obviously too large for practical experimentation, even during the construction of the dam. Therefore special procedures are required to overcome this problem. The methods generally are based on "size effect" models which have been developed from different points of view - a comprehensive review and discussion concerning the range of validity of the various concepts may be found e.g. by Di Tommaso (1989) and Elices et al. (1990). A fracture toughness assessment of dam concrete based on results of size-effect investigations for normal concrete was carried out by Brühwiler (1988). Recently Saouma et al. (1989) presented a hydraulic fracturing based field-test-method.

Within this contribution the problematic nature of G_F-interpretation on the basis of wedge splitting tests on drilling cores of dam concrete with diameters of 150 mm and 190 mm (determined by economics and statics) is pointed out.

2 Testing procedure and results

The tests were carried out with an improved version (minimization of friction - Tschegg, 1990) of the wedge splitting test method which was originally developed by Linsbauer and Tschegg (1986), (Fig.1) and has often been used in slightly modified form by Brühwiler (1988).

The testing procedure included three series, of ten drilling cores each. Two series of different diameters (150 mm and 190 mm) were chosen to show up the sample-size-influence (size effect) on the fracture energy. For quality assessment tests on drilling cores drawn from the construction joint interface were performed.

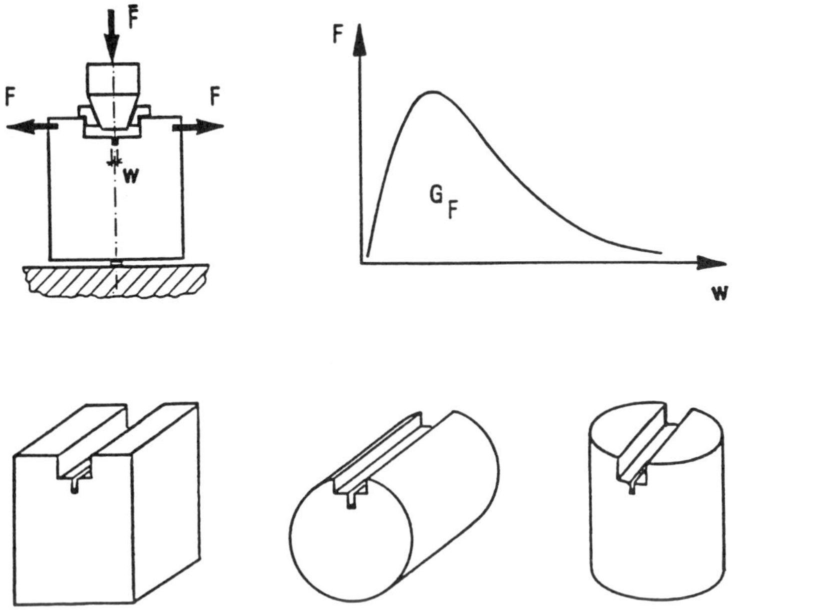

Fig.1. Wedge splitting test (Linsbauer-Tschegg 1986)

The spread of the test results for cores with 150 mm and 190 mm diameters is shown in Fig.2 and Fig.3 respectively. Typical GF-test results carried out on drilling cores with 190 mm diameter are represented in Fig.4, Fig.5 and Fig.6, where test sample No. 282 (Fig.4) may be regarded as representative of areas with middle-sized aggregates. The sample is characterized by an almost "homogeneous" surface of combined inter- and transgranular fracture processes.

In contrast to this, sample No.278 (Fig.5) and sample No.283 (Fig.6) show a typical transgranular (energy consuming) and intergranular (mica-gneis) fracture, respectively. The test results obtained from the drilling cores with 150 mm diameter have a similar characteristic.

A clear statement concerning the fracture energy investigations of the construction joint interface could not worked out because of an isufficient correspondence of the interface and the fracture area. This requirement was met in four tests only, where a drop in fracture energy was to recognize.

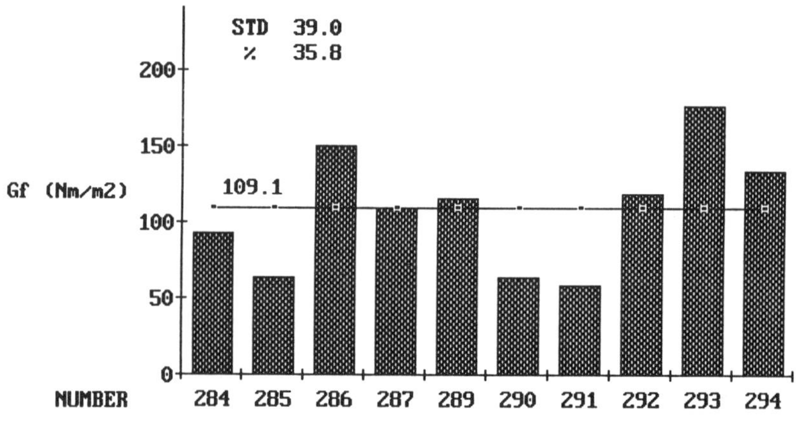

Fig.2. Fracture energy (G_f) test results **(D=150 mm)**

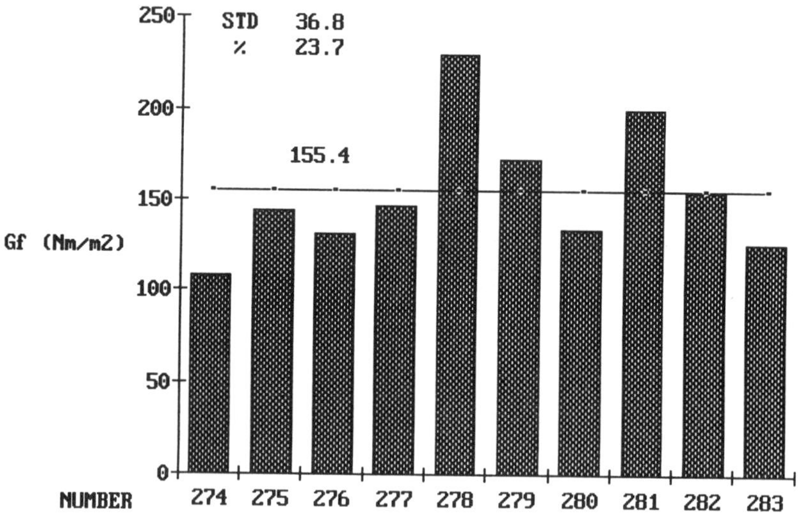

Fig.3. Fracture energy (G_f) test results **(D=190 mm)**

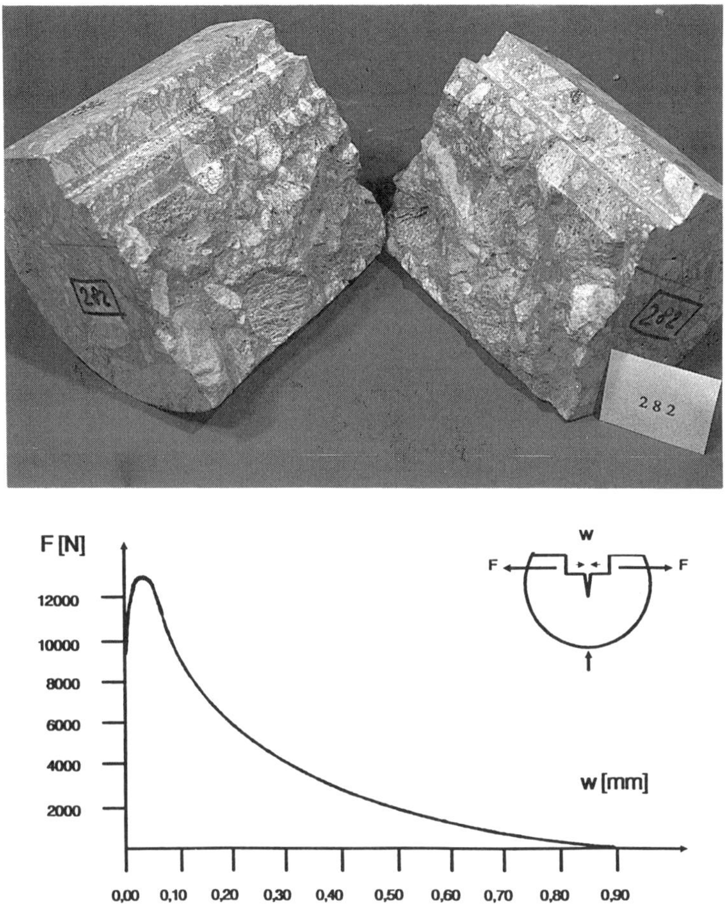

$$G_f = 155\ \mathrm{Nm/m^2}$$

Fig.4. Fracture surface and load-deformation diagram of sample No.282 (190 mm diameter)

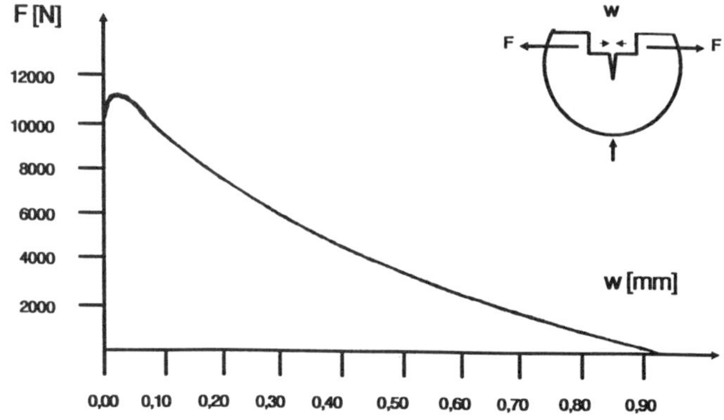

$G_f = 231\ Nm/m^2$

Fig.5. Fracture surface and load-deformation diagram of sample No.278 (190 mm diameter)

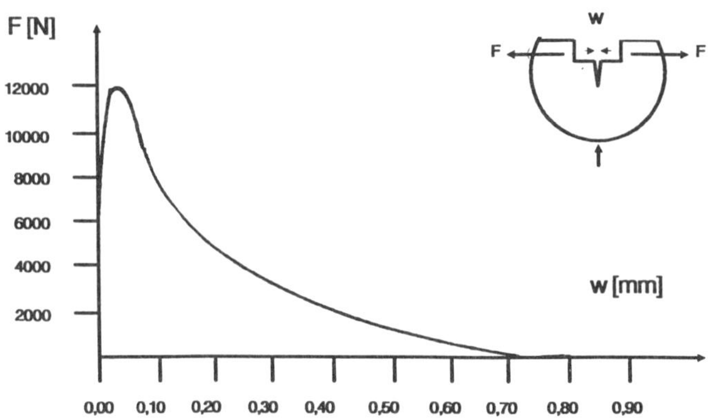

$$G_f = 126\ \mathrm{Nm/m^2}$$

Fig.6. Fracture surface and load-deformation diagram of sample No.283 (190 mm diameter)

3 Discussion

As has been demonstrated already, fracture energy determination of dam concrete is a very nasty problem. This has also been proved by standard deviation values of 36% and 24% for drilling cores with diameters of 150 mm and 190 mm, respectively (Fig.2., Fig.3 and Fig.8).

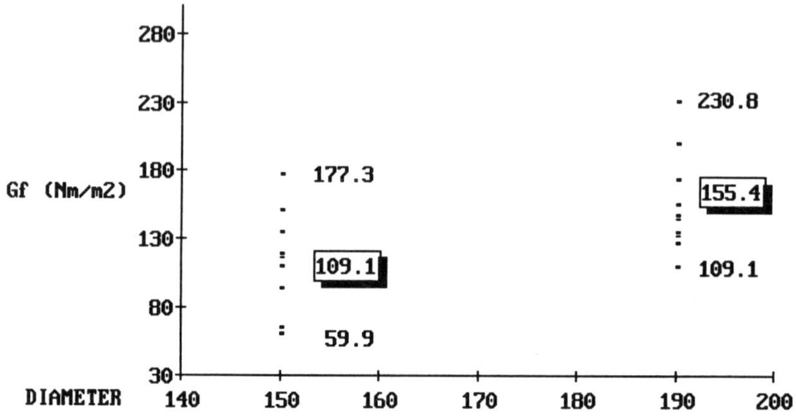

Fig.7. Range of scattering (D=150 mm and D=190 mm)

An attempt to gain valid values for the fracture toughness of dam concrete by applying size effect models (e.g. Bazant et al. 1982) must fail because of the inhomogenity of sample material.

In spite of the unsatisfactory results of tests for accurate material parameters, some useful statements may be made:

- the mean value of the fracture energy determined on drilling cores with larger diameters is significantly higher than the corresponding value for samples with smaller diameter (Fig.8)

- drilling cores with larger diameters show a smaller percentage of standard deviation

- construction joint interfaces (horizontal layers) represent areas of weakness

4 References

Bazant, Z.P., and Pfeiffer, P.A. (1987) Determination of Fracture Energy from Size Effect and Brittleness Number. **ACI Materials J.**, 10/11, 463-480

Brühwilwer, E. (1988) Fracture Mechanics of Dam Concrete Subjected to Quasi-Static and Seismic Loading Conditions. **Doctoral Thesis No.739** (in German), EPFL, Lausanne

DIN 19702 (1966) Berechnung der Standsicherheit von Wasserbauten. **Deutscher Normenausschuß** (German Standards)

Di Tommaso, A. (1989) Size Effects and Brittleness, in **Fracture Mechanics of Concrete Structures** (RILEM Report), (ed. L.Elfgren), Chapman and Hall, pp. 191-202

Elices, M. and Planas, J.(1990) Size Effect and Experimental Validation of Fracture Models. **Draft of the RILEM-Report TC89-FMT.**

Hillerborg, A., Modeer, M. and Petersson, P.E. (1976) Analysis of a crack formation and crack growth in concrete by means of fracture mechanics and finite elements. **Cem. Concr. Res.**, 6, 773-782

Hilsdorf, H.K. (1979) Sense and Limits of the Applicability of Fracture Mechanics in Concrete Technology (in German). **Forschungsbeiträge für die Baupraxis.** W.Ernst&Sohn, pp. 59-73.

Linsbauer, H.N., and Tschegg,E.K. (1986) Fracture Energy Determination of Concrete with Cube Shaped Specimens (in German). **Zement und Beton, 31,** 38-40

Linsbauer, H.N., Ingraffea, A.R., Rossmanith, H.P. and Wawrzynek, P.A. (1989) Simulation of cracking in large arch dam: Part I+II. **J. Struct. Eng.,** 115, 1599-1630

Tschegg, E.K. (1990) Paper to be published

Saouma, V.E., Broz, J.J., Boggs, H.L., and Brühwiler, E. (1989) A Comprehensive Investigation of Fracture Mechanics of Concrete Dams, **ICOLD**, Copenhagen, pp. 118-127

5 Acknowledgement

The author would like to express his thanks to the "Österreichische Draukraftwerke AG" both for placing the drilling cores at his disposal and for financial support and to Prof. E.K. Tschegg for experimental support.

70 STUDY ON FRACTURE BEHAVIOUR WITH WEDGE SPLITTING TEST METHOD

ZHAO GUOFAN
Dalian University of Technology, People's Republic of China
JIAO HUI
Shengyang Civil and Architectural Design Institute, People's Republic of China
XU SHILANG
Dalian University of Technology, People's Republic of China

Abstract
The wedge splitting test method proposed by Prof. Linsbauer has been adopted to investigate the fracture toughness K_{Ic} ,fracture energy G_F and crack tip opening displacement CTODc . Two sets of loading devices with sufficient stiffness are made. In order to study the size effect of specimen on K_{Ic} and G_F, four series of specimen are made. It is found that the values of K_{Ic} and G_F increase with ligament length of specimen. A formula for evaluating the K_{Ic} values of large size specimens from that of small size specimens is proposed. It is also found that the CTODc is independent of specimen size. Some proposals about design of specimen are proposed.
Keywords: Wedge Splitting Test, Fracture Toughness, Fracture Energy

1 Introduction

The wedge splitting test is a test method on fracture mechanics of concrete which is now being developed. With this test method, the effect of specimen weight can be neglected which is a big advantage over other test method especially compared with the three point bending beam test. A smooth load deflection curve can be obtained with this method. By taking cores from the existing concrete structures to prepare the specimens, the fracture toughness of such structures can also be measured, which is very useful in practical engineering. Using a simple, convenient and less expensive clamping apparatus specially designed, a small vertical load can be converted into a big horizontal splitting load, and the wedge splitting test can be carried out.Therefore the stiffness requirment for the test machine is less critical. Because of the advantages of this method, it is hopeful that this method might become a standard test method of measuring fracture energy of concrete.

Based on the research works [2], the research with the wedge splitting test method on the fracture energy and

Fracture Processes in Concrete, Rock and Ceramics Edited by J.G.M. van Mier, J.G. Rots and A. Bakker.
 Published by E. & F.N. Spon, 2–6 Boundary Row, London SE1 8HN. ISBN 0 419 15870 7

fracture toughness of concrete specimen of various sizes was carried out.

2 Test Programs

2.1 Design of the Loading Device

Using the wedge splitting method to carry out stable test on fracture mechanics of concrete, it is necessary to develope a specially loading device which can convert the vertical load from the test machine into horizontal splitting force. In the loading device designed by the authors, two loading plates mounted with weels were manufactured by welding 20mm thick steel plate. During the test the loading plates were symmetrically fixed on the specimen. For the convenience of mounting the specially designed extensometer to measure the crack mouth opening displacement, the slots were made in the middle of the loading plates. In order to coordinate various specimen sizes, two sets of loading device with sufficient stiffness were made.

During the test, the horizontal splitting force applied on the specimen can be determined with the following formula:

$$P_v=2P_h tg\alpha \qquad (1)$$

where: P_v-- vertical load applied by the test machine; P_h --the horizontal splitting force subjected on the specimen; α -- angle of the iron wedge (see Fig.1).

If the angle of the wedge a is small, a very big horizontal splitting force can be obtained with a small vertical load, and the displacement of the test machine is large compared with the crack mouth opening displacement of the specimen.

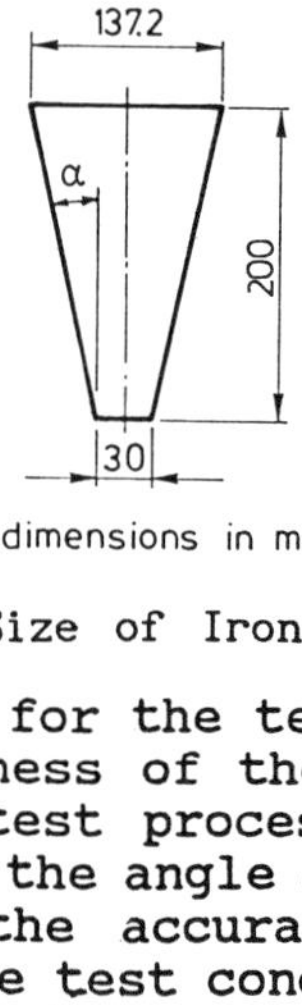

Fig. 1 Size of Iron Wedge

In this way the stiffness requirment for the test machine will be less critical. If the stiffness of the clamping apparatus is sufficient, the whole test process will be stable and easy to control. Therefore the angle a should be as small as possible. Considering the accuracy of the manufacture of the loading device, the test conditions and the basic requirments of the experiment, α=15 is choiced in this investigation.

2.2 Preparation of the Specimen

The specimen manufacture directly relates to the success or failure of the experiment, which should be considered carefully. In this investigation, cubic specimen as shown in Fig.2 was adopted. According to different test conditions, other specimen shape can also be used, for example, if the fracture parameters of rock or existing concrete structures should be determined by taking cores from the rock or the concrete structures, it is practical to use the cylinder specimen instead of cubic specimen.

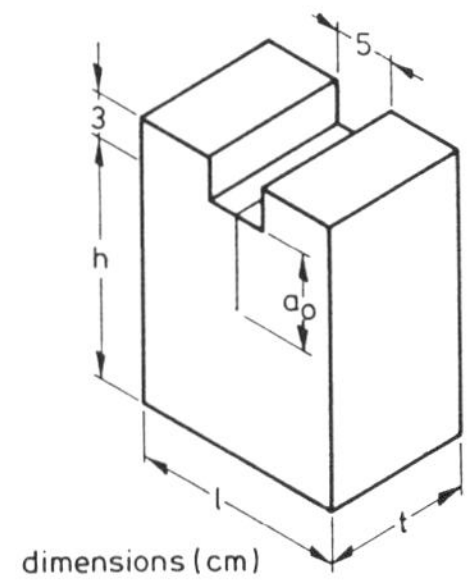

Fig.2 Specimen Shape

The initial crack on the specimen can be made either by sawing the specimen after the concrete hardening to form a post-initial crack or by inserting a steel plate inside the specimen during the concrete casting and taking the plate out after a period of curing to form a pre-initial crack in the specimen.

In order to study the effect of the specimen volume and thickness on the fracture energy G_F, two series of specimen of different size were designed. The mixes for all the specimen were equal with cement:sand:aggregate:water= 1:1.87:3.36:0.57. The grade of the cement is 425#. The maximum diameter of the aggregate is 20mm. The cubic compressive strength (150x150x150mm), tensile strength and the modulus of elasticity at test were 30.47Mpa, 1.7Mpa and 30.6 Gpa respectively.The slot on the specimen was designed to exert load to the specimen. The sizes of the slot were related to the sizes of the loading plate. The thickness of the loading plate used was 20mm , the width and depth of the slot were designed to be 50mm and 30mm respectively. After 24 hours curing of the specimen concrete, the steel plates in the concrete used to form the initial crack were taken out.

2.3 Instruments Used in the Experiment

All the tests were performed with the 5000KN compressive machine. The measuring range of the load cell was 0 -- 30 KN. The crack mouth opening displacement (CMOD) and crack tip opening displacement (CTODc) were measured with extensometer and recorded with X-Y plot. The extensometer used to measure the CMOD was fixed within the same plane as that of the splitting force applied on the specimen.

3 Fracture Energy G_F and Fracture Toughness K_{IC}

3.1 The Effect of the Supporting Conditions

To study the effect of supporting conditions on the test results, two types of support-- single line support and double line support as shown in Fig.3 were adopted. The supports were fabricated by welding a ϕ 20 steel rod on a 20 mm thick steel plate. For single line support, the supporting point should be within the plane of the initial crack (i.e. in the center of the specimen). For double line support, the supporting points should be carefully choiced to offset the effect of the specimen weight and the vertical load from the clamping apparatus exerted on the specimen. Various supporting points for the specimens of different size are shown in Table 1.

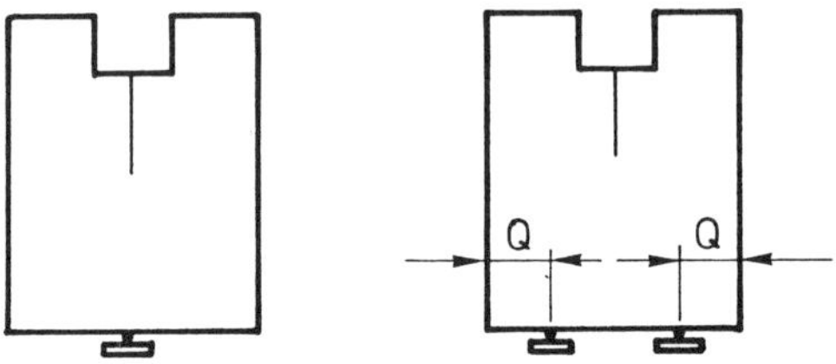

Fig.3 Supporting Conditions

Table 1. Details of the Supporting Conditions

Specimen	Size (lxtxh)mm	Q (mm)
V_1 — WS	150x150x150	37.5
V_2 — WS	200x200x200	50
V_3 — WS	300x300x300	90
V_4 — WS	400x400x400	150

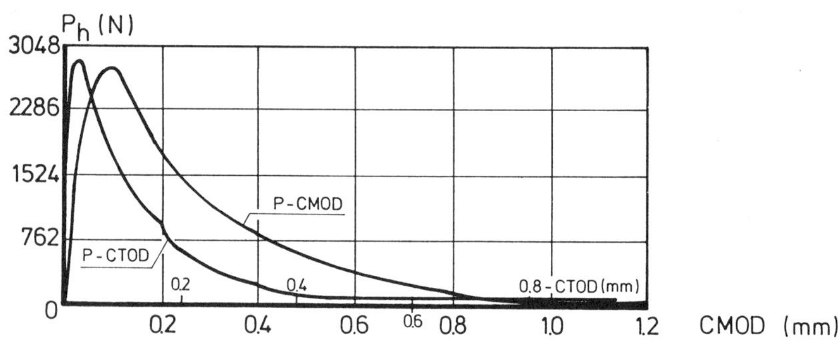

Fig.4 Measured Curve of P-CTODc and P-CMOD

During the test the curves of load-CMOD and load-CTODc were measured as shown in Fig.4. It can be observed that the test process carried out with the specially designed loading device was stable.

Based on the measured load-CMOD curve, the fracture energy G_F can be evaluated with the following formula:

$$G_F = W_0/A + mgV_0/(A \circ tg\alpha) \quad (2)$$

where: W_0 -- the area under the curve of load-CMOD; A -- the area of ligament; m -- the weight of the loading device, the weights of those two sets of loading device used were 23kg and 32kg respectively; V_0 -- the maximum crack mouth opening displacement.

Because the weight of the specimen has no contribution to the work of the load with this test method, the effect of the specimen weight is neglected in the calculation of the fracture energy G_F.

The loading condition of the wedge splitting test is similar to the small scale compact tension test. The vertical load does not affect the crack propagation. Therefore the fracture toughness K_{IC} can be calculated using the same formulas as that for the compact tension test. For the specimen of standard size (l/h=1.2), K_{IC} can be obtained with the following formula [1]:

$$K_{IC}=P_{max}/(t\circ h^{1/2})\circ F \quad (3)$$

where: $F=29.6(a/h)^{1/2}-185.5(a/h)^{3/2}+665.7(a/h)^{5/2}-1017(a/h)^{7/2}+638.9(a/h)^{9/2}$

P_{max} -- the maximum splitting force; the other notations are shown in Fig.5.

For the specimen of nonstandard size (l/h=1.0), the following formula to approximately evaluate K_{IC} was proposed [3]:

$$K_{IC}\circ t\circ h^{1/2}/P_{max}=3.675(1-a/h)^{-3/2} \quad (0.2\le a/h\le 0.8) \quad (4)$$

With formulas (2),(3) and (4), the fracture energy and fracture toughness for the specimen of different supporting conditions are analyzed. The calculated values of G_F and K_{IC} are shown in Table 2. It is observed that the difference between the values G_F and K_{IC} for those two supporting conditions is little. Therefore the effect of supporting conditions of single line or double line supporting on the fracture toughness and fracture energy can be neglected if the wedge splitting test method is adopted.

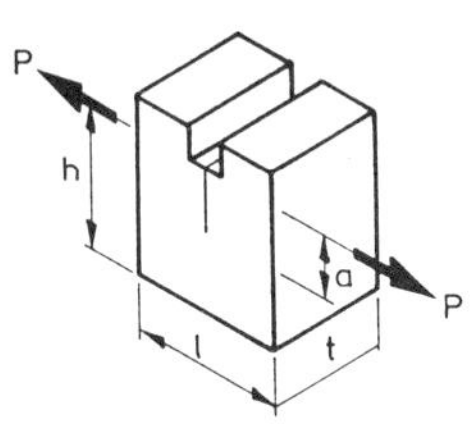

Fig.5 Size of Specimen

For double line support, the wedge splitting test method is similar to the compact tension test, the effect of specimen weight is eliminated and the specimen will not incline during the test, therefore the double line support is recommended for the test of cubic specimen. But for the test of specimen manufactured by taking cores from rock or the existing concrete structures, it is convenient to use the cylinder specimen and only the single line support can be used.

Table 2 Comparison of K_{IC} and G_F of Different Supports

Specimen	Sizes(mm)	a_0/h	Supports	G_F(N/M)	K_{IC}(MPa$\sqrt{m}$)
V2-WS(2,4)	200x200x200	0.5	S*	203.6	0.6898
V2-WS(3,5)	200x200x200	0.5	D*	225.2	0.7405
V3-WS(2,3)	300x300x300	0.5	S	263.8	1.0138
V3-WS(5,6)	300x300x300	0.5	D	261.1	1.0459

* S -- Single line support D -- Double line support

3.2 The Effect of Specimen Thickness

For the specimens of different thickness, the measured fracture energy and fracture toughness are shown in Table 3. Comparing the values G_F and K_{IC} in Table 3, we can see that the fracture energy and fracture toughness are independent of the specimen thickness, which is in accordance with other test results obtained with three point bending beam test and compact tension test.

Table 3 Effect of Specimen Thickness

Specimen	Size (mm) (lxtxh)	a_0/h	G_F(N/M)	K_{IC}(MPa$\sqrt{m}$)
T1-WS	200x150x200	0.5	226.7	0.7441
T2-WS	200x200x200	0.5	214.4	0.7152
T3-WS	200x300x200	0.5	228.9	0.8354
T4-WS	200x450x200	0.5	211.8	0.6295

3.3 The Effect of Specimen Volume

To study the effect of the specimen volume, four series ofspecimen with different size were designed and tested. Details about the specimen size and the calculated G and K with formulas (2) and (5) are listed in Table 4. The relationships between G_F, K_{IC} and the side length of the specimen are shown in Fig.6. Analysis of table 4 and Fig.6 shows that significant size effect on G_F and K_{IC} measured with wedge splitting test is also existed, which is in agreement with previous research results.

As mentioned in Ref.[3], the linear elastic fracture mechanics is applicable in the crack stability analysis for massive concrete structures and the fracture toughness K_{IC}

Fig. 6 Relationships of G_F, K_{IC} and Side Length of Specimen

can be used as the fracture criterion. The fracture toughness of specimens of different size can be calculated

Table 4 Effect of Specimen Volume

Specimen	Sizes(mm)	a_0/h	G_F(N/M)	K_{IC}(MPa$\sqrt{m}$)
V1-WS	150x150x150	0.5	196.7	0.6449
V2-WS	200x200x200	0.5	214.4	0.7152
V3-WS	300x300x300	0.5	226.4	1.0296
V4-WS	450x450x450	0.5	294.5	1.2583

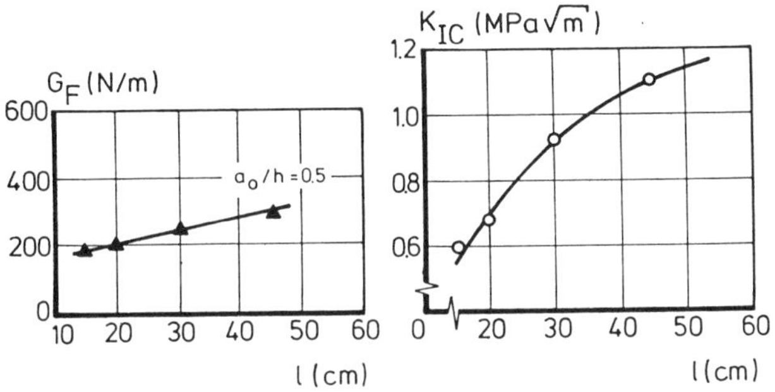

Fig.6 Relationship of G_F, K_{IC} and Side Length of Specimen

with the following empirical formula:

$$K_{IC}=K_{ICS}(h/h_s)^{0.45} \qquad (h<2m) \qquad (5)$$

where: h_s, K_{ICS} -- the height and fracture toughness of standard specimen (150x150x150mm is used) respectively; h, K_{IC} -- the height and fracture toughness of nonstandard specimen respectively.

If the fracture toughness of standard specimen is measured,the fracture toughness K_{IC} of large specimen can be evaluated with formula (5). For h>2m, h=2m should be used in the calculation. As the specimen thickness has little effect on K_{IC}, the specimen thickness is not considered in the formula. The loading conditions for wedge splitting test and compact tension test are same. Therefore formula (5) can also be used to evaluate the fracture toughness of compact tension specimen. With this method, the fracture toughness of the specimens reported in this paper and those reported in Ref.[3] which had the same mix design was analyzed. The comparison of the calculated and tested fracture toughness is shown in Table 5, which shows that the results of calculated and test measured fracture toughness are in good agreement.

For the stability analysis and hazard predictation of cracks of large, mass concrete structures, the linear elastic fracture mechanics is applicable and the fracture

Table 5 Comparison of the Calculated and Measured K_{IC}

Spec.	Sizes(mm)	K^t_{IC} (MPa$\sqrt{m}$)	K^c_{IC} (MPa$\sqrt{m}$)	K^t_{IC}/K^c_{Ic}
V1-WS	150x150x150	0.645	0.645	
V2-WS	200x200x200	0.715	0.734	0.974
V3-WS	300x300x300	1.030	0.881	1.170
V4-WS	450x450x450	1.258	1.057	1.190
CT-1	500x500x200	1.206	1.108	1.088
CT-2	1000x1000x200	1.383	1.514	0.913
CT-3	1500x1500x200	1.760	1.818	0.968
CT-4	2000x2000x200	2.018	2.068	0.976
CT-5	2500x2500x200	2.086	2.068	1.009
CT-6	3000x3000x200	2.030	2.068	0.982
			$\bar{X}$ =1.03	Cv=0.089

* K^t_{Ic}, K^c_{IC} — test and calculated values respectively; $x=K^t_{Ic}/K^c_{IC}$

toughness can be adopted as the fracture criterion. Therefore the size effect of fracture toughness should be considered. But for the analysis of small scale concrete structures, the linear method can not be used and the fictitious crack model and the fracture G_F will be used in the analysis. Because the structure is not large, the size effect of the fracture energy G_F can be neglected, which will not cause big error and is also conservative in practical analysis.

4. The Critical Crack Tip Opening Displacment of Concrete

The typical load-CTODc curves measured are shown in Fig.4 , At the beginning of the experiment ,a linear relationship between load and CTODc is obtained up to 0.55 of the maximum splitting force P_{max} . Then the nonlinearity in the relationship up to P_{max} is observed. When the splitting force reaches P_{max} , the crack will propagate at the tip of the initial carck and the corresponding crack tip opening displacement is the critical crack tip opening displacement (CTODc). In the post peak region,the carck propagates steadly and the load decreases slowly until finally the unstable damage happend.

The measured CTODc values are shown in Table 6, it can be observed that the value of CTODc is independent of the specimen size. Therefore CTODc can be regarded as a material constant and might be used as a crack criterion in the analysis of crack propagation of concrete.

Table 6 Measured CMODc and CTODc

Spec.	Sizes(mm)	CMODc(mm)	CTODc(mm)	n
V_1-WS	150x150x150	0.0909	0.02111	2
V_2-WS	200x200x200	0.1219	0.02115	3
V_3-WS	300x300x300	0.1624	0.02831	5
V_4-WS	450x450x450	0.1857	0.02191	6
T_1-WS	200x150X200	0.1319	0.0220	4
T_2-WS	200x200x200	0.1219	0.02115	3
T_3-WS	200x300x200	0.1219	0.02214	3
T_4-WS	200x450x200	0.1296	0.04617	3

* n — Number of specimens tested

5. Some Suggestions About the Specimen Preparation

As mentioned in the previous section, a initial crack and a slot should be prepared on the specimen during the specimen manufacture. If the wood mould is used to cast the concrete specimen, the wood piece used to make the slot will expand because it will absorb water after the concrete casting. After the concrete hardening, it will be very difficult to take the wood piece out. This problem can be solved by covering the wood poece with plastic film. But in this way another problem will arise, that is the foam will exist between the plastic film and the concrete.The fixing of the extensometer during test will be affected. If the trapezoid shape of the slot is adopted instead of the rectangular shape, the difficult of taking out the wood piece can also be overcomed. In this way, a small modification in the design of loading plates as shown in Fig.7 should be carried out.

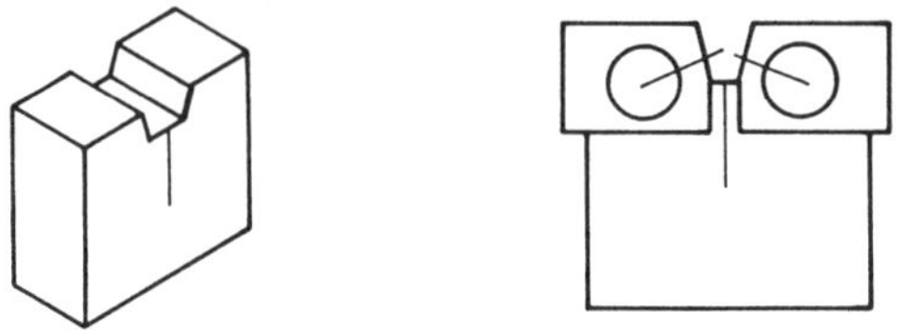

Fig.7 Specimens with Trapezoid Shape Slot

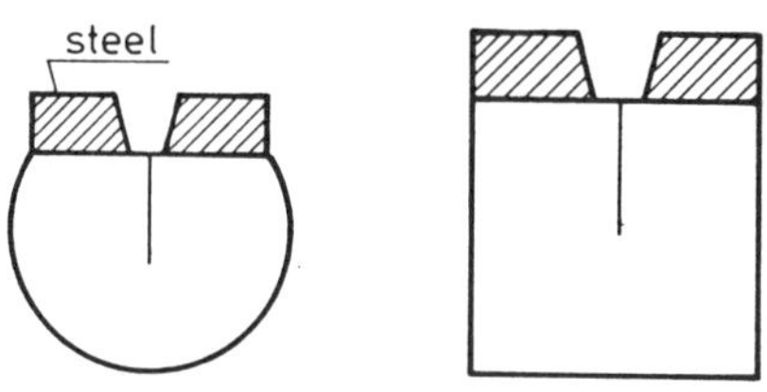

Fig.8 Preparation of the Slot by Sticking Metal Parts

For the specimens casted on site, it is easy to prepare a slot used to exert splitting load. But to saw such slots in the specimens manufactured by taking cores from rock or existing concrete structures is difficult. In this case, a easy way to prepare the slot is to grind a plane surface on the specimen first and then stick metal parts on the plane to form the slot as shown in Fig.8. This method of preparing slot can also be used for the specimens casted on site.

6. Conclusions

With the wedge splitting test method, the experiments on fracture parameters of concrete were carried out and the following conclusions can be obtained:

(1). Because the wedge splitting test method has many advantages over other test methods such as the less critical requirment for the stiffness of test machine, smooth curves measured during test, independence of the measured G_F on specimen weight and so on, the wedge splitting test method is a hopeful method which should be highly recommended.

(2) Test results show that significant size effect of fracture energy G_F and fracture toughness K_{IC} exists.

(3) If the specimen height is not high, the effect of the weight of loading device should be considered. Therefore the height of the specimen should be high enough during the test of fracture energy.

(4) The critical crack tip opening displacement (CTODc) can be regarded as a material constant independent of the specimen size. This material constant CTODc might be used as a fracture parameter in the analysis of concrete cracks. For this purpose, the relationship between CTODc and CMOD should be further studied in the future research activities.

7. REFERENCE

Beijing Research Institute of Steel and Iron (1977) Engineering fracture mechanics, Publishing House of National Defence Industry (in Chineses).

H.N. Linsbauer,E.K. Tschegg (1968) Fracture energy determination of concrete with cube-shaped specimens, Zement u. Beton 31,38-40.

Xu Shilang, Zhao Guofan (1988) Fracture toughness criterion of crack assesment of large scale specimen of concrete and concrete dams, Dalian University of Technology.

71 SEMI-DESTRUCTIVE TESTS AND FRACTURE MECHANICS OF CONCRETE

K. ROKUGO, Y. UCHIDA and W. KOYANAGI
Department of Civil Engineering, Gifu University, Japan

Abstract
The fracture mechanics parameters, G_F and f_t, were determined from semi-destructive tests, the bend-off test (similar to the break-off test) and the pull-off test. The fracture energy which was measured from the bend-off test was in good accordance with G_F obtained through the RILEM method. For the purpose of explaining the difference between the flexural strength from the bend-off test and that from the standard third-point bending test, the effects of the size and the cross-sectional shape (circular or square) on the flexural strength were investigated. The flexural strength decreased with the specimen size, in proportion to $H^{-1/7}$, where H is the specimen depth. The effects of the cross-sectional shape on the flexural strength were small. The mechanism of the effects of the size and shape on the flexural strength were investigated through the numerical analysis considering the tension softening behavior of concrete.
Keywords: Concrete, Fracture Mechanics, Semi-Destructive Tests, Fracture Energy, Flexural Strength, Bend-Off Tests, Pull-Off Tests.

1 Introduction

The numerical simulations of fracture behavior due to cracking and the evaluations of size-depending strengths have been attractive subjects in the field of the recent Fracture Mechanics of Concrete. Cohesive Crack Model with softening, Hillerborg (1978), Size-effect Law, Bazan and Oh (1983), and Two-Parameter Model, Jenq and Shah (1985), have been proposed as tools to deal with the size-effects on strengths of concrete. Hillerborg's Cohesive Crack Model can give results through numerical analysis and is effective to simulate failure behavior of concrete, where the fracture energy G_F, the tensile strength f_t, and the Young's modulus E are needed.

Semi-destructive tests, such as the break-off test, the pull-off test, and the pull-out test, have been used to estimate in-situ concrete strength. These methods are effective for large concrete structures such as pavements and dams. The break-off test, Johansen (1979), is performed by applying bending moment on a concrete core in a cantilever condition. In the pull-off test, a concrete core is pulled off in the direction normal to the concrete surface.

Fracture Processes in Concrete, Rock and Ceramics Edited by J.G.M. van Mier, J.G. Rots and A. Bakker.

Published by E. & F.N. Spon, 2–6 Boundary Row, London SE1 8HN. ISBN 0 419 15870 7

The purposes of this contribution are to determine fracture mechanics parameters G_F and f_t from the semi-destructive tests and to apply the fracture mechanics to discuss the test results. Loading test similar to the break-off test was performed to determine G_F. We call this testing method "bend-off test", because the size and procedures of the break-off test have been well specified. The tensile strength f_t was determined through one kind of the pull-off tests. Third-point bending tests were done to discuss the effects of the specimen size and the cross-sectional shape (circular or square) on the flexural strength (modulus of rupture). The mechanism of these effects was explained by means of the numerical simulation with the Hillerborg's cohesive crack model.

2 Semi-destructive tests to determine fracture mechanics parameters

2.1 Outlines of semi-destructive tests

Two kinds of semi-destructive tests, namely the pull-off test and the bend-off test as shown in Fig. 1, were performed to measure the tensile strength f_t and the fracture energy G_F of concrete.

In the pull-off test, core slits (diameter: 100 mm, depth: 100 mm) were drilled around each of embedded anchor bolts (depth: 50 mm) on a concrete block (200x200x1500 mm). The bolts were pulled off through a hydraulic jack. The loading tests were done at the ages of 7 and 27 days.

In the bend-off test, a larger core slit (diameter: 100 mm, depth: 200 mm) and a smaller core slit (diameter: 25 mm, depth: 160 mm) with the same center were drilled. A screw rod (diameter: 24 mm, length: 250 mm) was fixed into the smaller hole with epoxy resin. The concrete core with the rod was bent through a hydraulic

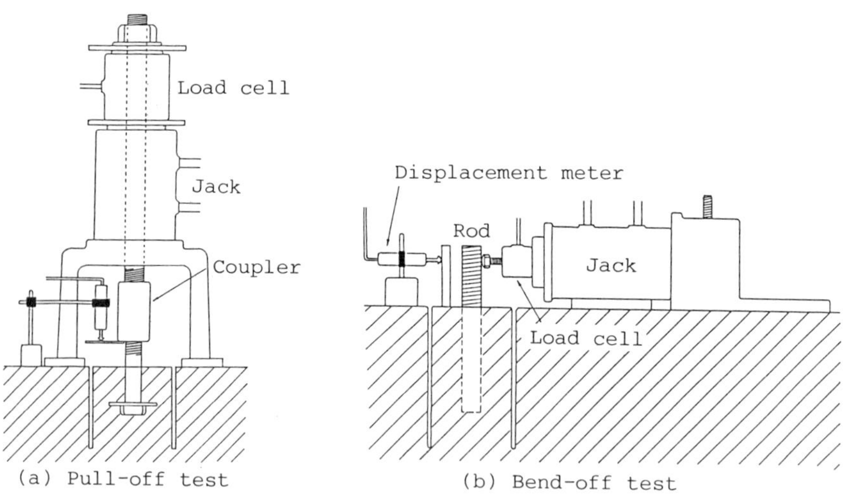

Fig. 1. Semi-destructive tests.

Table 1. Properties of concrete for semi-destructive tests

Testing age (days)	Compressive strength (MPa)	Tensile strength (MPa)	Flexural strength (MPa)	Fracture energy (N/m)
7	24.8 [0.73]	2.22 [0.23]	3.68 [0.37]	137 [19.]
28	39.4 [1.69]	3.16 [0.07]	4.49 [0.35]	157 [21.]

[ ]:Standard deviation

Details of concrete mix:

Cement:Water:Sand:Gravel=1.00:0.53:2.79:2.81 (weight ratios)

Maximum aggregate size: 15 mm, Cement: Normal Portland cement

Admixture: AE water reducing agent

jack. The displacement at the same level with the loading position was measured. Complete load-displacement curves were recorded. The loading tests were done at the age of 28 days.

Plain concrete was used. The compressive strength (from ϕ10x20 cm specimens), the splitting tensile strength (ϕ15x15 cm), flexural strength (10x10x40 cm, third-point bending test) and the fracture energy (RILEM method) of the concrete at the ages of 7 and 28 days are tabulated in Table 1. The standard deviations are indicated in the bracket. The details of the concrete mix are given below Table 1. The concrete were cured in wet condition until testing.

2.2 Results of semi-destructive tests and discussions

The tensile strength (maximum load/cross-sectional area) by means of the pull-off test is given in Table 2. The flexural strength (maximum moment/section modulus) and the fracture energy (area under the complete curve/cross-sectional area) by means of the bend-off test are also shown in Table 2.

In the pull-off test, all core specimens fractured at the tip of the core slit. The tensile strengths at the ages of 7 and 28 days through the pull-off test were in good accordance with those from standard splitting test.

In the bend-off test, the complete load-displacement curves were able to be measured through manual controlling as shown in Fig. 2. Loading and unloading operations were repeated, in case of necessity, after the peak load. There was no residual deformation in the screw rods. This means that the dissipated

Table 2. Results of semi-destructive tests

	Pull-off test	Bend-off test			
Testing age (days)	Tensile strength (MPa)	Max. load (kN)	Arm length (mm)	Flexural strength (MPa)	Fracture energy (N/m)
28	3.04 [5, 0.16]	1.99	258	5.23 [7, 0.34]	146 [5, 25.]

[ ]:Number of specimens, Standard deviation

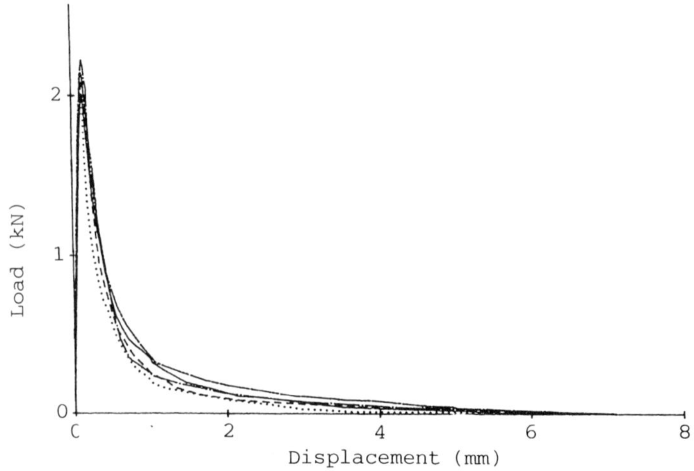

Fig. 2. Load-displacement curves at bend-off test.

energy was consumed only at the concrete fracture zone. The fracture energy from bend-off tests was almost the same with that from RILEM three-point bending tests on notched beams.

The flexural strength from bend-off tests of core specimens (circular cross-section, cantilever) was about 16 % higher than that from the standard third-point bending test on square beams. It has been reported that the break-off strength is higher than the beam strength by 20 % or more, Yener and Chen (1985). There would be several possible causes to explain this difference.

(a) Weakest link theory. In the bend-off test, the fracture zone is forced at the tip of the slit.
(b) Effects of shear force. In the bend-off test, there are shear stress distribution even in the fracture zone.
(c) Effects of section-shape. The cross-section of the cored specimens is circle.

The FE analysis with the cohesive crack model is effective to discuss the failure behavior of concrete due to cracking. The effects of the shape and size of specimens on the flexural strength are investigated in the next chapter.

3 Third-point bending tests on beams differing in size and shape

3.1 Outlines of third-point bending tests and analysis

Third-point bending tests on beam specimens with square or circular cross-section were carried out in order to investigate the effects of the shape of the cross-section and the size of the specimen on the flexural strength (modulus of rupture).

Table 3 shows the geometry and the number of the specimens.

Table 3. Test conditions of third-point bending tests

Square beams		Circular beams	
Size (mm) Depth, Width, Span	Number of specimens	Size (mm) Diameter, Span	Number of specimens
50x50x150	8	ϕ50x150	8
100x100x300	8	ϕ100x300	6
200x200x600	5	ϕ200x600	5
300x300x900	4	ϕ300x900	4
400x400x1200	4	-	-

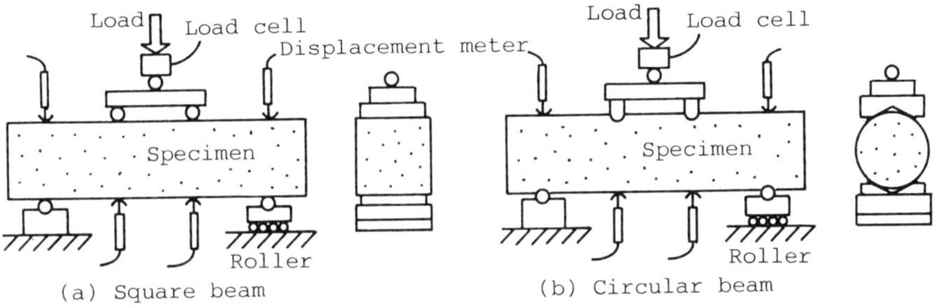

Fig. 3. Test configuration of third-point bending test.

Five sizes were chosen for square beams (depth: 50, 100, 200, 300, 400 mm). Four sizes were chosen for circular beams (diameter: 50, 100, 200, 300 mm). The span was taken to be three times of the depth or the diameter as illustrated in Fig. 3. The specimen length was about four times of the depth or the diameter. The sides of the specimens when concrete was cast in the mold were taken as the top and bottom surfaces at the loading test. Watertight steel molds and watertight paper molds were used for the square beams and circular beams, respectively. No diameter differed from the average diameter by more than 3 % for each circular beam.

Loading tests were performed in wet condition at the age of 27 through 30 days. Rollers were used at one of the supports. The load and the displacement at loading points were measured. The settlement at the supports was eliminated from the measured displacement at the loading points. The loading time from the beginning to the peak load was about 1 minute.

Normal concrete with the maximum aggregate size of 15 mm was adopted. The compressive strength f_c and the Young's modulus E_c which were measured from ϕ10x20 cm specimens were 32.9 MPa and 27 GPa, respectively. The splitting tensile strength from ϕ15x15 cm specimens was 2.79 MPa.

Fig. 4 shows the tension softening diagram and the finite element mesh, which were used in the FE analysis. The splitting tensile strength was used as f_t in the softening diagram. The value of G_F was estimated to be 0.132 based on our past experimental data.

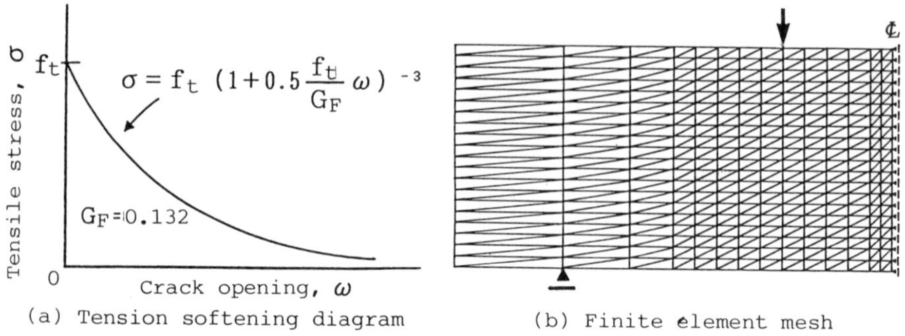

Fig. 4. Tension softening diagram and finite element mesh.

The reduced Young's modulus (2/3 of E_c) was adopted for the moment span in the analysis.

3.2 Results of third-point bending tests and discussions

Results of the third-point bending tests and the numerical analysis are given in Table 4. The values in the parentheses indicate the flexural strength modified with the specimen's own weight (free from weight). The calculated strength was obtained from the peak load on the simulated load-displacement curve.

Fig. 5 shows the measured load-displacement curves for the square beams with depth of 400 mm. Fig. 6 indicates the relation between the scatter band of the curves in Fig. 5 and the simulated load-displacement curve. Snap-back phenomena were observed in Fig. 5.

As seen from Table 4, in the case of square beam specimens, the calculated strengths are in good accordance with the measured flexural strengths. In the case of circular beams, the differences

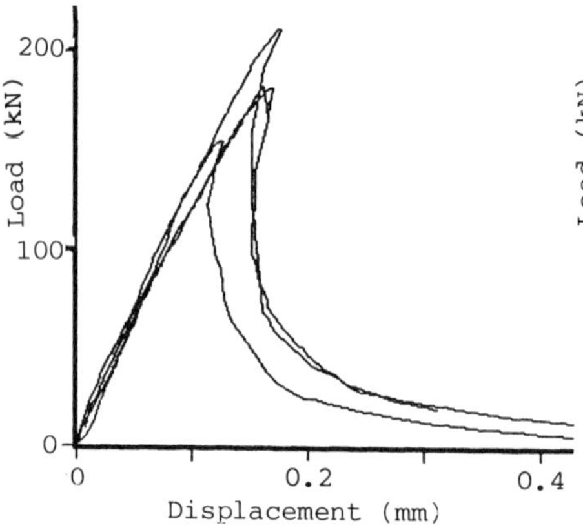

Fig. 5. Load-displacement curves from experiment (400-mm).

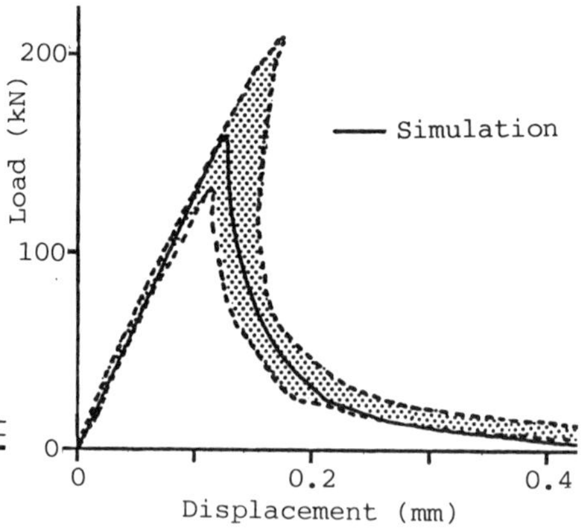

Fig. 6. Load-displacement curve from simulation (400-mm).

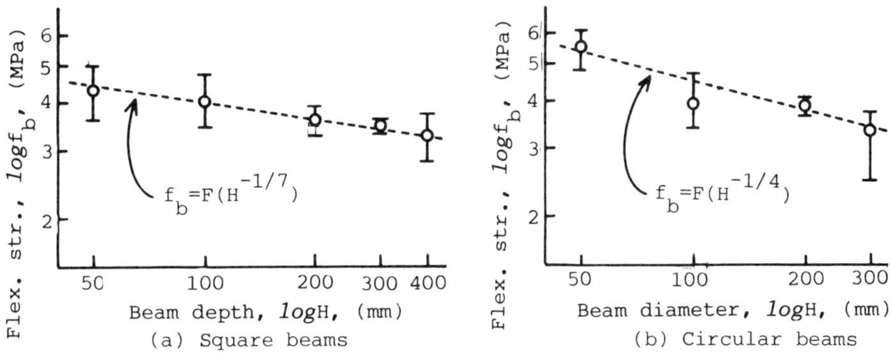

Fig. 7. Flexural strength and beam depth.

between the calculated and the measured strengths are large as compared with the square beams. As seen from Fig. 7, the flexural strength of square beams decreased with the specimen height H, that is, in proportion to $H^{-1/7}$. The flexural strength of circular beams is approximately proportional to $H^{-1/4}$. Shioya et al. (1989) have reported that flexural strength of concrete decreases in inverse proportion to the cubic to fourth roots of H ($H^{-1/3} \sim H^{-1/4}$). Hillerborg (1989) has reported numerical results about the size effects on the flexural strength.

The numerical results in Table 4 indicate that the flexural strength of the circular beams is higher than that of the square beams by more than 20 %. In the experimental results, however, the effects of the section shape on the flexural strength were small.

Fig. 8 shows the stress distributions at the calculated maximum load. Table 5 shows the resultant internal compressive force C_m, the internal lever arm z_m and the moment M_m for specimens with

Table 4. Results of third-point bending tests and numerical analysis

Specimen size (mm)		Maximum load in exp. (kN)	Flexural strength (MPa)	
			Experiment	Calculation
Square	50x50x150	3.73 [0.30]	4.34 (4.35)	4.49 (4.50)
	100x100x300	13.13 [1.19]	4.03 (4.04)	3.94 (3.95)
	200x200x600	49.9 [1.97]	3.63 (3.66)	3.37 (3.40)
	300x300x900	105.9 [3.03]	3.42 (3.46)	3.07 (3.13)
	400x400x1200	174.4 [19.5]	3.22 (3.29)	2.95 (3.01)
Circular	ϕ50x150	2.74 [0.19]	5.57 (5.58)	5.57 (5.58)
	ϕ100x300	7.74 [0.97]	3.92 (3.94)	4.85 (4.86)
	ϕ200x600	29.6 [0.96]	3.85 (3.89)	4.18 (4.23)
	ϕ300x900	56.4 [8.77]	3.30 (3.37)	3.84 (3.91)

( ):Flexural strength free from specimen's weight
[ ]:Standard deviation

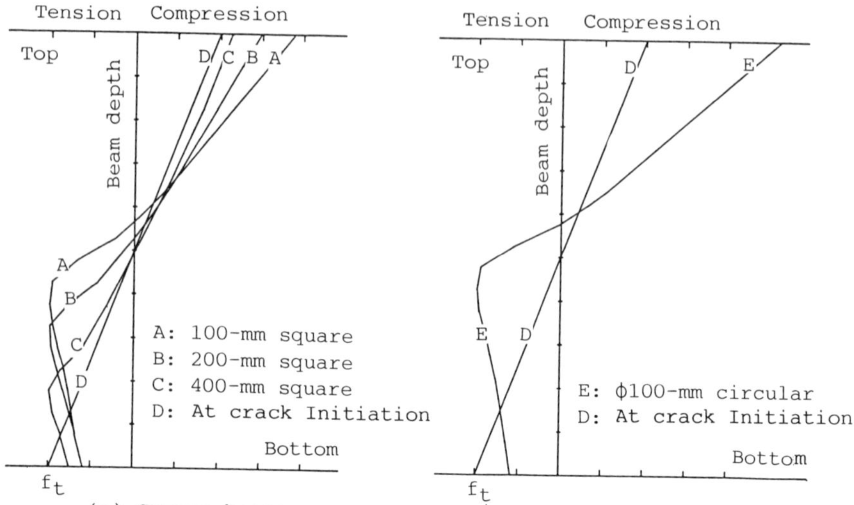

(a) Square beams (b) Circular beam

Fig. 8. Stress distributions at maximum load.

Table 5. Resultant compression C, lever arm z and moment M at maximum load (subscript m) and at initiation of crack (subscript c)

Specimen size (mm)	C_m/C_c	z_m/z_c	M_m/M_c	f_b/f_t
100x100x300 (100-mm square)	1.57	0.87	1.37	1.37
200x200x600 (200-mm square)	1.31	0.91	1.20	1.20
400x400x1200 (400-mm square)	1.11	0.97	1.08	1.08
φ100x300 (φ100-mm circular)	2.00	0.85	1.69	1.69

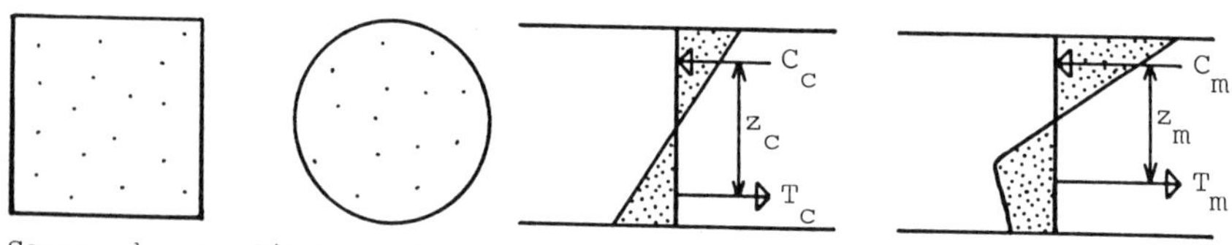

Square beam Circular beam At crack initiation At maximum load

different geometries at the maximum load. C_m, z_m and M_m are normalized with the corresponding values when the stress in the extreme tension fiber reached the tensile strength f_t.

When the specimen size increases or the shape becomes circular, the resultant internal force C_m increases and the internal lever arm z_m decreases because of the non-linear stress distribution due to the tension softening of concrete. As the result, the maximum moment M_m and therefore the flexural strength f_b increase. This is the mechanism of the size effects and the shape effects on the flexural strength.

4 Conclusions

(1) The complete load-displacement curves during the bend-off test (similar to the break-off test) were recorded. The fracture energy G_F which was calculated from the area under the curves was in good accordance with G_F obtained through the RILEM bending test on notched beams.

(2) The tensile strength f_t obtained from the pull-off test of drilled core specimens agreed with the value obtained from the standard splitting test.

(3) The flexural strength decreased with the specimen size, in proportion to $H^{-1/7}$, within the limits of this investigation (between 50 mm and 400 mm in the specimen depth H).

(4) The effects of the cross-sectional shape (square or circular) on the flexural strength were small in the experiments but large in the numerical analysis.

(5) The mechanism of the effects of the size and shape on the flexural strength was described through the numerical analysis considering the tension softening behavior of concrete. In the case of circular beams and smaller beams, the resultant internal force increases due to the tension softening of concrete, then the flexural strength increases.

References

Bazant, Z.P. and Oh, B.H. (1983) Crack band theory for fracture of concrete. Materials and Structures, **16**, 93, pp.157-177.

Hillerborg, A. (1978) A model for fracture analysis. Report TVBM-3005, Division of Building Materials, Lund Institute of Technology, Sweden.

Hillerborg, A. (1989) Fracture mechanics and the concrete codes, in **Fracture Mechanics: Application to Concrete** (eds V.C. Li and Z.P. Bazant), ACI SP-118, pp.157-169.

Jenq, Y.S. and Shah, S.P. (1985) Two parameter fracture model for concrete. J. Eng. Mech., ASCE, **111**, pp.1227-1241.

Johansen, R. (1979) In situ strength evaluation of concrete - The break-off method. Concrete International, ACI, **1**, 9, pp.45-51.

Shioya, T., Iguro, M., Nojiri, Y., Akiyama, H. and Okada, T. (1989) Shear strength of large reinforced concrete beams, in **Fracture Mechanics: Application to Concrete** (eds V.C. Li and Z.P. Bazant), ACI SP-118, pp.259-279.

Yener, M. and Chen, W.F. (1985) Evaluation of in-place flexural strength of concrete, ACI J., **82**, 6, pp.788-796.

72 ADHESIVE POWER OF BONDED CONCRETE

E.K. TSCHEGG
Technical University, Vienna, Austria
S.E. STANZL
University of Agriculture, Vienna, Austria

Abstract
Adhesive power and fracture behavior of old-new concrete bonds with different adhesives and with different pretreatments of the old concrete surface have been tested with a recently developed new method. Load-deformation curves were measured for stable crack propagation. From this, the specific fracture energy and the maximum load were determined. These results allow to judge the quality of different old-new concrete bonds. Consequences for testing the adhesive power of cement bonded materials are discussed.
Keywords: Concrete Bond, Interface Crack, Adhesive Power, Load-deformation Curve, Specific Fracture Surface Energy, Bending Tensile Strength, Adhesives, Surface Treatments.

1 Introduction

Pavement and preparation of bonds of cement-like materials are basic problems of the building industry. Various parameters influence endurance and mechanical load capacity of such bonds. These properties are mainly determined by physical and chemical characteristics as well as by shape and surface preparation of the base materials. If for example, emulsions or epoxi glues are used as adhesives, these materials themselves influence the adhesive power of the bond.

In order to characterize and judge bonds of cement-like materials, different testing methods have been used in the past, like for example tear-off test, tensile test, shear test and others. Comprehensive overviews and classifications of these testing methods are given by Hilsdorf (1981) and Schulz (1984). From such proof tests, a maximum stress (load per area) is usually obtained, which is the adhesion tensile stress for tensile loading and the shear stress for shear loading. However, it is almost impossible to decide if bonds have been separated in a brittle or ductile manner. The way of interface separation on the other hand is of considerable relevance for structural engineering purposes. All above mentioned testing procedures could not

Fracture Processes in Concrete, Rock and Ceramics Edited by J.G.M. van Mier, J.G. Rots and A. Bakker.
 Published by E. & F.N. Spon, 2–6 Boundary Row, London SE1 8HN. ISBN 0 419 15870 7

contribute to this problem and have been criticized therefore (e.g. Schulz 1984). They can only be used for comparison purposes of the adhesive power of different cement bonded materials, whereas determination of material-characteristic values is most problematic.

Fracture mechanical methods improved the possibilities of measuring the adhesive power of concrete bonds. Hillemeier and Hilsdorf (1976) have done pioneer work in this research field. They used CT-specimens and the LEFM in order to characterize the fracture properties of interfaces of cement bonded materials. In the following years LEFM methods however, have not been further developed in characterizing problems of concrete construction. Today, the "fracture-energy concept" (Proc. 1986) is discussed seriously. In this concept, the strain softening curve represents a kind of material law. It is an important base for a numerical simulation of fracture phenomenons in concrete.

In the following, it is tried to use the fracture energy concept to characterize the adhesive power of old-new concrete bonds. Different pretreatments of the old concrete surface and different adhesives are discussed.

2 Principle of the testing method

For this work, a splitting test method has been used, as patented by Tschegg and Linsbauer (1986) for concrete fracture tests and as adopted and used frequently by Wittmann and co-workers (Brüwiler 1990) later. First tests to characterize old-new concrete bonds have been performed and published by Tschegg and Stanzl (1989). Therefore, this method is described here only briefly.

The principle is shown in fig. 1. A cubic specimen with a rectangular groove and a starter notch is splitted with a wedge splitting device (wedge angle about 5-8°) along the

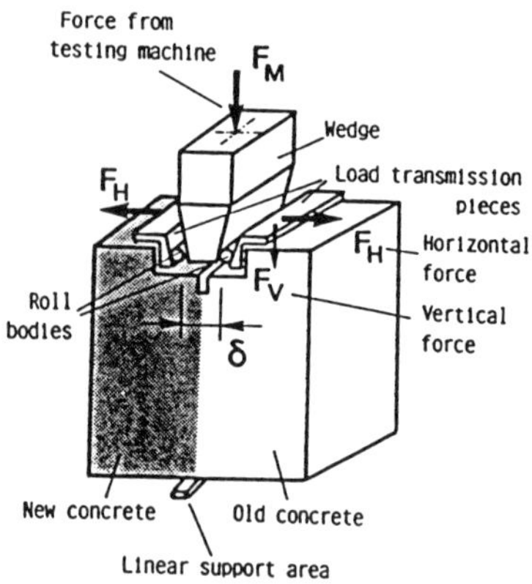

Fig.1. Principle of the testing method

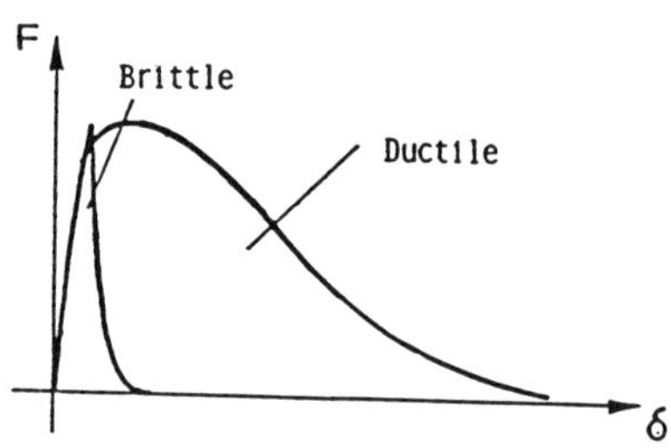

Fig.2. Schematic load-deformation curve

interface of the old-new concrete at a constant crack growth rate. For this, good alignment of interface, starter notch and linear support area is necessary.

The wedge is pressed into the load transmission pieces with a compression test machine, with a force F_M (which is measured with a load cell) thus producing high horizontal forces F_H, which split the cube. The horizontal forces may be determined easily by measuring the forces F_M and the wedge angle. The vertical forces F_V are small due to the small wedge angle and therefore do not influence the results; measurements on concrete have verified this result (Linsbauer and Tschegg 1989).

Inductive displacement gages are mounted on both ends of the starter notch in order to measure the displacement δ of the lines of force application (crack opening displacement). Force F_H and displacement δ are registered with an X-Y1,Y2 recorder or an electronic data logger; from this the load-deformation curve is obtained. In fig.2, such a curve is shown schematically for "brittle" and "ductile" separation of the bonded material. Notch-tensile strength or notch-bending tensile strength may be calculated from the maximum load. The area under the F-δ curve corresponds to the energy which has been spent to separate the specimen. Dividing this energy by the fracture area, yields the specific fracture energy G_F.

In order to reduce friction between wedge and load transmission pieces, roll bodies may be introduced, as described earlier (Tschegg 1990a). With these, the influence of friction becomes small enough to be neglected.

Such a wedge loading device together with a mechanic spindle machine or a hydraulic machine (in stroke or strain control) is appropriate to produce stable crack growth even in brittle material bonds. In another work (Tschegg 1990b) a table machine is described which fulfils the above mentioned requirements.

Cubes as used in civil engineering may be chosen as specimen shape, as their production is easy, inexpensive and quick in a laboratory. If however specimens have to be taken from constructions, drill cores may be used for testing purposes (Tschegg and Stanzl 1990). Starter notch and groove are introduced by a rock-saw either parallel or perpenticular to the cylinder area. As an example, a specimen with a groove and a starter notch parallel to the cylinder axis is shown in fig. 3.

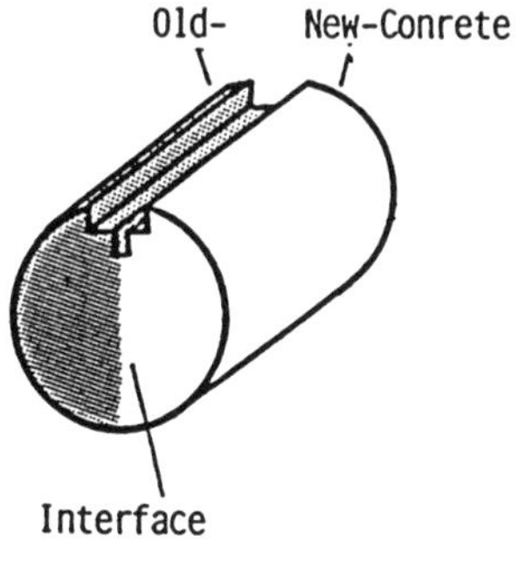

Fig.3. Drill core specimen

3 Specimen preparation

In this work cubes with an edge length of 150 mm have been used where the interface runs in the center line between old and new concrete (see fig.1). The procedure of specimen preparation as performed in this work was the following:

A) Production of old concrete: Conventional forms were subdivided into two identical halves with a sheet metal and were grouted with concrete of B400 quality (for chemical composition and mechanical behaviour see table 1) subsequently. The semi-cubes were removed from the forms after two days and stored in water for two months. Then, one area of each semi-cube was prepared in the following manner before reinforcing with new concrete:

* grinded superficially with a hand wire brush and then moistened; this treatment is called "no" in this paper. An old concrete surface is shown in fig. 4a after this treatment.
* "water blast" (see fig. 4a) with a water pressure of 800 bar.
* "sand blast" (see fig. 4a) and moistened afterwards .
* "grinded" (see fig. 4a) with a cup-grinding wheel and moistened afterwards.
* "fracture surface" means, that the fracture surface of a broken homogeneous old concrete cube served as bonding surface; it was moistened before bonding.
* "corrugated" specimens were produced with a corrugated sheet metal as one side of the form. The wavy surface remained untreated ("no") or has been "sand blast" and moistened then.

Fig.4. Surface of old concrete a) after pretreatments
b) after fracturing

Table 1. Concrete composition and compressive strength of B400 after 28 days storage in water

Max. grain size : 16 mm	Water/cement ratio: 0.53
Cement content: 380 kg/m^3	Compressive strength: 41.4 N/mm^2

B) Application of adhesives: The following adhesives have been applied to the old concrete surfaces with "no" and "sand blast" pretreatment (whereby preparation instructions of the manufacturers were regarded carefully).

* "epoxi adhesives (tough)", which remain ductile in the as-hardened condition.
* "epoxi adhesives (hard)", which are hard and form-closed after hardening.
* "emulsion" (plastic dispersion); an emulsion has been applied as "adhesive bridge" to the "sand blast" pre-treated old concrete surface and moistened (50g water:50g dispersion:50g cement) afterwards.
* "cement glue"; cement glue was applied with a water-cement ratio of w/c=0.5 to the moistened old concrete surface with pretreatment "no" and "sand blast".

C) New concrete: New Concrete B400 (chemical composition and mechanical behavior see table 1) was concreted to the old concrete part in usual cube-forms. The cubes were removed from the forms after two days and stored in a water-bath. After 26 more days the specimens were tested with the above described method. In order to determine the scatter of the results, four cubes of each specimen type were tested.

In addition, four homogeneous cubes (without an interface) of new concrete B400 were prepared. After storing them in water 28 days, they were tested in the same way as the bonded specimen.

4 Results

Maximum load F_{max} and specific fracture energy G_F are summarized in table 2 for all tests. The standard deviation (SD) of four measurements for each kind of specimen as well as relative values of G_F and F_{max} (i.e. ratio of G_F of old-new concrete bond and G_F-value of the homogeneous concrete, G_F/G_{con}; F_{max}/F_{con} is defined correspondingly) are added. Relative G_F and F_{max} values are plotted in a block diagram in figs. 5a and 5b.

In fig. 5a, the results of different old concrete surface treatments are summarized. The relative G_F-values go beyond 40% for the pretreatments "fracture surface" and "corrugated surface + sand blast". The relative F_{max} values correspondingly are in the range of 70% to 87%. For pre treatments "no" and grinded", the rel. G_F values are lowest, i.e. 10% and 7%; the rel. F_{max} are 45% to 55%. With other words, the rel. F_{max} values are reduced to their half, and the rel. G_F values to between a fourth and a sixth of their value in homogeneous concrete.

Correlating roughness of the old concrete fracture surface and G_F values, demonstrates that a macroscopically rough surface influences G_F- and F_{max}-values strongly. The influence of surface roughness in a microscopic scale, as

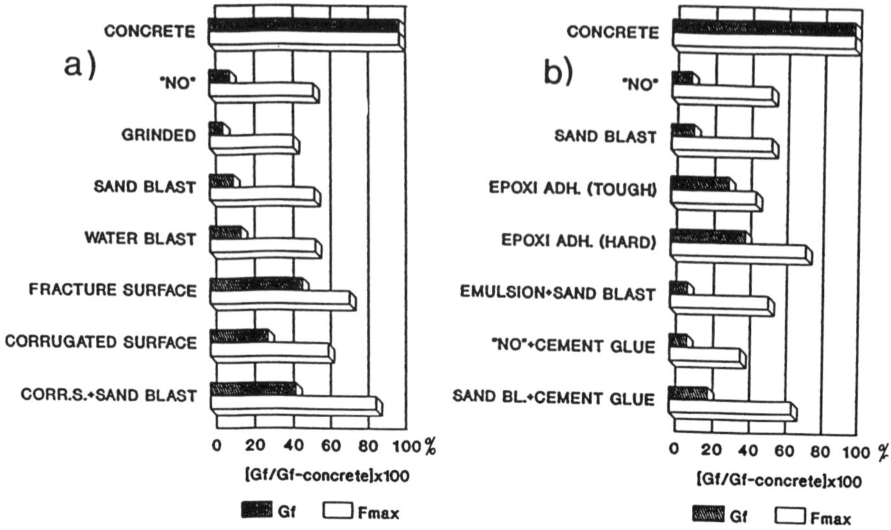

Fig.5. Comparison of G_F and F_{max}-values for
a) different surface pretreatments
b) different adhesives

formed by "grinding", "no", "water blast" and "sand blast" is smaller. The main reason for this probably is, that the crack leaves the interface of the old-new concrete interface if the surfaces are macroscopically rough, and then propagates with in the new or old concrete area partly or completely. As an example, the old concrete surfaces of "no", "sand blast", "water blast" and corrugated" treatment are shown in fig. 4b. It is obvious that old and new concrete are not fully separated along the interface, but partly within either the new or the old concrete area; more energy is necessary for fracturing with-in new or old concrete.

In fig. 5b, the rel. G_F and F_{max} values of different adhesives are summarized. For comparison purposes, the values of "no" and "sand blast" treatment are included. The rel. G_F values exceed 20% for "epoxi adh.(tough)", for "epoxi adh.(hard)" and for "sand blast+cement glue". Contrary to this, G_F of "emulsion+sand blast" and "no+cement glue" are only 10%. The rel. F_{max} values are approximately in the same range for all adhesives, i.e 40 -70%. Use of "emulsion" does not improve the adhesive power. Therefore, it seems worth while to think about the question, when and which adhesive should be used for producing an adhesive bond of old-new concrete for construction engineering purposes.

A main result of this work is the fact, that rel. F_{max} values between 40% and 80% are obtained for various old concrete pretreatments and adhesives, whereas the

Table 2. Maximum load and specific fracture energy of homogeneous and interface specimens

Treatment	F_{max} (N)	S.D. (%)	F_{max}/F_{con} (%)	G_F (N/m)	S.D. (%)	G_F/G_{con} (%)
Homog.Concr.	1121	3,6%	100%	90	2.5%	100%
"No"	605	7.4%	54.0%	9.7	12.0%	10.8%
Grinded	495	5.1%	44.2%	6.5	13.0%	7.2%
Sand blast	615	6.8%	54.9%	11.0	11.0%	12.3%
Water blast	617	7.5%	55.0%	14.4	7.5%	16.0%
Frac.surface	822	6.8%	73.4%	43.3	9.5%	48.1%
Corrug.surface	695	4.5%	62.0%	27.0	6.2%	30.0%
Corrug.surface + sand blast	981	4.0%	87.6%	40.0	3.2%	44.4%
Epoxi adhesive (tough)	517	3.5%	46.2%	28.0	6.5%	31.1%
Epoxi adhesive (hard)	825	5.6%	73.6%	36.4	4.9%	40.5%
Emulsion + sand blast	597	5.2%	53.3%	8.7	8.1%	9.6%
Cement glue / "no"	426	7.0%	38.0%	8.5	10.1%	9.4%
Cement glue / sand blast	747	8.1%	66.7%	18.7	8.1%	20.8%

Values are mean values of 4 measurements.
F_{max}= maximum load, S.D. = standard deviation, F_{max}/F_{con}= maximum load / maximum load of homog. concrete, G_F= specific fracture energy, G_F/G_{con}= specific fracture energy / specific fracture energy of homog. concrete.

rel.G_F values, which characterize the crack growth resistance, are only 7% to 50%.

The load-deformation curve reveals extensive information on the fracture behavior of concrete bonds . It is the most important base for a quantitative characterization of interface fractures in concrete bonds. In fig. 6, several typical load-deformation curves are plotted (note that all diagrams are drawn in the same scale). Comparing homogeneous concrete and bond-specimens with "no" and "water blast" pretreatment shows that the softening region of bonded specimens is substantially shorter. This shortening is more pronounced for "no" than for "water blast" treatment. For "epoxi adh.(hard)" and "(tough)" on the other hand, the softening region is more pronounced. For "epoxi adh. (hard)", a higher F_{max} value is obtained, which implies also a higher G_F value. Extensive studies on the reasons for the softening behaviouur of bonds are on the way. First numerical simulations of interface problems have revealed promising results.

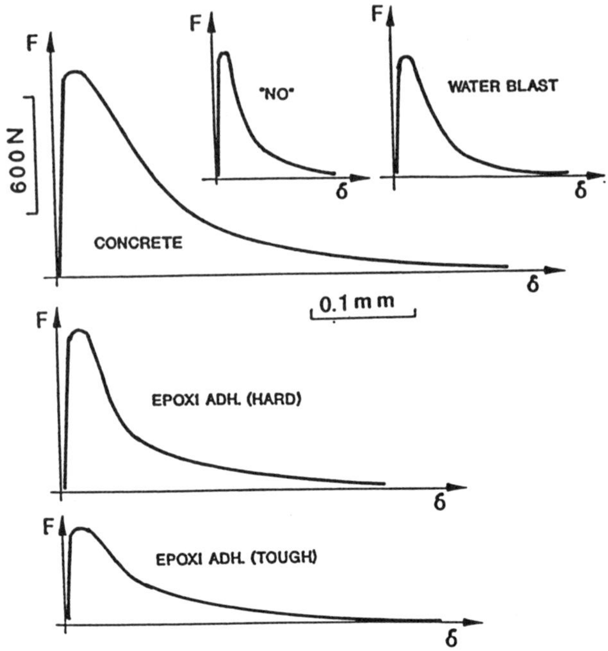

Fig.6. Typical load-deformation diagrams

5 Discussion

Different old-new concrete bonds have been tested with a recently developed testing method. From the resulting load-deformation curves, specific fracture energy G_F (which characterizes the crack growth resistance of an interface under mode I loading) and maximum load F_{max} (from which the bending tensile strength may be derived) are obtained. With conventional methods, like the tear-off test, only maximum load and from this the adhesive power can be determined. The crack growth resistance of the interface is in now way characterized. Identical maximum loads are obtained for separation of "brittle" and "ductile" bonds for example (see fig. 2), though the crack growth resistance is quite different.

With the new method however, it is possible to characterize the crack propagation properties of an interface (see fig. 2). During "brittle" separation, no forces are transmitted from one fracture surface to the other at small crack opening displacements δ; therefore, the strain softening regime is rather small (the curve declines rapidly). For "ductile" separation at same crack opening displacements δ, load-transmission between touching fracture surfaces takes place (large strain-softening area). Therefore, the crack prefers to propagate along the interface, in the case of "brittle" separation. This result is practical importance.

Advantages and disavantages of the fracture energy concept, as applied to bond problems of cement bonded materials, have been discussed extensively in a former work (Tschegg and Stanzl 1990). Specimen-size influence and formation of different fracture process zones have been treated there too.

From the results of this work some important general conclusions may be drawn, which are of relevance for construction engineering purposes: Bending tensile strength (calculated from the F_{max} values) and adhesive strength (obtained from a tear-off test) alone cannot characterize bonds of cement-bonded materials adequately. This becomes obvious, if for example "no" and "grinded" pretreatments are compared with "fracture surface" and "corrugated" pretreatments. For both types of bonds, F_{max} is approximately the same; the fracture behaviour however, is not characterized in any way by F_{max}. The ratio G_F/F_{max} on the other hand, is low for bonds with "no" or "grinded" pretreatment, and it is high for "fracture surface" and "corrugated" pretreatment. This demonstates, that at least one more controlling factor must be introduced, which is not directly correlated to F_{max} and which characterizes the energy consumption during crack propagation.

In this work, G_F has been found as most appropriate parameter for characterizing the crack growth resistance of cement-like material bonds. This specific fracture energy may be determined easily from the load-deformation curve according to the procedure of Tschegg and Stanzl(1990).

6 Summary

The present work has shown the following:

1. A recently developed method (Tschegg and Stanzl 1990) is appropriate to characterize the mechanical properties of bonds between cement-like materials.
2. For this characterization, the load-deformation curve is used, from which the specific fracture energy G_F and the maximum load F_{max} are determined.
3. The F_{max} and G_F values of all tested bonds are in the range of 40% - 87% (for F_{max}) and between 8% - 48% (for G_F) in comparision with homogeneous concrete.
4. Macroscopically rough surfaces of the old concrete part (e.g. "corrugated" and "fracture surface") give rise to high G_F values. Microscopically rough surfaces of the old concrete (e.g. "sand blast" or "water blast") on the other hand, do not increase the G_F values much in comparison with "no" pretreatment.
5. Expoxi adhesives lead to high G_F values at high F_{max} values, whereas "emulsion" as adhesive, brings about low G_F values at moderate F_{max} values.

References

Brühwiler E.and Wittmann F.H. (1990) The wedge splitting test, a new method of performing stable fracture mechanics tests, Eng. Frac. Mechanics, Vol. 35 No.1/2/3, p 117.

Linsbauer H. and Tschegg E.K.(1989) Determination of fracture energy of cement-like materials subjected to environmental influences, Final Report A1, COST 502, January 1889, Technical University Vienna, Austria.

Hillemeir B. (1976) Bruchmechanische Untersuchungen des Rißfortschrittes in zementgebundenen Werkstoffen,thesis, University (TH) Karlsruhe.

Hilsdorf H.K. (1981) in: "Forschung Straßenbau und Strassenverkehrstechnik", (herausgegeben vom Bundesministerium für Verkehr, BRD) Heft 342, p 47.

Proc. (1986) "Fracture toughness and fracture energy of concrete", (ed.F.H. Wittmann) Oct. 1-3. 1985, Elsevier, Amsterdam.

Schulz R. (1984) "Beton als Beschichtungsuntergrund - Über die Prüfung des Festigkeitsverhaltens von Betonoberflächen mit dem Abreißversuch" thesis 1984, RTWH Aachen.

Tschegg E.K. and Linsbauer H. (1986), Patent No. 233/86, No. 390328, Vienna, Austria, inventor E.K.Tschegg.

Tschegg E.K. (1990a) Patent 408/90, Vienna, Austria.

Tschegg E.K. (1990b) New equipment for fracture tests on concrete, submitted to Materials & Structures.

Tschegg E.K. and Stanzl S.E. (1990) Adhesive power measure ments of bonds between old and new concrete, submitted to J.of Mat.Science.

Acknowledgement: Thanks are due to Mag. W. Zikmunda for his experimental assistance and Mr. A. Pötzl (BPV, Lanzendorf) for machining the specimens.

73 CRACK GROWTH RESISTANCE EVALUATION FOR BRITTLE NONMETALLIC MATERIALS

G.S. VOLKOV
Institute for Problems of Strength, Academy of Sciences of the UkrSSR, Kiev, UkrSSR

Abstract
A method for estimating the crack growth resistance of brittle nonmetallic materials (BNM) at various stages of the subcritical crack growth has been developed. Its essence consists in loading a DCB specimen with tensile and compressive forces applied to specimen's ends at different distances from the crack tip. The load ratio is maintained constant in the course of loading and can be more than, or less than, or equal to unity. This makes it possible to provide at the tip of a propagating crack an increase, or a decrease, or a constancy of the stress intensity factor K_I at a fixed level of the external load. The method allows the crack growth rate ϑ to be controlled at the stages of its nucleation, stable subcritical growth, and retardation right up to the arrest.
Keywords: Fracture, Crack, Specimen, Brittle Nonmetallic Material, Stress Intensity Factor.

1 Introduction

Brittle nonmetallic materials find an ever increasing application in various industries. Owing to a higher corrosion resistance, low specific weight and ability to retain their physico-mechanical properties at high temperatures, in a number of cases these materials are preferable to metals. Composite materials, various types of ceramic materials and rocks are widely used.

Most of such materials are polycrystalline aggregated structures containing defects in the form of cracks, pores, foreign inclusions, interlayers of a lower strength, etc. The existence of lower strength surfaces of a geologic, production-process, service, or another origin affects appreciably the behavior of such materials under load. They generally exhibit dissimilar resistance to the action of tensile and compressive loads, with both strength and elastic characteristics being significantly dependent on the type of loading, as well as on the number and orientation or existing defects.

Fracture Processes in Concrete, Rock and Ceramics Edited by J.G.M. van Mier, J.G. Rots and A. Bakker.
 Published by E. & F.N. Spon, 2–6 Boundary Row, London SE1 8HN. ISBN 0 419 15870 7

The fracture of BNM is generally preceded by a preparatory period involving nucleation and development of a system of microcracks which form a lower-strength process zone, whose strength and elastic characteristics may differ substantially from those of the initial material.

In view of specific features of the BNM structure and behavior under load it is difficult to apply to them the ASTM E-399 standard fracture mechanics methods developed for metallic materials. The difficulties may be both methodical and fundamental because of the complexity of determining the true position of the crack tip. In view of this fact the development of methods for estimating the crack growth resistance characteristic of such materials, which are often employed for manufacturing critical structures, is of great practical importance.

2 Test Procedure

This report presents a test procedure providing for estimation of the BNM crack resistance at various stages of the subcritical growth of a crack, including its nucleation, extension, and arrest. The essence of the procedure consists in a simultaneous loading of a double cantilever beam (DCB) specimen with tensile P_t and compressive P_c forces applied to the specimen's ends at different distances from the crack tip. The ratio between the forces is maintained constant in the course of specimen loading and, depending on the test purposes, can be less than, or more than, or equal to unity.

A special loading device has been developed to implement said method (Volkov, 1986). **Its kinematic scheme is** shown in Fig. 1 and the general view in Fig. 2. It comprises housing 1 accommodating the loading system. The latter comprises angular levers 2 whose arms adjoining specimen 3 carry movable thrust pieces 4 for interaction with side surfaces of the specimen. By means of pivot pins 5 opposite arms of levers 2 are coupled to tie-rods 7, adjustable in length with the aid of turnbuckle 6, which are coupled to each other and to a loading device by means of hinge 8.

The specimen with notch 12 made in its end face and crack 13 is installed between grips 11 and thrust pieces 4, which are positioned at a given distance from spindles 9 mounted in bearings 10. To ensure perpendicularity of loads to the crack line, the lever arms adjoining the specimen are installed parallel to the specimen side surfaces by adjusting the height of thrust pieces 4. Housing 1 is attached to the upper support of the test rig. The applied load P acts on levers 2 through tie-rods 7. Levers 2 transmit the compressive load P to the specimen via thrust pieces 4 and the tensile load P via spindles 9 and grips 11. For the adopted loading scheme

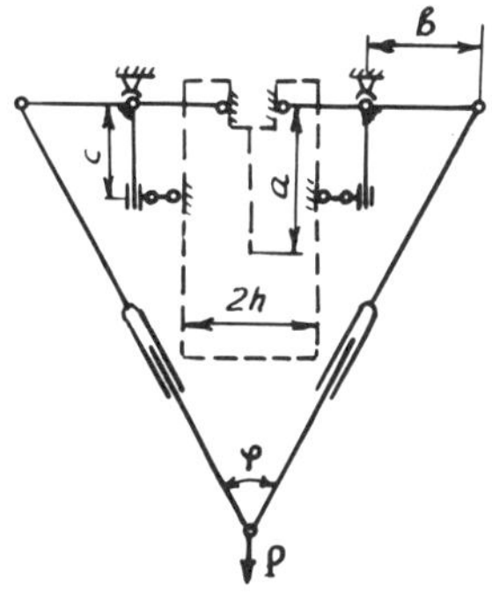

Fig. 1. Kinematic diagram of the loading device

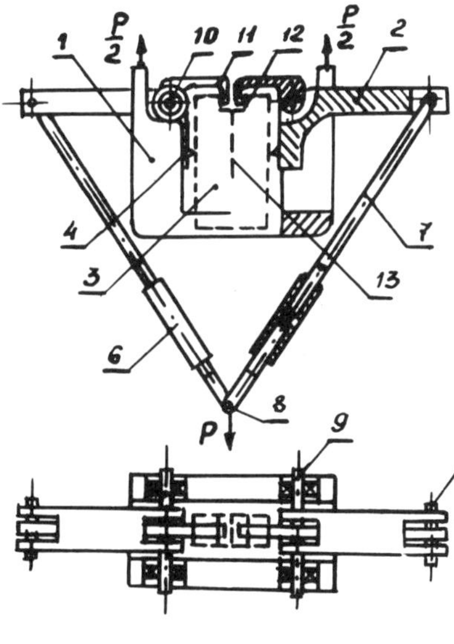

Fig. 2. General view of the loading device

(Fig. 1) the loads applied to the specimens are given by the following expressions:

$$P_t = \frac{P}{2}\left(\frac{b}{c} - \tan\frac{\varphi}{2}\right) ; \qquad P_c = \frac{P}{2}\cdot\frac{b}{c} \quad , \tag{1}$$

where P is the total load applied to the loading system;
- b is the length of the horisontal arm of lever 2 (Fig. 1);
- c is the distance between lines of action of loads P and P ;
- φ is the angle between tie-rods 7.

From expression (1) it follows that

$$\frac{P_t}{P_c} = 1 - \frac{c}{b}\cdot\tan\frac{\varphi}{2} \quad .$$

The analysis of this relation reveals that the ratio between the tensile and compressive loads depends solely on geometric parameters of the loading system and remains constant in the process of the test. The ratio can be less than unity with converging tie-rods of the loading device, equal to unity with parallel tie-rods, and more than unity when the tie-rods diverge. The three cases provide for three types of experimental problems corresponding to various stages of crack evolution: a stable growth, an accelerated propagation, and retardation right up to its arrest.

The calculation of crack growth resistance characteristics is carried out with the well-known relation for determining the stress intensity factor in a DCB specimen loaded only with tensile forces:

$$K_I = \frac{2P_t}{B}\cdot\left(\frac{3}{h}\right)^{1/2}\cdot\left(\frac{a}{h} + \frac{\Delta a}{h}\right) \quad , \tag{2}$$

where B is the specimen thickness;
- a/h is the correction for the relative crack length allowing for a nonrigid fixation of specimen's ends on the crack tip line.

It was ascertained by Fisher (1983) that the value of the correction varies within 0.64 $< a/h <$ 0.70 and that its mean value of 0.67 practically coincides with an exact solution for crack lengths $a/h >$ 1.0 (Foote, 1983). Assuming the correction to be independent of the crack length, by analogy with expression (2) we obtain for the case of the specimen loaded only with compressive forces:

$$K_I = \frac{2P_c}{B}\cdot\left(\frac{3}{h}\right)^{1/2}\cdot\left(\frac{c - a - \Delta a}{h}\right) \quad , \tag{3}$$

where a is the crack length from its tip to the line

of action of tensile loads.

Since an elastic problem is being considered, then, using the superposition principle and proceeding from relations (1), (2), and (3), we obtain the following formula for calculating the stress intensity factor for a DCB specimen loaded according to the scheme shown in Fig. 1:

$$K_I = \frac{P}{B} \cdot \left(\frac{3}{h}\right)^{1/2} \cdot \left[\frac{b}{h} - \tan\frac{\varphi}{2}\left(\frac{a}{h} + 0.67\right)\right] . \qquad (4)$$

The analysis of formula (4) reveals that as the crack in the specimen grows at a constant level of external load P, the stress intensity factor at the crack tip decreases. It means that tensile stresses at the crack tip decrease in the course of testing which results in the crack arrest without reduction of external load.

The proposed crack resistance test method allows a more correct determination of the BNM crack resistance characteristics at the crack arrest stage by direct load monitoring from the moment of the crack growth onset to its arrest. This specimen loading procedure better corresponds to service conditions of many real products, since the load acting on the specimen remains constant over the whole stage of crack extension up to its arrest.

The above loading scheme allows growing of an initial crack in the DCB specimen. To do this, a more rapid decrease of the stress intensity factor with the growth of the crack should be provided by increasing the angle between the loading tie-rods. Moreover, crack resistance characteristics at the moments of the crack growth onset and arrest are determined simultaneously in one test. To calculate the former the initial crack length should be used in formula (4), while to calculate the latter the final crack length should be used.

If = 0 is set in the loading device then, as follows from expressions (1), the compressive and the tensile forces will be equal, with the result that the specimen will be loaded with a constant moment. Expression (4) for calculating the stress intensity factor will take the form

$$K_I = \frac{2}{bh} \cdot \left(\frac{3}{h}\right)^{1/2} \cdot M \quad , \qquad (5)$$

where M = (P/2)•c is the moment applied to the specimen.

From equation (5) it follows that the stress intensity factor is independent of the crack length. Consequently, the constancy of the stress intensity factor at the tip of both stationary and propagating crack is provided at a constant loading level. This makes it possible, on the one hand, to determine the BNM crack resistance without measuring the crack length in the specimen and, on the other hand, to ensure constancy of the crack growth rate. The magnitude of the stress intensity factor and hence the crack growth rate can be varied by changing the total load on the specimen. This allows the BNM crack resistance to be estimated at different stages of the subcritical crack growth, including sustained fracture under conditions of creep. The latter fact allows this method to be used to plot K - v diagrams and the BNM crack resistance to be evaluated over the whole range of the subcritical crack growth.

If the ratio between the tensile and compressive loads is more than unity ($\varphi > 0$), then at a constant external load the stress intensity factor at the tip of the growing crack will grow. The crack will accelerate, which will allow the material crack resistance to be investigated at the stage of a fast crack.

The method employs DCB specimens of a simplified configuration (Fig. 1) without holes drilled commonly for the loading bars. This simplifies preparation of specimens from hard-to-work materials which include most ceramic ones. Owing to a vertical arrangement of the specimen in the loading device, the proposed method is easy to apply to low-temperature tests, as well as to tests in corrosive media. In this case the loading elements and the crack tip opening displacement gauges can be positioned outside the corrosive medium with the bottom part of the specimen with the crack tip being inside it which will simplify the procedure of recording the force and strain parameters in the course of an experiment.

The proposed method was tested with model materials, such as soda-lime glass and polymethyl methacrylate (PMMA), where the crack nucleation and growth processes are easy to observe. Other types of specimens of such materials, in particular compact and V-type ones, were tested. A fair agreement of the crack resistance estimation results was obtained, which allowed the method to be applied to other BNMs. In particular, ceramic refractory materials and rocks were tested. Their tests demonstrated the possibility of a deviation of the crack extension from a rectilinear one and even of a break-off of the specimen end in the process of loading. It is therefore recommended to make side notches in the specimen's side surfaces to create conditions for plane strain on the one hand, and on the other hand, to guide the crack propagation. Pisarenko (1977) recommended to take their depth from the condition $B/B_n = 1.5...2.5$, where B_n is the thickness of the net cross-section of the specimen. Larger values are recommended for coarse-aggregated materials, as well as for low-strength ones. With the side notches the stress intensity factor is calculated from the following relationship:

$$K_I^* = K_I \cdot \left(\frac{B}{B_n}\right)^m ,$$

where K_I is the nominal value of the stress intensity factor calculated from relation (4) for the conditions of the absence of the side notches;
m is the coefficient depending on the specimen type; its value is little dependent on the crack length and for a DCB is close to 0.65.

In view of relatively low BNM crack resistance characteristics, as compared to metals, an UP-8 bench press with a maximum force of 5 kN was used for loading. It provided for testing specimens with control of both the load and the crosshead displacement. In testing small-size specimens of BNM with a low crack resistance the loading can be effected by suspension of a weight. In testing specimens of relatively large sizes or of materials with a high crack resistance the device proposed for the implementation of the above method should be coupled to the power circuit of the set up of the required power.

3 References

Annual Book of ASTM Standards, E-399 (1980), pt. 10, pp. 580-601.

Fisher, W.B. (1983) The stress intensity factor for the double cantilever beam. Int. J. Fracture, 22, 2, 133-143.

Foote, R.M.L. and Buchwald, V.T. (1983) An exact solution of the stress intensity factor for a double cantilever beam. Ibid., 29, 3, 125-134.

Pisarenko, G.S., Naumenko, V.P. and Volkov, G.S. (1977) On determining the stress intensity factor in specimen with side notches. Problemy Prochnosti, 10, 5-10.

Volkov, G.S. (1986) USSR Inventor's Certificate No 1, 262, 335. Method and Apparatus for Determining Crack Resistance of Brittle Nonmetallic Materials. Bull. No 37.

PART TWELVE

RATE EFFECTS, SUSTAINED LOADING AND HYGRAL EFFECTS

74 INFLUENCE OF SUSTAINED LOAD ON THE FRACTURE ENERGY AND THE FRACTURE ZONE OF CONCRETE

E. AASSVED HANSEN
SINTEF, Cement and Concrete Research Institute, Trondheim, Norway

Abstract
This paper describes the results from tests on notched concrete beams. In addition to measure the fracture energy, the fracture zone was monitored by displacement transducers. In order to reveal the influence of sustained load, the concrete response due to a procedure including partially sustained load, was compared with the response due to a procedure with constant deformation rate. In beams of age 28 days the fracture energy was influenced by sustainment. It increased by 24% when the period of sustainment was about two hours. The results also showed that the timedependency was considerably reduced, and the fracture zone more narrow, at age 90 days compared with age 28 days.
Keywords: Fracture energy, fracture zone, microcracking, CTOD, timedependency, sustained load, holography.

1 Introduction

The objectives of this investigation are to reveal the influence of a short period of sustained load on the development of the fracture zone and the fracture energy. Although, they are important properties, the knowledge on how the fracture energy and the fracture zone are influenced by factors like time, are still limited. The relation between them are neither fully understood.

To study the timedependency, tests are made on notch concrete beams in a fracture energy determination unit. The development of the fracture zone is monitored by displacement transducers and observed by TV-holography. The TV-holography observations have been reported earlier, Aassved Hansen (1989) and (1990).

2 Materials and specimens

The specimens were made of concrete with a water-cement ratio of 0.48, and 1:2.25:2.4 ratios of cement-sand-gravel. Five slabs of dimension 100·600·1200 mm, and 26 beams of dimension 100·100·1200 mm were cast in a horizontal position and stored in water. After 14 days, five beams (100·100·1200 mm) were cut from each slab and restored in water until testing. The cutting was primarily done to provide more representative concrete surfaces for the TV-holography observations.

Fracture Processes in Concrete, Rock and Ceramics Edited by J.G.M. van Mier, J.G. Rots and A. Bakker.
 Published by E. & F.N. Spon, 2–6 Boundary Row, London SE1 8HN. ISBN 0 419 15870 7

The beams were used in identical test programmes at ages 28 and 90 days: (1) the dynamic modulus of elasticity was determined on each beam, (2) all beams were cut in two halves of length 600 mm, (3) the tensile strength and secant modulus were determined on eighteen 600 mm beams, (4) the fracture energy was determined on the remaining 600 mm beams. The microcrack formation during the fracture energy tests were monitored by TV-holography and displacement transducers.

The compressive 100 mm cube strengths were 65.3 and 74.2 MPa at 28 and 90 days respectively. The tensile strengths determined at 45 and 90 days were both 4.0 MPa. The secant (50% of f_t) and dynamic moduli were 33.0 and 43.5 GPa respectively at 90 days.

3 Experimental technique

3.1 Fracture energy determination unit

The fracture energy was determined in the unit shown in Fig 1. As can be seen, half the beam is clamped, while the applied load acts upwards. A notch, 50 mm deep and 2 mm wide, was sawn in each beam.

3.2 Fracture zone displacement recordings

The fracture energy beams were monitored with three inductive displacement transducers, Fig 2. The transducers were attached by a light aluminum frame. Their purpose and position were:

CMOD: Crack Mouth Opening Displacement. The transducer is thread trough a hole in the frame at the underside of the beam. The tip rests against a steel plate on the opposite side of the notch.
CTOD: Crack Tip Opening Displacement. Transducer with reference length 25 mm and three sensing probes, clamped to the concrete surface across the notch tip.
CPOD: Crack Path Opening Displacement. A transducer similar to the CTOD, positioned 25 mm ahead of the notch tip.

3.3 Loading procedure

Two different loading procedures, termed A and B, were used:

A. Normal loading. The load is applied at a constant deformation rate, so the maximum load is reached within two-three minutes.

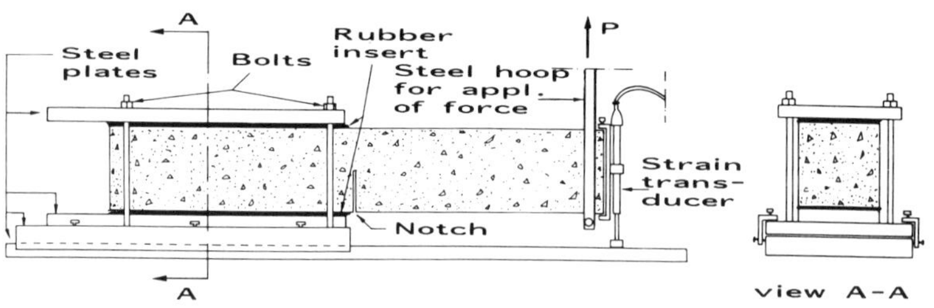

Fig 1. Sketch of fracture energy determination unit

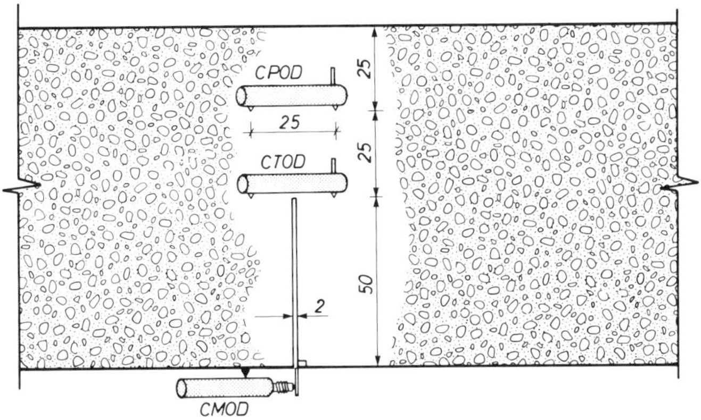

Fig 2. Position of transducers

B. Partially sustained load. Initially, a constant deformation rate is applied, as in A. At about 90% of the maximum load, the deformation is halted. The load level is then maintained by deformation control. That is, to compensate for relaxation due to the progressive microcracking in the fracture zone, a small increase in deformation is frequently applied. When the microcracking reaches some critical level, the beam can no longer support the load. The load therefore declines. At this stage, the deformation rate is again increased to correspond with the rate used in procedure A.

The period in which the load was kept constant according to procedure B, varied from two to 240 minutes. Fig 3 shows the recorded load history for two beams tested according to procedure A and B respectively. It is seen that due to a very precise deformation control, it is possible to keep the load almost constant, i.e., within 0.1% or ± 1 N.

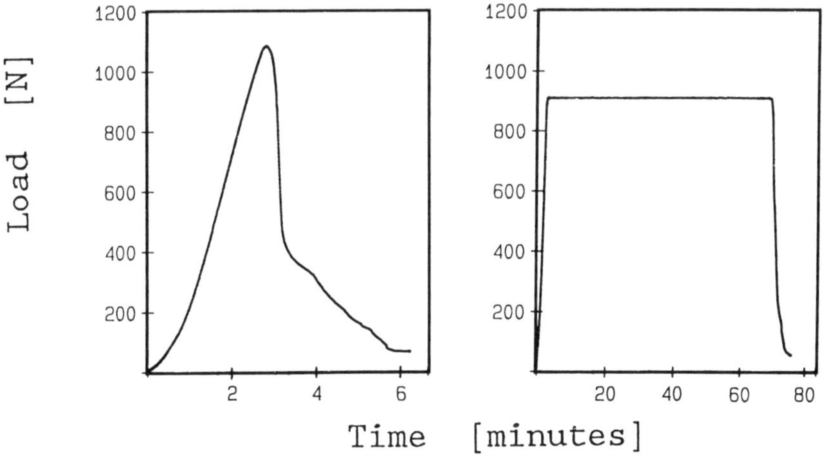

Fig 3. Recorded load histories according to procedures A and B

4 Results

4.1 Fracture energy

The fracture energy results given in Table 1, were group with respect to time under sustained load. Test procedure A makes the zero group. No difference was found between cut beams and beams cast in mould.

Table 1. Average fracture energy (G_F) with standard deviations (SD)

Test age (days)	Constant load (min)	Number of beams	G_F (Nmm)	SD (%)
28	0	11	150	14
	2-6	7	156	7
	20-120	6	185	4
90	0	10	147	17
	1-3	6	133	6
	20-40	7	153	7
	70-240	4	157	11

4.2 Fracture zone displacements under normal loading

Examples on recorded displacements from the three transducers are given in Fig 4 for one beam of age 28 days (denoted A-28), and in Fig 5 for one beam of age 90 days (denoted A-90).

The curves obtained at 28 and at 90 days are almost identical. In the given examples, slightly steeper ascending branches and more smooth peaks at 90 days, seem to depend more on individual variations than on age. In all beams the CTOD is approximately linear up to 30% of the maximum load (P_{max}). The deviation from linearity at this level is probably due to some minor bond cracking. A second more distinct deviation from linearity occurs at about 70-75% of P_{max}. This second deviation seem to coincide with an arise in the CPOD, which up to this level is about zero. Thus, indicating that a linear stress distri-

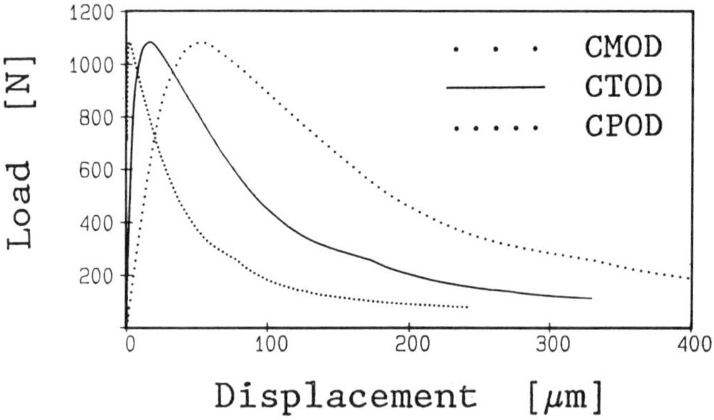

Fig 4. Load versus CMOD, CTOD and CPOD for beam A-28, age 28 days

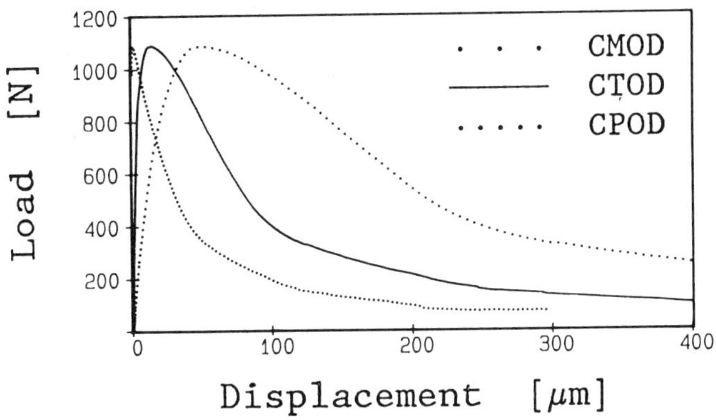

Fig 5. Load versus CMOD, CTOD and CPOD for beam A-90, age 90 days

bution is no longer present, and, most likely, a fracture zone emerges at the notch tip.

At 70-75% of P_{max}, the CTOD for A-28 and A-90 is in the range 3.9-4.2 μm. This equals an average tensile strain of $160\text{-}170 \cdot 10^{-6}$. In comparison, the ultimate tensile strains measured on specimens in the direct tensile test varied between $150 \cdot 10^{-6}$ and $180 \cdot 10^{-6}$.

The increase in the CPOD and CTOD rates as the load approaches 70% of P_{max} is easily seen when plotted as in Fig 6. The ratio CMOD to CTOD seems to change from one constant value (about five) to an other (about two). This change occur between 70% of P_{max} and P_{max}, which corresponds to CTOD values in the range 4 to 15 μm.

A quite similar change is observed for the CPOD-CTOD ratios, which increase from zero to about 0.5. The 0.5 value is, however, not obtained until the load has passed its maximum, and CPOD is about 15 μm.

Fig 6 also shows that the ratios CMOD to CTOD is almost identical in the two beams up to P_{max}. When the load declines, the ratio for the A-90 beam increases more than for the A-28 beam. The CPOD-CTOD curves are, on the other hand, almost identical.

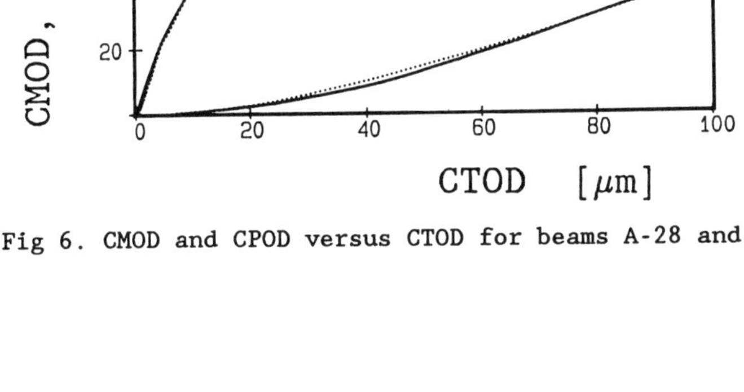

Fig 6. CMOD and CPOD versus CTOD for beams A-28 and A-90

4.3 Fracture zone displacements under sustained loading

Examples on the recorded displacements are given in Fig 7 for one beam of age 28 days (B-28), and in Fig 8 for one of age 90 days (B-90). The sustained period is about 70 minutes for both, confer Figs 10 and 11.

In B-28, the CTOD is 5.8 μm at the beginning, and 33 μm at the end of the sustained period. In B-90, these CTOD values are 5.5 and 18 μm. The average values for all beams of age 28 days were 7.5 μm (range 5.4-9.6 μm) and 29 μm (21-40 μm), respectively. While for the beams of age 90 days, they were 6.4 μm (5.3-7.6 μm) and 25 μm (18-33 μm).

In the 28 days old beams the CTOD value at the end of the sustained period seemed to increase with the length of the period, while no such correlation is present for the more mature beams.

The CMOD-CTOD and CPOD-CTOD ratios shown in Fig 9, are quite similar to those in Fig 6. A major difference is a steeper CMOD-CTOD curve for B-90 compared with A-90. The CPOD-CTOD ratios for B-28 and B-90 are also slightly higher than for A-28 and A-90.

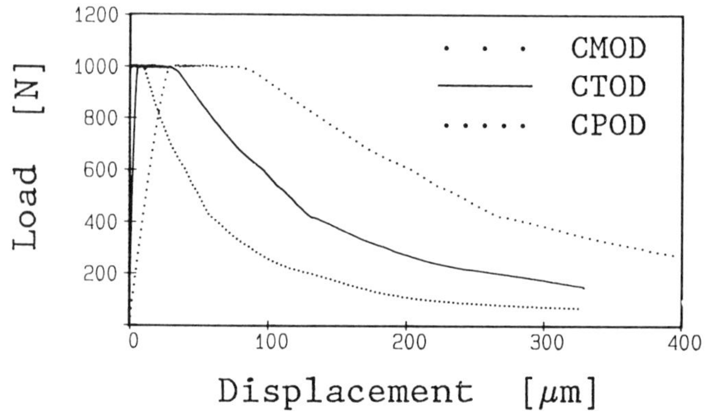

Fig 7. Load versus CMOD, CTOD and CPOD for beam B-28, age 28 days

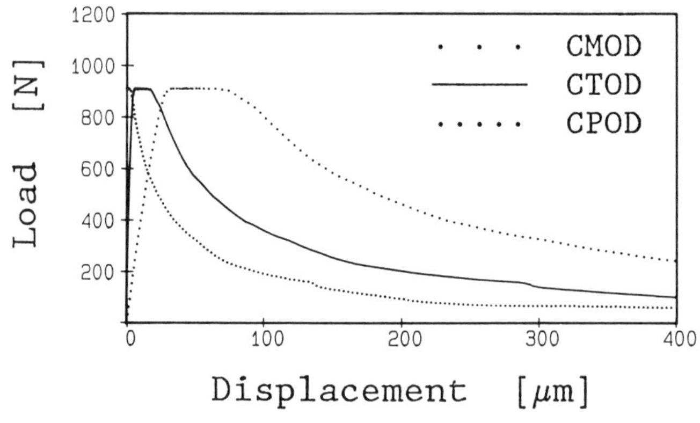

Fig 8. Load versus CMOD, CTOD and CPOD for beam B-90, age 90 days

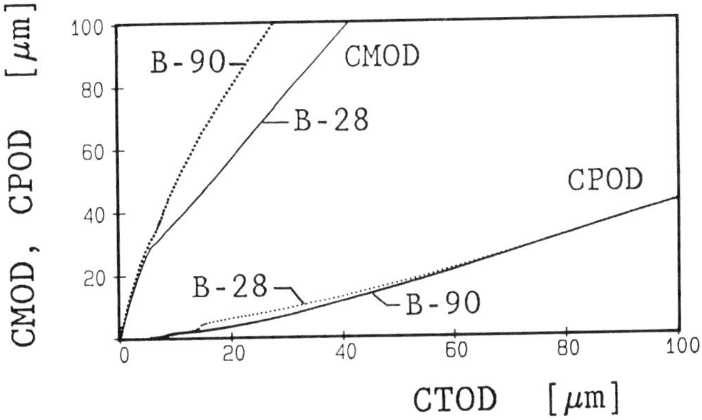

Fig 9. CMOD and CPOD versus CTOD for beams B-28 and B-90

The influence of age and sustainment can also be studied by calculating the area under the various load-displacement curves, Table 2. Although, these values have no physical meaning, they reveal some interesting information when compared with each other.

As for the fracture energy, the timedependency seems to be present in the 28 days old beams only. When the concrete age increases, the results show significant reductions in CTOD and CPOD areas. CMOD does not seem to depend on concrete age. The differences in CMOD-CTOD and CMOD-CPOD ratios at 28 and 90 days imply, that the size of the fracture zone is reduced with increased concrete age.

The increased gap between CMOD and CTOD with increased age is also evident from the curves shown in Figs 7 and 8. As can be seen, the development in CMOD is almost identical in B-28 and B-90, while CTOD and CPOD are significantly lower at age 90 days compared with 28 days.

When plotting the displacement histories as in Figs 10 and 11, the reduction in CTOD and CPOD rates with increased maturity is evident. In Figs 10 and 11 the three typical stages of a creep curve can be observed in the displacement recordings, i.e., the primary creep with declining displacement rate, the secondary creep with constant rate, and the final stage with accelerated displacement.

Table 2 Area (Nmm) under load-displacement curves average values

Age (days)	Constant load (min)	Number of beams	CMOD Mean	CMOD SD(%)	CTOD Mean	CTOD SD(%)	CPOD Mean	CPOD SD(%)
28	0	11	255	13	116	11	55	18
	2-6	7	244	12	113	11	55	16
	20-120	6	293	9	136	9	67	24
90	0	11	255	19	95	18	42	22
	1-3	7	251	17	91	14	42	17
	20-40	7	243	12	97	7	44	10
	70-240	5	241	15	99	12	47	9

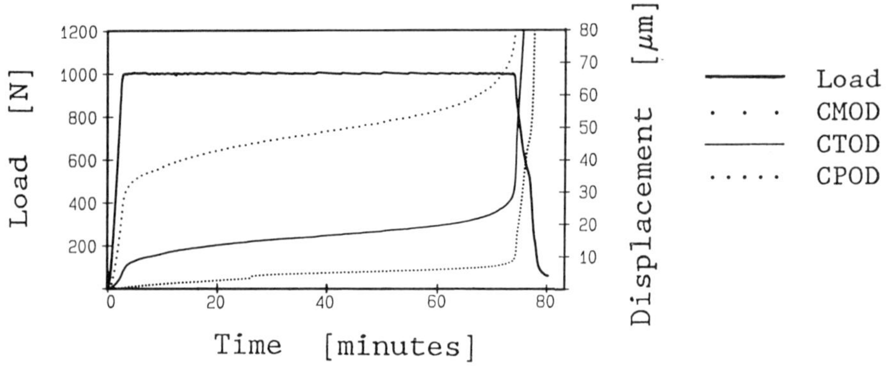

Fig 10. Load and displacements versus time for beam B-28

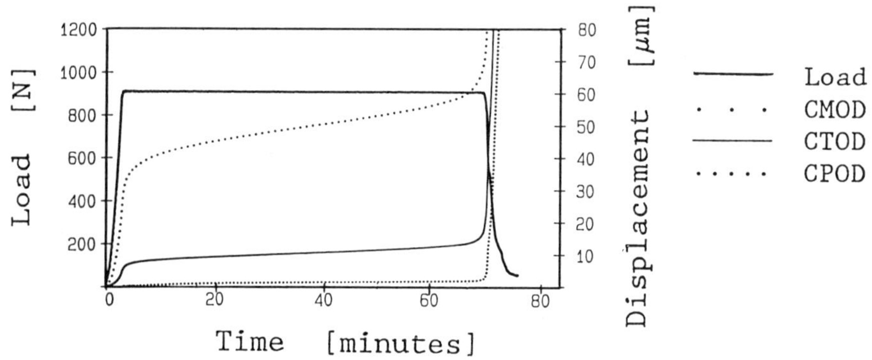

Fig 11. Load and displacements versus time for beam B-90

5 DISCUSSION

5.1 Influence of sustained load

The fracture energy results indicate a dependency of time under sustained load. In beams of age 28 days, the average increase in fracture energy due to a sustained period of 20 to 120 minutes only, is as high as 24%. In spite a longer period of sustainment, the increase is considerably lower for the more mature beams of age 90 days. In fact, the average fracture energy increase of 7% at 90 days, is not statistically significant, due to a large scatter in the results.

Wittmann et al. (1987) observed influence of time on the fracture energy by varying the loading rate on 28 days old notched beams. They found a minimum value at intermediate deflection rates with an increase towards lower rates, i.e., increased test duration.

In the present tests, the higher fracture energy under sustained load, most likely results from increased widening of the fracture zone. The slow crack growth promotes increased microcracking and more tortuous crack paths, and thus, it increases the true fracture area.

An extended widening under sustained load was, as previously shown, confirmed by the displacement recordings on the 28 days old beams, while no significant influence was present at 90 days.

5.2 Influence of concrete age

The fracture zone displacement recordings indicate that the fracture zone becomes smaller at increased concrete age. Although the CMOD values are almost identical, the CMOD-CTOD ratios, Figs 6 and 9, increase with increased concrete age. This can also be seen from Fig 12, which compares the CMOD and CTOD curves for beams B-28 and B-90.

The apparently more rigid behaviour with increased age may result from changes in the relative strength between the cement paste constituents and increased bond strength.

Berger et al. (1973) and Grudemo (1979), among others, have observed that cracks in young cement paste develop through the high porosity C-S-H gel, deviating away from the rigid CH crystals. In more mature paste the cracks tend to progress through the CH crystals in a less tortuous path. Two reasons for this transition have been proposed. With increased hydration a CH free crack path becomes extremely tortuous and energy consuming. In addition the difference in strength and fracture energy between CH and C-S-H decreases.

A similar hypothesis may be applied to concrete. Increased bond strength makes a interfacial crack more energy consuming, also because it results in a larger fracture area than a crack through an aggregate. Thus, a transaggregate path may give the lowest total fracture energy. SEM observations by Tait and Garrett (1986) confirm that cracking through aggregate particles may occur without any debonding, especially for well-hydrated mortar. The TV-holography observations within this investigation also revealed a clear tendency for increased transparticle crack growth in the mature beams, Aassved Hansen (1990).

In addition to explain the age effect, this could also explain the reduced timedependency in the mature beams. Due to reduced difference in fracture strength between C-S-H gel and CH-crystals, and between the interface and aggregates in mature concrete, a reduction in crack propagation rate will not have the same influence on the crack path. At age 28 days, a slow crack propagation rate will favour interparticle paths, and thus, increasing the total fracture energy.

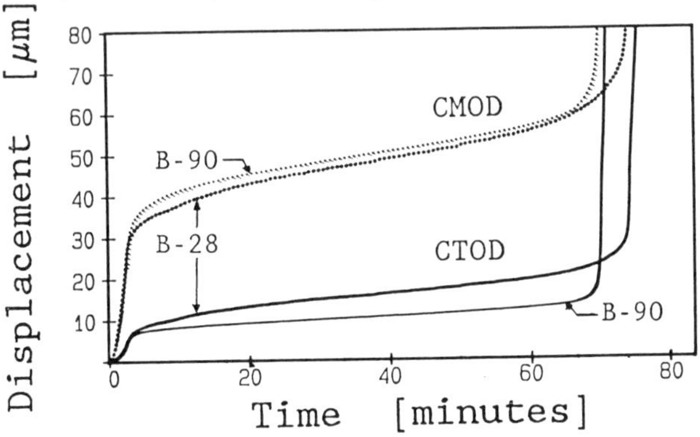

Fig 12. CMOD and CTOD versus time for beams B-28 and B-90

6 Conclusion

The following conclusions may be drawn from the test on notched beams with partially sustained load:

The fracture energy was influenced by a short period of sustained load, which imply that the fracture energy is timedependent.
The influence of sustained loading decreased with increased concrete age.
In 28 days old beams the average increase in fracture energy was 24% when the period of sustainment lasted for 20 to 120 minutes.
The width of the fracture zone increased significantly during a period with sustained load.
While the total width of the fracture zone increased in beams of age 28 days, due to the sustained load, no such influence seemed to present in the 90 days old beams.
The width of the fracture zone seemed to depend on concrete age. It became more narrow as the age increased from 28 to 90 days.

7 References

Aassved Hansen, E. (1989) A holographic real time study of crack propagation in concrete. **Cem. Con. Res.,** 19, 611-620.

Aassved Hansen, E. (1990) Real time TV-holography observations of the fracture process zone, in **Micromechanics of Failure of Quasi-Brittle Materials** (eds S.P. Shah, S.E. Swartz and M.L. Wang), Elsevier Science Publ., London, pp. 504-513.

Berger, R.L. Lawrence, F.V. and Young, J.F. (1973) Studies of the hydration of tricalcium silicate pastes - II. Strength development and fracture characteristics. **Cem. Con. Res.,** 3, 497-508.

Grudemo, Å. (1979) Microcracks, fracture mechanism, and strength of the cement paste matrix. **Cem. Con. Res.,** 9, 19-34.

Tait, R.B. and Garrett, G.G. (1986) In situ double torsion fracture studies of cement mortar and cement paste inside a scanning electron microscope. **Cem. Con. Res.,** 16, 143-155.

Wittmann, F.H. Roelfstra, P.A. Mihashi, H. Yiun-Yuang, H. Xin-Hua, Z. and Nomura, N. (1987) Influence of age of loading, water-cement ratio and rate of loading on fracture energy of concrete. **Materials and Structures,** 20, 103-110.

75 EXPERIMENTS ON CONCRETE UNDER LATERAL COMPRESSION AND TENSILE IMPACT LOADING

J. WEERHEIJM
TNO Prins Maurits Laboratory, Rijswijk, The Netherlands
H.W. REINHARDT
Institute for Engineering Materials, Stuttgart University, Germany
S. POSTMA
TNO Prins Maurits Laboratory, Rijswijk, The Netherlands

Abstract
Laterally compressed concrete specimens were tested on axial impact response in a modified Split Hopkinson bar apparatus. The residual impact strength and deformation capacity at various levels of compression were determined. The influence of the lateral compression device on the axial response was estimated and incorporated in the data. The rate effect on the residual strength appears to depend on the level of compression.
Keywords: Concrete, Dynamic Loading, Biaxial Testing, Material Properties, Experimental Investigations.

1 Introduction

The rate effect on concrete response is most dominant under uniaxial tensile loading. The impact tensile strength as a function of the loading rate has been studied and determined in many research programmes.

Programmes on concrete under multiaxial, dynamic loading are very scarce. As a first step in understanding and gathering data on the dynamic response of concrete under multiaxial loading, TNO-PML and Delft University of Technology (DUT) performed a research programme on the biaxial loading condition of static lateral compression and axial impact tensile loading.

In the experimental part, the attention was mainly focussed on the question whether the shape of the failure envelope is affected by the loading rate. From static loading it is known that the fracture process, which determines the ultimate strength, depends on the loading condition. Therefore, it might be expected that rate effects, which depend also on the fracture process, will affect the shape of the failure envelope. Hitherto, none of the scarce dynamic test programmes has proven or refuted this dependency.

To determine the residual impact strength at various levels of lateral compression and to gather data on the deformation capacity, experiments were performed using a modified Split Hopkinson bar (SHB) device at the DUT.

2 Theory and data from literature

As a reference for the dynamic tests, the failure envelope under static loading has to be known. Experimental data for this loading region are characterized by scatter, because the almost inevitable interaction between both loading devices is combined with the

Fracture Processes in Concrete, Rock and Ceramics Edited by J.G.M. van Mier, J.G. Rots and A. Bakker.
 Published by E. & F.N. Spon, 2–6 Boundary Row, London SE1 8HN. ISBN 0 419 15870 7

scatter of the material properties under compression as well as tension. However, the extensive programmes of, among others, Kupfer, Nelissen and Gerstle give a good average feature of the failure envelope, which will be used as a reference for the dynamic tests, see Figure 3. Suaris and Shah (1985) proposed a curve based on a cumulative damage model. This envelope corresponds with the empirical data of Kupfer and Gerstle (1973).

The general feature is a continuously decreasing (residual) tensile strength with increasing compression level. The strength decrease is coupled to a change in failure mechanism, from tensile failure to shear failure at high compression levels.

Under uniaxial impact tensile loading the strength increase can be attributed to the changing fracture plane and increasing energy demand for crack extension, see Weerheijm (1989 and 1990) and Zielinski (1982). With increasing static lateral compression, the level of internal damage will also increase, which changes the energy demand for axial tensile failure. The question is whether the rate effect on the energy demand to form a final failure plane for axial tension will be independent of the level of compression. Takeda (1974) and Zielinski (1985) concluded from their experiments that the rate effect is unique, which means independent of the compression level. This conclusion will be considered next, with reference to some theories on the fracture process.

When the failure process is simplified and schematized, the applied energy in the tensile loading can be absorbed in two ways. Firstly, a single failure plane is formed and all the applied energy is used to create this failure plane. Due to lateral compression and the associated internal damage, the energy demand will decrease with increasing compression level. When the loading rate is increased, the change in geometry of the failure plane and the corresponding change in energy demand will depend on the compression level. This will not result in a unique formulation of the loading rate effect, the rate effect will decrease with increasing compression level.

The second option is that the applied energy is initially distributed over multiple cracks and, at a later stage, the final failure or shear planes are formed. With increasing loading rate the multiple crack extension will last longer, resulting in an increasing energy demand with increasing loading rate. The geometry of the failure plane(s) will depend on the compression level as it does in option 1. In this situation, the rate dependency of the failure envelope cannot be predicted.

Considering both options, it is unlikely that a unique formulation can be used to describe the rate effects. This corresponds with the model proposed by Suaris and Shah (1985) who concluded, based on the accumulating damage principle, that the failure envelope expands non-isotropically with increasing loading rate. Note that Suaris applied the model for proportional compression-tension loading, but the coupling of the rate effect to the specific failure mode leads, also for static prestressing, to the conclusion of a non-unique rate effect.

3 The experimental programme

In the experimental programme a low and a high quality concrete were tested at four levels of lateral compression. The identification and test conditions of the series are given in Table 1. The static material properties are summarized in Table 2. The uniaxial impact tests were a reference for the biaxial tests. The compression levels of 50% and 75% roughly indicate the average level in the experiment.

The specimen geometry (Figure 1) was prismatic and notches were applied to fix the failure plane. The axial and lateral strains outside the failure zone were measured with strain gauges, while the total axial deformation was recorded with displacement gauges. These recordings enabled us to determine the deformation of the failure zone.

The experiments were performed in a modified SHB apparatus at the DUT. The impinging and transmitted pulse are traditionally recorded with strain gauges at the lower and upper bar of the SHB. The modification consisted of an additional lateral compression device. The limited available space led to a compact equilibrium system of two hydraulic jacks, the specimen and two tension rods, see Figure 2.

In Weerheijm (1989) the biaxial test set-up itself and the evaluation are described extensively, here just a few remarks will be made.

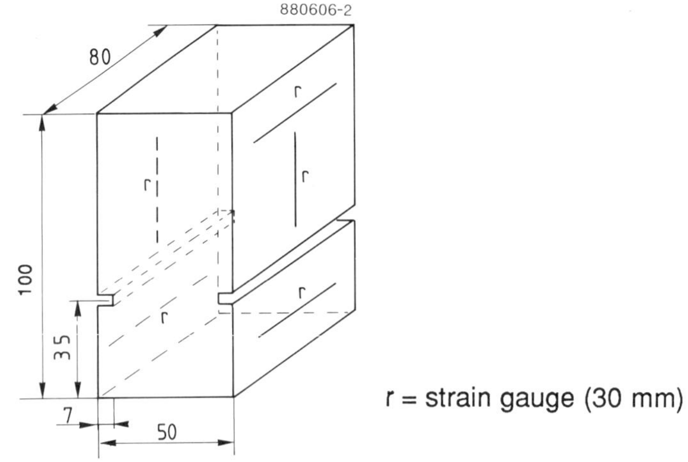

Fig. 1. Geometry of specimen and location of strain gauges.

Table 1. The test series and test conditions

Compression level		0%		50%	75%	100%
Notch depth	0	5	7	7	7	7
Quality A		A5	A7	A50	A75	A100
Loading rate [GPa/s]		17.7	12.0	15.0	13.4	-
Quality B	B0	B5	B7	B50	B75	B100
Loading rate [GPa/s]	10.0	13.7	13.4	10.8	11.0	-

Table 2. Static material properties and mix proportions

		Mix A	Mix B
Compressive strength	[MPa]	41.4	23.6
Tensile strength	[MPa]	3.0	1.6
Young's modulus	[GPa]	33.9	30.9
Portland cement	[kg/m^3]	375	200
Water-cement ratio	[-]	0.5	0.6
Air content	[%]	2.1	2.5

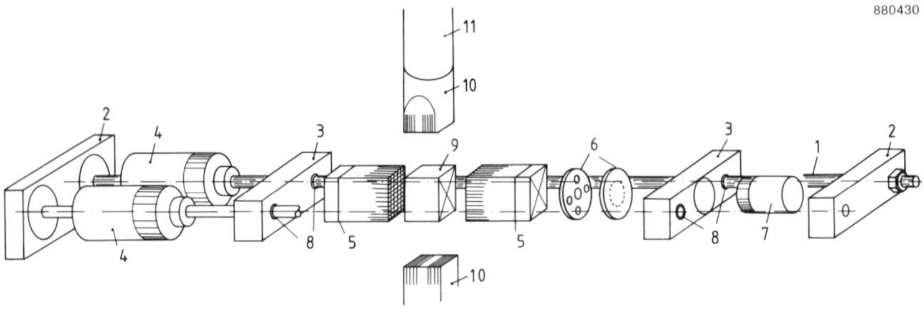

1 tension rod	7 force measuring device
2 fixed platen	8 bronze bearing
3 guiding platen	9 specimen
4 jack (200 kN)	10 aluminium adapter
5 brush-bearing platen	11 transmitter bar of SH equipment
6 spherical seat	

Fig.2. Exploded view of the lateral compression device.

The geometry of the brush bearing platens satisfies the stiffness requirements with respect to the deformation of the loaded faces of the specimen, see Nelissen (1972). The saw cuts in the adapters of the SHB were necessary to reduce the lateral stiffness. In combination with a thickened glued seam, a sufficient lateral deformation capacity of the transition zones was realized ($\varepsilon_{lat} > 0.1$ %). The final remark concerns the recording of the compressive force. The force was recorded by means of strain gauges on the tension rods between the bearing platens to preclude friction effects on the recording.

Considering the biaxial set-up, the question had to be answered as to what extend the axial and lateral devices interact during the impact tests and disturb the data of the material response. This aspect was examined during the evaluation of the set-up with an aluminium specimen.

It appeared that the uniaxial wave reflection and transmission were disturbed. Fifteen per cent of the load was transmitted to the lateral device. The pulse recorded in the upper bar of the SHB equalled the pulse in the specimen. It was concluded that for the concrete tests, the transmitted pulse could be used to determine the load and thus the residual strength.

The preliminary tests also showed that the influence of the lateral device increased during an impact test, due to the axial and lateral deformation. Consequently the biaxial set-up can only be applied before the interaction process between both devices becomes dominant. This demand was verified and fulfilled during the biaxial concrete tests. However, the possible influence of the static device on the dynamic response necessitates careful interpretation of the test results.

4 Test results

4.1 The failure envelope

The uniaxial impact tests showed that the rate effects on the uniaxial tensile strengths and Young's moduli corresponded with earlier DUT test programmes (Weerheijm, 1991). To define the compression level, the compression strength of the specimen in the lateral loading device was determined. Due to the limited stiffness of the device, the reference compressive strength was slightly smaller than derived from cube testing. The reference points for the failure envelope are given in Table 3.

Table 3 Reference points for the failure envelope.

		Mix A	Mix B
Static tensile strength	[MPa]	3.0	1.6
Dynamic tensile strength	[MPa]	4.9	2.7
Static compressive strength	[MPa]	39.6	18.6
Dynamic Young's modulus	[GPa]	40.7	37.1

As reference for the data, the static failure envelope derived by Kupfer is used and the same envelope extended uniformly with the rate effect as followed from the uniaxial impact tests. These curves are depicted in Figure 3, along with two sets of data points. The first set (depicted as crosses) represents the residual strength as followed directly from the tests. However, the stress-strain relation for the material outside the failure zone showed a varying, and mostly too high, Young's modulus, exceeding the dynamic modulus for uniaxial impact. The regular wave transmission from specimen to upper bar, as observed in the tests with aluminium, apparently did not occur in the concrete tests. Part of the observed loading pulse was transmitted via the brushes. To estimate this part, the following procedure was followed. The experiments were divided into groups with a small variance in compression level. Foı each group the minimum of the observed Young's modulus was taken as the reference modulus. The reference values varied from the dynamic Young's modulus to ninety per cent of the static Young's modulus. The actual load level for a specific test was estimated by using the ratio of the observed Young's modulus and the reference value. This led to the second set of data points in Figure 3. The reference points and data were used for third order polynomial fits, which are also given in the figure.

In spite of the fact that the measured data seems to correspond rather well with the uniformly extended failure envelope, it was decided to reject this correspondence because of the mentioned role of the brushes. The adjusted data are regarded as the best representation of the residual strength that can be derived from the current tests.

Contrary to the test results of Zielinski and Takeda, a definite reduction of the rate effect is observed with increasing lateral compression. It seems that the rate effect vanishes and the dynamic failure envelope shrinks to the static one. Referring to both extreme failure processes given in section 2, this result corresponds with the feature that the rate effect for moderate loading rates is governed by the changing geometry of the fracture plane. Due to the compression, the geometry is almost prescribed and the rate effect vanishes. This phenomenon is less pronounced for the low quality concrete. The rate effect reduces from 1.7 at uniaxial loading to 1.3 beyond the level of fifty per cent compression. Probably the second mechanism, multiple crack initiation, is more important. This should be supported by a relatively large deformation.

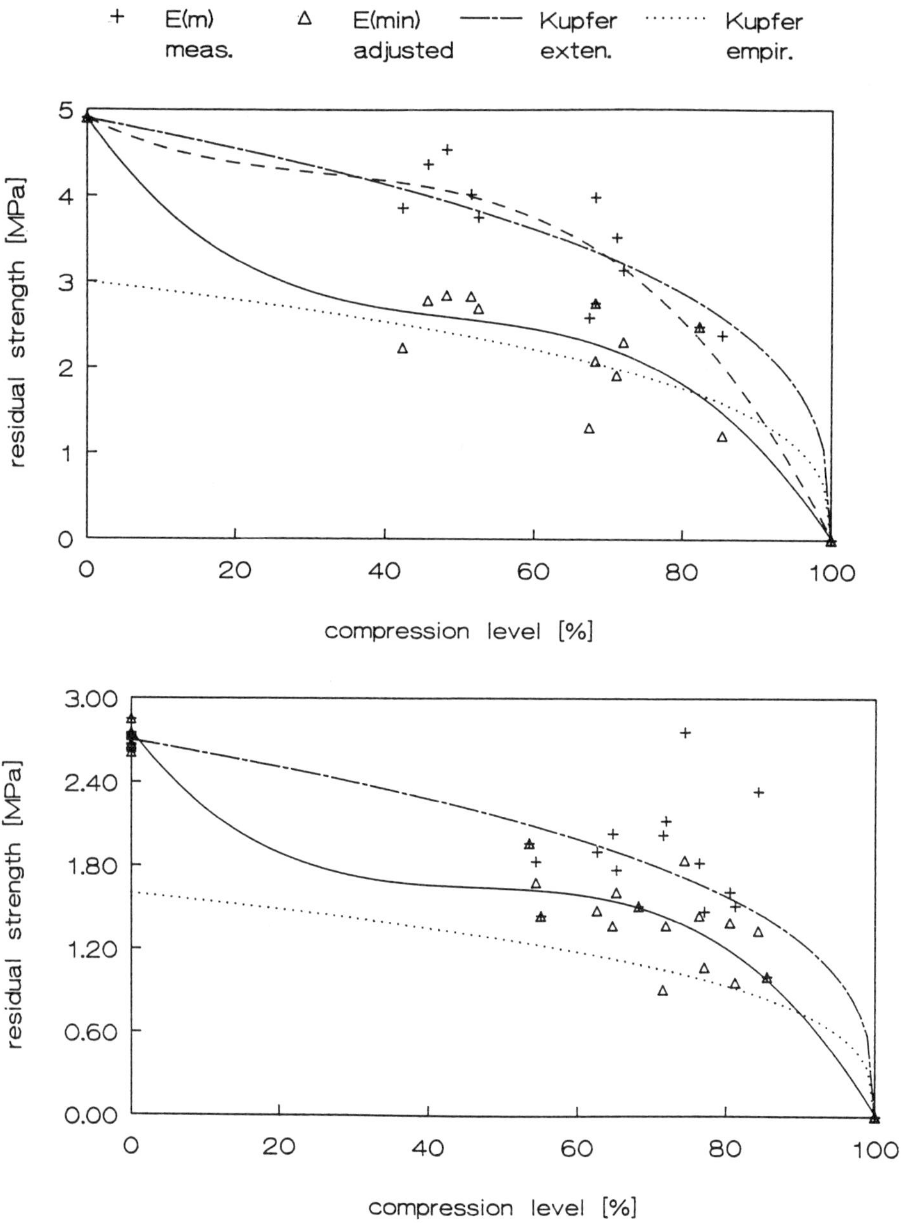

a) Mix A, high quality
b) Mix B, low quality
Fig.3. Reference failure envelopes and test results.

It is concluded that the rate effect is not unique under combined compression and tension for loading rates in the moderate region, i.e. up to 10 [GPa/s]. For higher rates, the steep strength increase, observed in uniaxial tests, occurs due to the dominant inertia effects on crack extension (Weerheijm, 1990). Therefore, it is expected that for high loading rates, the rate effect on residual strength does not depend on the compression level.

4.2 The fracture energy

The total fracture energy Gf could be derived accurately from the uniaxial impact tests (Weerheijm,1991). For the high and low quality concrete the Gf values were 100 and 72 [N/m] respectively. The dissipated energies at ultimate strength were 10 and 9 [N/m]. The latter numbers are less accurate because of the structural response of the specimen which obscured the true load-deformation relation of the material.

The other reference, the Gf values under static loading, is unknown. A few test results are mentioned by Van Mier (1984), but these are insufficient to draw conclusions for the influence of increasing compression level.

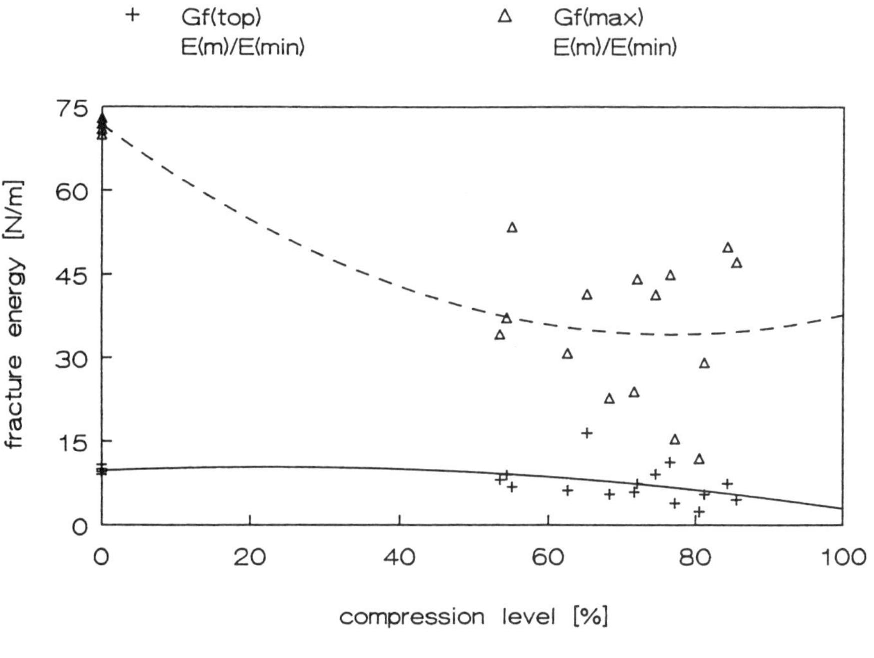

Fig.4. The Gf values at ultimate load and unloading with polynomial fits.

In the biaxial tests, the influence of the lateral device increases with increasing axial deformation. The load transfer through the brushes to the upper bar becomes more important but also more unpredictable. The specimen response in a specific test could be reconstructed using the recorded deformations. An average response or failure process, however, could not be deduced because of the unpredictable inter-action between specimen, lateral and axial loading device. The best results were obtained for the B-quality. The Gf values at ultimate strength (Gf(top)) and at the moment of unloading are given in Figure 4. The almost constant fracture energy at ultimate load could indicate a contribution of multiple crack initiation at higher compression levels. Because of the scatter and the mentioned uncertainties, no definite conclusion can be drawn.

4.3 The deformation capacity.

The material outside the failure zone was loaded at about 75% of the ultimate strength. The stress-strain relations showed that the material response was almost elastic. This observation might be partly incorrect because of the fact that, during the descending branch, a part of the observed load was transmitted through the brushes.

The load-deformation curves for several compression levels are depicted in Figure 5. Note that the reliability of these curves decreases with increasing axial deformation.

The general feature is an increasing deformation with decreasing residual strength, while the softening branch was definitely influenced by the lateral device. The latter resulted in too high stress levels in the softening branch.

The results of the low-quality concrete are the most reliable, as stated in the former section. In this paper, the curves will be considered up to ultimate strength.

From all the curves it emerges that before ultimate strength, non-linear deformation occurred. With increasing compression level the deformation increased almost proportionally to the decrease in residual strength. The magnitude of the observed deformation is of the same order as under static loading. This seems to contradict the general feature of dynamic test results which shows an increase in strength and deformation. The explanation for this apparent discrepancy is probably the fact that in the current programme, just a single failure zone is considered due to the applied notched specimen. In most former programmes on dynamic response, multiple fracture zones could develop outside the final failure zone, which resulted in the observation of a larger deformation capacity. To enable the characterization of dynamic material properties uniquely, it is very important to define the size of the considered fracture zone.

The failure envelopes of the high and low quality concrete indicated that for the low quality, a part of the applied energy might be dissipated in multiple crack initiation. Comparison of the deformation increases for both qualities give no answer because of the unreliable results for the A-quality. From the proportional increase of the deformation, it is concluded that the failure mechanism in the impact tests is not affected by the level of compression.

Besides the effect of multiple crack initiation, it is possible that for the low quality concrete, the inertia effects on crack extension are already more important than for the high quality concrete. The applied loading rate is beyond the theoretical threshold for the low quality concrete where the inertia effects become dominant. This might explain the observed difference in rate effect.

At this stage of the research programme it is not yet possible to explain the differences between the data on the high and low quality concrete. Further research is necessary on the failure mechanisms and the possibilities to eliminate the influence of the lateral compression device on the data.

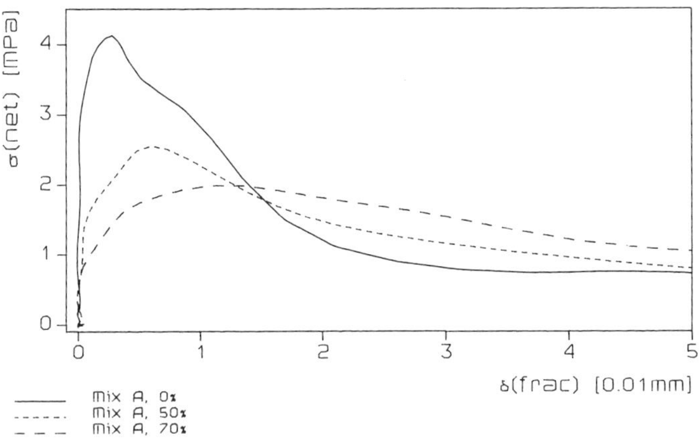

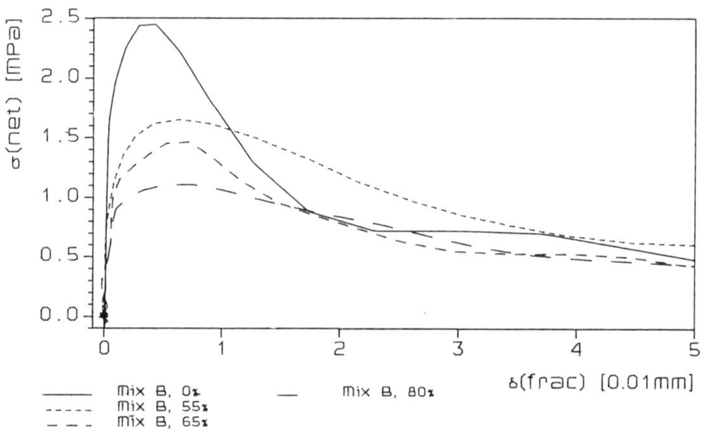

a) A-mix, high quality concrete
b) B-mix, low quality concrete
Fig.5. The load-deformation curves for the low and high quality concrete.

6 Concluding remarks

The current tests with the modified SHB apparatus emphasized the difficulty of testing concrete under the combined loading of tension and compression. Improvements and additional measurements are necessary to limit or measure the interaction between the axial and lateral loading devices. However, from the current test results the following conclusions can be drawn.

With the modified set-up and the notched specimen, data on the residual strength at various levels of lateral compression can be obtained.

The structural response of specimen and loading device obscured the true material response. In the uniaxial impact tests the structural response could be quantified. For the biaxial tests, the strength data could be adjusted whereas the relation between load and deformation are less reliable due to the unpredictable interaction between loading devices and specimen when the axial deformation increases.

Because of the notched specimen, failure was limited to a single zone which resulted in a smaller deformation capacity at high loading rates than observed in most impact tests.

The main goal of the experimental programme was achieved. The failure envelope for the residual impact strength was derived. The rate effect on the envelope appears to be non-unique. For the tested high quality concrete, the rate effect had vanished at compression levels beyond 50% of the compressive strength. The rate effect for the low quality concrete reduced from 1.7 at uniaxial loading to about 1.3 beyond the level of 50% compression.

Based on the failure mechanism under uniaxial tensile loading, it is expected that at high loading rates, beyond 15 [GPa/s] where inertia effects become dominant, the rate effect on the failure envelope will be unique.

7 References

Kupfer, H.B. and Gerstle, K.H. (1973) Behaviour of Concrete under Biaxial Stresses, Journal of the Engineering Mechanics Division, ASCE, Vol. 99, No. EM4, August 1973, pp. 583-866.

Mier, J.G.M. van (1984) Strain-Softening of Concrete under Multiaxial Loading Conditions. Wibro, Helmond, 1984 (thesis).

Mlakar, P.F. et al. (1985) Dynamic Tensile-Compressive Behavior of Concrete, ACI Journal July-August, 1985, pp. 484-491.

Nelissen, N.L.M. (1972) Biaxial Testing of normal Concrete, HERON Vol.18, 1972, No.1.

Suaris, W. and Shah, S.P. (1985) Constitutive Model for Dynamic Loading of Concrete. Journal of Structural Engineering, Vol. 111, No.3, March, 1985, pp. 563-576.

Takeda, J. et al. (1974) Mechanical Behaviour of Concrete under Higher Rates of Loading than in Static Tests. Proc. Symposium: Mechanical Behaviour of Materials, Kyoto 1974, Vol.II, pp. 479-486.

Weerheijm, J. (1989) Properties of Concrete under Dynamic Loading 2: Biaxial Loading Device. PML report PML1989-59, May 1989.

Weerheijm, J. (1990) Properties of Concrete under Dynamic Loading 3: Fracture Model for Brittle Materials. PML report PML1990-65, October 1990.

Weerheijm, J. (1991) Properties of Concrete under Dynamic Loading 4: Uniaxial Impact Tests, to be published.

Weerheijm, J. and Reinhardt, H.W. (1989) Modelling of Concrete Fracture under Dynamic Tensile Loading. Proc. Symp. Recent Developments in the Fracture of Concrete and Rock. Cardiff, Wales, September 1989, pp. 721-729.

Zielinski, A.J. (1982) Fracture of Concrete under uniaxial Impact Tensile Loading. Delft University Press, Delft, 1982 (thesis).

Zielinski, A.J. (1985) Concrete under Biaxial Loading: static compression-impact tension. Stevin report 5-85-1, Delft, 1985.

76 FAVORABLE AND UNFAVORABLE RATE EFFECTS ON FRACTURE PROCESS OF CONCRETE

J. TAKEDA, H. KOMOTO and T. TANIKAWA
Kumamoto Institute of Technology, Japan

Abstract
Considering the practical application of fracture mechanics to the general reinforced concrete structures, the development of dynamic fracture mechanics on concrete is in urgent need, in which several rate effects, namely the changes of the mechanical properties, fracture toughness or propagation speed of cracks in concrete etc., by the loading rate are brought about. In the rate effects on concrete, some unfavorable effects for dynamic design of structures are contained. The prediction of brittle fracture of concrete are also studied.
Keywords; Rate Effects, Dynamic Fracture Mechanics, High Rate Loading Test, Impact Test, Explosion, Brittle Fracture of Concrete.

1 Introduction

It is really impressive that the fracture mechanics on concrete is now assidiously being studied by world-wide investigators. And it is also proper, these many studies are based on the statical application of loads, for making its fundamental developments.

There are, however, many accidents and disasters which daily fall on cities and structures (Fig. 1), and impose the investigators of concrete engineering to develope the protection techniques for these disasters, in which the dynamic fracture mechanics of concrete might be needed.

A evident distinction between static loadings and dynamic loadings

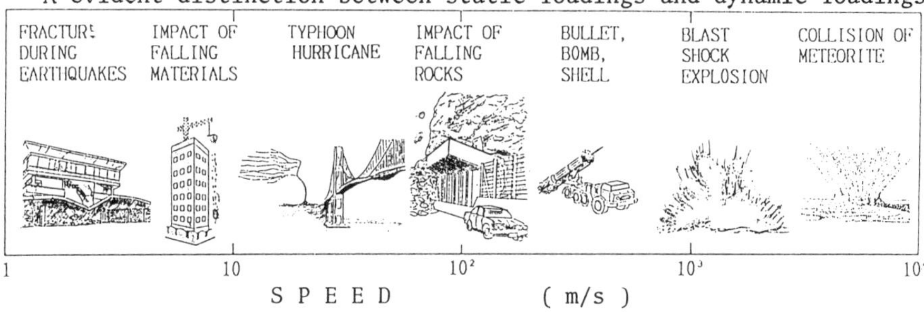

Fig. 1. Various accidents and disasters and approximate speed.

Fracture Processes in Concrete, Rock and Ceramics Edited by J.G.M. van Mier, J.G. Rots and A. Bakker.
 Published by E. & F.N. Spon, 2–6 Boundary Row, London SE1 8HN. ISBN 0 419 15870 7

is that in the former, the increasing rate of applied load on specimens (in the case of the tests) is quite small so that the tests are an isothermal changes in thermal equilibrium state, though the latter is inversely just an adiabatic process in unequilibrium state and is infallibly accompanied with stress waves generated during the tests. In higher loading rate than that in static loading tests, the responses of the specimens are governed by the new constitutive equations which changed by the loading rate namely rate effects.

The above rate effects occur also in the tests in fracture mechanics on concrete and make the changes not only on fracture toughness and various constants in fracture mechanics, but also on the propagation speed of cracks generated in the specimen during loading stage.

This paper deals with the summary of the rate effects on concrete based on the experiments carried out by the authors by this time.

2 Rate effects on concrete

A significant distinction between static loading test and dynamic loading test is that in the former, the increasing rate of applied load on specimens is entirely small so that the tests are as a matter of fact isothermal changes in thermal equilibrium state and all the process is expressed by the difference of Helmholtz free energy during the test, though the latter is an adiabatic process in unequilibrium state inversely.

Even in the case of the test carried out in quasiequilibrium state, however, if a movement of material elements occurs in the specimens and it has velocity v, then the stress (zv) should be yielded in that portion and propagate immediately to the other part as the stress wave, in where z is the mechanical impedance of the specimen. Therefore it is clear that the loading rate even in the test done in quasiequilibrium state has to be limited within the allowable range and also the use of load-cell for the measurement of applied load must be controlled within some range of the loading rate.

Addition to these circumstances, under actual intense dynamic loadings, for example, severe earthquakes, disastrous explosion or impact of falling rocks, etc., prominent vibrations and stress waves might be also added. Moreover, the responses of the specimens, tested with higher rate of loading than in static loadings, are governed by new mechanical properties and constitutive equations which are changed by the loading rate concern, namely affected by rate effects.

Accordingly, the true nature of the rate effects are quite thermodynamic effects and they infallibly occur in all high rate loading tests and impact tests, namely dynamic loading tests.

Controlling load application, however, so as not to make any disturbance and so as not to yield any significant stress waves, a kind of tests can be carried out in quasiequilibrium state, which are usually called high rate loading tests. It is important to know that some rate effects on concrete can be seen even in high rate loading tests on concrete or reinforced concrete structures, though influential rate effects not yet appear in these tests.

The reason of the occurrence of the rate effects which occur on

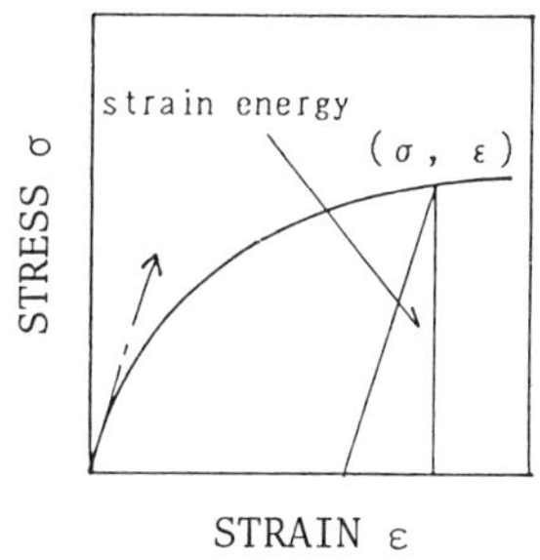

Fig. 2. Stress-strain curve and strain energy.

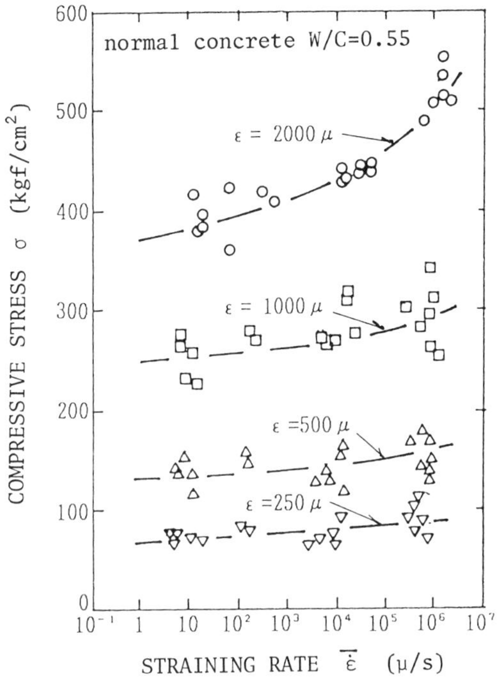

Fig. 3. Relations between straining rate $\bar{\dot{\varepsilon}}$ and compressive stress σ at the certain values of strain.

materials in high rate loading tests (quasiequilibrium state) has been explained by A. M. Freudenthal by means of Statistical Physics (Freudenthal, 1950). He derived equations describing inelastic deformation in terms of atomic structure which are based on the assumption that the energy distribution over atoms can be expressed by the Boltzman-Maxwell distribution function and the inelastic deformation of materials may be brought about by thermodynamical changes of equilibrium position within atomic structure, which will require definite activation energy in order to be able to across the barrier of potential energy. If an external force is applied in one direction, the energy level required for activation in the direction of the applied force will be reduced to $(q-\varepsilon_s)$ from the initial value q (where ε_s denotes the potential energy of the applied force), whereas in the opposite direction, it will be increased to $(q+\varepsilon_s)$ with respect to the possible movement. Hence, the rate of deformation in the direction of the applied force u is given by the difference between the rate of activation in the direction of the applied force u_1 and in opposite direction u_2,

$$u=u_1-u_2=C_1\frac{kT}{h}\exp(-(q-\varepsilon_s)/nRT)-C_1\frac{kT}{h}\exp(-(q+\varepsilon_s)/nRT$$

$$=C_1\frac{kT}{h}\exp(-q/nRT)\sinh(\varepsilon_s/nRT) \quad (1)$$

where C_1 is a constant, k is Boltzman's constant, h is Planck's constant, T is absolute temperature, R is gas constant and n denotes a number of moles.

From this equqtion, Freudenthal has given explicit explanations on viscosity, relaxation or the relation between straining rate and temperature, etc.

Now, examin the above theory of Freudenthal on to the experimental data of high rate compressive test on concrete, carried out by the authors (Takeda, 1971). It is assumed that the potential energy of applied load ε_s can be expressed by the energy induced by elastic strain and expressed by the next equation, at the arbitrary poit (σ,ε) on the stress-strain curve in Fig. 2,

$$\varepsilon_s = \frac{1}{2} C_i V \sigma^2 \tag{2}$$

where C_i is the initial compliance of the σ-ε curve and V denotes the volume of the specimen. Hence, from the equations (1) and (2), replacing u by straining rate $\bar{\dot{\varepsilon}}$,

$$\bar{\dot{\varepsilon}} = C_1 \frac{kT}{h} \exp(-q/nRT) \sinh(C_i V \sigma^2/2nRT) = C\sigma^2, \quad C = C_1 C_i V/2nhN \tag{3}$$

where N is Avogadro number and C is a constant at a certain value of strain ε. In this place, the function sinh is approximated, because ε_s is sufficiently small in comparison with thermal energy RT.

Fig. 3 shows the relation between the straining rate $\bar{\dot{\varepsilon}}$ and the stress σ at the various values of strain ε, obtained from the test results. If the above theory of Freudenthal is valid in this test, the relation between $\bar{\dot{\varepsilon}}$ and σ at a certain value of strain ε must be expressed as a straight line in this figure by equation (3), though such tendency can be seen only within small range of both ε and $\bar{\dot{\varepsilon}}$. The data points in Fig. 3 have considerably large deviation, nevertheless, the exceedingly plain relation between the mean stressing rate $\bar{\dot{\sigma}}$ and the mean straining rate $\bar{\dot{\varepsilon}}$, averaged in the range from the beginning to the maximum stress point was obtained in the high rate compressive test of concrete and shown in Fig. 4.

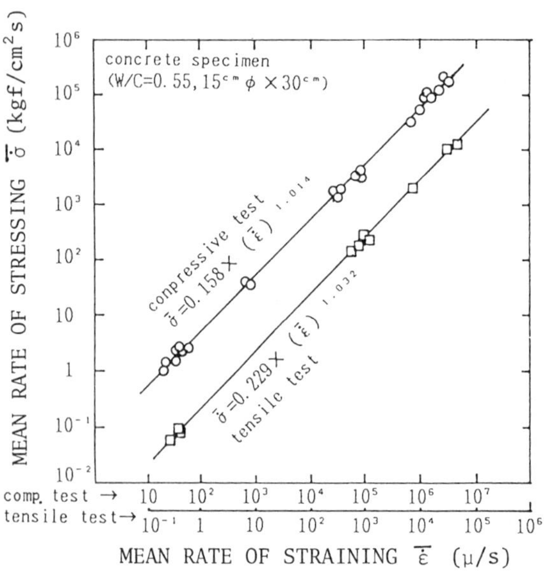

Fig. 4. Relations between mean rate of stressing $\bar{\dot{\sigma}}$ and mean rate of straining $\bar{\dot{\varepsilon}}$.

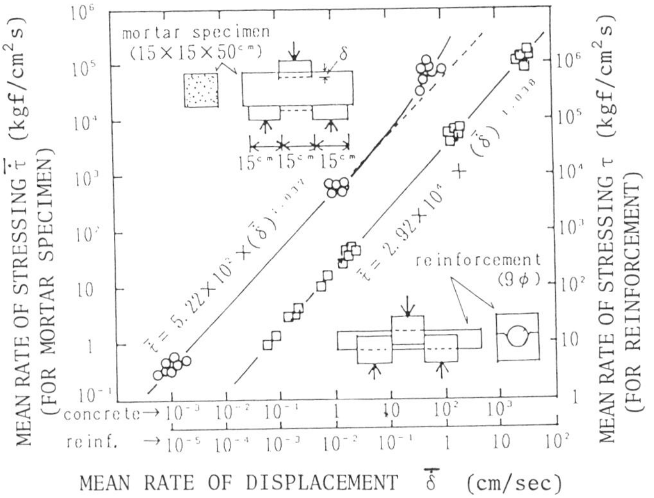

Fig. 5. Relations between mean rate of stressing $\bar{\dot{\tau}}$ and mean rate of displacement $\bar{\dot{\delta}}$ in shear tests.

From the relation,

$$\bar{\dot{\sigma}} = \alpha\,(\bar{\dot{\varepsilon}})^{\beta} \qquad (4)$$

is obtained, where α and β are both the constants which are influenced by the kinds of the tests and specimens, and β is usually a bit larger than 1.00. The similar relations to Fig. 4 have been found in all kinds of high rate loading tests not only on many kinds of concrete with various mix proportions and with various types of loading (Fig.5), but also reinforced concrete structural members (Takeda,1971), wood (Takeda, 1962), etc. Therefore it is considered that these relations must be fundamental relations over whole range of high rate loading tests.

3 Results obtained in preceeding studies

Summarizing the test results on concrete and reinforced concrete structural members done with various loading types, the rate effects on concrete are divided into two effects, the first is on the relation between stress and strain in loading process before fracture begins and the second is on fracture criterion, expressed by the equation (4) and the fracture process thereafter. Concerning the first effect, the similar equation to (4) can be also obtained (Takeda, 1971). Supporse now, concrete specimens are being tested with high rate compressive loading, and the stress- strain curve like as Fig. 6 was obtained in the test. From the equation (4),

$$\bar{\dot{\sigma}} / \bar{\dot{\varepsilon}} = \sigma_m/\varepsilon_m = \alpha\,(\bar{\dot{\varepsilon}})^{\beta-1} \qquad (5)$$

where σ_m, ε_m are the stress and strain respectively at the maximum stress point B in the figure. This equation gives the inclination of

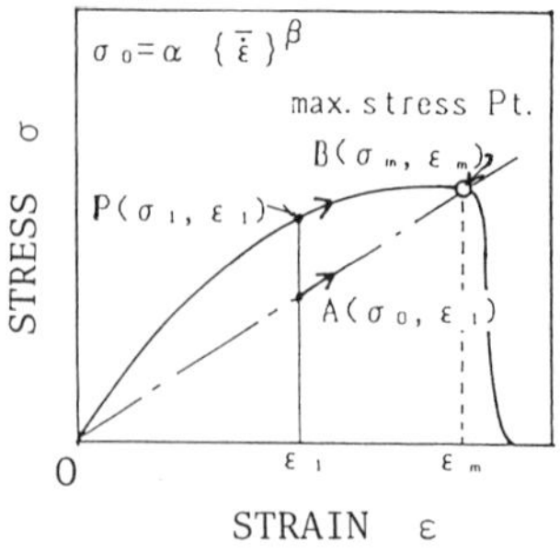

Fig. 6. Maximum stress point on stress-strain curve.

the straight line OB, which is determined by the mean straining rate $\bar{\dot{\varepsilon}}$ concern, in where point O is the origin. Suppose, further, that the stress and strain at a moment during the test are expressed by the point P (σ_1 ,ε_1). Taking point A to just under the point P and on the straight line OB, the value of the stress at the point A can be obtained by the equation (6).

$$\sigma_0 = \alpha\ (\bar{\dot{\varepsilon}})^{\beta-1}\ \varepsilon_1 \qquad (6)$$

It is known that both the points P and A will move forward respectively in the loading process of the test, and at the instant the point P and A coincide each other, the fracture of specimen will begin. Moreover, it is really impressve that not the stress σ_m or the strain ε_m alone, rather only the ratio (σ_m/ε_m), which is obtained from the equation (6), can make the start of the fracture of the specimen.

The value of (β-1) in the equation (5) and (6) is usually positive, as mentioned above. Therefore, it is clear that the inclination of the straight line OB in Fig. 6 increases gradually with the increment of the mean straining rate $\bar{\dot{\varepsilon}}$ (Fig. 7). There are two cases, however, as seen in (a) and (b) in the figure, reflecting both the two effects, the first and second rate effects stated above, which work independently respectively. In the case of the figure (a), the strain (or displacement) at the maximum stress (or load) increases with the increment of the straining rate (or displacement rate), and rather favorable rate effect for dynamic design of reinforced concrete structures. In this case, comparatively small value of (β-1), generally

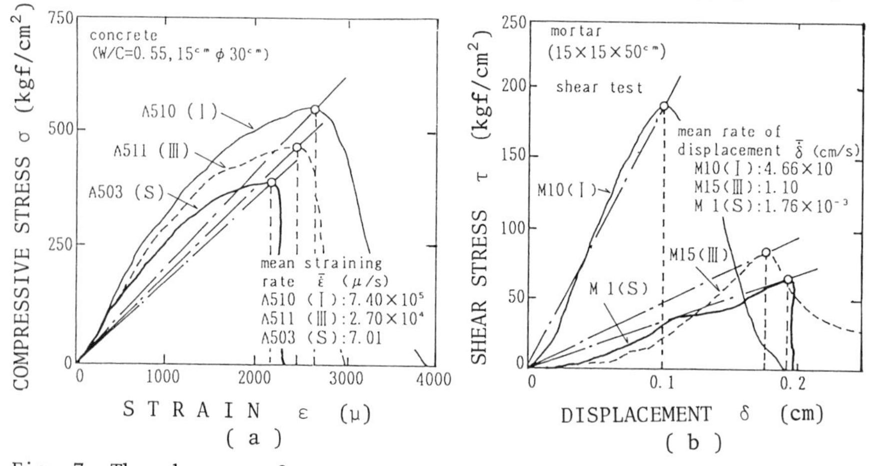

Fig. 7. The changes of stress-strain and stress - displacement curves with the increment of mean rate of straining or mean rate of displacement.

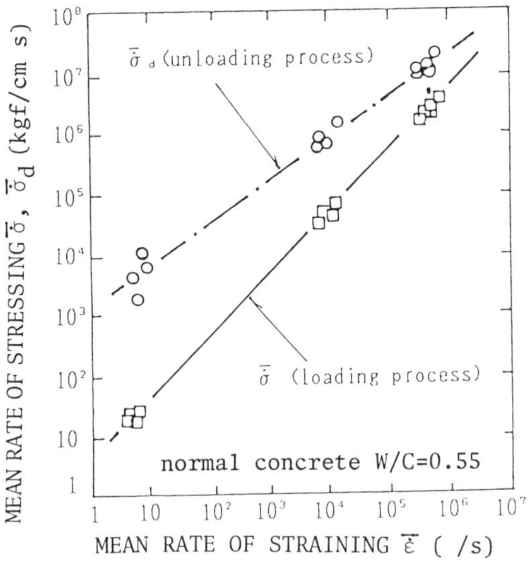

Fig. 8. Comparison of the relations between mean stressing rate and mean straining rate, and mean stress decreasing rate and straining rate in unloading process.

less than 0.02 is obtained in the tests (Takeda, 1971). This was found in the high rate compressive, tensile and bending tests on concrete and reinforced concrete structural members.

On the contrary, there is the case in which the strain (or displacement) at the maximum stress (or load) decreases with the increment of mean straining rate (or displacement rate) as seen in the figure (b), and in the highest rate of loading, exceedingly brittle fracture is brought about on the specimens. This is very dangerous and unfavorable rate effect for dynamic design of structures. It was seen in shear tests on concrete, mortar (Fig. 7 (b)), reinforced concrete beam and even in the case of steel reinforement only or bolt (Takeda, 1969, 1971).

The stress decreasing rate in unloading process after fracture of specimen started in the test, is also influenced by the mean straining rate $\bar{\dot{\varepsilon}}$ in loading process, shown in Fig. 8, and this also one of the second rate effect, though it is not favorable for dynamic design of structures.

4 Rate effects in fracture mechanics on concrete

The second rate effect, above mentioned, on concrete is concerning the start of practical fracture of specimens, expressed by the equation (6) and (7), which is just a start of unstable propagation of cracks generated newly or grown from micro-flaw in heretofore loading stage. Hence, it is considered that the similar equations to (4) and (5) may be found also on the fracture toughness or critical energy release rate in fracture mechanics.

Then, based on the data of Mode I test carried out with various loading rates by the authors (Takeda, 1982, 1985), the relation between the mean loading rate $\dot{P}$ and the mean straining rate near the crack tip $\bar{\dot{\varepsilon}}_a$ in the loading stage was examined and shown in Fig. 9. The mean straining rate $\bar{\dot{\varepsilon}}_a$ is calculated from the measurement of " strain gage G_a" which is stuck perpendicular to the initial crack line at the distance of 8mm from the crack tip. In the figure, the similar tendencies to Fig. 4, 5 are also seen, except only a part in the high-

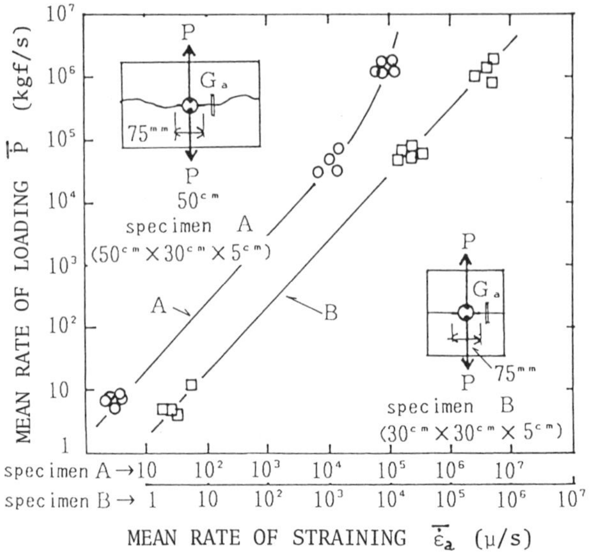

Fig. 9. Relations between mean loading rate and mean straining rate in Mode I test.

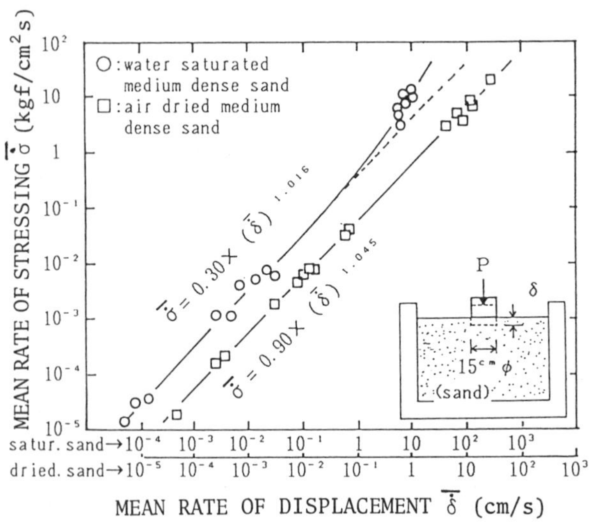

Fig. 10. Relations between mean stressing rate and mean displacement rate in compressive test of sand layer

est loading rate of "specimen A", just in which curved propagation of cracks in the test, shown in the figure, was found (Takeda, 1982). It is already proved theoretically by Yoffe that at higher velocity the running crack in a media tends to become curved. Therefore, it is inferred that the above tendencies recognised in the test done with the highest loading rate on the "specimen A" were brought about by the large acceleration which is yielded in near part to the cracks rapidly propagating in the specimen, by the load application in this test. The above inference, concerning especially the bend of the line of log $\dot{P}$ - log $\overline{\dot{\varepsilon}}_a$ shown in Fig. 9, is supported also by the compressive

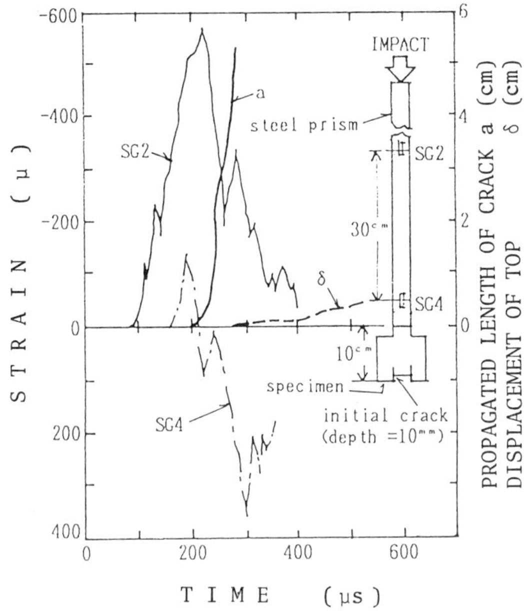

Fig. 11. Example of measurement in impact test of Mode II mortar specimen.

test of sand layer (Fig. 10), in which the influences of acceleration of sand particles may intrinsically be intensified and severe liquefaction of sand layer occurs in high rate loadings, reflecting that the sand layer consists of independent solid particles. It is widely known that the liquefactions of sand layer have ever been frequently brought about the severe subsoil disaster during earthquakes (Takeda, 1969).

It is considered that the relations deduced from the result of Fig. 9, which are similar to that obtained from the equations (4) and (5), may give the threshold value of applied load namely fracture toughness influenced by the loading rate concern, at which the initial crack begins unstational propagation. Therefore, it is clearly known that the fracture toughness in fracture mechanics on concrete also changes with the loading rate or the straining rate concern, and being governed by the similar equations (4) and (5), and eventually violent brittle fracture of concrete may occur by high speed propagatio of cracks. It has been already reported that the stress intensity factor, etc. also change with the loading rate before fracture begins, and the propagation speed of cracks in concrete under high rate loading is larger than that in static loading and it is nearly 1000 m/s in the test of highest loading rate (Takeda, 1988, 1988, 1989).

The change of the propagation of cracks is also one of the rate effects in the fracture mechanics on concrete and in Mode II test, the surface of propagated cracks in impact test is completely different from that in static loading test, as reported before (Takeda, 1988). Fig. 11 shows the example of measurements in impact test of Mode II mortar specimens and the surprisingly high speed of crack propagation (a in the figure) can be seen (Takeda, 1988).

5 Conclusion

(1) For the dynamic fracture mechanics on concrete, the study on the rate effects holds the vital key to the developement.
(2) The rate effects are one of the thermodynamic effects, though the explanation based on the Statistical Physics in thermal equilibrium state is yet insufficient to the whole range of the rate effects caused in various dynamic loadings (high rate loading and impact loading).
(3) The rate effects, obtained in many kinds of dynamic tests on concrete and reinforced concrete structural members, is divided into two effects, the first is on the response before the fracture begins and the second is on the start of fracture and thereafter. The second rate effect eventually works as the unfavorable effect for the dynamic design of structures.
(4) The above unfavorable rate effect can be predicted by the relation between the mean stressing rate and mean straining rate. This is the same as in the test of fracture mechanics.
(5) Not only the stress intensity factor. etc., but also the speed of crack propagation are affected significantly by the loading rate in various dynamic loads.
(6) The propagation of cracks reaches nearly 1000 m/s in high rate loading test and impact test, which is one of the second rate effect.

6 References

Freudenthal, A.M. (1950) The inelastic behavior of engineering materials and structures, John Wiley and Sons, 109-140.
Takeda, J. et al. (1962) High rate loading test on wood, Kanto-district, AIJ (in Japanese).
Takeda, J. et al. (1969) Effect of displacement rate of holizontal soil reaction in sand, Soils and Foundation, 11, 15-34.
Takeda, J. et al. (1971) Deformation and fracture of concrete subjected to dynamic load, ICM1, 4, 267-277.
Takeda, J. et al. (1982) Basic concept on the responses of structural members and structures under impact or impulsive loading, and other papers, RILEM-CEB-IABSE-IASS Int. Sym. Berlin,S1:13-18,S2:83-91,S3: 185-187.S4:293.
Takeda, J. et al. (1985) Strain rate effects on concrete and reinforcements and their contributions. MRS Sym. Boston, 64, 15-20.
Takeda, J. et al. (1988) Rate effects on crack extension in Mode II specimens of mortar, Int. Workshop, Sendai, 386- 392.
Takeda, J. et al. (1989) High speed propagation of cracks in concrete or reinforced concrete structural members under impact load, JSMS Sym. on impact problems, 69-72 (in Japanese).

77 THE INFLUENCE OF CONCRETE HUMIDITY ON ITS DESTRUCTION PROCESS

J. HOŁA
The Institute of Building Engineering, Technical University of Wrocław, Poland

Abstract
The present paper submits an experimental determination of the influence concrete humidity exerts upon the process of destruction in concrete. The accomplished studies concerned normal concrete varying in humidity. The concrete was charged by immediate loads, being compressed quasi-axially. Acoustical methods, particularly those of acoustic emission, gave us rise to stating the humidity of concrete substantially influenced the destruction process course. Above all, this is proved by diversified values of the critical stresses, σ_I and σ_{II}, distinguishing between the particular stages of the destruction process, as well as by the differences in the totting sum of the acoustic emission recorded while the material was being destroyed.
Keywords: Normal concrete, Humidity, Acoustical methods, Destruction course, Critical stresses.

1 Introduction

Since constructions made of concrete work, depending on their assignment, in various humidity conditions, it is essential to determine the influence of the material's humidity on the course of its destruction caused by a load. However, exhaustive empirical data in this scope has not yet been obtained. It is also important to determine the influence of the humidity concrete contains on acoustically-measured parameters, especially those obtained owing to the acoustic emission method, which serve as the basis for determining the scale or the course of load-evoked destruction in this material's structure.

In order to explain the above-cited issues, what we carried out were the investigations for various humidity concretes subjected to immediate loads when compressed quasi-axially. The wide range of the concrete's humidity was achieved due to its appropriate storage, which rendered it possible for the material to remain in a dry state, the state of maximal sorption humidity and the state of full saturation. The ultrasound and the acoustic

Fracture Processes in Concrete, Rock and Ceramics Edited by J.G.M. van Mier, J.G. Rots and A. Bakker.
 Published by E. & F.N. Spon, 2–6 Boundary Row, London SE1 8HN. ISBN 0 419 15870 7

emission methods were made use of as the investigation techniques.

2 Description of the tests

The tests referred to normal concrete of the B20-class range compression strength, widely applied to concrete structures. The tests were carried out after a 90-day curing of cube samples 10 x 10 x 10 cm and 15 x 15 x 15 cm in size, made of a concrete mix, the formula of which was, in terms of 1 m^3, as follows:

portland cement '35' from Groszowice Cement Plant	- 320 kg,
natural gravel 'Proszowice'	- 1188 kg,
river sand 'Wrocław'	- 716 kg,
municipal water	- 177 l.

Depending on the way the samples were stored, various humidities of the concrete were obtained at the test time, i.e.:

concrete humidity W = 0.00%	- the samples had been stored for 86 days in a climatic chamber at a temperature of +18°C (±1°C) and relative air humidity of ca. 95%, then dried at a temperature of 105°C to reach constant weight (dry state).
concrete humidity W = 2.28%	- the samples had been stored for 28 days in a climatic chamber at a temperature of +18°C (±1°C) and relative air humidity of ca. 95%, then, till the 90th day, in the laboratory at a temperature of +18 +18°C (±3°C) and relative air humidity of ca. 65% (sorption humidity state).
concrete humidity W = 2.56%	- the samples had been stored for 90 days in a climatic chamber at a temperature of +18°C (±1°C) and relative air humidity of ca. 95% (maximal sorption humidity).
concrete humidity W = 5.84%	- the samples had been stored for 86 days in a climatic chamber at a temperature of 18°C (±1°C) and relative air humidity of ca. 95%, then water-saturated to reach the full saturation state.

Depending on concrete humidity, the samples were divided into 4 groups named series. The following series notations were accepted:

the 'As'-series, W = 0.00%,
the 'Ap'-series, W = 2.28%,
the 'Ak'-series, W = 2.56%,
the 'Aw'-series, W = 5.84%.

Each of the series included 24 10 x 10 x 10 cm samples and 12 15 x 15 x 15 cm samples. The accomplished tests employed the ultrasound and the acoustic emission methods while the samples were being subjected to quasi-axial compression. The axial compression of the samples was done having withdrawn the friction at the contact spots of the samples' surfaces and the tester's pressure plates. To achieve this, the surfaces had been grinded and then lubricated with cup grease.

The stand to measure the acoustic emission included the elements shown in Fig. 1. The ultrasound tests employed a Polish-made tester, the Unipon 541-type, with 500 kHz-frequency heads. While testing the 10 x 10 x 10 cm samples, the totting sum of the acoustic emission, RMS voltage of the acoustic emission, as well as the transition time of the longitudinal ultrasound waves along the direction perpendicular to that of the force action were recorded as functions of stress increment. The tests were carried on at the air temperature of 20°C (±2°C) and relative air humidity of 55% (±5%).

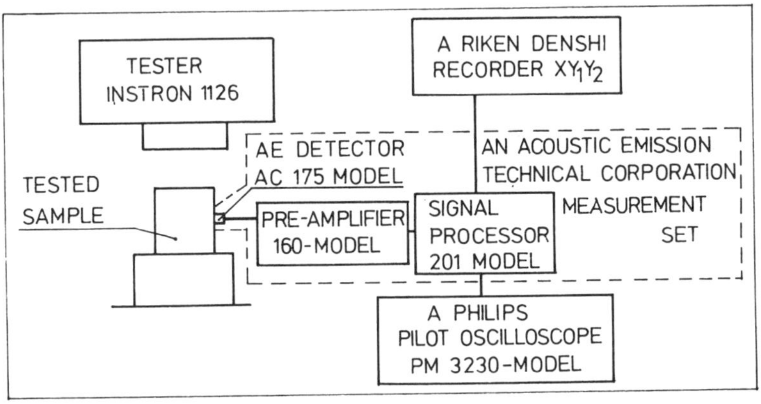

Fig. 1. A diagram of the test stand for acoustic emission measurements

3 Test results and their analysis

On the grounds of the classical strength tests of the concrete in question, having tested the 15 x 15 x 15 cm samples, determined was mean compression strength after a 90-day curing. The obtained results are displayed in Table 1.

Table 1. Averaged compression strength of concrete, determined in 15 x 15 x 15 cm samples after a 90-day curing.

Concrete series denotation	Humidity condition of concrete	Concrete humidity W (%)	Mean compression strength R (MPa)	Softening coefficient during compression
As	dry	0.00	31.50	1.00
Ap	humid	2.28	27.00	0.86
Ak	humid	2.56	27.30	0.87
Aw	saturated	5.84	23.49	0.75

According to the test results obtained with the help of the ultrasound and the acoustic emission methods, the following relationships were determined:

- a velocity change of the longitudinal ultrasound waves for various humidity concretes, the concretes compressed quasi-axially, as a function of stress increment (Fig. 2),
- variability course of the acoustic emission totting sum for varied humidity concrete compressed quasi-axially, as a function of stress increment (Fig. 3).

The obtained test results imply that, for each of the series of the concrete in question, the velocity of the longitudinal ultrasound waves decreases as the stress level increases, whereas in the case of the particular series, the stress values at which a distinct decrease of the velocity is marked tend to vary. As also implied by the ultrasound tests, the measurement of the longitudinal ultrasound wave velocity is only possible up to a certain stress level, different for the particular series of the concrete tested and lower than destructive stresses. The results obtained with the help of the acoustic emission method clearly indicated the totting sum of the acoustic emission unquestionably depended upon the humidity condition of the concrete. This dependence points out that both the totting sum of the acoustic emission yielded at the particular stages of the destruction and the overall totting sum of the acoustic emission is the smaller the more humid is the concrete, which is exemplified in Figs 3 and 4. We are justified to think concrete humidity appears to have an influence on the value of the acoustic emission

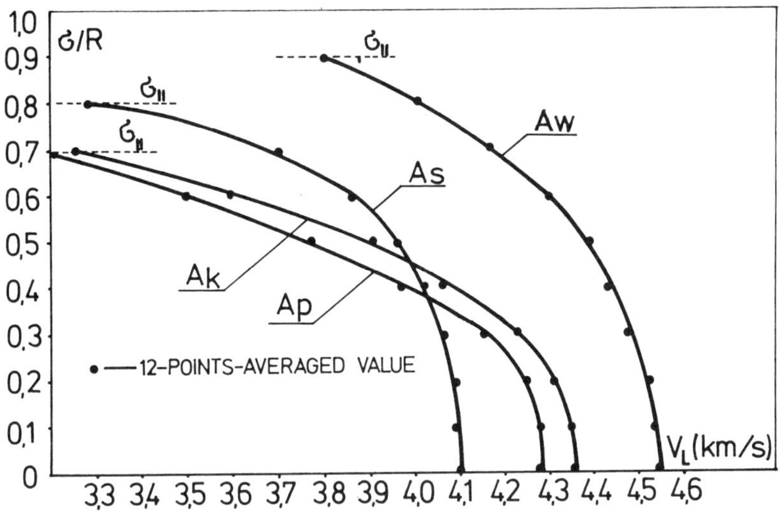

Fig. 2. Velocity change of longitudnial ultrasound waves for the 'As', 'Ap', 'Ak' and 'Aw"-series concrete compressed quasi-axially, as a function of stress increment.

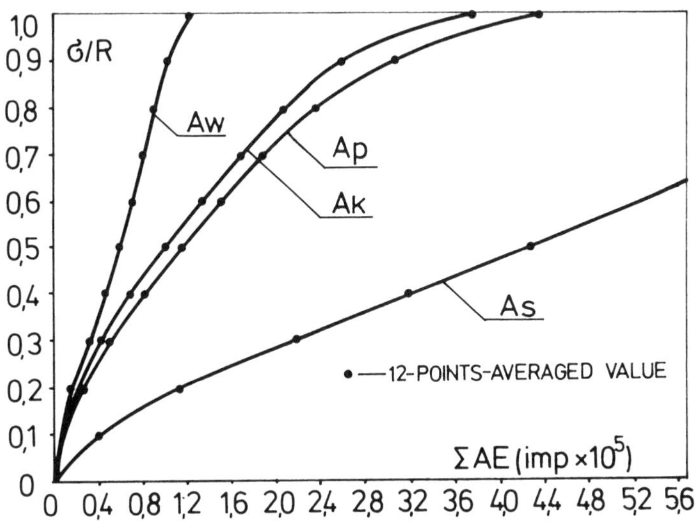

Fig. 3. Variability course of the acoustic emission totting sum, for the 'As', 'Ap', 'Ak' and 'Aw'-series concrete compressed quasi-axially, as a function of stress increment.

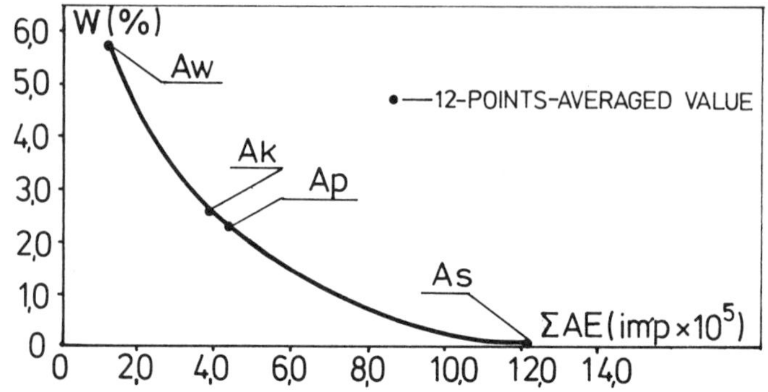

Fig. 4. Relationship between humidity of concrete and the overall totting sum of the acoustic emission yielded during the quasi-axial compression.

totting sum, the emission yielded during the destruction of concrete. The influence is thought to be associated with, above all, a weakening of bonds between the structural lattice crystals. This weakening follows the fact that, as the material's humidity is increasing, the bonds get partially dissolved, resulting in a gradual decrease of the concrete's compression strength. What we also observed while testing the concrete series was a decrease of, characterized by the values of the softening coefficient, drawn up in Table 1. It is worth mentioning the softening coefficient is defined as the ration of compression strength of a concrete fully water-saturated to its dry state strength. The softening of the concrete structure and the subsequent compression strength loss immediately influence the number of the spots exposed to potential structural destruction, as well as the character of the stress-evoked destruction, which is manifested by the number of the recorded acoustic emission impulses.

To make the image of the concrete destruction process course changes, corresponding to concrete humidity, more complete, we determined the intensity of the acoustic emission totting sum growth as a function of stress increment. For instance, this variability course for the concrete series subjected to a quasi-axial compression test is presented in Fig. 5. Figure 5 implies the course of the variability of the acoustic emission totting sum is three-staged depending on stress increment. The lengths of the particular stages, however, vary for the series tested and depend upon concrete humidity. Thus, for water-saturated concrete (the 'Aw'-series), the range of 0-0.2 σ/R exhibits a constant growth of the acoustic emission totting sum, the range of 0.2-0.9 σ/R shows its stabilization, and over 0.9 σ/R its repeated growth is visible.

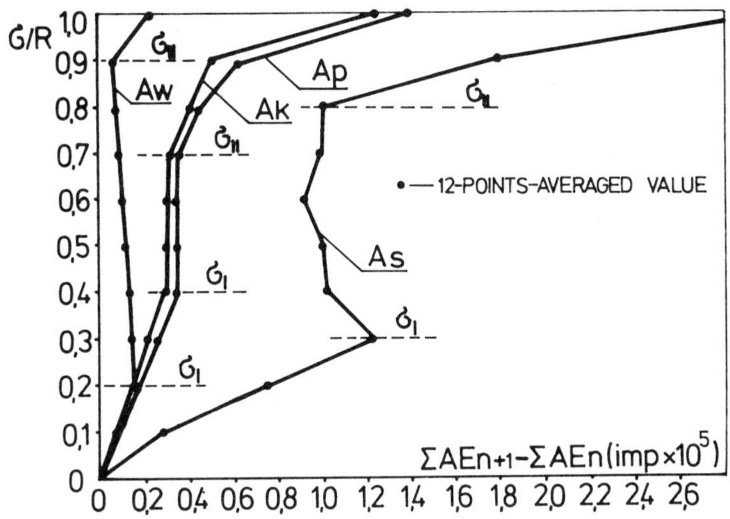

Fig. 5. Variability intensity course for the acoustic emission totting sum growth, the 'As', 'Ap', 'Ak' and 'Aw'-series concrete compressed quasi-axially, as a function of stress increment.

In the case of the concretes of similar humidity (the 'Ak' and 'Ap'-series), this variability course is alike. It shows a constant growth of the acoustic emission intensity within the range of 0-0.4 σ/R, the intensity stabilization marked within the range of 0.4-0.7 σ/R, and then the repeated growth. On the other hand, for a dry concrete (the 'As'-series) the growth of the acoustic emission totting sum intensity occurs within the range of 0-0.3 σ/R the range of 0.3-0.8 σ/R displays, though less clearly than in the case of the humid concretes, the stabilization of its growth, and up to the level of 0.8 σ/R - observed is a rapid intensity growth. The above-described regularities imply three characteristic stages to be distinguished during the process of concrete destruction, the lengths of the stages undoubtedly depending on the material's humidity. According to (2), the first stage is the one of stable crazing initiation, characterized by that some isolated spots of the structure exhibit the occurrence of microcrazes initiated at the stage of forming as microslots and pores. The peculiarity lies in that, at this stage of concrete destruction, the microcrazes fail to develop. Nevertheless, what we are to face here is the phenomenon of their multiplication, as was proved by the present paper's authors, which is indicated by a slight velocity drop of the longitudinal ultrasound waves and a constant growth of the acoustic emission totting sum intensity. An increase of the load makes the concrete destruction process enter the stage of stable crazing propagation, during which we are to face the propagation of the microcrazes formed during the first stage as well as the formation of new stable microcrazes - resulting from either the loss of adhesion between the aggregate grains and the slurry or within the slurry itself (2). This condition manifests itself in a considerable attenuation of the ultrasound waves and the stabilization of the intensity growth of the acoustic emission totting sum. Provided the load further increases, the concrete destruction process enters the stage of catastrophic destruction. This stage exhibits the formation of distinct, wide cracks that spread out unstably to reach the eventual destruction of the material. This phenomenon is made visible by the total loss of measurability of the longitudinal ultrasound wave velocities, as well as by a rapid growth of the acoustic emission totting sum intensity.

The boundaries between the particular stages of concrete destruction are determined by the critical stresses, σ_I and σ_{II}, which according to among others (1, 3) are identified with fatigue strength and long-term strength of concrete, respectively. These stress values, which characterize the qualitative changes occurring in the structure of the material examined as a response to load increase, were determined for the particular concrete sereis tested with the help of the ultrasound and the acoustic emission methods. Noticeably, when utilizing the ultrasound method, determined were only the values of the critical stresses σ_{II} since they had been assumed to correspond to the stress level at which the possibilities to measure the running time of the longitudinal ultrasound waves are totally lost. In the case of utilizing the acoustic emission method, the subject to determination were both the values of the critical stresses σ_I and σ_{II}, on the grounds of the intensity of the acoustic emission totting sum determined as a function of stress increment. We accepted the critical stresses σ_I to correspond to the stress level at which the stages of constant growth and stable growth of the acoustic emission totting sum are clearly distinguishable. On the other hand, the critical stresses σ_{II} correspond to the stress level at which the stages of stable growth and rapid growth of the acoustic emission totting sum intensity. The values of the critical stresses, determined according to the principles given here, are marked in Figs 2 and 5. Moreover, Fig. 6 submits a matching to show the variability course of the critical stress values, σ_I and σ_{II}, in relation to the humidity condition of a concrete under investigation. The analysis of Fig. 6 implies concrete humidity is of a crucial influence on these stress values. Thus, as concrete humidity is increasing, the values of the critical stresses, σ_I and tend to grow initially and then decrease down the humidity value of ca. 2.60%. For the critical stresses σ_{II}, the relationship is reverse.

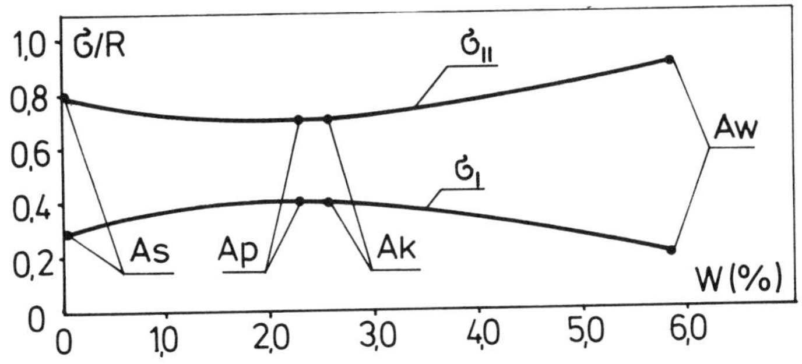

Fig. 6. Variability course of the critical stress values, σ_I and σ_{II}, dependind on concrete humidity.

4 Conclusions

The application of the acoustic methods, i.e. ultrasound and acoustic emission, to the research has pointed out concrete humidity to have a substantial influence on the process of concrete destruction. The results obtained thanks to these methods, especially the acoustic emission method, clearly prove distinguishable are three characteristic stages during the process of concrete destruction, the duration of which depends largely upon humidity of concrete. The values of the critical stresses σ_I and σ_{II}, constituting boundaries between the particular phases of concrete destruction, were found out to be related to the humidity of the material in question. The values of the critical stresses, σ_I, were stated to grow initially together with the growth of the tested material's humidity, the growth from 0.3 to 0.4 σ/R, and then, starting from the humidity of ca. 2.60%, to decrease to 0.2 σ/R. For the critical stresses σ_{II} this relationship looks reverse, yet initially the values decrease from 0.8 down to 0.7 σ/R and then grow up to 0.9 σ/R. The occurrence of differences in the course of the process of concrete destruction, the differences connected with concrete humidity, is also proved by the differences in the totting sum of the acoustic emission recorded during concrete destruction. It was also revealed that the acoustic emission totting sum recorded while destroying concrete depends substantially on the humidity of the material tested. The sum is the smaller the greater humidity of concrete.

5 References

(1) Flaga, K. and Furtak, K. (1982) Critical stresses in concrete as dependent on the type of aggregate, **AIL**, 4, 653-666.

(2) Newman, K. and Newman, J.B. (1971) Failure theories and design criteria for plain concrete, **Civil Engineering Materials Conference**, Wilay-Interscience, London, pp. 963-995.
(3) Shah, S.P. and Chandra, S. (1968) Critical stress volume change and microcracking of concrete. **J. ACI.**, 9, 770-781.

PART THIRTEEN

APPLICATIONS: STRUCTURAL ANALYSIS

INVITED PAPER

78 FRACTURE MECHANICS APPROACH TO REINFORCED CONCRETE DESIGN

A. CARPINTERI
Department of Structural Engineering, Politecnico di Torini, Italy

Abstract
Tensile and flexural strength both appear experimentally to decrease with size, for plain as well as for reinforced concrete. Two different fracture mechanics models – the ideal material with a random distribution of microdefects and the initially uncracked material with a growing cohesive macrocrack – both give account for such a decrease. In the real cases – e.g. shear resistance – a combined effect may arise. As regards the minimum amount of reinforcement, it can be provided by the transitional condition of simultaneous concrete cracking and steel yielding. The failure mechanism changes completely when the beam depth is varied, the steel percentage being the same. Only when the steel percentage is inversely proportional to the square root of the beam depth, the transitional behaviour is tendentially reproduced.
Keywords: Size Effects, Reinforced Concrete, Fracture Mechanics, Shear Resistance, Minimum Reinforcement, Brittleness Number, Transitional Condition, Defect Size Distribution of Self-Similarity.

1. Introduction

Several aspects of reinforced concrete design may be improved by the application of fracture mechanics concepts. Specially for high strength concrete, which is a very brittle material, and in the case of large structural members, the fracture mechanics approach could provide rational design rules, in a compromise between structural reliability and economical aspects.

Two specific topics are selected, for which a rational approach could introduce variations in the reinforced and prestressed concrete design codes. The former is the strength size effect; the latter is the minimum reinforcement. As strength dangerously decreases with size, so the minimum percentage of reinforcement conservatively does.

While the modern fracture mechanics was developed in the 1950's, mainly for metals, we should recall the early attempt by Westergaard (1934) specifically devoted to reinforced concrete. Then, in the 1980's, the contributions due to Bazant and Cedolin (1980), Carpinteri (1981, 1984), Saouma and Ingraffea (1981), Hawkins (1985), Gustafsson and Hillerborg (1985), Tassios (1985), Rokugo et al. (1986a), Bosco et al. (1989; 1990), should be mentioned. Very recently, a workshop specifically devoted to the applications of fracture mechanics to reinforced concrete, has been held at the Politecnico di Torino, Italy, with the sponsorship of RILEM and CEB.

2. Size effects for tensile and flexural loading

2.1. Influence of defects and porosity on microcracking initiation and tensile strength

In the present section, two-dimensional structures with a multitude of microcracks or voids of a

Fracture Processes in Concrete, Rock and Ceramics Edited by J.G.M. van Mier, J.G. Rots and A. Bakker.
 Published by E. & F.N. Spon, 2–6 Boundary Row, London SE1 8HN. ISBN 0 419 15870 7

given size-distribution are considered. Three hypotheses are assumed:

(1) the structure is macroscopically homogeneous;

(2) the structure is macroscopically isotropic;

(3) the interaction among the imperfections is negligible.

As a first case, let us consider a set of similar structures, where a multitude of cracks and/or polygonal voids of constant size a are embedded (Fig. 1-a). They can be considered as specimens of the same material, and their failure occurs when the fracture condition at the imperfection of the most critical orientation is realized. Since such imperfections are all the same size, we can assert that any size effect is absent in this ideal case.

Let us consider now a set of similar structures, where the imperfections have a constant size which is proportional to the size of the structure (Fig. 1-b). In this case, they cannot be considered as specimens of the same material. The failure occurs at the imperfection of the most critical orientation, which is of a size proportional to the size of the structure. Since the influence of

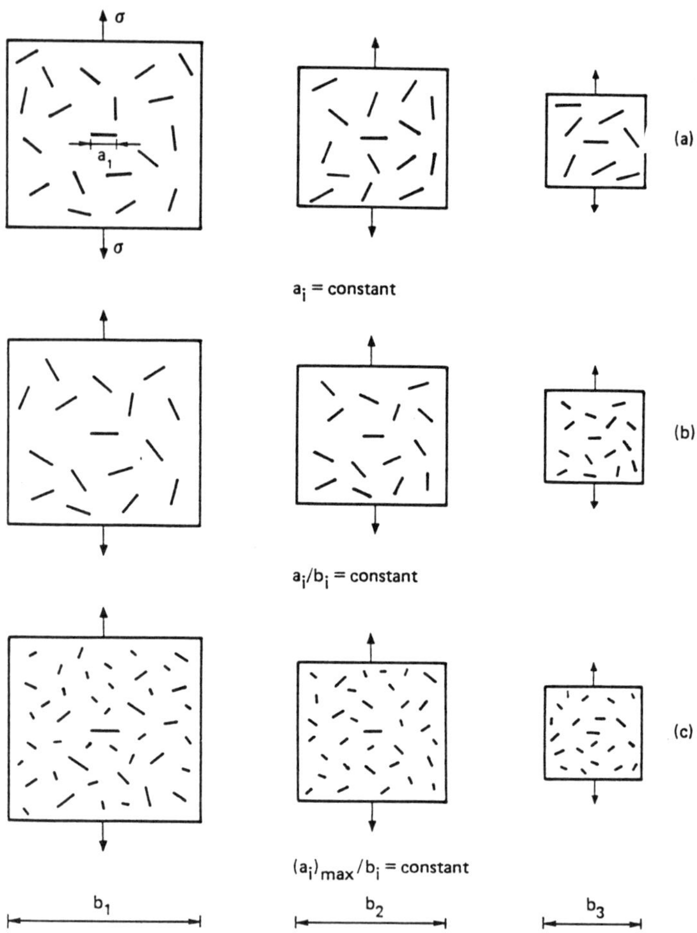

Fig. 1. Geometrically similar structures with random defects: (a) constant defect size; (b) defect size constant and proportional to the structural size; (c) defect size distribution of self-similarity.

the other imperfections is assumed negligible, this case is completely equivalent to that of a unique imperfection of size proportional to that of the structure. Thus, the strength size effect is represented by a linear ln $\sigma_f - \ln b$ diagram with slope $-\alpha$, where α is the power of the stress-singularity produced by the imperfections. If we have imperfections of the same size, but with different shape (e.g. cracks and circular pores), the fracture condition has to be evaluated only for the imperfections with the maximum value of $\alpha(0 \leq \alpha \leq 1/2)$.

Let us consider finally a set of similar structures where the imperfections with the most dangerous shape ($\alpha = \alpha_{max}$) have a probability density $p(a)$ of size distribution (Figs. 1-c and 2-a). We can assert that if the size distribution $p(a)$ is such that the maximum size a_{max} is proportional to the linear scale b, then the strength size effect will be represented by a linear ln $\sigma_f - \ln b$ diagram with slope $-\alpha_{max}$. The above hypothesis is very restrictive and is valid only when the probability density of size distribution $p(a)$ presents particular properties. If ρ is the density of imperfections (number of imperfections per unit area), the maximum size a_{max} can be defined as follows (Carpinteri 1986; 1989a):

$$\rho b^2 p(a_{max}) \frac{1}{2\pi} \Delta a \Delta \omega = 1, \tag{1}$$

where the factor $(1/2\pi)$ is due to the fact that all the imperfection orientation angles ω are equally likely. If a geometrically similar structure of characteristic size kb is considered and the above hypothesis is assumed (Fig. 1-c), we can write:

$$\rho(kb)^2 p(ka_{max}) \frac{1}{2\pi} \Delta a \Delta \omega = 1. \tag{2}$$

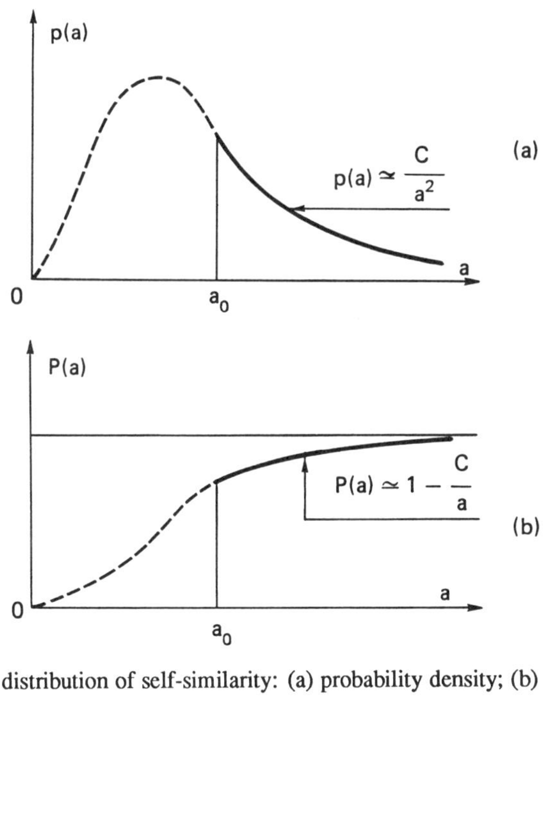

Fig. 2. Defect size distribution of self-similarity: (a) probability density; (b) cumulative distribution.

Since a_{max} is a function of ρ and b, besides the ranges Δa and $\Delta \omega$, it follows that eqns (1) and (2) must be valid for any defect size a:

$$\rho b^2 p(a) \frac{1}{2\pi} \Delta a \Delta \omega = 1, \tag{3}$$

$$\rho k^2 b^2 p(ka) \frac{1}{2\pi} \Delta a \Delta \omega = 1. \tag{4}$$

From eqns (3) and (4) it follows that:

$$p(a) = k^2 p(ka), \quad \forall a \gg \bar{a}, \quad \forall k \in R^+, \tag{5}$$

and then function p assumes the form:

$$p(a) = \frac{C}{a^2}, \quad \forall a \gg \bar{a}, \tag{6}$$

where C is a constant with the physical dimension of a length and $\bar{a}$ is the average defect size. Equation (6) will be referred to as the *defect size distribution of self-similarity.* The related cumulative distribution function P is:

$$P(a) = \int_0^a p(x)\,\mathrm{d}x = \int_0^{a_0} p(x)\,\mathrm{d}x + \int_{a_0}^a \frac{C}{x^2}\,\mathrm{d}x, \tag{7}$$

where $a_0 \gg \bar{a}$ is the value beyond which the decreasing branch of function p can be approximated by eqn (6). Carrying out the integrations in eqn (7), we obtain:

$$P(a) = P_0 + \left[-\frac{C}{x}\right]_{a_0}^{a} = P_0 + \frac{C}{a_0} - \frac{C}{a}. \tag{8}$$

Since for $a \to \infty$ the cumulative distribution $P(a) \to 1$,

$$P_0 + \frac{C}{a_0} = 1, \tag{9}$$

and then (Fig. 2-b):

$$P(a) = 1 - (C/a), \quad \text{for} \quad a > a_0, \tag{10}$$

with

$$C = (1 - P_0)a. \tag{11}$$

Generally speaking, the cumulative distribution function P will have the following form:

$$P(a) = 1 - \frac{C}{a^N}, \quad \text{for} \quad a > a_0, \tag{12}$$

with

$$C = (1 - P_0) a_0^N. \tag{13}$$

In this case, the strength size effect can be represented by a linear $\ln \sigma_f - \ln b$ diagram with slope $-\alpha_N$ (Carpinteri 1986; 1989a):

$$\alpha_N(\gamma, N) = \frac{\alpha(\gamma)}{N^\zeta}, \tag{14}$$

where γ is the reentrant angle of the regular voids and the exponent ζ depends on the secondary features of the material (e.g. density of imperfections, size distributions of the less dangerous defects, etc.). The probability density of size distribution in the general case – see eqn (12) – is:

$$p(a) = \frac{\mathrm{d}P}{\mathrm{d}a} = N\frac{C}{a^{N+1}}, \quad for \quad a > a_0, \tag{15}$$

which becomes eqn (6) when $N = 1$.

Equation (14) shows that the size effect vanishes when $\gamma = \pi$ (circular pores) and/or when $N \rightarrow \infty$ (nearly constant defect size). On the other hand, the size effect becomes enormous when $N \rightarrow 0$ (very large dispersion in the imperfection size distribution). The treatment of 2-D structures can formally be extended to 3-D structures with polyhedral voids.

If the preceding assumptions are valid, the experimental $\ln \sigma_f - \ln b$ strength diagram must appear linear with a negative slope α_N, according to eqn (14). When the dispersion in the imperfection size distribution is not very high $(N \geq 1)$, we have the theoretical upper bound $\alpha_N \leq 0.50$, which is rarely exceeded in the experimental results reported in the relevant literature.

The results of Strange and Bryant (1979) are plotted in Fig. 3. The graph of flexural tensile strength against the beam width shows a very regular linear decrease in the bilogarithmic diagram for all the four cement composites investigated. The results of Sabnis and Mirza (1979) show the same trend for the flexural tensile strength (Fig. 4) and for the indirect tensile strength (Fig. 5).

2.2. Apparent decrease in the ultimate bending strength of initially uncracked materials

In the present section, two-dimensional structures, initially uncracked and with a unique growing cohesive macrocrack, are considered (Carpinteri 1989a; 1989b). A three point bending slab (Fig. 6) is numerically simulated by varying the *brittleness number* (Carpinteri, 1985):

$$s_E = \frac{\mathcal{G}_F}{\sigma_u b}, \tag{16}$$

where $\mathcal{G}_F$ is the fracture energy, σ_u the ultimate tensile strength and b the slab depth.

Some dimensionless load-deflection diagrams for a concrete-like material are plotted in Fig. 7, with ultimate tensile strain $\varepsilon_u = 0.87 \times 10^{-4}$, Poisson ratio $\nu = 0.1, t = b, l = 4b, a_0/b = 0.0$. The specimen behaviour is brittle (snap-back) for low s_E numbers – i.e. for low fracture toughnesses $\mathcal{G}_F$, high tensile strengths σ_u and/or large sizes b. For $s_E \lesssim 10.45 \times 10^{-5}$, the $P - \delta$ curve has a positive slope in the softening branch, and a catastrophic event occurs if the loading process is deflection-controlled. Such an indenting branch is not virtual only if the loading process is controlled by a monotonically increasing function of time – for example, the displacement discontinuity across the crack (Fairhurst *et al.*, 1971; Rokugo *et al.*, 1986b; Biolzi *et al.*, 1989). When the post-peak behaviour is kept under control up to complete structure separation, the area delimited by the load-deflection curve and the deflection axis represents the product of the fracture toughness $\mathcal{G}_F$ and the initial cross-sectional area bt.

The maximum loading capacity $P_{max}^{(1)}$ of initially uncracked specimens with $l = 4b$ is ob-

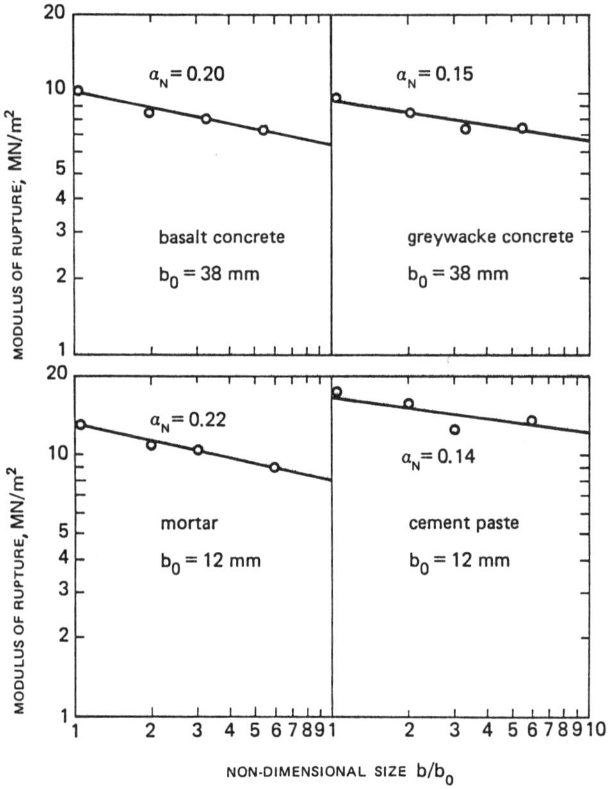

Fig. 3. Effect of size on flexural tensile strength of cementitious materials (Strange and Bryant, 1979).

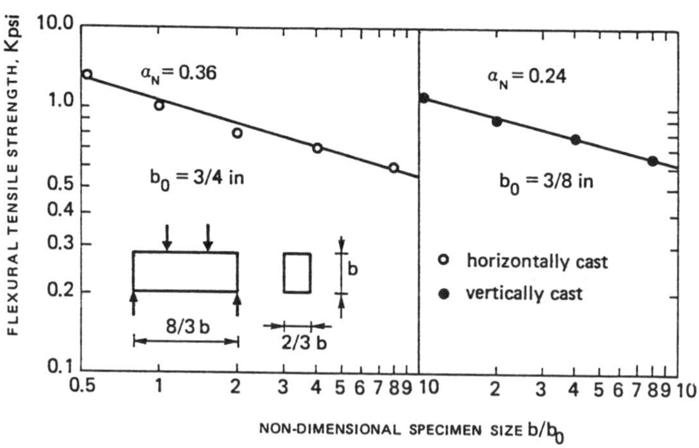

Fig. 4. Effect of size on flexural tensile strength of concrete (Sabnis and Mirza, 1979).

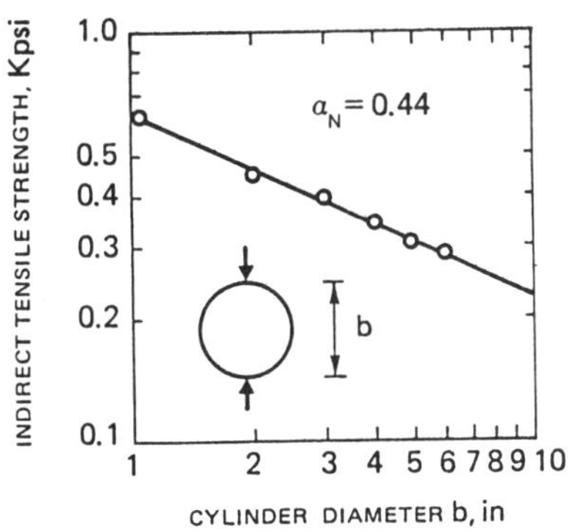

Fig. 5. Effect of size on indirect tensile strength of concrete (Sabnis and Mirza, 1979).

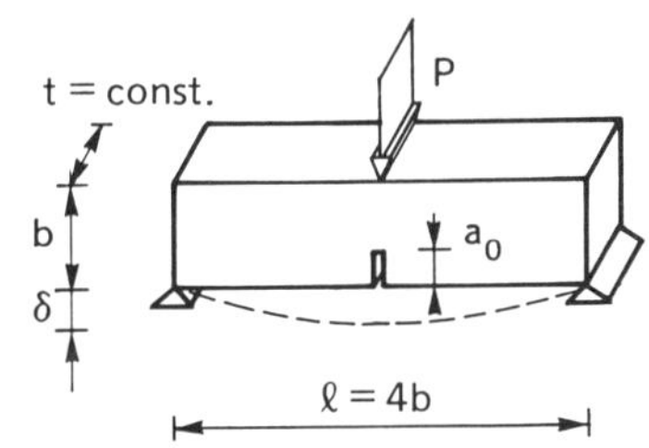

Fig. 6. Three point bending specimen.

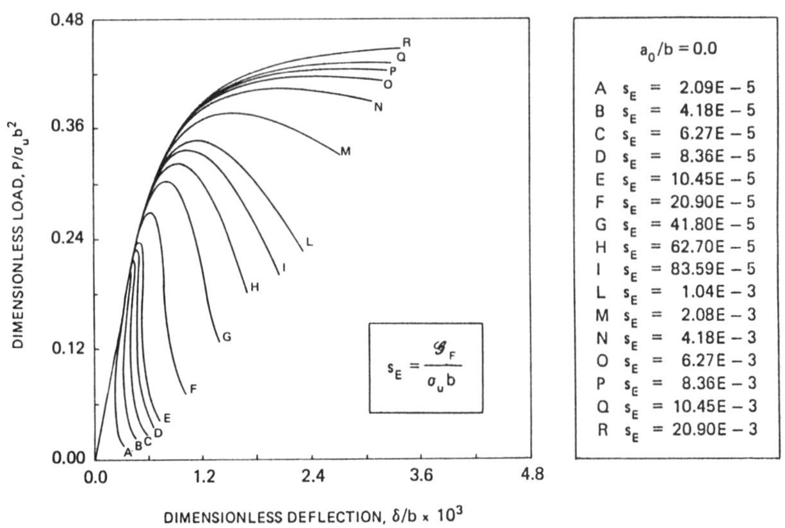

Fig. 7. Dimensionless load vs. deflection diagrams with various brittleness numbers $s_E = G_F/\sigma_u b$.

tained from Fig. 7. On the other hand, the maximum load $P_{max}^{(3)}$ of the ultimate strength is given by:

$$P_{max}^{(3)} = \frac{2}{3}\frac{\sigma_u t b^2}{l}. \tag{17}$$

The values of the ratio $P_{max}^{(1)}/P_{max}^{(3)}$ may also be regarded as the ratio of the apparent tensile strength σ_f (given by the maximum load $P_{max}^{(1)}$ and by applying eqn (17)) to the true tensile strength σ_u (considered as a material constant). It is evident from Fig. 8 that the results of the cohesive crack model tend to those of the ultimate strength analysis for low s_E values:

$$\lim_{s_E \to 0} P_{max}^{(1)} = P_{max}^{(3)}. \tag{18}$$

Therefore, for comparatively large specimen sizes only, the tensile strength σ_u can be obtained as $\sigma_u = \sigma_f$. With the usual laboratory specimens, an apparent strength higher than the true one is always found (Petersson, 1981).

As a limiting case, for the size $b \to 0$ or fracture energy $\mathcal{G}_F \to \infty$ (elastic-perfectly plastic material in tension), i.e. for $s_E \to \infty$, the apparent strength $\sigma_f \to 3\sigma_u$. In fact, in the center of the beam, the uniform stress distribution (Fig. 9) produces a plastic hinge with a resistant moment M_{max} which is twice the classical moment of the bi-rectangular limit stress distribution (elastic-perfectly plastic material in tension and compression).

2.3. Interaction between growing macrocrack and inherent microcracks

Two different models have been considered:

(1) the ideal material with a random distribution of microdefects (Section 2.1);
(2) the initially uncracked material with a growing cohesive crack (Section 2.2).

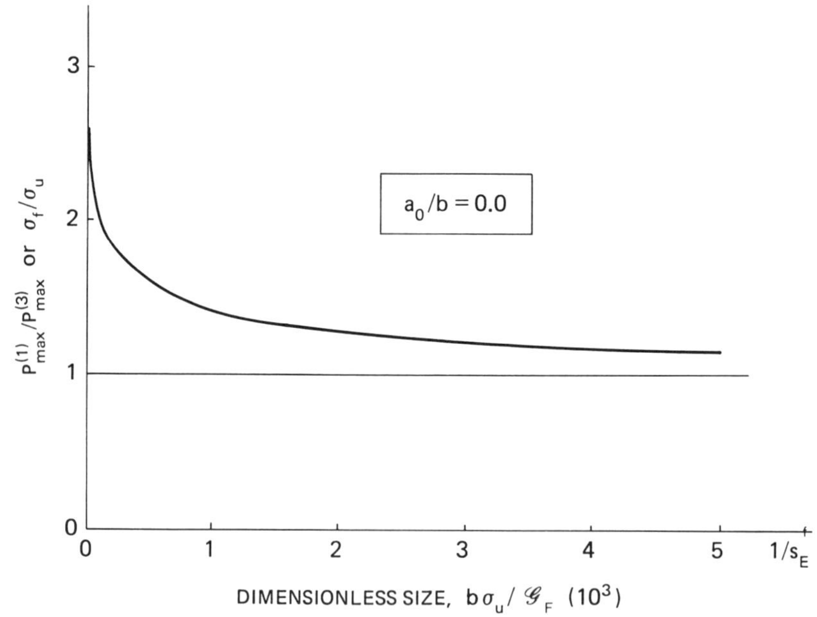

Fig. 8. Decrease in apparent strength by increasing specimen size.

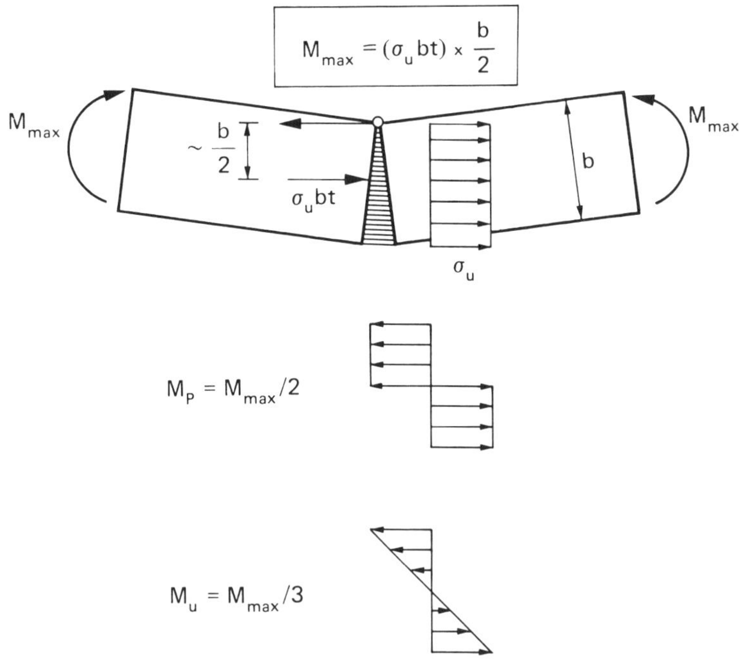

Fig. 9. Constant distribution of cohesive stresses.

Both such models lead to a *decrease in the apparent strength* with an increase in the size of the specimen.

In the former idealization, the apparent strength (stress of microcracking initiation) tends to zero when the size tends to infinity, while it tends to the true value σ_u (constant material property) when the size tends to zero.

On the other hand, with the latter model, the apparent ultimate bending strength tends to the value σ_u when the size tends to infinity, whereas it tends to the limit $3\sigma_u$ when the size tends to zero (Figs. 8 and 9).

The two models may describe different real situations, according to the geometry and failure mechanism of the specimen. If the microdefects pre-existing in the material produce very severe stress conditions and the failure mechanism is tensile and brittle, the former model is likely to be applicable. On the other hand, if the failure mechanism is stable or ductile and is produced by the slow propagation of a unique crack in a partially compressed material (in bending), the latter model is more suitable for interpreting the ultimate strength decrease with size.

In some cases, for example the shear resistance of concrete beams, a combined effect of the two mechanisms may arise. The Eurocode EC2 standards take into account of this effect through the relation:

$$\frac{V_u}{bh} = 0.25 f_{ct} k \left(1.2 - 40 \frac{A_s}{A} \right), \tag{19}$$

where V_u is the ultimate shear force, b the beam thickness, h the beam depth, f_{ct} the concrete tensile strength, A_s/A the steel percentage, and $k = 1.6 - h > 1$ (metre) is the size effect factor, decreasing up to the beam depth of 1 metre. The EC2 requirements, together with those of ACI and the experimental results obtained by Bosco, Carpinteri and Debernardi (1989, 1990) are shown in Fig. 10.

3. Minimum reinforcement

3.1. Crack growth stability in reinforced concrete

Crack formation in reinforced concrete beams provokes unstable behavior. If the loading process is performed by controlling deformation or crack opening, the unstable phenomenon can be followed in its actual evolution. Otherwise, a positive jump of deformation or a negative jump of loading capacity, occurs, according to the loading control, i.e. dead load or fixed grip condition. A qualitative load vs. deflection diagram is represented in Fig. 11, where the snap-back and snap-through actual behavior is by-passed along the short cuts AB (load-control) or AC (deflection-control).

The analysis of such experimental behavior is useful in the determination of rational criteria to compute the minimum amount of reinforcement for concrete members in flexure. The minimum amount of reinforcement could be given by the condition for which first concrete cracking and steel yielding are simultaneous. Several national and international standard codes provide

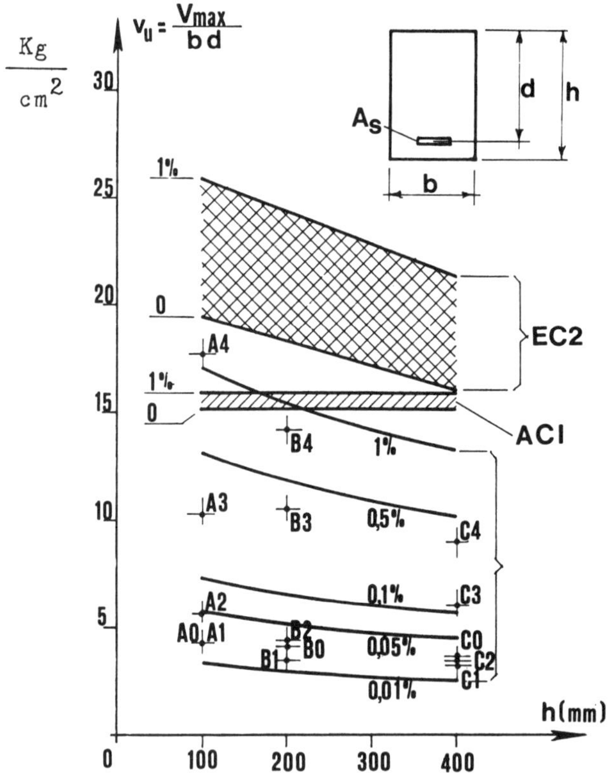

Fig. 10. Shear resistance against the beam depth for various steel percentages (Eurocode and ACI requirements compared with experimental results).

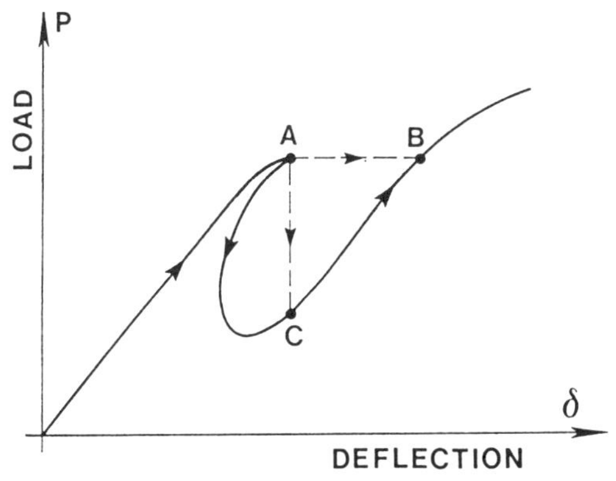

Fig. 11. Snap-through and snap-back behaviours, respectively with load-control (AB) and deflection-control (AC).

empirical formulas for the determination of the minimum amount of reinforcement. At the most, two are the parameters taken into account in these formulas: the concrete tensile strength and the steel yield strength, whereas other important features are neglected, such as the size-scale of the concrete member or its fracture energy. Moreover, the current codes establish minimum amounts which, sometimes, are very different.

In some codes (France, Switzerland, Eurocode EC2) the minimum amount is established on the basis of the ratio between the computed stresses in concrete and steel; other codes (Great Britain, American Concrete Institute) take into account only the steel yield strength, while in Italy and URSS a minimum percentage of steel is fixed independently of any geometrical and mechanical feature (Szalai, 1987).

It is possible to take into consideration the beam size effect on the minimum steel percentage through the concepts of fracture mechanics and propose a dimensional analysis criterion to determine the minimum amount of reinforcement for concrete beams in flexure. To this purpose a *brittleness number* is defined (Carpinteri 1981; 1984):

$$N_P = \frac{f_y h^{1/2}}{K_{IC}} \frac{A_s}{A}, \tag{20}$$

f_y being the steel yield strength, K_{IC} the concrete fracture toughness, A_s/A the steel percentage and h the beam depth. The brittleness of the structural member increases by increasing the size and/or decreasing the steel content. On the other hand, a physically similar behavior is revealed in the cases where the brittleness number N_P is the same. The experimental investigation carried out by the author shows that the failure mechanism changes completely when the beam depth is varied, the steel percentage being the same. Only when the steel percentage is inversely proportional to the square root of the beam depth, the mechanical behavior is reproduced (Bosco, Carpinteri & Debernardi 1989; 1990).

3.2. Transitional brittleness number N_{PC}

If the experimental tests are carried out with a direct applied load, it is possible to obtain different kinds of load-deflection or moment-rotation diagrams, which, with reference to Fig. 12, represent:

(a) the behaviour of beams with a very low content of steel, which, when the moment reaches the first cracking value, cannot sustain the external load if it is not reduced. Therefore a sudden

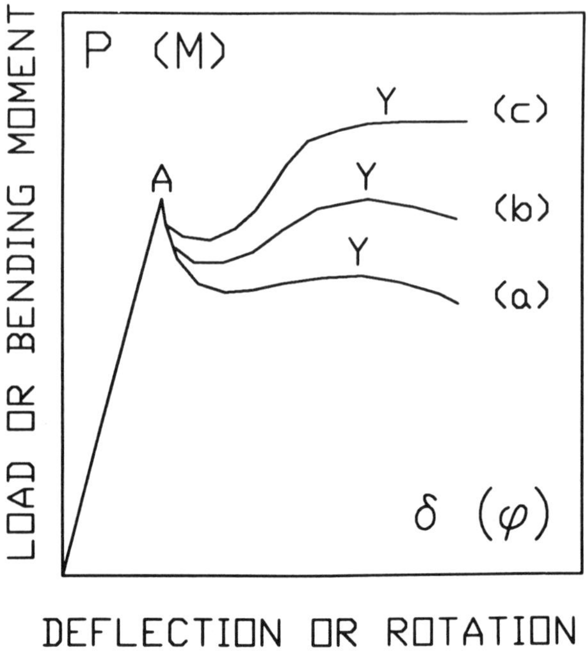

Fig. 12. Unstable (a), transitional (b) and stable (c) structural response of reinforced beam.

failure occurs, without warning signs. Experimentally it is possible to observe the softening branch by controlling the deformation;

(b) the transitional condition, when the moment of concrete cracking is equal to the moment of steel yielding;

(c) the condition for which, after cracking, the beam is able to sustain further increase in applied load.

Fracture mechanics is able to represent the actual behaviour of the cross-section of a reinforced concrete beam, using a simple model, which gives theoretical indications in good agreement with the experimental results (Carpinteri 1981; 1984). As a matter of fact, it is possible to define a dimensionless parameter, the brittleness number, which can represent different behaviours, varying both steel percentage and/or cross-section size. For different values of this parameter, it is possible to obtain qualitative diagrams representing unstable behaviour as indicated in Fig. 12 by the curve (a), or stable behaviour, as indicated by the curve (c), or transitional behaviour, as represented by the curve (b).

An experimental investigation was carried out on initially uncracked R.C. beams, by means of three-point bend tests, by controlling the tensile strain on the lower edge of the beam or, after cracking, the crack mouth opening displacement. The analysis of the experimental behaviour supplies useful indications for the determination of the minimum steel percentage to be provided in reinforced concrete structures in flexure. 76 unnotched beams were tested, subdivided in:

– Five concrete classes, with a cylindrical compressive strength varying between 16 and 80 N/mm^2.

– Four dimensional classes between 10 and 80 cm of beam depth.

Concrete class	Cylindrical strength (N/mm^2)
I	16
II	30
III	40
IV	65
V	80

Dimensional class	Beam depth (mm)
A	100
B	200
C	400
D	800

The experimental results obtained so far show that the transitional value of the brittleness number, N_{PC}, is tendentially constant for each concrete grade, by varying the beam depth h. As a consequence, the determination of the minimum steel percentage, is allowed on the basis of the brittleness number N_{PC}.

To represent the transitional brittleness number N_{PC} we can refer to the experimental diagram shown in Fig. 13, where N_{PC}, as an average, increases monotonically against the compressive strength of concrete. It is even possible to obtain a linear relationship, that, for safety reasons, is assumed to bound the maxima of the distribution:

$$N_{PC} = 0.1 + 0.00233 f_{cm}. \tag{21}$$

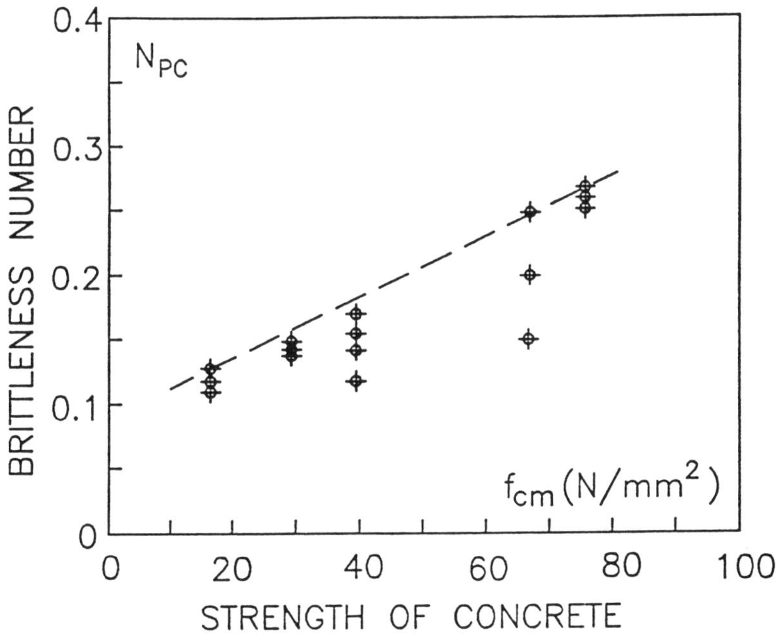

Fig. 13. Transitional brittleness number N_{PC} against the compression strength of concrete.

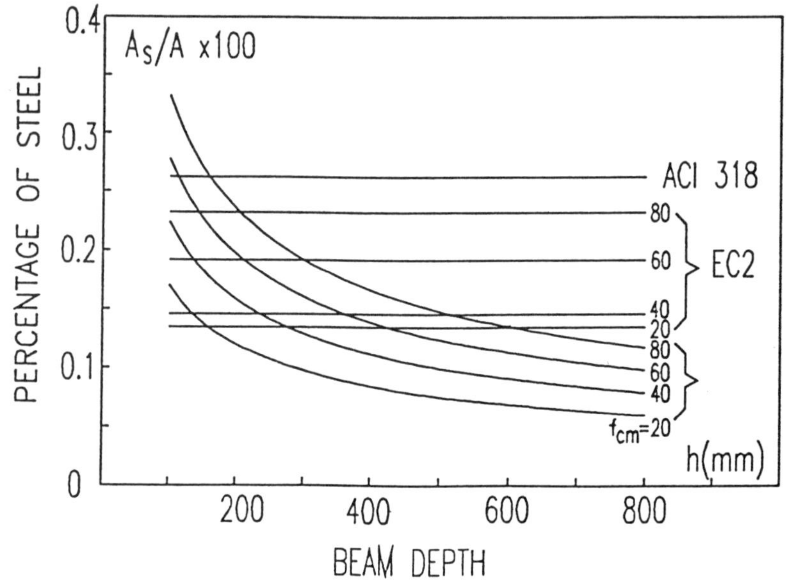

Fig. 14. Minimum steel percentage against the beam depth for various concrete grades (Eurocode and ACI requirements compared with experimental results).

Fig. (14) provides the minimum reinforcement percentage, A_s/A, obtained from relation (21), as a function of beam depth h, for common concrete strengths in structural applications. In the same diagram, the minimum steel percentages required by Eurocode EC2 and ACI 318, are reported. While the values supplied by both this standards and by other international codes, are independent of the beam depth h, the relationship which is established by the brittleness number N_{PC}, calls for decreasing minimum steel percentages with increasing beam depths.

On the other hand, it is necessary to point out the following aspects.

- The diagrams in Fig. 14 only compare the minimum percentage of steel content obtained from the experimental results to the requirements of widely used Codes, without taking into account any safety criteria.
- Relationship (21) is valid for ordinary values of steel strength.
- Relationship (21) is obtained for concrete aggregates with maximum diameter of 15mm and effective beam depth to total beam depth ratio equal to 0.9.
- The fracture energy $\mathcal{G}_F$ was obtained using the RILEM procedure, while the fracture toughness K_{IC} derives from the well-known relation:

$$K_{IC} = \sqrt{\mathcal{G}_F E}, \tag{22}$$

E being the Young's modulus of concrete.

4. Conclusions

(1) Tensile and flexural strength both appear experimentally to decrease with size, for plain as well as for reinforced materials.

(2) For the ideal material with a random distribution of microdefects, the apparent strength tends to zero when the size tends to infinity, while it tends to the true value σ_u when the size tends to zero.

(3) For the initially uncracked material with a growing cohesive macrocrack, the apparent bending strength tends to the true value σ_u when the size tends to infinity, whereas it tends to the limit $3\sigma_u$ when the size tends to zero.

(4) The Eurocode EC2 standards take into account of a combined effect of the two above-described mechanisms in the case of the beam shear resistance.

(5) The minimum amount of reinforcement for beams in flexure can be provided by assuming simultaneous concrete cracking and steel yielding (transitional condition).

(6) A physically similar flexural behaviour is obtained in the cases where the brittleness number N_P is the same. The failure mechanism changes completely when the beam depth is varied, the steel percentage being the same. Only when the steel percentage is inversely proportional to the square root of the beam depth, the mechanical behaviour is reproduced, see eq. (20).

(7) The experimental results show that the transitional value of the brittleness number, N_{PC}, is tendentially constant for each concrete grade, by varying the beam depth.

(8) While the minimum steel percentages provided by Eurocode and ACI are independent of the beam depth, the relationship established by the brittleness number, calls for decreasing values with increasing beam depths.

Acknowledgments
The financial support of the Department for University and Scientific and Technological Research is gratefully acknowledged by the author.

5. References

Bazant, Z.P. and Cedolin, L. (1980). Fracture mechanics of reinforced concrete, *Journal of the Engineering Mechanics Division* (ASCE), **106**, 1287-1306.

Biolzi, L., Cangiano, S., Tognon, G. and Carpinteri, A. (1989). Snap-back softening instability in high strength concrete beams, *Materials and Structures* (RILEM), **22**, 429-436.

Bosco, C., Carpinteri, A. and Debernardi, P.G. (1989). Size effect on the minimum steel percentage for reinforced concrete beams, in *Recent Developments on the Fracture of Concrete and Rock* (eds. S.P. Shah, S.E. Swartz and B. Barr), Elsevier Applied Science, pp. 672-681.

Bosco, C., Carpinteri, A. and Debernardi, P.G. (1990). Minimum reinforcement in high strength concrete, *Journal of Structural Engineering* (ASCE), **116**, 427-437.

Carpinteri, A. (1981). A fracture mechanics model for reinforced concrete collapse, in *Advanced Mechanics of Reinforced Concrete* (IABSE), Delft University Press, pp. 17-30.

Carpinteri, A. (1984). Stability of fracturing process in R.C. beams, *Journal of Structural Engineering* (ASCE), **110**, pp. 544-558.

Carpinteri, A. (1985). Interpretation of Griffith instability as a bifurcation of the global equilibrium, in *Application of Fracture Mechanics to Cementitious Composites* (ed. S.P. Shah), Martinus Nijhoff Publishers, pp. 287-316.

Carpinteri, A. (1986). *Mechanical Damage and Crack Growth in Concrete: Plastic Collapse to Brittle Fracture*, Martinus Nijhoff Publishers, Dordrecht.

Carpinteri, A., (1989a). Decrease of apparent tensile and bending strength with specimen size: two different explanations based on fracture mechanics, *International Journal of Solids and Structures*, **25**, pp. 407-429.

Carpinteri, A. (1989b). Size effects on strength, toughness and ductility, *Journal of Engineering Mechanics* (ASCE), **115**, pp. 1375-1392.

Fairhurst, C., Hudson, J.A. and Brown, E.T. (1971). Optimizing the control of rock failure in servo-controlled laboratory tests, *Rock Mechanics*, **3**, pp. 217-224.

Gustafsson, P.J. and Hillerborg, A. (1985). Improvements in concrete design achieved through the application of fracture mechanics, in *Application of Fracture Mechanics to Cementitious Composites* (ed. S.P. Shah), Martinus Nijhoff Publishers, pp. 667-680.

Hawkins, N.M. (1985). The role of fracture mechanics in conventional reinforced concrete design, in *Application of Fracture Mechanics to Cementitious Composites* (ed. S.P. Shah), Martinus Nijhoff Publishers, pp. 639-666.

Petersson, P.E. (1981). Crack growth and development of fracture zones in plain concrete and similar materials, *Report TVBM 1006, Lund Institute of Technology.*

Rokugo, K., Iwase, H. and Koyanagi, W. (1986a). Application of fracture energy of concrete in compression to section design of reinforced concrete members, in *Fracture Toughness and Fracture Energy of Concrete* (ed. F.H. Wittmann), Elsevier, pp. 587-596.

Rokugo, K., Ohno, S. and Koyanagi, W. (1986b). Automatical measuring system of load-displacement curves including post-failure region of concrete specimens, in *Fracture Toughness and Fracture Energy of Concrete* (ed. F.H. Wittmann), Elsevier, pp. 403-411.

Sabnis, G.M. and Mirza, S.M. (1979). Size effects in model concretes, *Journal of the Structural Division* (ASCE), **105**, 1007-1020.

Saouma, V.E. and Ingraffea, A.R. (1981). Fracture mechanics analysis of discrete cracking, in *Advanced Mechanics of Reinforced Concrete* (IABSE), Delft University Press, pp. 413-436.

Strange, P.C. and Bryant, A.H. (1979). Experimental tests on concrete fracture, *Journal of the Engineering Mechanics Division* (ASCE), **105**, pp. 337-342.

Szalai, K. (1987). Principles of dimensioning slightly-reinforced concrete structures, *CEB Bulletin d'Information, 25th Plenary Session, Treviso,* **185**, pp. 117-134.

Tassios, T.P. (1985). Design codes and fracture mechanics, in *Application of Fracture Mechanics to Cementitious Composites* (ed. S.P. Shah), Martinus Nijhoff Publishers, pp. 681-693.

Westergaard, H.M. (1934). Stresses at a crack, size of the crack, and the bending of reinforced concrete, *Journal of the American Concrete Institute*, **30**, pp. 93-102.

79 ANCHOR BOLTS ANALYZED WITH FRACTURE MECHANICS

U. OHLSON and L. ELFGREN
Luleå University of Technology, Sweden

Abstract

Non linear fracture mechanics makes it possible to study size effects and design criteria for anchor bolts. In the last few years many studies have been carried out in this field. A Round Robin Analysis has also been initiated by RILEM TC90-FMA "Fracture Mechanics of Concrete - Applications".

In the paper theoretical results are analyzed and compared to experimental values.

Keywords: Anchor bolt, Brittleness number, Size effects, Fracture energy, Design criteria, Characteristic length

1. Introduction

Anchor bolts, see Fig. 1, embedded in concrete or grouted in concrete or rock are very important structural details. They are used to transfer local loads into structures e.g. from cantilevers to beams and walls, from beams to columns and from columns to slabs and foundations.

Various design procedures have been developed over the years based on simplified assumptions and empirical test results, see e.g. Klinger and Mendonca (1982) and Rehm et al (1988). One general way of writing the capacity F_u has been proposed by Eligehausen and Sawade (1989).

$$F_u = a_1 \; f_c^{a2} \; d^{a3} \quad \text{or with numerical values} \tag{1.1}$$

$$F_u \approx 15 \; f_c^{0.5} \; d^{1.5} \; [\text{kN}] \tag{1.2}$$

where a_1, a_2 and a_3 are empirical constants
f_c = concrete compressive strength (N/mm^2)
d = embedded depth (mm)

The factor a_1 is used to assure dimensional correctness of the formulae and to calibrate measured loads to predicted values. In Eligehausen and Sawade (1989) a_1 = 15.5 is proposed.

The expression f_c^{a2} represents an approximate value of the tensile strength of the concrete derived from the compressive strength. Values of a_2 = 0.5 and 0.66 are used.

Fracture Processes in Concrete, Rock and Ceramics Edited by J.G.M. van Mier, J.G. Rots and A. Bakker.
 Published by E. & F.N. Spon, 2–6 Boundary Row, London SE1 8HN. ISBN 0 419 15870 7

The influence of the embedment depth d is given by $a_3 = 1.5$ to 1.54 by Eligehausen and Sawade (1989). This means that the failure load does not increase in proportion to the surface of the failure cone. However, in the US code ACI (1978) direct proportionality between the failure load F_u and the size of the failure cone surface is assumed by a choice of $a_3 = 2$. This assumption can thus be unsafe for bolts with large embedment depths as it tends to overestimate the influence of the depth compared to test results.

Contributions to the development of fracture mechanics methods for anchor bolts have among others been given by Ottosen (1981), Elfgren et al. (1980), (1982), Peier (1983), de Borst (1986), Ballarini et al. (1986), Hellier et al. (1987), Rots (1988), Eligehausen and Sawade (1989), Ohlsson (1990), Eligehausen and Ozbolt (1990), Cervenka et al. (1990), Ballarini and Shah (1991) and Ohlsson and Elfgren (1991). In order to compare the various methods, RILEM TC 90-FMA (1990) initiated a Round Robin Analysis of anchor bolts.

One way to model the failure is to study the **elastic energy** and the **fracture energy** as proposed by Bache, see e.g. Elfgren (1989) and di Tommaso (1989).

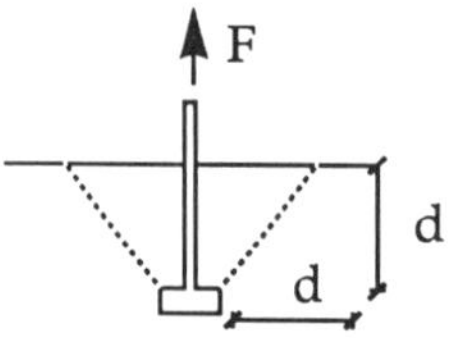

Fig. 1. Anchor bolt embedded in concrete

The elastic energy is the sum of the strain energy $0.5\,\sigma\varepsilon = 0.5\,\sigma^2/E$ over the volume of the anchor bolt. If we approximate the volume with $V = \pi \cdot d^2 \cdot d$, see Fig. 1. we may write

$$W_{el} \sim \frac{1}{2}\frac{\sigma^2}{E}\pi d^3 \sim \frac{\sigma^2 d^3}{E} \qquad (1.3)$$

The fracture energy needed to create a crack is proportional to the crack area.

$$W_{cr} \sim G_F \pi d^2 \sim G_F d^2 \qquad (1.4)$$

A **brittleness number** B can now be defined as the ratio of elastic to fracture energy at failure

$$B = \frac{\text{Elastic energy}}{\text{Fracture energy}} \sim \frac{\sigma^2 d^3}{E G_F d^2} = \frac{\sigma^2 d}{E G_F} \qquad (1.5)$$

or

$$B = \frac{f_t^2 d}{E G_F} \qquad (1.6)$$

(Here EG_F/f_t^2 is often called the characteristic length l_{ch} and $B = d/l_{ch}$, see e.g. Hillerborg (1985))

The failure load F can now be studied as a function of the brittleness number B. This will be done in this paper.

2. Plane stress specimen

2.1 Experimental work

Test specimens according to Fig. 2 were cast. The test configuration was proposed by Arne Hillerborg. A similar test set up was earlier used by Shah and Ballarini, see Ballarini et al. (1986). The specimens were cast in steel moulds and were made with two different geometries, here denoted series 1 and series 2. The geometry in series 2 is according to the RILEM, TC 90-FMA (1990) Round Robin Analysis of Anchor bolts. Test series 1 has earlier been presented in Ohlsson (1990) and in Ohlsson and Elfgren (1991) and is only summarized here.

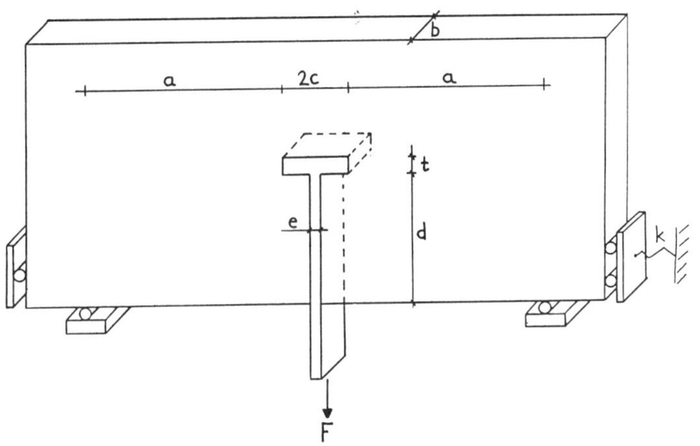

Fig. 2. Test specimens. In series 1, b = 50 mm, 2 c = 100 mm, t = 25 mm, e = 15 mm. The ratio a/d was 300/200 = 1.5, 150/100 = 1.5 and 200/100 = 2.0. In series 2, b = 50 mm, 2c = 15 and 45 mm, t = 5 and 15 mm, e = 5 and 15mm. The ratio a/d were 100/50 = 2.0,.150/150 = 1.0, and 300/150 = 2.0. The stiffness k was obtained by the tension in two 12 mm reinforcement bars.

The concrete was made with Portland cement and natural aggregates, see table 1.

Table 1. Concrete mixes.

Concrete	Cement (kg/m^3)	Water (kg/m^3)	Aggregate 0-8 mm (kg/m^3)	Aggregate 8-12 mm (kg/m^3)	Micro silica (kg/m^3)	Super plastisizer (kg/m^3)
C40	237	177	1064	934	13	
C45	374	173	933	859		
C80	450	156	770	1149	70	10

Table 2. **Material properties.**

Concrete	f_{cc} (MPa)	f_{cspl} (MPa)	E_c (GPa)	G_{FE} (N/m)
C40	44q 1 (12)	4.1 0.5 (6)	-	151 23 (5)
C45	55 2 (3)	3.4 0.2 (3)	38.2 0.9 (3)	91 7 (3)
C80	85 7 (3)	5.9 0.3 (6)	40.4 0.9 (4)	121 1 (3)

Numbers in parenthesis denote number of tests; ± values denote standard deviations.

Cubes (150 mm) were cast in order to determine the splitting strength f_{cspl} and the compressive strength f_{cc}. The modulus of elasticity, E was determined on drilled concrete cores, diameter 70 mm, length 150 mm. The fracture energy, G_{FE}, was determined with the RILEM three point bending test on notched beams (840 x 100 x 100 mm), see Hillerborg (1985).

The tests were performed in a servohydraulic testing machine. The C45 concrete specimens in series 1 were loaded by pulling the anchor rod with a constant velocity. The other specimen were tested in displacement control with the anchor head displacement (the displacement of the anchor head relative to the supports) as the feedback signal to the servohydraulic system.

The failure generally consists of three main cracks, see Fig. 3. Besides the pull out cone, a bending crack is formed. The bending crack can be avoided if the specimen is very high compared to the embedment depth, but the specimen would then be very difficult to handle.

Fig. 3. Anchor bolt No 5 after testing.

Fig. 4 show examples of load-displacement curves. The first part of the curves is linear elastic with no crack growth. At a certain load level cracking starts. The specimen weakens. The load-displacement diagram has a peak and an unloading part. At the end of this second part the crack system is visible. As the deformation increases the stiffness of the specimen also increases. The stiffening is caused by shear stresses in the existing tensile cracks. Some of the specimen with small embedment depths do not show this stiffening effect.

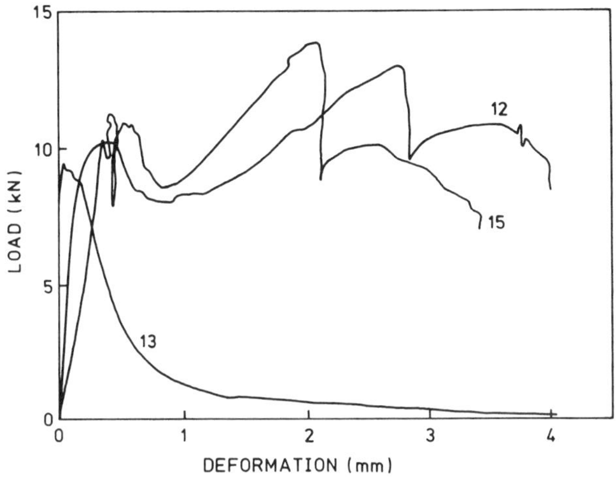

Fig. 4. Load-displacement curves for test No 12, 13 and 15.

Table 3 shows the peak loads obtained.

Table 3. Peak loads

Test No	Concrete type	d (mm)	a (mm)	F_{max} (kN)	Test No	Concrete type	d (mm)	a (mm)	F_{max} (kN)
Series 1					Series 2				
1	C45	200	300	16.0	12	C40	50	100	12.9
2	C45	200	300	20.1	13	C40	50	100	9.4
3	C45	100	150	15.2	14	C40	50	100	11.4
4	C45	100	150	15.4	15	C40	50	100	13.8
5	C80	200	300	27.5	16	C40	150	150	51.3
6	C80	200	300	31.8	17	C40	150	150	33.5
7	C80	200	300	25.8	18	C40	150	300	17.0
8	C80	100	150	38.1	19	C40	150	300	16.5
9	C80	100	150	23.6					
10	C80	100	200	24.8					
11	C80	100	200	21.3					

2.2 Finite Element Analysis

The pullout test of the series 1 specimen was analyzed with the finite element method. The analysis has focused on the influence of different embedment depths and material properties.

The pull-out test were modelled using the finite element program ABAQUS. Two embedment depths, 100 and 200 mm, were studied.

A discrete cracking model was used. The crack path was modelled as a straight line from the upper edge of the anchor head to the support on the top surface, see Fig. 5. The crack path is marked as a solid line in the figure.

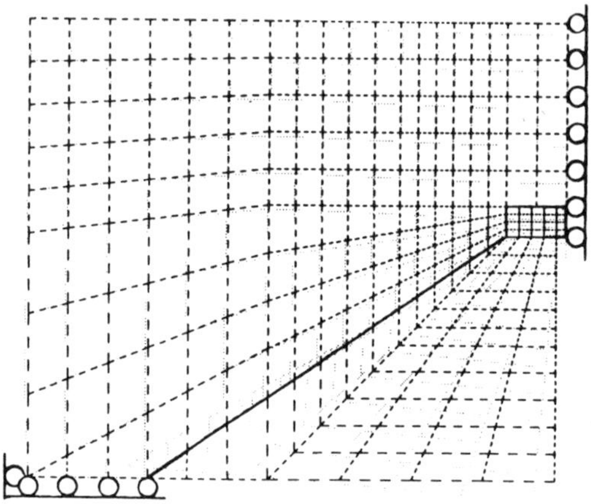

Fig. 5. Finite element model. Embedment depth d = 100 mm and d = 200 mm.

Four-noded plane stress elements, type CPS4 with 4 integration points were used. The fracture zone was modelled with spring elements, type SPRING2. A load-displacement relation in the 1-direction (crack opening direction) was chosen to give a linear stress-crackwidth relation. In the 2-direction (crack sliding direction) the stiffness was zero. Bond between steel and concrete was modelled only on the horisontal parts of the anchor head which transferred compressive stresses to the concrete.

The material properties was varied according to table 4.

Table 4. Material properties in the analysis.

Series	G_F (N/m)	f_{ct} (MPa)	E_c (GPa)	d (mm)
A	50,100,200,400	3.0	30	200
B	100	0.75,1.5,3.0,6.0	30	200
C	100	3.0	15,30,60,120	200
D	50,100,200,400	3.0	30	100
E	100	0.75,1.5,3.0,6.0	30	100

Load-displacement curves showing the influence of the parameters G_F, f_t, E_c and d are shown in Fig. 6. A high fracture energy gives a higher peak load and a more ductile behaviour after peak load. The peak load increases with a higher tensile strength but the brittleness increases too. A low tensile strength gives a very ductile behaviour. A high modulus of elasticity gives a higher peak load and a stiffer initial behaviour.

The influence of different embedment depths, d = 100 and d = 200 mm, has also been studied. The peak load is higher for the larger embedment depth. The ratio between the peak loads is ≈ 1.3. The studied pull-out test with two different embedment depths is not a true size effect problem. The size of the anchor bolt is the same for the both embedment depths. The smaller specimen will be stiffer compared to the larger specimen. This stiffness increase will increase the peak load.

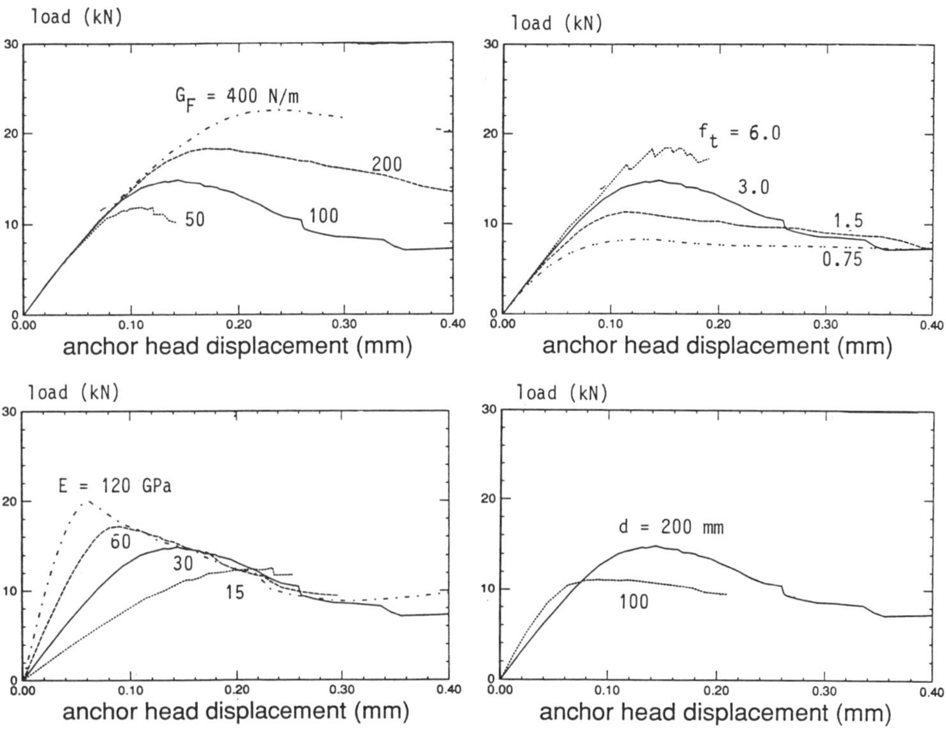

Fig. 6. Load-displacement curves from Finite Element Analysis. Material properties according to Table 4.

2.3 Brittless number

The brittleness number is the ratio between elastic energy and fracture energy in a structure, as shown in section 1.

The fracture energy is the amount of energy that the structure can absorb in the fracture zone. When the elastic energy in a structure is large compared to the fracture energy, the structure will have a brittle behaviour, compare Gustafsson and Hillerborg (1985), Bazant and Pfeiffer (1987) and Di Tommaso (1989).

The elastic energy in a specimen is correlated with the stiffness of the specimen. An effective modulus of elasticity, E_{eff}, proportional to the stiffness of the specimen is therefore introduced. E_{eff} is determined from linear elastic finite element calculations. A Finite Element calculation with $E_{steel} = E_{concrete}$ is made. The stiffness F/S obtained is denoted K_{ref}. In the second calculation E_{steel} is raised to its real value. The stiffness obtained is denoted K_i. The effective modulus of elasticity is then calculated as:

$$E_{eff} = E_{concrete} \cdot \frac{K_i}{K_{ref}} \qquad (2.1)$$

E_{eff} is therefore a function of $E_{concrete}$, E_{steel} and the geometry of the specimen.

The brittleness number can now be written $B = f_t^2 d/(E_{eff} G_F)$

Investigating the influence of the brittleness number, B, on the peak load, F_{max}, we plot a dimensionless diagram with the brittleness number, B, versus a nondimensional load $F_{max}/d \cdot b \cdot f_t$, see Fig. 7.

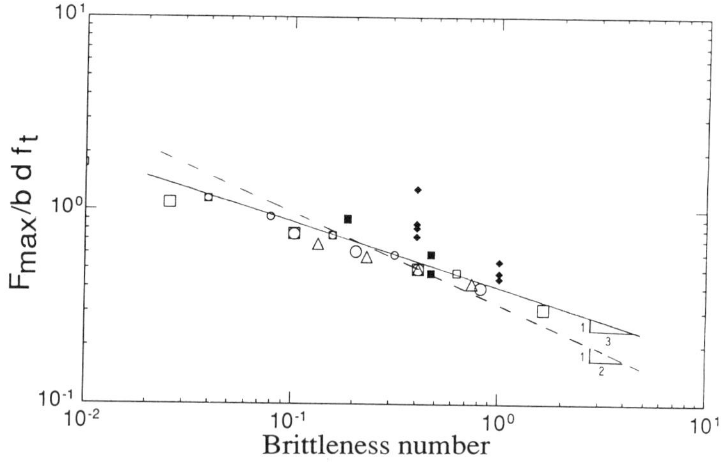

Fig. 7. Results from finite element calculations and laboratory tests. Open symbols denote results from Finite Element Calculations and filled symbols denote results from laboratory tests.

Fig. 7 shows a correlation between the brittleness number and the peak load in the Finite Element Calculations. The laboratory tests also show this correlation, but the scatter is very large. The Finite Element Calculations seems to predict a lower limit for the bearing capacity. A curve between the plotted results from the finite element calculations will have a slope of $k \approx -1/3$. This means that

$$\frac{F_{max}}{b \cdot d \cdot f_t} \sim \left(\frac{1}{B}\right)^{1/3} \quad \text{or} \quad F_{max} \sim b\,(E_{eff} G_F f_t \cdot d^2)^{1/3} \tag{2.2}$$

Thus, according to the finite element calculations, E_{eff}, G_F, and f_t has equal influence on the bearing capacity of the anchor bolt.

The slope of the $B - F_{max}/b \cdot d \cdot f_t$ curve may be a starting point when considering design formulas

$k = 0$	will give	$F_{max} \sim b \cdot d \cdot f_t$ (no size effect)
$k = -1/4$	will give	$F_{max} \sim b \cdot d^{3/4} \cdot f_t^{1/2} \cdot E^{1/4} \cdot G_F^{1/4}$
$k = -1/3$	will give	$F_{max} \sim b \cdot d^{2/3} \cdot f_t^{1/3} \cdot E^{1/3} \cdot G_F^{1/3}$
$k = -1/2$	will give	$F_{max} \sim b \cdot d^{1/2} \cdot E^{1/2} \cdot G_F^{1/2}$ (linear fracture mechanics)

3. Axi symmetric specimen

Some test results are given in Table 5 and in Fig. 8. Analytical results presented by Eligehausen and Özbolt (1990) indicate a slope changing from $k = -1/2$ to $k = -1/3$ for decreasing brittleness numbers. A slope of $k = -1/2$ (corresponding to a linear elastic fracture mechanics analysis) will give $F_{max} \sim d^{3/2}\, E^{1/2}\, G_F^{1/2}$, compare with Eq. (1.1). More results are needed and will hopefully be provided by the ongoing Round Robin Analysis and Tests, RILEM TC-90 FMA (1990).

Table 5. Test results

Test	d mm	f_t MPa	E GPa	G_F N/m	F_u kN
Barr (1990)	50	1.5	20	100	22.5
"	50	1.5	20	100	29.75
"	150	1.5	20	100	159
Eligehausen &	120	1.8	23.5	70	97.2
Sawade (1989)	250	1.8	23.5	70	290
"	500	1.8	23.5	70	885
Ohlsson (1991)	50	4.1	30	150	26

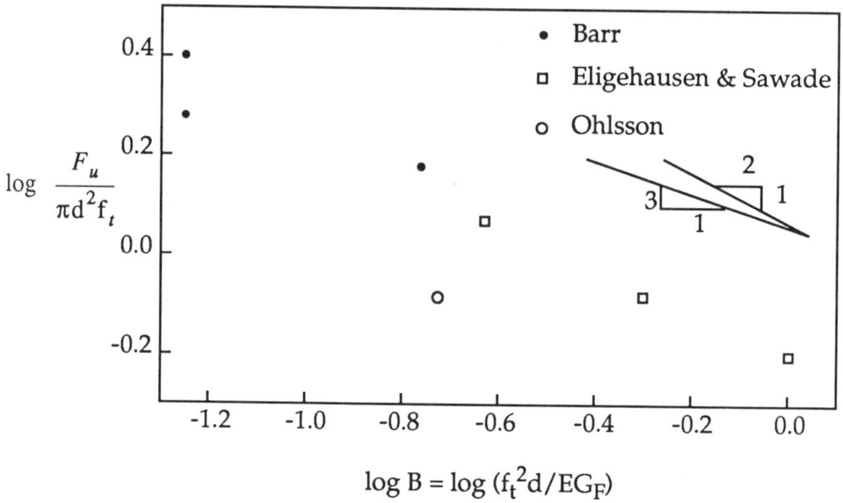

Fig. 8. Failure load F_u as function of brittleness number for axisymmetric specimen.

4. References

ACI 349-76 (1978) Steel embedments. Proposed addition to code requirements for nuclear safety related concrete structures. **ACI Journal** (Detroit), Vol 75, No 8, August 1978, pp 329-347.

Ballarini, R., Shah, S.P. and Keer, L. M. (1986) Failure characteristics of short anchor bolts embedded in a brittle material. **Proceedings of the Royal Society of London**, Vol A404, pp 35-54.

Ballarini, R. and Shah, S.P. (1991) Fracture mechanics based analysis of pull-out tests and anchor bolts. **Analysis of Concrete Structures by Fracture Mechanics** Proceedings of the RILEM workshop dedicated to Professor Arne Hillerborg (edited by Lennart Elfgren and Suru P Shah) Chapman & Hall, London, pp 245-280.

Barr, B.I. (1990) Contribution to Round-Robin Analysis of Anchor Bolts, Personal Communication.

Bazant, Z.P. and Pfeiffer, P.A. (1987) Determination of fracture energy from size effect and brittleness number. **ACI Materials Journal**, (Detroit), Vol 84, No 6, pp 463-480.

de Borst, R. (1986) **Non-linear analysis of frictional materials.** Proefschrift ter verkrijging van de graad van doctor in te technische wetenshappen aan de Technische Hogeschool Delft... Delft, 140 pp.

Cervenka, V. Puli, R. and Eligehausen, R. (1991) FEM simulation of concrete failure. **ECF 8 Fracture Behaviour and Design of Materials and Structures**, Vol II (editor D Firrao), EMAS, Warley, pp 728-733 (ISBN 0 947817 43 3)

Elfgren, L., Broms, C.E., Johansson, H. och Rehnström, A. (1980) **Anchor bolts in reinforced concrete foundations. Short time tests**, Division of Structural Engineering, Luleå University of Technology. Research Report TULEA 1980:36, 117 pp.

Elfgren, L., Broms, C.E., Cederwall, K. and Gylltoft K. (1982) Fatigue of anchor bolts in reinforced concrete foundations. IABSE Colloquium **Fatigue of steel and concrete structures**, Lausanne March 1982, Proceedings, IABSE report, Vol. 37, Zürich, pp 105-117.

Elfgren, L. (editor) (1989) **Fracture Mechanics of Concrete Structures. From theory to applications**, A RILEM Report by TC 90-FMA. Chapman & Hall, London, 410 pp (ISBN 0 412 30680 8).

Eligehausen, R. and Sawade, G. (1989) Analysis of anchorage behaviour. A fracture mechanics based description of the pull-out behaviour of headed studs embedded in concrete (chapter 13.2) in **Fracture Mechanics of Concrete Structures. From theory to applications** (edited by Lennart Elfgren). Chapman & Hall, London, pp 263-280.

Eligehausen, R. and Ozbolt, J. (1990) Size effect in anchorage behaviour (chapter 13.1). **ECF 8 Fracture Behaviour and Design of Materials and Structures**, Vol II (editor D Firrao), EMAS, Warley, pp 721-727 (ISBN 0 947817 43 3).

Gustafsson, P.J. and Hillerborg, A. (1988) Sensitivity in shear strength of longitudinally reinforced concrete beams to fracture energy of concrete. **ACI Structural Journal** (Detroit), Vol 85, No 3 May-June, pp 286-294.

Hellier, A. K., Sansalone, M., Ingraffea, A.R., Carino, N. J. and Stone, W.C. (1987) Finite element analysis of the pullout test using a nonlinear discrete cracking approach. **Cement, Concrete and Aggregates**,(ASTM), Vol. 9, No 1, Summer 1987, pp 20-29.

Hillerborg, A. (1985) The theoretical base of a method to determine the fracture energy G_F of concrete. **Materials and Structures**, Vol 18, No 106, pp 291-296.

Klinger, R.E. and Mendonca, J.A. (1982) Tensile capacity of short anchor bolts and welded studs: A literature review. **ACI Journal** (Detroit) Vol 79, No 4, July-August, pp 270-279.

Ohlsson, U. (1990) **Fracture Mechanics Studies of Concrete Structures**. Division of Structural Engineering, Luleå University of Technology, Licentiate thesis 1990:07L, 8 + 12 + 18 + 7 +21 pp.

Ohlsson, U. and Elfgren, L (1991) Anchor bolts in Concrete Structures. Two dimensional modelling. **Analysis of Concrete Structures by Fracture Mechanics.** Proceedings of the RILEM Workshop dedicated to Professor Arne Hillerborg (edited by Lennart Elfgren and Suru P Shah). Chapman & Hall, London, 1991, pp 281-301.

Ottosen, N.S. (1981) Non-linear finite element analysis of pull-out tests. **Journal of the Structural Division** (New York), ASCE, Vol 107, No ST4, April 1981, pp 591-603.

Peier, W. H. (1983) Model of pull-out strength of anchors in concrete. **Journal of Structural Engineering,** (New York) ASCE, Vol. 109, No 5, May 1983, Paper No. 17949, pp 1155-1173.

Rehm, G., Eligehausen, R. and Mallee, R. (1988) Befestigungstechnik (Fastening technology). In German. In **Beton Kalender 1988**, Ernst & Sohn, Berlin 1988, pp 569-663.

RILEM TC 90-FMA (1990) Round-Robin Analysis of Anchor Bolts - Invitation **Materials and Structures**, Vol 23, No 133, Jan 1990, pp 78. Revised invitation in **RILEM News** (Cachan Cedex), No 1991.1, January 1991.

Rots, J. (1988) **Computational modelling of concrete fracture**, Dissertation, Delft University of Technology, Department of Civil Engineering, Delft, 132 pp.

Di Tommaso, A. (1989) Size effect and brittleness. Chapter 7 **in Fracture Mechanics of Concrete Structures. From theory to Applications**, (edited by Lennart Elfgren). Chapman & Hall, London, pp 191-207.

80 FRACTURE ANALYSIS OF CONCRETE PLANE-STRESS PULL-OUT TESTS

V. ČERVENKA and R. PUKL
Klokner Institute of the Czech Technical University, Prague, Czechoslovakia
R. ELIGEHAUSEN
Institute for Building Materials, Stuttgart University, Germany

Abstract
This paper reports on a parameter study which was performed for the Round Robin Analysis Of Anchor Bolts organized by RILEM Committee TC-90 FMA (Fracture Mechanics Applications). 18 plane-stress pull-out specimens were analyzed by the computer program SBETA which is based on the finite element method and takes into account nonlinear fracture mechanics. A simplified formula for the design of such anchors is derived.
Keywords: Pull-Out Test, Anchor Bolt, Non-linear-fracture Mechanics, Finite Element Method.

1 Introduction

The authors have performed a numerical study in connection with the RILEM Round Robin Analysis Of Anchor Bolts [1]. A part of this study, which concerns the plane stress solution, is described in this paper. The plane stress anchor represents a linear anchoring element which can be a steel profile or a row of anchoring bolts. The failure mechanism of anchors subjected to a pull-out load is a complex problem which is typically investigated experimentally. Recent advances in the computational fracture mechanics make it possible to perform a computer simulation of such tests.

2 Description of pull-out tests

A two-dimensional specimen with a steel anchor as shown in Fig.1 is considered. The specimen is supported by two roller bearings and an axial loading force is applied to the anchor. Two cases of lateral constraints were considered, namely, with rigid constraint and without constraint. The thickness of the specimens is 100 mm. The material parameters are specified in Fig.1. The aim of the test was to determine the maximum load which can be transmitted by the anchor under variable geometry and constraint conditions. Three values of the embedment depth d and three ratios of the support span a to the embedment depth are considered. By varying the geometrical parameters and two lateral constraint conditions 18 cases are obtained. All these tests were simulated by computer analysis. This

Fracture Processes in Concrete, Rock and Ceramics Edited by J.G.M. van Mier, J.G. Rots and A. Bakker.
 Published by E. & F.N. Spon, 2–6 Boundary Row, London SE1 8HN. ISBN 0 419 15870 7

numerical study is in detail described in the report [4].

3 Method of FE-fracture analysis

The analysis was performed by the computer program SBETA which is based on the finite element analysis and a non-linear-elastic constitutive model of concrete. A smeared material approach based on the Bazant's crack-band model [5] is used. The stress-strain law for concrete in tension is bi-linear with a linear part up to the tensile strength and linear softening, Fig.2. In order to model correctly the discrete nature of cracking the softening modulus E_t is related to the fracture energy parameter G_f, the crack band width h and the tensile strength f_t by the formula

$$E_t = -f_t^2 \frac{h}{2G_f} \tag{1}$$

The crack band width is related to the mesh size by $h = \sqrt{A}$, where A is the finite element area. Two types of crack model are used. In the fixed-crack model the crack direction is determined at the instant of crack initiation and kept fixed during the analysis. The shear strain on the crack plane can arise and thus the shear stiffness as defined by Eq.(2) is used. The shear stiffness is modeled by Kolmar's relation [6] in which the shear modulus G declines with the crack band strain ϵ_u:

$$G = r_s\ G_c\ , \qquad r_s = 0.0923\ ln\left(\frac{\epsilon_u}{5.33}\right) \tag{2}$$

In this formula G_c is the initial shear modulus of concrete and r_s is the reduction coefficient. In the rotated-crack model the crack direction coincides with the principal strain axes. The material properties are modeled only by the normal stress components and the shear stiffness is not used.

The model contains also other properties to describe the concrete behavior, namely, a bi-axial failure function, a non-linear response in compression and a reduction of the compressive strength in cracked concrete. However, these aspects of the SBETA constitutive model are not important for the brittle mode of failure treated here and therefore shall not be discussed in detail. A four-node quadrilateral finite element is used. It is composed of two four-node subtriangles with an exact stiffness integration. The nonlinear solution is performed by the arch-length method which enables the analysis of the post-peak behavior. Details about the program SBETA can be found in references [2,3]).

4 Results of parameter study

The subject of the parameter study was to simulate 18 cases of pull-out tests described in Section 2. An example of a finite element mesh for the pull-out specimen is shown in Fig.3. Note that only half of the specimen was analyzed taking advantage of the symmetry. Similar meshes were used for all specimens. All cases were calculated for fixed and rotated crack models. The calculated peak loads are given in Table 1. It contains also the data from an analysis based on the

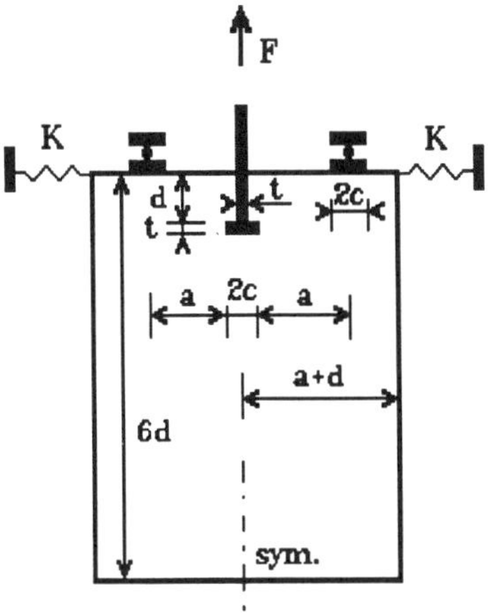

Geometry:
$d = 50, 150, 450$ mm
$a = d/2,\ d,\ 2d$
$2c = 3d/10,\ t = d/10$
thickness = 100 mm

Lateral constraint: $K = 0, \infty$

Material properties:
$f_t = 3$ MPa, $f_c = 40$ MPa,
$E = 30$ GPa, $\nu = 0.2$
$G_f = 100$ N/m

Fig.1 Geometry and material properties of the pull-out specimens.

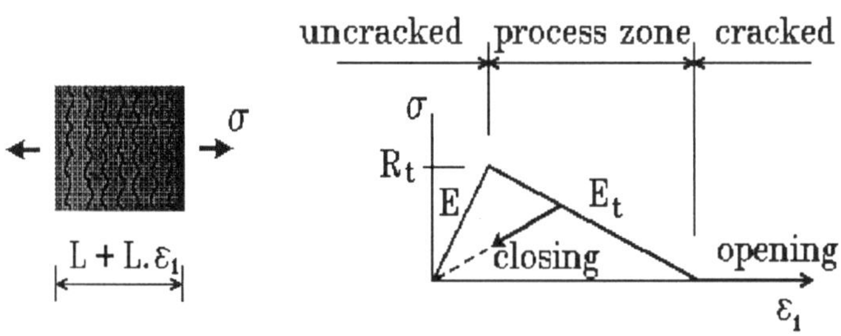

Fig.2 Stress-strain law for concrete in tension.

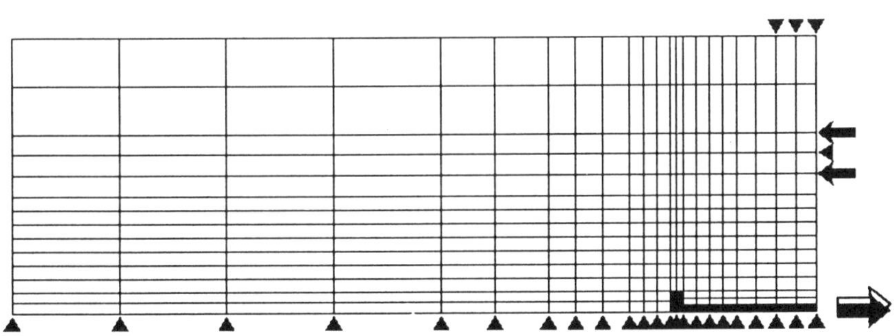

Fig.3 Finite element mesh for typical specimen.

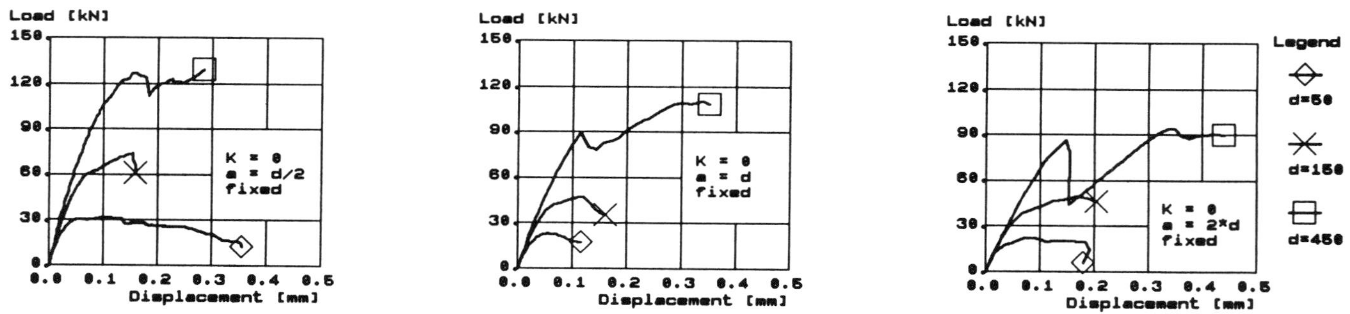

Without lateral constraint.

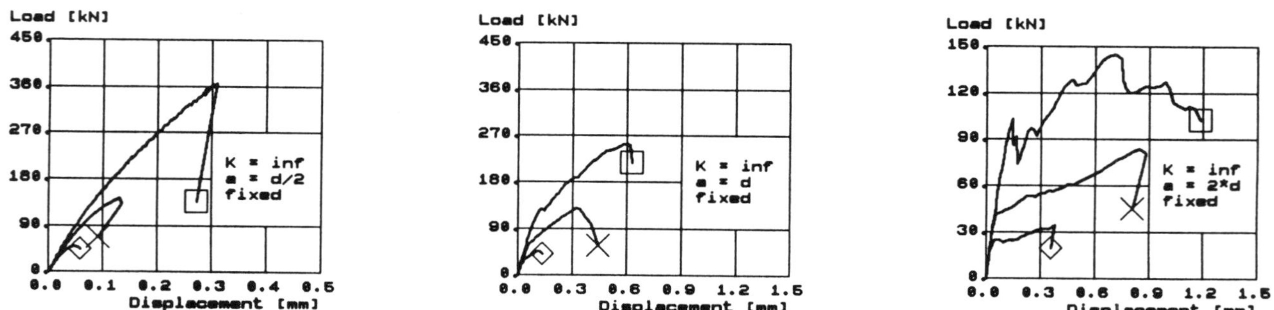

With lateral constraint.

Fig.4 Load-displacement diagrams of 18 specimens analyzed by fixed crack model.

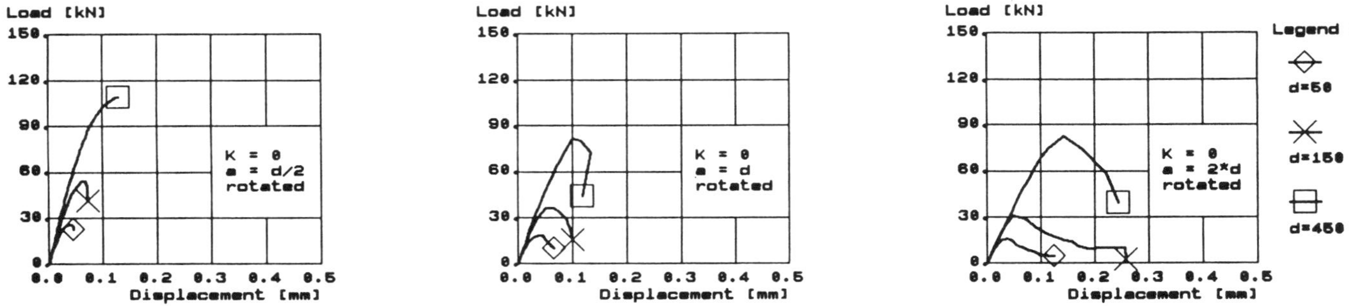

Without lateral constraint.

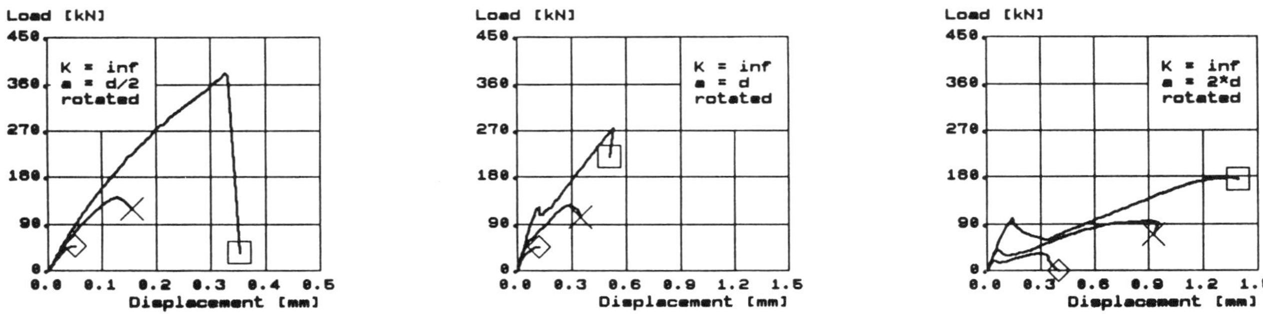

With lateral constraint.

Fig.5 Load-displacement diagrams of 18 specimens analyzed by rotated crack model.

microplane model which will be discussed later. Table 2 summarizes displacements of the anchor head at peak load. The load-displacement diagrams for fixed-crack analysis of all specimens are shown in Fig.4 and for rotated-crack analysis are shown in Fig.5. (Note that different scales of the diagrams are used. This had to be accepted because of the wide range of the loads, varying from 16.6 to 382.7 kN.)

Table 1 Peak loads [kN]

lateral constraint		0			∞		
d [mm] =		50	150	450	50	150	450
a =	material model						
d/2	fixed cracks	31.9	74.3	129.4	49.8	144.6	366.5
	rotated cracks	26.0	54.3	109.0	49.5	142.7	382.7
d	fixed cracks	23.1	47.2	110.2	47.6	130.7	254.9
	rotated cracks	18.8	36.5	81.9	47.1	129.3	276.1
	microplane	17.5	42.7	93.4	-	-	-
2 d	fixed cracks	22.2	49.5	94.3	34.6	84.0	145.5
	rotated cracks	16.5	31.1	82.9	34.3	96.8	179.6

Table 2 Displacements [mm] of anchor head at peak load

lateral constraint		0			∞		
d [mm] =		50	150	450	50	150	450
a =	material model						
d/2	fixed cracks	0.102	0.154	0.284	0.047	0.130	0.308
	rotated cracks	0.035	0.067	0.123	0.045	0.125	0.325
d	fixed cracks	0.053	0.107	0.338	0.111	0.316	0.585
	rotated cracks	0.035	0.050	0.100	0.104	0.287	0.527
	microplane	0.034	0.087	0.186	-	-	-
2 d	fixed cracks	0.071	0.164	0.336	0.378	0.843	0.704
	rotated cracks	0.034	0.046	0.141	0.280	0.911	1.322

The most diagrams have only one peak at the maximum load. However, some specimens reached first local instability before maximum load, then, after forming a new stable crack pattern the load could be further increased up to the peak. The post-peak descending branch could be calculated until a kinematically unstable system was reached (pulling-out of the anchor).

Each diagram containing three curves shows the effect of the embedment depth (size effect) on the pull-out behavior of the specimens. It can be observed that the ratio of the failure loads for the specimens with the geometrical scaling factor 3 is much smaller than this value. This effect will be quantitatively evaluated in Section 5. From comparison of the different diagrams the effect of the various parameters can be observed. The tests with lateral constraint gave always greater maximum loads than tests without constraint. With increasing support span the maximum load decreases and the effect of lateral constraint also decreases.

The effect of the crack model can be found from the comparison of Fig.4 and 5. A comparison for a particular case (d=450 mm, a=d, k=0) is shown in Fig.6. Here the fixed crack model gives a higher maximum load than the rotated crack model. This could be expected because the rotated-crack model is by it's nature more flexible. It is also interesting to compare these models with an analysis based on the microplane model which was performed by Ozbolt for the same Round-Robin Analysis [1]. The microplane model considers a discrete material structure [7]. It appears that the microplane model gives an intermediate solution between fixed and rotated crack models.

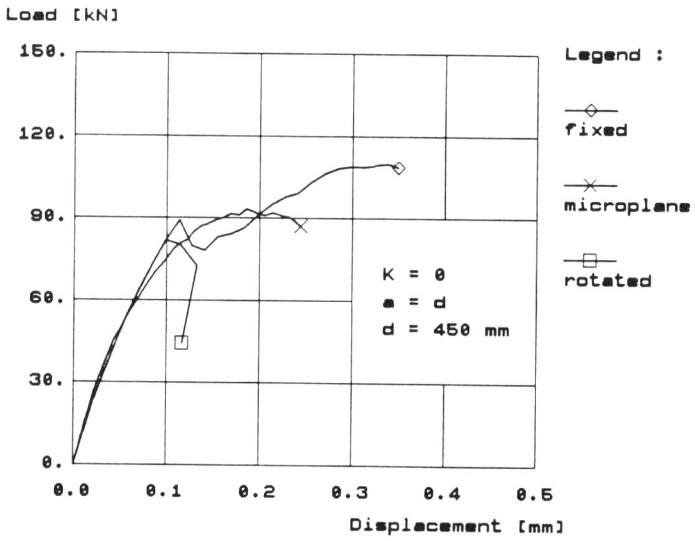

Fig.6 Comparison of load-displacement diagrams for three crack models.

However, the fixed-crack model is not always stronger (see Table 1). E.g. in constrained specimens with d=450 mm the rotated crack analysis gave greater maximum loads than the fixed crack one. Thus a general conclusion about the relation between fixed and rotated crack analysis with respect to the maximum load cannot be made.

The computer simulation provided also a large amount of information about the deformed states, crack patterns and stress fields. They were evaluated using the SBETA graphical system and are comprehensively described in the report [4]. Here one example of the crack propagation is shown only. Fig.7 illustrates the crack development for a typical specimen. In the bottom the crack process zone is depicted by crack-lines, while in the upper part of the figure the crack is indicated by means of strain isolines. Both of these graphical forms are useful for the

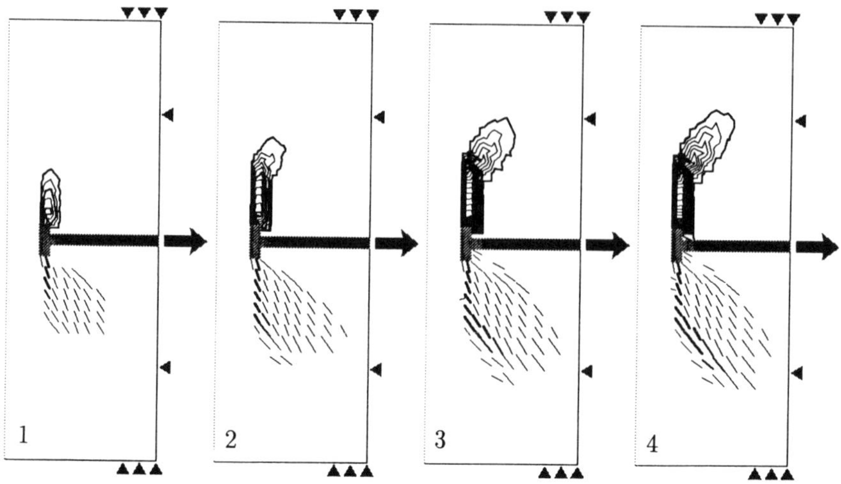

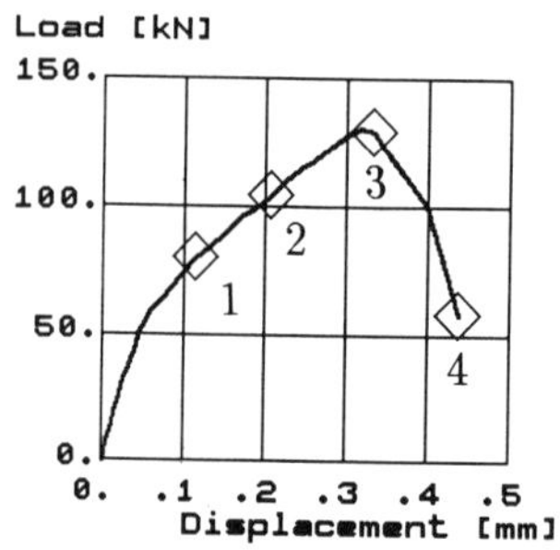

Fig.7 Crack propagation calculated by SBETA for the specimen d=50 mm, a=d, fixed crack model.

Fig.8 Load-displacement diagram for the specimen from Fig.7.

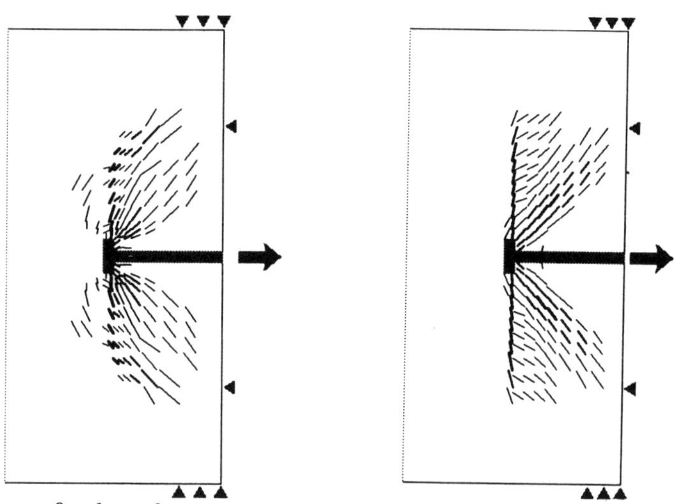

fixed cracks rotated cracks

Fig. 9 Comparison of the final crack patterns for two crack models ($K = \infty$).

investigation of the crack localization. Fig.8 shows the corresponding load-displacement diagram with locations of points for crack patterns (total number of load steps was 26). In Fig.9 the final crack patterns for fixed and rotated crack analyses are compared.

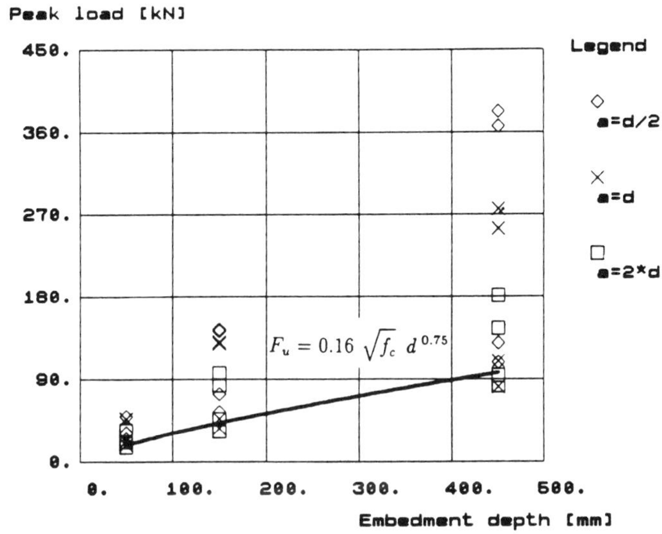

Fig.10 Comparison of Eq.(4) with all simulated tests.

5 Size effect

Some practical conclusions can be drawn from this parameter study for the design of linear anchors in plane stress state. A formula for the maximum load can be proposed in the following form:

$$F_u = k\sqrt{f_c}\, d^{\alpha}\left(\frac{a}{d}\right)^{\beta} \tag{3}$$

Failure load F_u is in kN. The embedment length d and the support span a are in mm. The thickness is considered as $b = 100$ mm. In the calculation $f_c = 40$ MPa. In Eq.(3) the effect of concrete strength on F_u is represented by the expression $\sqrt{f_c}$. This is based on the generally accepted assumption that the the pull-out failure load is directly related to the tensile strength of concrete which is approximately proportional to this expression $\sqrt{f_c}$. However, recent investigation on the behavior of anchors indicate that the concrete cone failure load is proportional to the parameter $\sqrt{EG_f}$ [8]. A more general expression could be derived by a similar numerical study with variation of the parameters E and G_f.

Constants k, α, β were derived using the least-square fit method. For specimens without lateral constraint they amount to: $k = 0.25, \alpha = 0.68, \beta = -0.28$ and for specimens with lateral constraint: $k = 0.31, \alpha = 0.80, \beta = -0.40$. The results indicate that for a large span the effect of lateral constraint vanishes. Using Eq.(3) and the above given parameters this occurs at $a/d \doteq 5$. In such a case the expression $k(a/d)^{\beta}$ is equal to 0.16. This corresponds to practical cases with large support spans. Thus Eq.(3) can be simplified as

$$F_u = 0.16\ \sqrt{f_c}\ d^{\,0.75} \tag{4}$$

The formula is compared with the simulated tests in Fig.10. The exponent $\beta = 0.75$ in Eq.(4) is half way between the values valid for linear fracture mechanics ($\beta = 0.5$) and theory of plasticity ($\beta = 1.0$). Before practical implementation of the above results an experimental verification should be made at least for some representative cases.

6 Conclusion

The computer simulation of pull-out tests based on the finite element method and nonlinear fracture mechanics was successfully applied. A simple formula for the maximum pull-out force of the linear anchors in plane stress state is derived.

References

[1] ELFGREN, L. (Editor) - Round-Robin Analysis of Anchor Bolts, RILEM TC-90 FMA Fracture Mechanics of Concrete-Application, Preliminary Report, Luleå University of Technology, September 1990.

[2] ČERVENKA, V., ELIGEHAUSEN, R., PUKL, R. - SBETA Computer Program For Nonlinear Finite Element Analysis of Concrete Structures, Part 1, Program Description, Part 2, User's Manual, Part 3, Examples of Applications, Mitteilungen des Instituts für Werkstoffe im Bauwesen No.1990/1, Universität Stuttgart.

[3] ČERVENKA, V., ELIGEHAUSEN, R., PUKL, R. - FEM Computer Models Of Concrete Structures, IABSE Colloquium Structural Concrete, April 10-12, 1991, Stuttgart, Germany.

[4] PUKL, R., ČERVENKA, V., ELIGEHAUSEN, R. - Study of pull-out tests of anchor bolts - 2D computer simulation, Institut für Werkstoffe im Bauwesen, Universität Stuttgart, 1991 (Report under preparation)

[5] BAŽANT, Z.P., OH, B.H. - Crack Band Theory for Fracture of Concrete, Mater. Struct. RILEM, Paris, France, 16, 1983, pp. 155-177.

[6] KOLMAR, W. - Beschreibung der Kraftübertragung über Risse in nichtlinearen Finite-Element-Berechnungen von Stahlbetontragwerken, Dissertation, Technische Hochschule Darmstadt 1985, p.94.

[7] BAŽANT, Z.P., OŽBOLT, J. - Nonlocal microplane model for fracture, damage, and size effect in structures, ASCE Journal of the Engineering Mechanics Division, 116, No.11, Nov. 1990, pp.2485-2505.

[8] OŽBOLT, J., ELIGEHAUSEN, R. - Numerical Analysis of Headed Studs Embedded in Large Plain Concrete Blocks. Institut für Werkstoffe im Bauwesen, Universität Stuttgart, Report No. 4/10-90/9, Oct. 1990.

81 CONSEQUENCE OF CONCRETE FRACTURE: BRITTLE FAILURE AND SIZE EFFECT OF R.C. BEAMS UNDER SHEAR

† Q.B. WANG and J. BLAAUWENDRAAD
Department of Civil Engineering, Delft University of Technology, The Netherlands

Abstract
Fracture of concrete leads to brittle failures and as a result causes size effect of decreasing strength in structures of increased sizes. These consequences are investigated numerically for slender reinforced concrete beams without shear reinforcement under shear load. In such beams, the shear resistance due to post-cracking shear mechanism of aggregate interlock is found to have minor significance and instead the softening tensile stress has a major significance because the deformation in the failing crack is governed by crack opening. So the shear failure can be considered as a matter of mode I fracture of concrete. A further investigation shows that the primary cause of size effect is concrete fracture of the localized failure crack. The deterioration of bond of reinforcement due to distributed cracking is a secondary one.
Keywords : Beam, Brittle failure under shear, Size effect, Discrete crack, Progressive concrete fracture, Aggregate interlock, Bond.

1. Introduction

Concrete beams without shear reinforcement, especially slender ones, often suffer from brittle failures when loaded under shear. An interesting phenomenon of the failures is that a single localized so-called diagonal crack arises and leads to the collapse of a beam. As far as the strength is concerned, the most interesting issue of the failure type is the size effect, i.e. the fact that the shear strength decreases as the depth of the beam increases (while geometrically it is properly scaled).

The mechanism of shear failure is not quite understood in the past and because of this, the size effect could not be explained clearly. In fact, diverse suggestions have been reported to this matter. Taylor [2], for instance, suggested that the shear resistance depends for an important part on the aggregate interlocking due to crack sliding and that the size effect is the result of decreasing interlock contribution in larger beams due to wider crack widths; Bazant and Kim [1] studied the size effect in shear strength by applying a size effect formulation which holds probably for the fracture of plain concrete structures. But they did not explain any direct relations between the failure mechanisms they used and the fracture property of concrete. The present paper presents the outcome of a numerical study on shear failures, revealing what is happening in the localized failure crack and on the basis of that linking the size effect and the failure mechanism together. The conclusion is: the shear failure is a matter of mode I fracture of concrete and the size effect is caused primarily by concrete fracture.

Fracture Processes in Concrete, Rock and Ceramics Edited by J.G.M. van Mier, J.G. Rots and A. Bakker.
 Published by E. & F.N. Spon, 2–6 Boundary Row, London SE1 8HN. ISBN 0 419 15870 7

2. Analysis methods

2.1. Crack approach

Diagonal cracking failures involve single localized cracks. Because of this, it requires a good crack model to simulate the shear transfer behavior in the cracks and to predict the ultimate load correctly. Unfortunately, it was found that the smeared crack approach did not fulfill this need [5b]. The reason is as follows. In the smeared crack approach, the shear stiffness of cracked concrete is modeled with a retained shear stiffness, usually expressed as βG (β is called the shear retention factor). Analyses using such a model showed that for values of β between 0.05 and 0.2, the brittle shear failure could not be simulated and instead only flexural failure was predicted. In fact, it was found that in the failure diagonal cracks the crack opening was far larger than the crack sliding. This required, from aggregate interlock theory [3], that β be arbitrarily small. However when β was further reduced, the analyses could not be continued at a certain stage due to convergence problem and could neither produce reliable results. It can be concluded [5b] that the smeared crack method treating shear transfer with a material stiffness is simply not suitable for shear critical discrete cracks. A crack sliding in a discrete crack can only occur parallel to the crack plane. But when it is transformed to a shear strain in a material model, it may occur both parallel and perpendicular to the crack plane. This causes an unnecessary stiffness reduction for the shear deformation perpendicular to the crack plane and may trigger additional convergence problems.

The discrete crack approach had to be used in the research. In absence of a program with automatic generation of discrete cracks, a predefined crack method was applied. By this it is meant that a line of interface elements is embedded in-between plane-stress elements. The interface elements have a high initial stiffness and are put cracked once a certain criterion involving the tractions is satisfied. Following [7], analyses were performed according to a "predictor-corrector" procedure. In a predictor analysis, the smeared crack approach was applied to predict the position of the critical (failure) crack. Further in a corrector analysis, a potential crack path was defined beforehand at the predicted position. For the system with only one discrete crack, the brittle shear failure could be predicted without numerical problems.

2.2. Constitutive models

Concrete was modeled with plane-stress elements. The smeared crack analyses were performed with the 8-noded square quadratic elements while for discrete crack analyses the linear triangular elements were used. The constitutive models for the materials are as follows:

1. Concrete before cracking or yielding: linearly elastic;
2. Crack initiation: $\sigma = f_t$ = tensile strength of concrete;
3. For the coexisting normal traction σ and tangential one τ in an interface element, a quadratic relation was used: $(\frac{\sigma}{f_t} + \frac{\tau}{f_s}) + ((\frac{\sigma}{f_t})^2 + (\frac{\tau}{f_s})^2) = 2$ following [8], with $f_s = 1.6\, f_t$.
4. Smeared cracks: linear tension-softening with a fracture energy $G_f = 0.07$ N/mm for a crack band;

5. Constant shear-retention ;
6. Discrete crack mode I: a Hillerborg's model with a linear stress-crack opening relation and a fracture energy $G_f = 0.07$ N/mm. The softening relation was applied to the resulant of σ and τ and the crack displacement in the corresponding direction;
7. Discrete crack mode II: aggregate interlock model [3];
8. Concrete yielding: Mohr-Coulomb criterion and perfect plasticity;
9. Steel: elastic-perfect-plasticity;
10. Bond-slip: An elastic-perfect-plastic relation with an initial stiffness of $E_s = 250$ N/mm^3 and a limit stress of $\tau_{su} = 5.0$ N/mm^2 ;
11. Dowel action: a non-linear spring, whose stiffness was obtained from simulation of the dowel specimens of Fenwick [6].

3. Brittle failure governed by mode I fracture

To understand the mechanism of diagonal cracking failures, some numerical results for a simply supported beam under concentrated loads are considered. For beams under distributed loads, numerical results [5b] showed a similar behavior. The beams were tested in [4] and the geometry and the crack pattern after failure can be seen in Fig.1.

The predictor analysis with the smeared crack approach made with a shear-retention factor of 0.2 produced a crack pattern as shown in Fig.1b. It was judged that the inclined crack band on the extreme left side corresponded to the failure crack. The criterion of the judgement was the inclination of the crack band and the fact that it meets the reinforcement line with a gentle slope. For the rather diffuse crack band, two cases of discrete cracks were investigated, Fig.1c. The crack path of case 1 appears to represent a flexural-shear crack with a bit long segment steeply across the reinforcement, while that of case 2 to represent a typical final collapse crack meeting the reinforcement with a gentle slope. In case 2 also a longitudinal discrete crack along the reinforcement was predefined. Fig.2 shows the load-displacement responses of the discrete crack analyses. It turns out that, in case 1 the load can be continually increased without brittle failure; In case 2, however, a peak load is predicted close to the test result, following which a brittle snap-back appears. Afterwards the load can increase slightly till the crack extends deeply into the compression zone (yielded finally). It is proved that, among the two crack paths, case 2 represents the failure crack but case 1 does not.

The deformations and shear stresses along the diagonal path plotted in Fig.3 (for case 1) explain the failure behavior of the beam. The basic characteristic of the deformation is that the two fragments separated by the crack rotate around the crack tip outwards. In the web nearby the crack tip, the deformation is dominated by a crack opening and shear stress is negligibly small. In the steep portion nearby the crack base, however, a large sliding occurs leading to a relatively large shear stress. In case 1, the shear force in the crack is so large as to remain the diagonal crack stable; In case 2, the situation is totally different. At the peak load the crack is very short, Fig.4a, and almost no aggregate interlock stresses (and dowel force) are developed (so not plotted). During the snap-back following the peak load, a large sliding displacement occurs at the crack base. But a high shear stress and dowel force are excluded by the longitudinal crack caused by themselves. In this case, the resistance of the crack is provided almost totally by the softening tensile stresses of concrete, see Fig.4b.

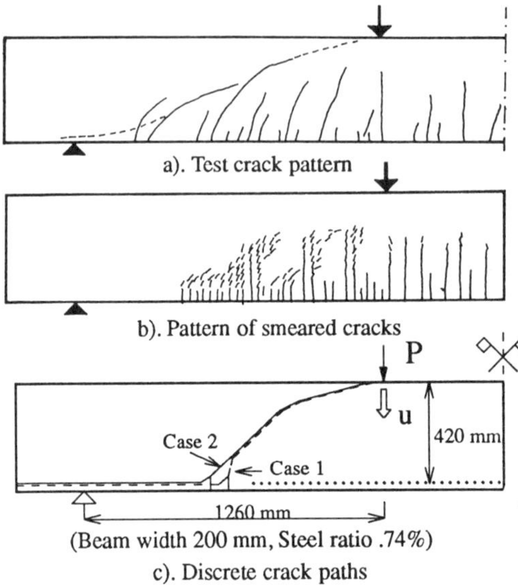

Fig.1. A simply supported beam

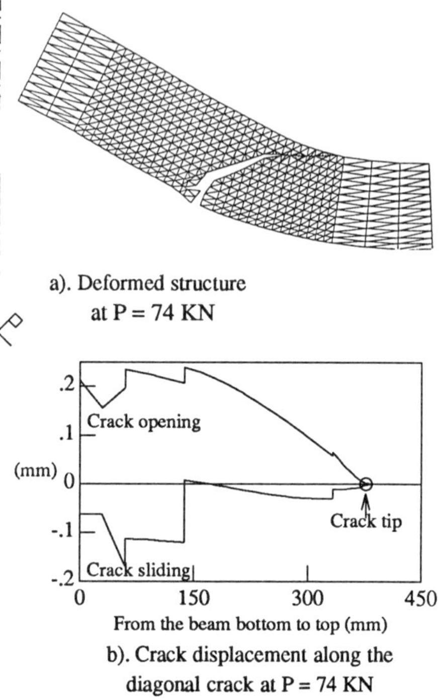

Fig.3. Case 1:Crack displacements & shear stress along diagonal path

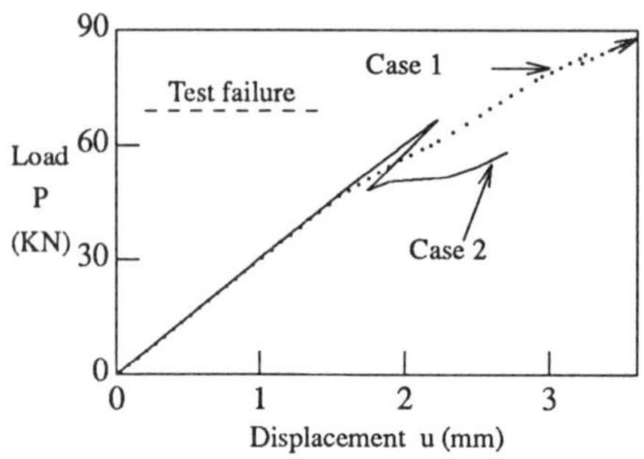

Fig.2. Load-displacement responses of discrete crack analyses

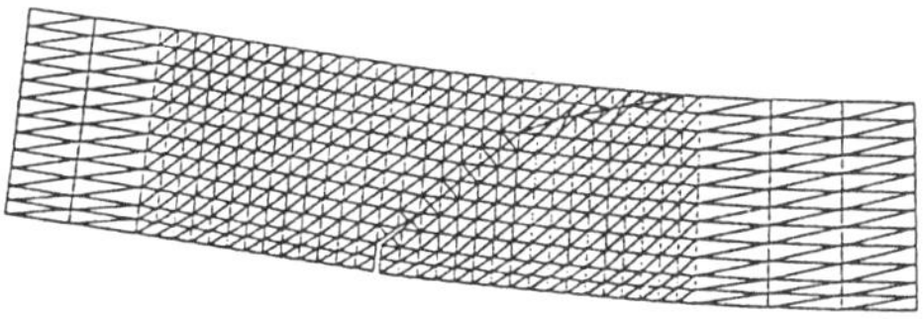

a). Deformed structure

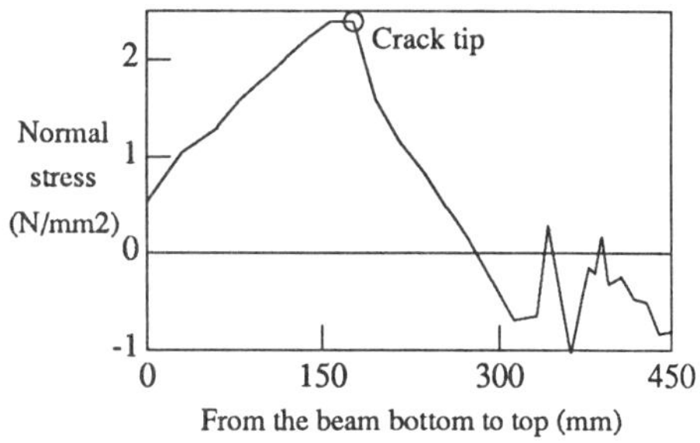

b). Normal stress along the diagonal cracks

Fig.4. Case 2: Normal stress along the diagonal crack (at peak load)

It turns out that, in the flexural-shear cracks of a beam the aggregate interlock mechanism may contribute much in resisting the external load. But in the flat failure crack interlocking is not activated before failure and instead, the crack is subjected primarily to a mode I deformation. Therefore, the shear failure of diagonal cracking can be considered as a matter of concrete fracture in the mode I condition.

4. Size effect

4.1. Size effect for perfectly scaled beams

The size effect was investigated from beams under distributed loads. Fig.5 shows the prototype beam with depth d = 150 mm (tested by [12]). Two larger beams are obtained by geometrically enlarging the prototype, having d = 300 mm and 600 mm respectively. The geometrical enlargement means a proportional multiplication of the depth d, the reinforcement area A_s, the bond-interface $\sum C$ of the circumference of the rebars. From smeared crack analyses it was found that the position of the critical inclined crack is approximately the same for all beams. So in the discrete crack analyses to determine the load capacity, a single element mesh was used (with different dimensions). The chosen discrete crack is shown in Fig.5c. Following the conclusions of the preceding section, the aggregate interlock and dowel action were all excluded in the analysis. In this way, there was no need to model the longitudinal crack along the rebar and numerical calculations became easier. Further, the material properties (elasticity modulus E_c, tensile strength f_t and fracture energy G_f) were kept constant.

The shear strength is plotted against the depth and compared with two empirical size effect formulas, Fig.6. Fig.7. shows the tensile stress along the diagonal path at the failure stage. The same figure also indicates the crack states at failure stage by the mark of the crack tips. It turns out that a considerable size effect is exhibited for the perfectly scaled beams. This size effect is all due to the fracture of concrete since no other nonlinearities were observed in the system (note that the bond stress was observed to be smaller than the limit bond strength and no plasticity was observed at the failure stage). Furtheron, the theoretical size effect is considerably weaker than the prediction of the empirical formulas. This difference will be discussed next.

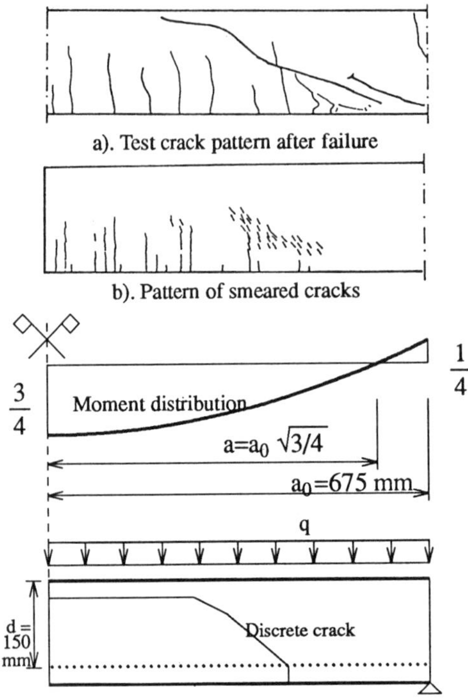

(Beam width 150 mm, Steel ratio 2%)

c). Geometry, loads and discrete path

Fig.5 The prototype beam (the same discrete crack path for beams of d = 300 and 600 mm)

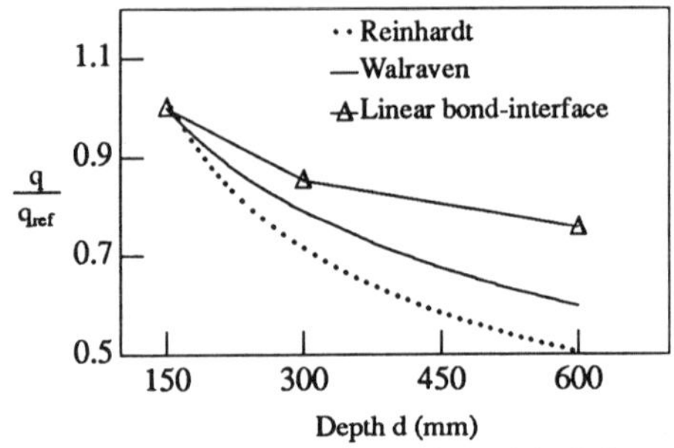

Fig.6. Calculated and empirical size effect, Reinhardt: $d^{-1/2}$ Walraven: $(1 + 10d)^{-1/2}$)

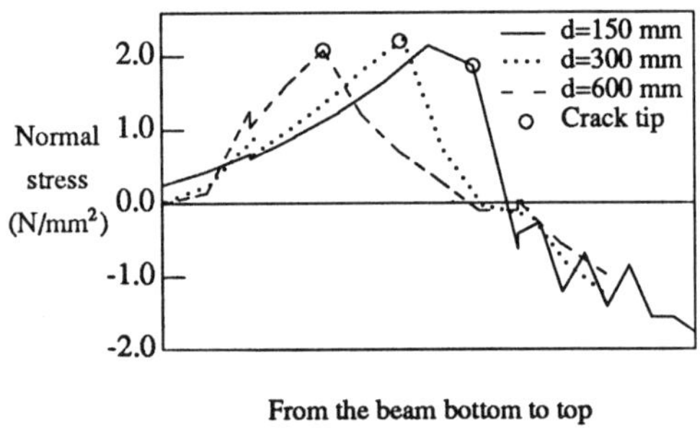

Fig.7. Normal interface stress along the diagonal path at failure stage

4.2. Effect of bond-slip due to distributed cracking

Bond-slip elements have been used in the analyses described above. But these may be not enough to take account of all the slip deformations in the case that only a single discrete crack is included. In reality, distributed cracks are often observed in the vicinity of the reinforcement especially when many rebars of small diameters are used. What effect would these distributed cracks in the plane of the beam have on the load carrying capacity? For the beam with d = 300 mm a calculation was made, in which the smeared crack model was applied to the area above and below the reinforcement, see the crack pattern shown in Fig.8a. The result is compared to that of the case with only discrete crack in Fig.8b. The smeared cracks cause indeed a considerable decrease of the strength. A closer examination of the deformations [5c] indicated that the bond-stress nearby the crack mouth was considerably lower in presence of the smeared cracks.

It is perhaps an appropriate way to modify the bond-interface $\sum C$ if the combination of smeared and discrete cracks should be avoided. One may imagine the whole group of rebars together with the concrete cover as one 'rebar'. Since this 'rebar' contacts the web only through the width of the beam, this width can be considered as an effective bond-interface in the plane stress condition. The three geometrically similar beams described in the last section were reanalyzed using a constant effective bond-interface equal to the beam width. The obtained size effect is shown in Fig.9. It turns out that an extra size effect is exhibited compared to the case with linearly scaled $\sum C$. Besides a better agreement with the empirical formulas is achieved.

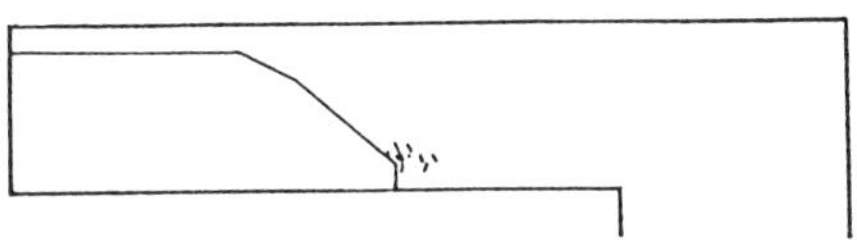

a). Smeared cracks taken into account

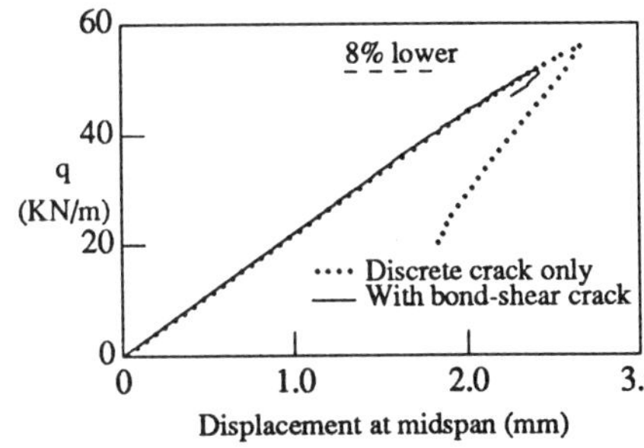

b). Load-displacement responses

Fig.8. Influence of distributed bond-shear cracking on the shear capacity (Smeared cracks for bond-shear cracks)

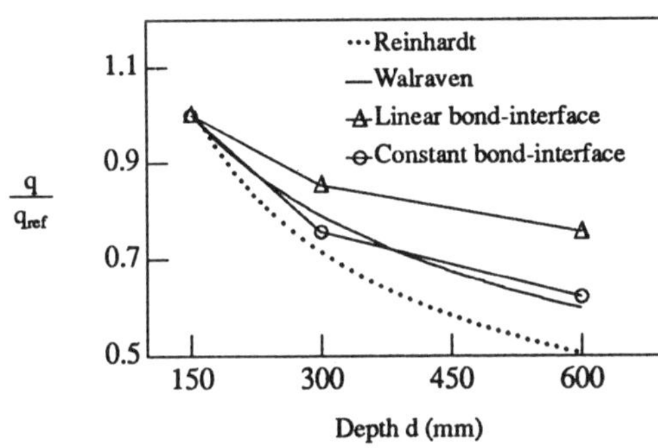

Fig.9. Calculated and empirical size effects

4.3. On the modeling of size effect

The size effect of shear strength is primarily the result of fracture of concrete. So it is appropriate to model it using the energy balance concept of fracture mechanics: fracture occurs when an equilibrium is reached between the release of the system potential energy and the fracture energy [10,11]. The classical linear elastic fracture mechanics, which applies for non-stress transferring cracks, gives the following well-known fracture strength for an infinitely large plate with a center crack of length $2a_{cr}$ and subjected to a remote uniform stress:

$$\sigma_u = \left[\frac{EG_f}{2\pi\, a_{cr}}\right]^{1/2} \tag{1}$$

where E is the elasticity modulus. Dividing the equation by the tensile strength f_t, one obtains,

$$\frac{\sigma_u}{f_t}=\left[\frac{EG_f/f_t^2}{2\pi a_{cr}}\right]^{1/2}=\left[\frac{l_{ch}}{2\pi a_{cr}}\right]^{1/2} \tag{2}$$

Here $l_{ch}=EG_f/f_t^2$ is the so-called characteristic length and can be considered as a material constant.

Yet, the above equation does not necessarily apply for concrete which has a plastic strength limit i.e. f_t and obeys progressive stress softening without a complete release of strain energy and dissipation of fracture energy, especially in case of non-precracked structures just as the beams studied in the preceding sections. Three situations are illustrative. (1). For an extremely small structure size d, eq.(2) shall give an unlimitedly large strength since a_{cr} approaches 0. But the actual structural strength should approach the perfectly plastic strength; (2). In case of an extremely large structure size, a_{cr} approaches 0 too. This trend can be observed from Fig.7 clearly. In that case, the structure fails at first appearance of the failure crack and the structure strength can be evaluated by the elastic brittle fracture approach; (3). Eq.(2) even does not apply for the flexural strength of a plain concrete beam subjected to bending in combination with normal tension or compression. In [9] it was shown numerically that for different magnitude of normal actions the size effect is different. A simple case is that when the beam is subjected to an axial stress of magnitude of f_t, the flexural strength does not change with the size of the beam. So for materials with progressive softening, the size effect is dependent on the global stress or strain gradient along the concerned crack, which determines (together with softening of concrete) the extent of stress redistribution. This point is of importance for dealing with the size effect of strength of R.C. beams subjected to shear, bending and axial action, because the development of the diagonal crack is strongly related to the bending deformation.

Another difficulty hindering the use of eq.(2) is that, in contrast to what is usually assumed [1,11], the crack length a_{cr} is not necessarily proportional to the structure size d. As can be concluded from Fig.7, a_{cr}/d decreases as d increases. Therefore it is, strictly speaking, not correct to replace a_{cr} by d in the equation. But as an engineering approximation, l_{ch}/d may be used as a size effect parameter and the strength should be a general function of it.

For non-precracked structures, it could be a good way to apply a size effect factor to the strength increment between the elastic brittle fracture strength σ_{el} and the perfect plasticity strength σ_{pl}. The nominal ultimate strength σ_u of a structure can be expressed as,

$$\sigma_u=\sigma_{el}+(\sigma_{pl}-\sigma_{el})\,f(l_{ch}/d) \tag{3}$$

where $f(l_{ch}/d)$ varies from 1.0 for zero-sized structures to 0.0 for infinitely large structures. In this way the effect of stress and strain gradient can be accounted for in the ultimate strength. Note that eq.(3) in fact gives a lower bound for the nominal strength, i.e. σ_{el}. Such a lower bound existed already in many design codes e.g. in the Eurocode [13].

For the size effect of shear strength shown in Fig.9, an exponential function may be adopted tentatively: $f(l_{ch}/d) = (d/l_{ch})^B$. The function does not have a upper bound of 1.0. This does not matter much because no extremely small sizes are considered. Fig.10. shows the relation between $\mathrm{Log}(q-q_{el})$ and $\mathrm{Log}(d/l_{ch})$ (note that for distributed loads q characterizes the nominal shear strength). The two almost straight lines imply that the adopted function is a good approximation. The slope of the lines indicates the value of B: $B \approx -0.52$ for linearly scaled bond-interface and -0.82 for constant bond-interface.

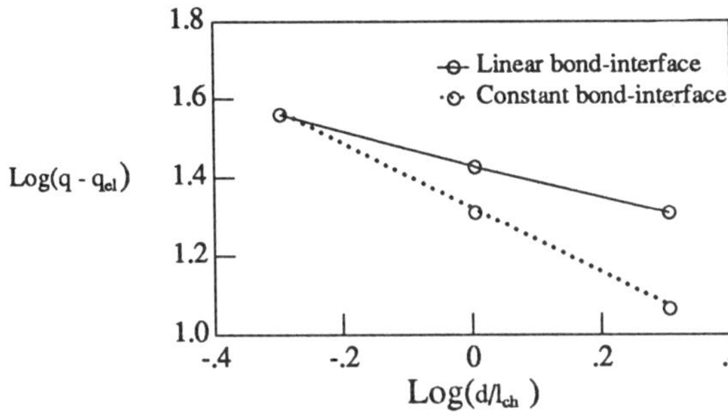

Fig.10. Size effect as a logarithmic relation between $q - q_{el}$ and the ratio d/l_{ch}

6. Conclusions

1. In a correctly chosen diagonal crack the crack opening dominates the deformation. The crack shear mechanism of aggregate interlock is not activated. Therefore, the failure due to diagonal cracking is a matter of mode I fracture of concrete;
2. The fracture of concrete along the failure diagonal crack causes directly a significant size effect of shear strength;
3. Bond transfer decreases in higher beams due to in-plane distributed cracking around the reinforcement. This may also cause certain size effects;
4. For non-precracked concrete with progressive tension-softening, it is suggested that a size effect function be applied to the increment between the elastic brittle strength and the perfect-plastic strength.

7. Acknowledgement

The support of prof. R. de Borst, prof. J.C. Walraven and dr. J.G. Rots in the research project is highly appreciated. To decide on the proper steps in the project and in interpreting the computational results, their contribution is of great help.

8. In commemoration of Q.B. Wang

It is for me, supervisor Johan Blaauwendraad, a sorrowful duty at the end of this paper to inform the reader that author Qingbiao Wang suddenly passed away on December 31, 1990, at the age of 28 years. He was a highly appreciated and promising doctoral student. His dead was a big loss for the Delft group working on concrete mechanics and concrete fracture. The preparation of this paper was practically completely done by mr. Wang.

9. References

1. Bazant, Z.P. and Kim, I.K.: "Size effect in shear failure of longitudinally reinforced concrete beams", ACI Journal Sept-Oct 1984, pp. 456-468.
2. Taylor, H.P.J.: "Shear strength of large beams", J. of Struc. Div. ASCE, Nov. 1972, pp. 2473-2490.
3. Walraven J.C.: "Mechanisms of shear transfer in cracks in concrete", Report A26/79-02, Delft Univ. of Tech., Nov. 1978.
4. Walraven J.C.: "The influence of depth on the shear strength of light-weight concrete beams without shear reinforcement", Report 5-78-4, Delft Univ. of Tech., 1978.
5. Wang Q.B.: "Shear strength of reinforced concrete concrete beams and slabs without shear reinforcement", reports of Delft University of Technology,
   a. part 3: Analysis of interlock and dowel action in a Stevin-beam, Reportnr. 25.2-89-2-18, Dec. 1989;
   b. part 4: Analysis of shear capacity of a series of IBBC-beams, Reportnr.25.2-90-2-08, Aug. 1990;
   c. part 5: Size effect analysis, Reportnr. 25.2-90-2-09, Sept. 1990.
6. Fenwick R.C. and Paulay T.: "Mechanisms of shear resistance of concrete beams", J. of Structural Division, ASCE, Oct. 1968, pp. 2325-1339.
7. Rots, J.G.: "Computational modeling of concrete fracture", Dissertation, Delft Univ. of Tech., 1989.
8. P.J. Gustafsson: Fracture mechanics studies of non-yielding materials like concrete, Report. TVBM-1007, Div. Build. Mat., Lund Inst. of Tech., 1985
9. Duda, H.:"Basic concepts for using concrete tensile strength", introductory reports, Proc. of IABSE Colloquium on "Structural Concrete", Stuttgart, 1991.
10. Griffith, A.A.: "The phenomena of rupture and flow in solids", Phil. Trans. Roy. Soc. (London), Series A, 221(1920), p.163-198.
11. Reinhardt, H.W.: Simplitude of brittle fracture of structural concrete, IABSE Delft 1981, pp. 175-184.
12. TNO-IBBC: Results of the first series of shear tests, report B-77-128 (in dutch).
13. Eurocode No.2. Common uniform code for concrete structures, EUR8848, 1984.

82 ANALYSIS OF REINFORCED CONCRETE BEAMS WITHOUT SHEAR REINFORCEMENT USING NON-LOCAL MICROPLANE MODEL

J. OŽBOLT and R. ELIGEHAUSEN
Stuttgart University, Germany

Abstract
The shear resistance of reinforced concrete beams without shear reinforcement is studied using the non-local microplane model and plane stress finite elements. The main objective of the present work is the study of the size effect. Calculated failure loads for geometrically similar specimens of four different sizes are compared with test data and the recently proposed size effect law. Results of the analysis as well as test results exhibit significant size effect. Observed failure is of the brittle type and is due to failure in tension-compression. Further studies with variations of the mesh size and load path demonstrated that results of the calculations using a non-local continuum are not influenced by the above parameters. However, the calculated failure loads depend on the characteristic length over which the strains are measured. In contrast to that, the calculated failure loads are inobjective when a local continuum is used and depend on mesh size, load path and convergence criteria. This is due to the stability of the numerical analysis, which may lead to an overriding of the critical failure mode and activating to a more stable one. In order to correctly predict failure load using local analysis and crack band approach it is necessary to check the stability of the numerical procedure.
Keywords: Size effect, Shear, Non-local Continuum, Microplane Model, Stability, Objectivity.

1 Introduction

Kani, (1967) was one of the first who demonstrated that the shear strength of identical concrete beams decreases with increasing beam depth and that the shear design provisions used at that time were unsafe for larger beams. Recently, several experimental studies confirm the size effect in a diagonal shear type of failure (Bazant and Sun, 1987; Walraven, 1990; Bazant and Kazemi, 1990; Walraven and Lehwalter, 1990). In the last two studies, the size effect law proposed by Bazant (1984) was confirmed. However, according to Walraven and Lehwalter, the aggregate size has no influence on the size effect law for a maximum aggregate size between 8 and 32 mm.

To improve the understanding of the mechanism of the diagonal shear failures and of the size effect on the failure load, four

Fracture Processes in Concrete, Rock and Ceramics Edited by J.G.M. van Mier, J.G. Rots and A. Bakker.
 Published by E. & F.N. Spon, 2–6 Boundary Row, London SE1 8HN. ISBN 0 419 15870 7

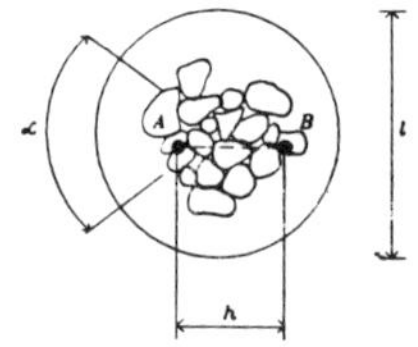

Fig. 1 Interaction among various orientations and at distance

geometrically similar specimen with different sizes are analyzed using the non-local microplane model and four node plane stress finite elements. Results of the numerical analysis are compared with above test results and size effect law.

In many applications of finite elements in failure simulations of concrete structures, a local continuum approach based on the crack band theory is used. Studies by Rots (1988) indicate that the results may be inobjective and may depend on the finite element size and orientation. To explore the reason for this inobjectivity, studies are performed using the microplane model and a local continuum. Varied are the mesh size and load path. Furthermore, it is checked weather objective results are reached when a non-local continuum is used.

In a numerical analysis based on a non-local continuum the key parameter is the characteristic length over which the strains are averaged. Therefore, the influence of this parameter on the diagonal shear failure load is studied.

2 Review of the non-local microplane model

The basic idea of the microplane models, that were initiated by G.I.Taylor (1938), is that the plastic slips in the modeling of the plasticity were calculated independently on various crystalographic planes, based on the resolved shear stress component, and were then superimposed to obtain the plastic microstrain. Recently (Bazant, 1984; Bazant & Gambarova, 1985; Bazant & Prat, 1988), this approach was extended to include strain softening of concrete, and was renamed more generally as the "microplane model", in recognition of the fact that the approach is not limited to plastic slip but can equally well describe cracking and strain softening damage. However, in order to prevent instability due to strain softening, the microplanes must be constrained kinematically rather than statically.

A basic requirement for a continuum model for a brittle heterogeneous material such as concrete is that it must correctly display the consequences of heterogeneity of the microstructure. A continuum constitutive model lumps the average response of a certain characteristic volume of the material (Fig. 1). In essence, one may distinguish two types of interactions among the particles or damage sites in the microstructure, which must be somehow manifested in the continuum model: (1) Interaction at **distance** among various sites (e.g., between A and B, Fig. 1); and (2) interaction **among various orientations** (see angle α in Fig. 1).

The interactions at distance control the localization of damage. They are ignored in the classical, local continuum models but are reflected in non-local models (Pijaudier-Cabot and Bazant, 1987). The non-local aspect is a requisite for a realistic description of the size effect, as well as for the modeling of fracture propagation in the form of a crack band.

According to the non-local concept, the stress at a point depends not only on the strain at the same point but also on the strain field in a certain neighborhood of the point (Kröner 1968; Krumhansl 1968; Levin 1971; Eringen & Edelen 1972). In the current study, an effective form of the non-local concept, in which all variables that are associated with strain softening are non-local and all other variables are local is used. The originally proposed non-local concept (Pijaudier-Cabot and Bazant, 1987) is here modified introducing additional weighting functions that control the averaging of the total (positive and negative) strain field into the directions of main principle stresses. An important advantage of this formulation, called non-local damage or non-local continuum with local strain, is that the differential equations of equilibrium as well as the boundary conditions are of the same form as in the local continuum theory, and that there exist no zero-energy periodic modes of instability.

The key parameter in the used non-local concept is the characteristic length l_c over which the strains are averaged and it has a significant influence on the results of the analysis. Bazant & Pijaudier-Cabot (1989), assumed that this length is a material parameter which can be taken as $3d_a$ (d_a = maximum aggregate size). However, in general 3D stress-strain situations the characteristic length is difficult to interpret as a material parameter depending on the concrete mix only, but may be influenced by other parameters as well such as stress-strain state and geometry of the particular problem. Principally, the characteristic length must be chosen so that together with the stress-strain law used in the smeared crack approach can correctly prescribe structural behavior.

The non-local microplane model as well as an effective numerical iterative algorithm for the loading steps, that is used in the finite element code is described in detail by Bazant and Ozbolt (1990).

3 Size effect study

3.1 Finite element meshes and material parameters

A size effect study of reinforced concrete beams, previously tested by Bazant and Kazemi (1990), is carried out. Four different sizes of geometrically similar specimen are analyzed. The geometry of the specimen used in the analysis are shown in Fig. 2. Test results for the specimens A - C where available.

Only one half of the specimen is analyzed i.e. a symmetric failure mode is forced. In the analysis four-node plane stress finite elements are used. The finite element meshes used in the analysis are shown in Fig. 3. Specimen A - C are analyzed using four integration points and specimen D is calculated by employing nine integration points.

The material parameters used in the analysis are chosen such that the uniaxial compression and tensile strength was about the same as in the experiment, (initial Young moduli E^0= 40000 MPa, Poisson's ratio ν= 0.18), and microplane model parameters taken such that the

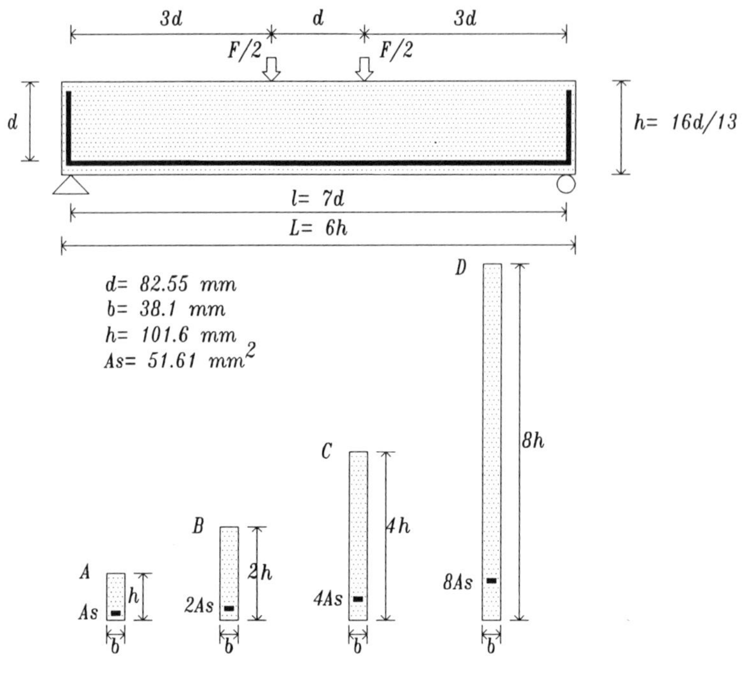

Fig. 2 Specimen geometry used in the analysis

calculated uniaxial concrete compression, and tensile strengths are f_c = 46 MPa and f_t = 2.9 MPa respectively. These values for the f_t and f_c are calculated using one finite element in a plane stress situation. The characteristic length of the non-local continuum was determined by fitting the average test results for the specimen B. It should be noted that this length, in the zone of interest, should not be smaller than at least two times the smallest finite element size since otherwise the analysis is equivalent to the classical local continuum analysis. In the present size effect study, the characteristic length was fixed as l_c = 40 mm. However, in the case of the smallest specimen (d= 82.55 mm) the characteristic length has been taken as l_c = 25 mm, because for l_c = 40 mm the strains over the half the beam depth would be averaged, what is not realistic.

Reinforced steel bars are modeled by introducing a corresponding smeared reinforcement area in the second row of the finite elements (see Fig. 3). The behavior of steel is taken as linear elastic, with an elasticity modulus E^s= 210000 MPa. It is assumed that there is no slip between concrete and steel. In the experiments this was assured by extending the steel bars at the beam ends over the beam depth (see Fig. 2).

The present numerical results are obtained by prescribing a force at the top of the specimen, i.e. by load control.

3.2 Results of the analysis

Load-displacement curves for four different specimen sizes are plotted

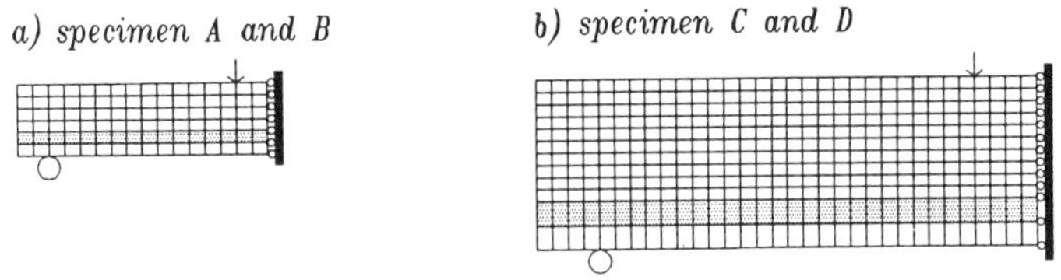

Fig. 3 Finite element meshes used in the analysis

in Fig. 4. Basically, the shape of the curves is bilinear. The first part is almost linear elastic up to the point where the first vertical bending crack is opening, following by the second, globally almost straight, but less steep line up to failure. The first part of the second line is consisting of a few local peaks that are due to the formation of the further bending cracks. The descending branch of the load-displacement curve could not be calculated probably because of the snap-back.

The maximum principle deformations of the specimen size C before failure are plotted in Fig. 5. Rectangulars plotted in this figure are proportional to the magnitude of the max. principle strains. The larger side of each rectangle is plotted in the direction perpendicular to the direction of the maximum principle strains. Fig. 5 indcates alarge tensile deformations at the compression part of the beam, in the direction perpendicular to the beam axis (horizontal cracking). Due to this, the concrete tensile strength is reduced and as a consequence, a critical diagonal crack, running from the bottom of the beam and approaching the compression zone at an angle (measured from the beam center line) of about $\alpha = 30^{\circ}$, causing brittle failure. Since before failure the critical crack is opening, there is almost no shear stress along the crack and failure is due to Mode I.

In Fig. 6 the nominal stresses at failure $\sigma_N = F_U/2bd$ (F_U = peak load), obtained in the numerical analysis and in the experiments (average values) are compared with the size effect law proposed by Bazant (1985):

$$\sigma_N = B\, f_t (1 + \beta)^{-1/2}, \qquad \beta = d/d_0 \tag{1}$$

The optimum values for the parameters B and d_0 are obtained by a linear regression analysis of the experimental results. It can be seen that the agreement between the results of the numerical analysis and the test data is good and that the failure loads must be described by the size effect law, i.e. the nominal stresses at peak load decrease with increasing specimen size. As expected, due to the fact that the failure type is brittle, the size effect is close to linear elastic fracture mechanics.

Bazant size effect law and size effect law proposed by Walraven and Lehwalter, (1990) are compared in Fig. 7. Values of the shear strengths are plotted relative to the beam depth d= 200 mm. Differences between these two predictions are large if the smaller beam depth is considered. This is probably due to the fact that the Walraven and Lehwalter (1990) proposed size effect law for the concrete mix with the maximum aggregate size between 8 and 32 mm.

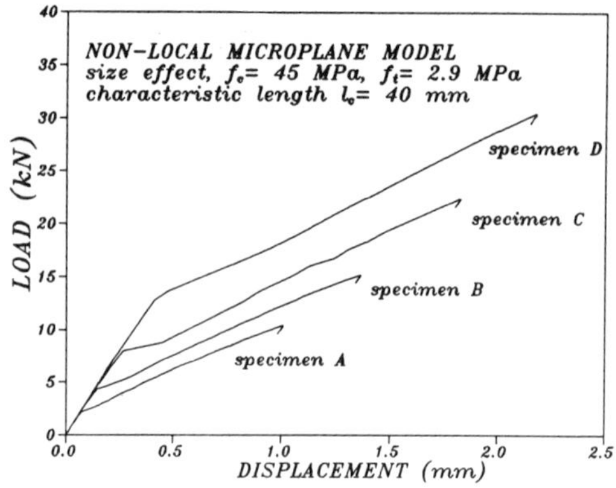

Fig. 4 Load-displacement curves for different specimen sizes

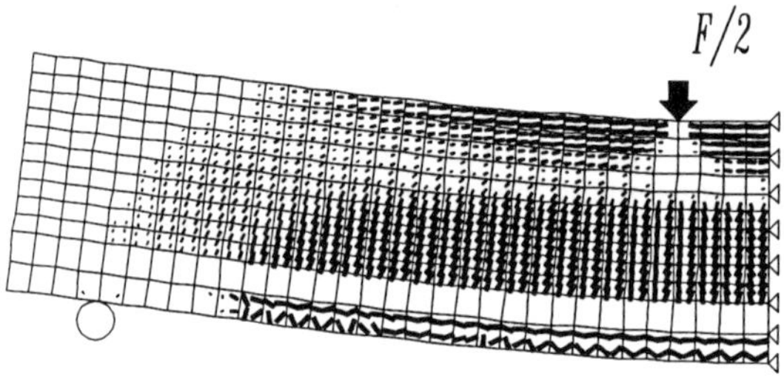

Fig. 5 Maximal principle deformations at peak load

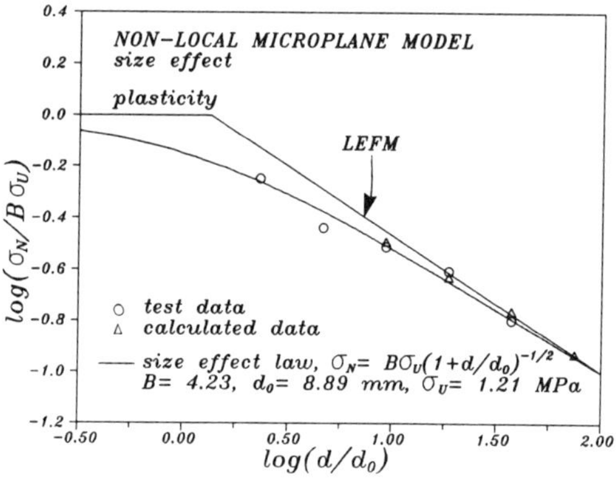

Fig. 6 Comparison between calculated and test data and size effect law

However, Bazant size effect law for the present example is calculated using the specimen with maximum aggregate size of 1.4 mm. Due to this, prediction employing Bazant size effect law is more close to linear elastic fracture mechanics.

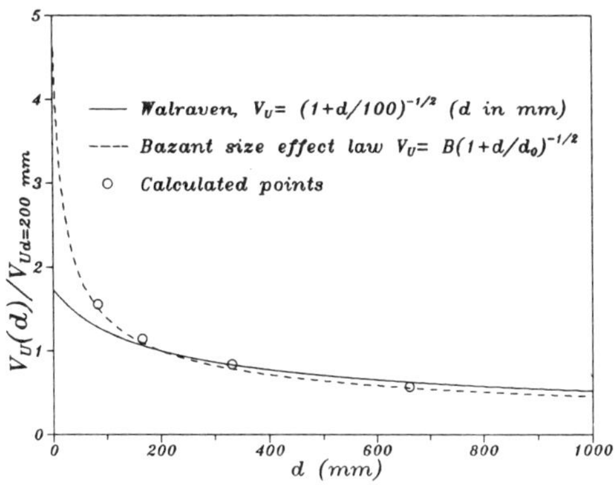

Fig. 7 Comparison between different size effect predictions

4 Objectivity and stability of the numerical analysis

Currently, failure simulations of reinforced concrete structures is often performed using the smeared crack approach and the crack band theory. However, studies by Rots (1988), indicate that in many cases results can be inobjective and depend on the finite element choice. In the analysis of the reinforced concrete beams without shear reinforcement using microplane model and local continuum the same problem arised.

In order to demonstrate the inobjectivity of the local analysis and make it more clear, specimen C has been analyzed using two different meshes. The same type of the mesh as shown in Fig. 3 have been used. The first mesh was relatively coarse, with 112 finite elements, and in a second mesh the number of elements was doubled, i.e. the size of the finite elements has been decreased by the factor of two. The resulting load-displacement curves are plotted in Fig. 8. This figure clearly demonstrates large mesh sensitivity. Namely, using a finer mesh, the failure load is much higher than in the case of a coarse mesh. Furthermore, if a coarse mesh is used, failure is due to diagonal shear, while the finer mesh specimen exhibits arc action and the calculated failure is due to failure of concrete in compression. It is interesting that refining the mesh, the real failure mode is somehow overridden by the arch action. For load up to approximately the failure load that is observed for the coarse mesh specimen, the convergency is relatively bad for both meshes. After that, for the fine mesh specimen suddenly the convergency becomes much better since a more stable mode is activated. It seems that one point exists (stability point), after which the diagonal shear failure mode is switched into a more stable failure mode - arch action. The same problem arises when in the coarse mesh instead of four integration points nine integration points is used.

To check this assumption, a stability analysis of the finer mesh specimen is carried out. Namely, the determinant of the symmetric tangent stiffness matrix, K_t^*, is calculated. The matrix K_t^* is calculated by symmetrization of the nonsymmetric tangent stiffness matrix K_t, obtained employing the microplane model, as follows:

$$K_t^* = \frac{1}{2} (K_t + K_t^T) \tag{2}$$

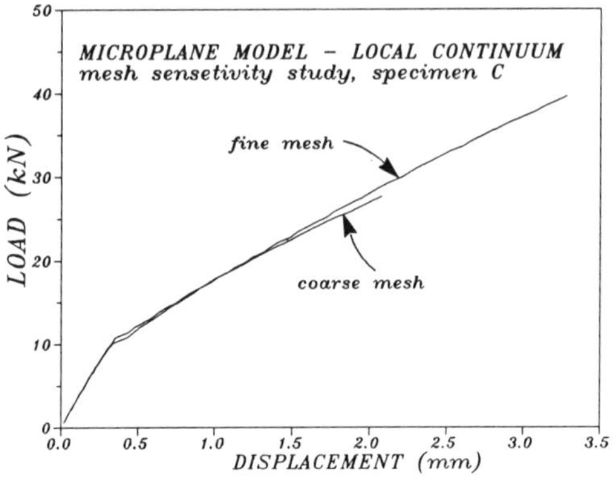

Fig. 8 Finite element mesh sensitivity in local analysis

The matrix K_t^T is transpose of the matrix K_t. Up to the load approximately equal to the failure load that is due to the coarse mesh specimen, $\det|K_t^*| > 0$, and $\det|K_t^*| < 0$ after that point. Since the determinant of the tangent stiffens matrix K_t changed the sign, it is clear that a bifurcation point is passed somewhere between these two calculations (de Borst, 1987). As shown by Bazant (1988), the path that occurs after the bifurcation point must minimize the second order work $\Delta W = \delta f \; \delta u/2$ where δf is the increment of the prescribed force and δu the calculated displacement increment. It is evident that in the present example the path taken does not minimize the second order work since the analysis indicates that an arch action is activated which is not the most critical path. This clearly shows that in order to detect the bifurcation (failure) point in the local continuum analysis, a stability analysis should be carried out.

A similar problem is present in the local analysis if different loading paths or different convergence criteria (Kazic, 1990) are prescribed. To demonstrate the influence of a loading path, in Fig. 9 load-displacement curves for the specimen size B, using the finite element mesh shown in Fig. 3a, are plotted. In the employed local continuum analysis, the calculated failure load is much higher if displacement control is used rather than load control. In contrast, using a non-local continuum, calculated failure loads are almost the same for both cases.

In many practical applications, where fracture of the structure is due to fracture in tension, local continuum analysis based on the crack band approach can correctly predict failure load. However, in the case of diagonal shear failure, where failure is due to failure in tension-compression, crack band concept can not objectively predict failure load. Reason is probably due to the fact that using this

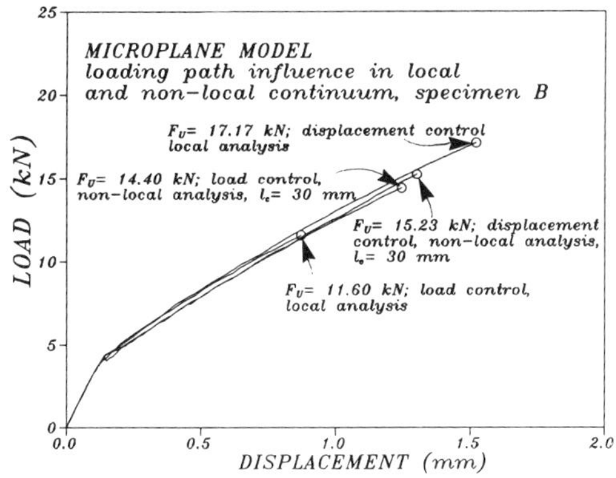

Fig. 9 Load path sensitivity

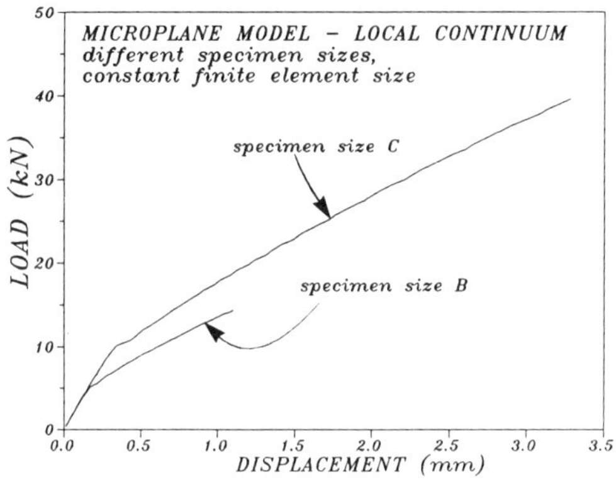

Fig. 10 Inobjectivity with respect of the specimen size

concept there is no interaction between a tension and a compressive part of the structure, i.e. there is no interaction over a distance. To demonstrate this, specimen of two different sizes (Specimen B and C, Fig. 2) are analyzed. In both cases the size of the finite elements were constant and the same, so that according to the crack band approach σ-ε curves are constant. Since the increase of the size is by a factor of two it would be expected that the failure load for the larger specimen is maximum two times larger (no size effect) or approximately 1.4 times larger with the size effect, than the failure load of the smaller specimen. However, the results show (see Fig. 10) that the failure load for the larger specimen is more than two times larger, due to the fact that in the case of the larger specimen instead of a diagonal shear failure, an arc action is formed. This indicates that a local analysis, even if it is used with the crack band approach, can not objectively predict diagonal shear failure.

It is interesting that in a local analysis, with increasing mesh refinement the failure loads increase. Further more, the failure mode is changed from a diagonal shear failure, to failure of the concrete in compression, caused by an arc action. This is in contradiction with experience from the local continuum analysis of the concrete structure where mesh refinement usually lead to strain localization and decrease of a failure loads. However, in present case mesh refinement increase the possibility that the most critical failure mode is override by another one, that is more stable. To detect a correct failure mode in the case of a diagonal shear using a local continuum approach, stability analysis should be carried out.

The demonstrated drawbacks of the local continuum analysis can be overcome if a non-local continuum approach is employed. Namely, in all analyzed examples where the non-local microplane model is used the calculated failure mode was of diagonal shear type. However, it is important that the finite element mesh is fine enough so that the characteristic length is at least two times the smallest size of the finite element. Otherwise, the continuum is too much local and the same problems, as demonstrated in the local analysis, may arise.

5 Influence of the characteristic length

In the non-local type of the finite element analysis the key parameter is the characteristic length, that define the volume of the material over which the strains are averaged. It is still not clear, whether this length is a material parameter or depends also on the geometry of the structure and the stress-strain state.

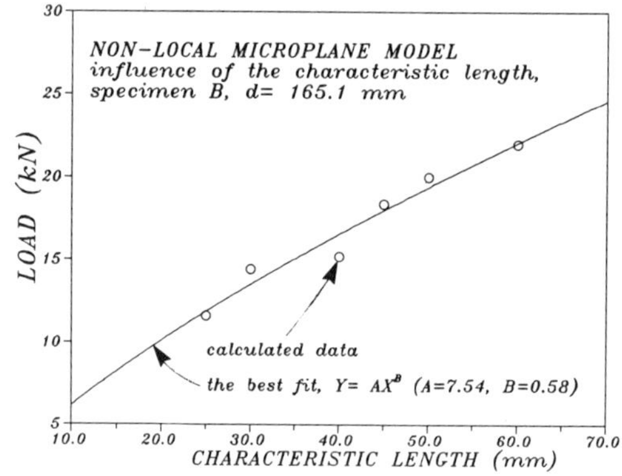

Fig. 11 Influence of the characteristic length

Here, the influence of the characteristic length on the diagonal shear failure load, of reinforced concrete beams, without shear reinforcement, is demonstrated. The study is performed using the specimen size B and the finite element mesh shown in Fig. 3a (four integration points). Geometry, finite element mesh and microplane material parameters are fixed. Six analysis are carried out with l_c = 20, 30, 40, 45, 50 and 60 mm. The smallest l_c value is about the finite element size. The largest characteristic length is approximately 1/3 of the beam depth. The load-displacement curves of all analyzed beams are similar and all fail in diagonal tension. However, the failure load increased with increasing characteristic length (Fig. 11). Comparison between the calculated data and the average test data indicate that the optimum value of l_c, for the present case, is between 40 and 45 mm.

Since Fig. 11 clearly indicates a significant influence of the characteristic length on the failure load, further studies are needed to investigate the correlation between basic concrete properties i.e. tensile strength, fracture energy, specimen geometry and strain state and this length. However, presently the characteristic length should be chosen such that together with the assumed stress-strain relations the fracture of concrete structures is correctly prescribed. As already mentioned, the size of the l_c should be at least two times the smallest finite element size. On the other side, l_c should not be to large relative to the specimen geometry, since in such a case the strain field from the bottom to the top of the beam is averaged, what is not realistic.

6 Conclusions

A analysis of reinforced concrete beams without shear reinforcement, using the non-local microplane model, indicates significant size effect. Comparison of calculated failure loads with test data and size effect law demonstrate good agreement.

For all analyzed specimen sizes, observed failure was of diagonal shear type, unexpected and brittle. Failure is due to failure of concrete in tension-compression.

Results of the finite element analysis are inobjective if local continuum approach is used. Large mesh sensitivity and load path dependency is observed. As a consequence of a local continuum, the most critical failure mode can be easily overridden by a more stable one. It is observed that local analysis, even if the crack band approach is employed, can not objectively predict the failure. In order to detect the real failure mode in local continuum finite element analysis, stability analysis should be carried out.

Using the non-local finite element analysis with characteristic length that is at least two times larger than the smallest finite element size, the critical failure mode was always obtained. However, as expected, significant influence of the characteristic length on the calculated results is demonstrated. Further studies are needed in order to clarify the relation between characteristic length and concrete fracture parameters, specimen geometry and stress-strain state.

7 References

Bazant, Z.P., Size Effect in Blunt Fracture: Concrete, Rock,Metal. J. of Eng. Mechanics, ASCE, 110, No. 4, Apr. 1984, pp.518-535.

Bazant, Z.P., and Sun H.-H., Size effect in Diagonal Shear Failure: Influence of Aggregate Size and Stirrups, ACI Material Journal, V. 84, No. 4, 1987, pp. 259-272.

Bazant, Z.P, and Kazemi, M.T., Size effect on diagonal shear failure of beams without stirrups. Internal Report, Center for Advanced

Cement-Based Materials, Northwestern University.
Bazant, Z.P., and Gambarova, P.G., Crack shear in concrete: crack band microplane model. J. Struc. Eng., ASCE,110(9), pp. 2015-2035, 1984.
Bazant, Z.P., Microplane model for strain-controlled inelastic behavior, Mechanics of Engineering Materials, C.S. Desai and R.H. Gallager, eds., John Wiley & Sons, Chichester and New York, chap. 4, 1984, pp. 45-59.
Bazant, Z.P., and Prat, P.C., Microplane model for brittle-plastic material - parts I and II. J. of Eng. Mech., ASCE, 114(10), pp. 1672-1702, 1988.
Bazant, Z.P., and Pijaudier-Cabot, G., Measurement of Characteristic Length of Nonlocal Continuum. J. of Eng. Mech, ASCE, 115 (4), 1989, pp. 755-767.
Bazant, Z.P., and Cedolin, L., Stability of structures: elastic, inelastic, fracture and damage theories and codes, Oxford University Press, New York, 1990 (in press).
Bazant, Z.P., and Ozbolt, J., Nonlocal microplane model for fracture, damage and size effect in structures. J. of Eng. Mech., ASCE, 116, (11), 1990, pp. 2485-2505.
R. de Borst, Stability and Uniqueness in Numerical Modeling of Concrete Structures, Colloquium on Computational Mechanics of Concrete Structures, IABSE, August 1987, pp. 161-176.
Eringen, A.C., and Edelen, D.G.D., On non-local elasticity. Int. J. of Eng. Sci. 10, pp 233-248, 1972.
Kani, G.N., How Safe are Our Large Concrete Beams ?. ACI Journal, No. 64-12, March 1967, pp. 128-141.
Kazic, M. and Eligehausen, R., Influence of Anchoring Technique on Shear Strength of Reinforced Concrete Beams. Report in Preparation, Stuttgart University, Institut für Werkstoffe in Bauwesen, December, 1990.
Kröner, E., Interrelations between various branches of continuum mechanics. Mech. of Generalized Continuua, E. Kröner, ed., Springer, W.Berlin, pp. 330-340, 1968.
Krumhansl, J.A., Some considerations of the relations between solid state physics and generalysed continuum mechanics. Mech. of Generalized Continuum, E. Kröner, Ed., Springer, W.Berlin, pp. 298-331, 1968.
Levin, V.M., The relation between mathematical expectation of stress and strain tensor in elastic micro-heterogeneous media. Prikl. Mat. Mehk. 35, pp. 694-701, 1971 (in Russian).
Pijaudier-Cabot, G., and Bazant, Z.P., Nonlocal damage theory. J. of Eng. Mech., ASCE, 113, pp.1512-1533, 1987.
Rots, J.G., Computational modeling of concrete fracture. Thesis presented to Delft University of Technology, at Delft, The Netherlands, September 1988.
Taylor, G.I., Plastic strain in metals. J. of Inst. of Metals, Vol. 62, pp. 307-324, 1938.
Walraven, J., Scale Effect in Beams with Unreinforced Webs, Loaded in Shear. Annual Report, Delft University of Technology, Delft, The Netherlands, V.1, 1990, pp. 103-112.
Walraven, J. and Lehwalter, N., Einfluss des Masstabs in schubbeanspruchten Bauteilen ohne Schubbewerung. Beton und Stahlbetonbau 85, 1990, H. 9, pp. 228-232.

83 USE OF A STOCHASTIC MODEL FOR MATERIAL BEHAVIOUR ANALYSIS AND APPRAISAL OF STRUCTURES

P. ROSSI and X. WU
Laboratoire Central des Ponts et Chaussées, Paris, France

Abstract
The description of the non-linear behavior of materials and structures is moreoften deterministic and the obtained results (ultimate loads, load-displacement response, patterns of cracking...) are generally supposed to be correct. Our purpose is to show that these results can be debated and that the deterministic models are anyway unsufficient if more accurate approach of safety is required. The paper shows how a model on a probabilistic basis can improve the quality of the numerical simulation of the loading of concrete specimens. It shows also that a stochastic approach can give very important information on the problem of the localization of cracks in an heterogeneous material like concrete.
Keywords: Stochastic model, concrete, steel fibre reinforced concrete, safety, localization, cracks, heterogeneity.

1. Introduction

The term "stochastic" covers very different approaches having the following in common:

- the desire to take into account the influence of local heterogeneities on the behavior of a structure,
- recourse to random variables (ultimate strength, Young's modulus, damage parameter, cracking threshold, etc.),
- the use of statistical techniques (simulation of chance, calculation of means and variances, etc.).

An exhaustive inventory of works on the subject is impossible, but most of them use:

- an analytical modelling (most often derived from Weibull's work: Jayatilaka (1979), Vardar (1975), Mihashi (1983), Sentler (1985),

Fracture Processes in Concrete, Rock and Ceramics Edited by J.G.M. van Mier, J.G. Rots and A. Bakker.
 Published by E. & F.N. Spon, 2–6 Boundary Row, London SE1 8HN. ISBN 0 419 15870 7

- a numerical approach, the application of which is limited to bar structures, Burt (1977), Krajcinovic (1982), Schorn (1896),
- or finite element analysis; heterogeneities are randomly distributed in the volume of the test specimen and its behavior simulated, Roelfstra (1986), Bazant (1986), Breysse (1988), Rossi (1987).

2. The proposed modellings

The numerical model developped by Rossi et al. (1987, 1988, 1989) at the Laboratoire Central des Ponts et Chaussées (L.C.P.C.) belongs to the family of statistically-based deterministic models and uses finite element method. It differs from other models of the same type in a greater consistency between the scale of modelling used and the cracking criteria chosen and in a better match to the actual process of cracking of concrete. As a result, it is today the only stochastic model in wich it is possible to enter cracks as discrete kinematic discontinuities (information about the length, opening width, and orientation of each crack spacing, is given explicitly). For this purpose use is made of special contact elements with no thickness that interface all of the "volume" elements. On these contact elements random tensile strengths are distributed (these elements open when the tensile strength is reached). The "volume" elements are equilateral triangles so that cracks can be created in three directions (in the case of a two-dimensional model). A procedure consisting of relating the distribution function of the tensile strengths to the size of the mesh elements (a relation detemined experimentally for each concrete) yields relatively objective calculations.

From this model of concrete without fibres, taking the fibres into account to model an SFRC is done very simply by introducing, just after the opening of a contact element that, in the case of an SFRC, simulates the cracking of the matrix, return forces that depend on the crack opening width. These forces model the forces taken out by the fibres across the macrocrack. The necessary experimental data are then obtained from uniaxial tensile tests on notched specimens.

Remark: The return forces act only above some minimum crack opening width, or threshold, and disappear when this crack opening width reaches a critical value that reflects failure or pull-out of the fibres.

3. Material behaviour analysed with the model

3.1 Tensile test on plain concrete

The tensile test analysed consists in creating assumed distributed damage obtained through aluminium bars glued on the lateral surfaces of the concrete specimen; the load is applied on the bars which impose an uniform strain state to concrete (figure 1). This type of test was performed by Bazant and Pijaudier-Cabot (1987), and more recently by Mazars et al. (1988).

The results of our numerical analysis are presented in figure 2 and 3. In figure 2, are compared the experimental results obtained on a concrete having, approximately, the same aggregate sizes than the one on which the distibution functions of strength have been determined. In figure 3, are shown the crack pattern before and after the peak load, before the rupture, and the stress-strain curves corresponding to each state of cracking.

Remark: from the global response, assuming that aluminium behavior is linear elastic and writing the equilibrium of the central section of the system, we can deduce the behavior of the concrete as follows:

$$F = F_A + F_C \Longrightarrow \sigma_C = \frac{F - E_A \times \epsilon_S \times S}{S_C} \qquad (1)$$

Where : F is the total force,
F_A is the force in the aluminium bars,
F_C is the force in the concrete,
E_A is the Young's modulus of aluminium,
ϵ_A is the strain in the aluminium bars,
S is the total section,
S_C is the section of the concrete specimen,
σ_C is the stress in concrete.

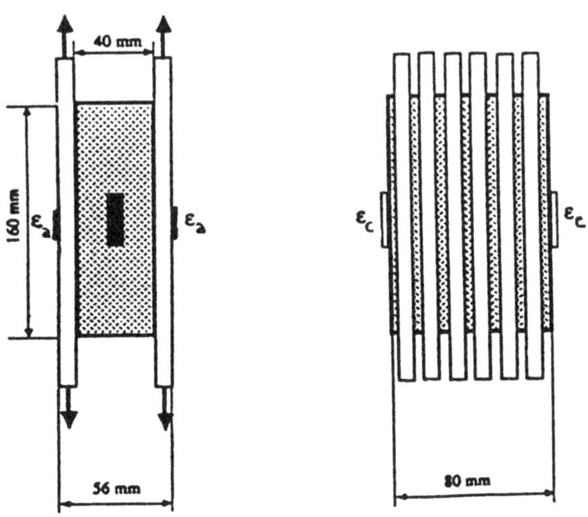

Fig. 1. Tensile test with aluminium bars (after Mazars et al.)

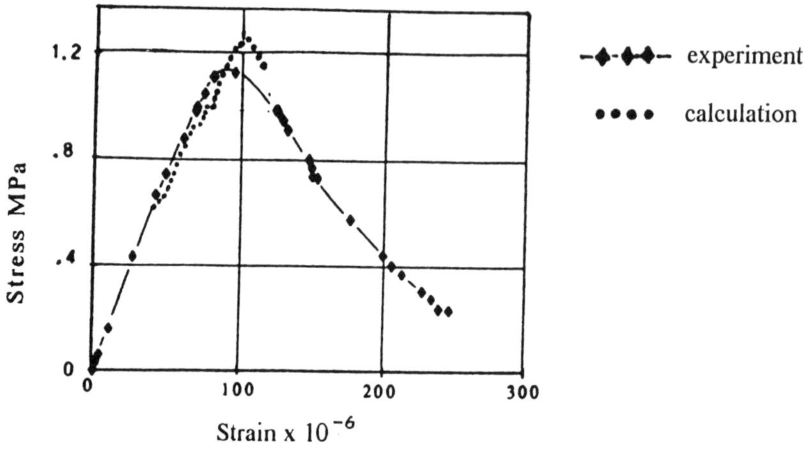

Fig.2. Tensile test with aluminium bars. Comparisons between experiment and calculation

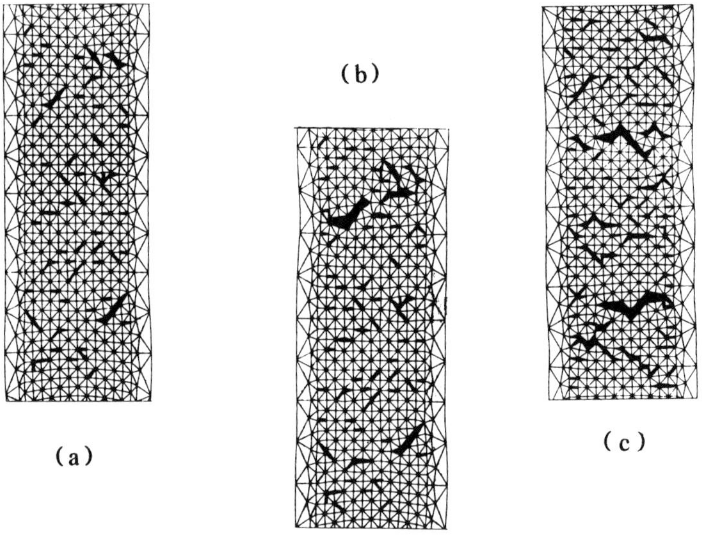

Fig. 3. Tensile test with aluminium bars. State off cracking before(a), and after (b) the peak load, and before the failure (c).

If we consider the results given in figures 2 and 3, we can make the following comments:

- we observe an enough acceptable aggreement between the experimental and the numérical results,
- the peak load corresponds to the localization of the cracks,
- if we consider the post peak behavior, we can observe the creation of severals macrocracks like the cracking process in a reinforced concrete beam (it is structural behavior).

3.2. fibre reinforced concrete behavior during a bending test

We have applied our model to analyse the difference of behavior of a plain concrete beam and a fibre reinforced concrete beam, this beam being loaded in four points bending. In figure 4 is presented the mesh used, and the dimensions of the specimen. In figure 5 are shown the crack patterns at the peak load, and the load-displacement curves concerning the two materials.

Remarks: - In these analyses the load is imposed.

- The one difference between the two calculations is the use of return-forces after cracking of the matrix. That means that the distribution function of the strength and the spatial repartition of these strength are the same.

Concerning the results presented in figure 5, we can make the following comments:

- at the peak load the density of cracks is more important in steel fibre reinforced concrete than in plain concrete,
- the mechanical behavior of the steel fibre reinforced concrete specimen is more ductile than the plain concrete specimen, and its peak load is 40 % higher.

All these results are coherent with that is known in literature.

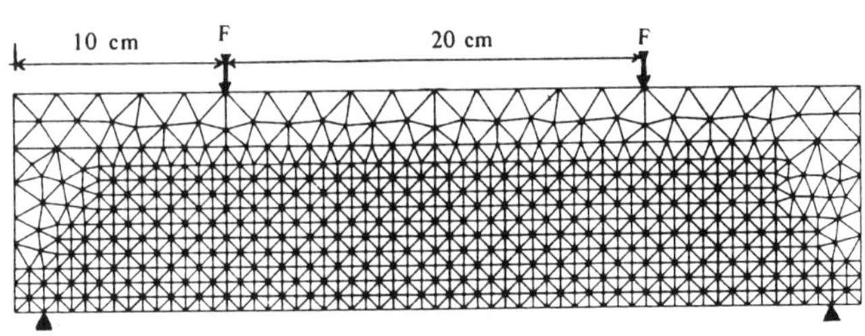

Fig. 4. Bending test analysis. Mesh and dimensions of the specimen.

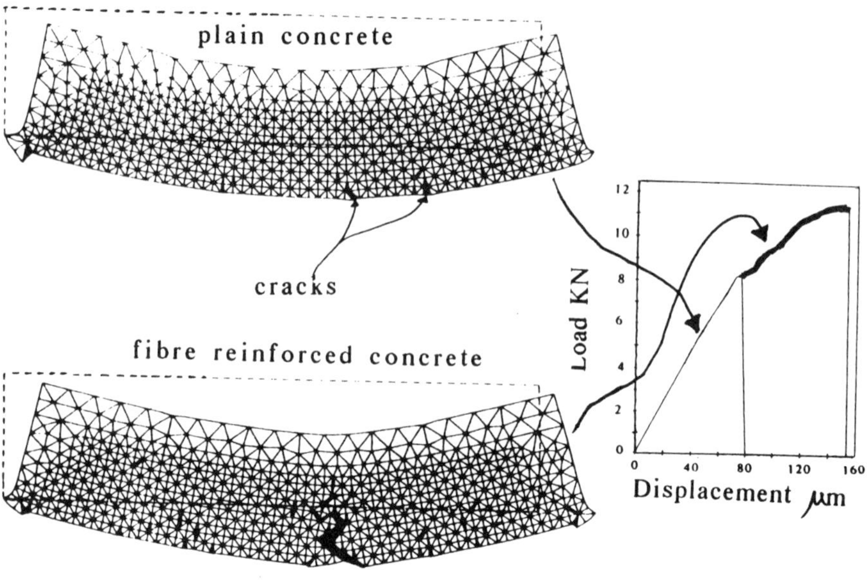

Fig. 5. Bending test analysis. Comparison between plain and fibre reinforced concrete.

4. The safety evaluation of structures.

Our purpose is to show that to obtain usefull information concerning the behavior of structures, like for example ultimate loads, or cracks patterns, deterministic results obtained by using deterministic models are anyway insufficient if the safety of these structures are involved.

For that, we propose three numerical analyses of three type of structures whose sensitivity to inhomogeneity within the material is supposed to be different. These structures are:

- 10x10x10 cm^3 plain concrete beam, in four point bending (plane stresses),
- 10x10x40 cm^3 steel reinforced concrete beam, in four point bending (plane stresses)
- 10x10x10 cm^3 cube in plain concrete, under an concentrated eccentric compressive load (plane stresses).

Ten simulations of each structure are realised.

Remark: concerning the modeling of steel reinforced concrete, we use a discrete approach consisting of separate models for the concrete, the steel bar, and the interface between the bar and the concrete. The steel bar can have an elasto-plastic behavior (Von Mises criterium for example) and the interface is modelized with the same contact element used in our stochastic modelling of plain concrete. A Coulomb law is then used.

In figures 6, 7, and 8, are respectively presented the results concerning the cube, the plain concrete beam and the reinforced concrete beam. These results are the load-displacement curves, and an exemple of pattern of cracking for each structures.

In figure 9, is proposed a statistic analysis of the results consisting in relating the cumulated probability of failure with the load of failure (which is normalised).

From this figure 9, we can make the following remark: it is clear that a deterministic approach would not give satisfactory information concerning the safety and the pattern of cracking of the structures studied.

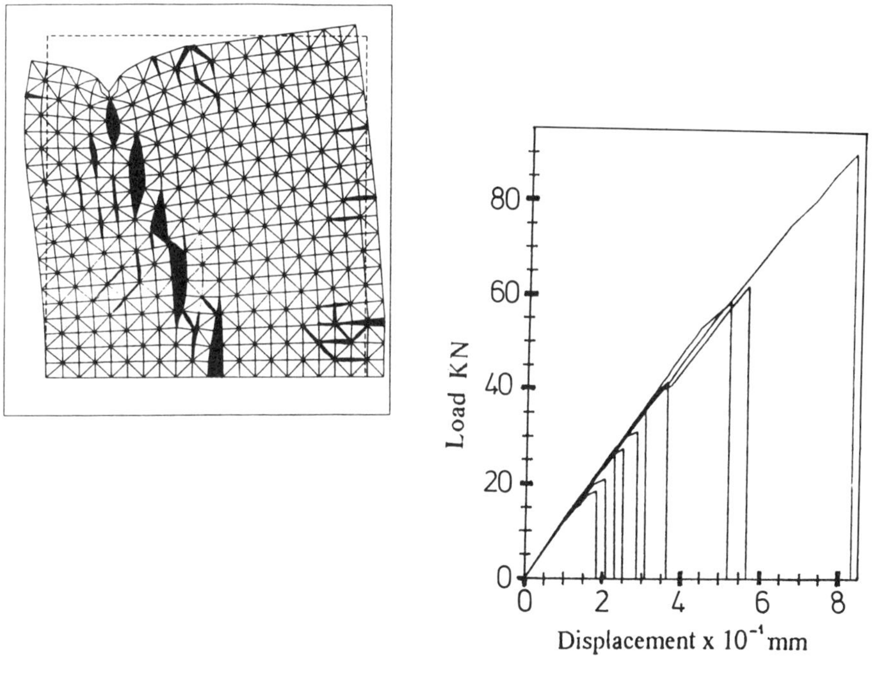

Fig. 6. Plain concrete cube loaded by under concentratred excentric compressive force.

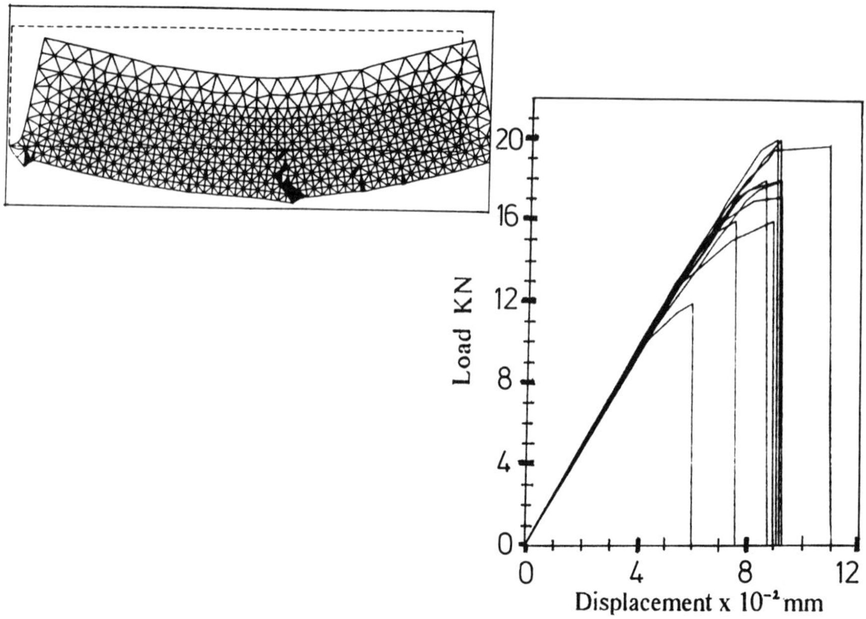

Fig. 7. Plain concrete beam loaded in four point bending.

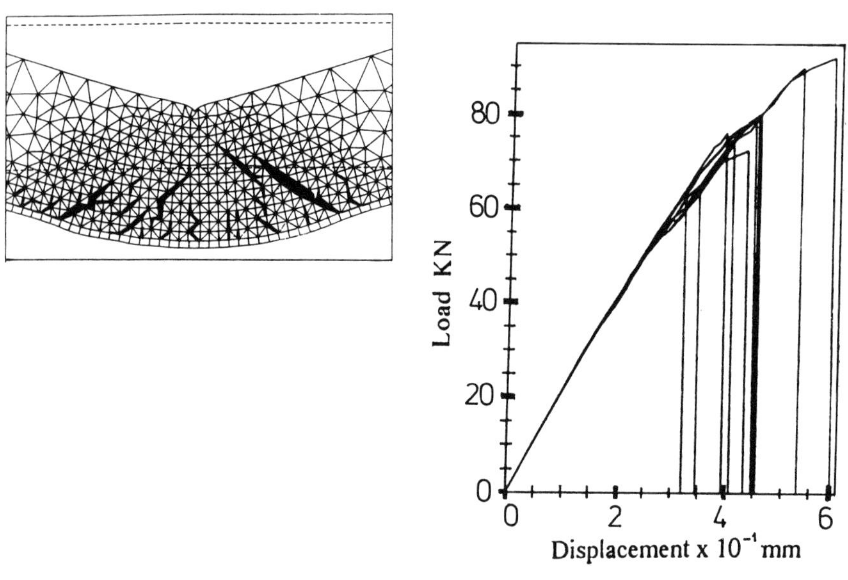

Fig. 8. Reinforced concrete beam loaded in three point bending.

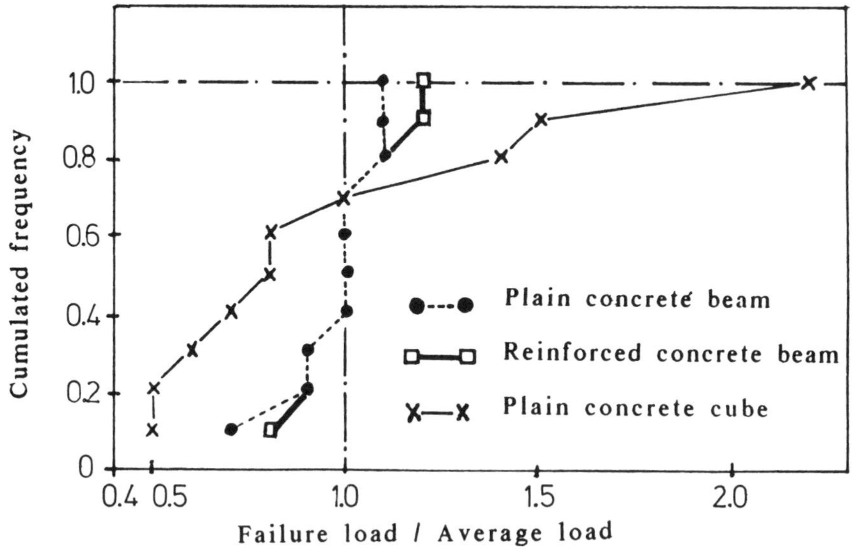

Fig. 9. Probability of failure for each structure

5. Conclusions

We have presented in this article new numerical modellings based on a stochastic approach. We think that we have shown that these modellings are very interesting to describe with good accuracy the mechanical behavior of very heterogeneous materials like concrete and fibre reinforced concrete, and also give better results than those given by deterministic models when a precise approach of safety is required.

6. Acknowledgement

A part of the results presented were carried out with the help of the Rheology of Geomaterials Joint Research Group (France), whom we thank.

7. References

Bazant, Z.P. (1986) Mechanics of distributed cracking. Applied Mechanics Review, 39, pp. 675- 705.

Bazant, Z.P. and Pijaudier Cabot, G. (1987) Measurement of characteristic length of non local continuum. **Report n° 87-12/498 m. Center for Concrete and Geomaterials, Northwestern University**; and also in **Journal of Engineering Mechanics,** ASCE.

Burt, N.J. and Dougill, J.W. (1977) Progressive failure in a heterogeneous medium. **Journal of Engineering Mechanics Division**, pp. 365-376.

Jayatilaka, A.S. (1979) Fracture of engineering brittle materials.**Applied Science Publishers, London.**

Krajcinovic, D. and Da Silva (1982) Statistical aspects of continuous damage theory. **International Journal of Solids and Structures**, vol. 18, n°7, pp. 551-562.

Mazars, J. Ramtani, S. and Berthaud, Y. (1988) An experimental procedure to delocalize tensile failure and to identify the unilateral effect of distribued damage. **Cracking and Damage, strain localization and size effect,** edited by Mazars and Bazant,Elsevier Applied Science,pp. 55-64.

Mihashi, H. (1983) A stochastic theory for fracture of concrete. **Fracture Mechanics of Concrete**, edited by F.H. Wittmann, Elsevier.

Roelfstra, P.E. and Sadouki, H. (1986) Fracture process in numerical concrete. **Fracture Toughness and Fracture Energy of Concrete**, edited by F.H. Wittmann, Elsevier.

Rossi, P. and Richer, S. (1987) Numerical modelling of concrete cracking based on a stochastic approach. **Materials and Structures**, vol. 21; pp. 3-12.

Rossi, P. and Piau, J.M. (1988) The usefulness of statistical models to describe damage and fracture in concrete. **Cracking and Damage, strain localization and size effect,** edited by Mazars and Bazant, Elsevier Applied Science, pp. 91-103.

Rossi, P. (1989) Numerical modelling of the cracking of concrete using a non deterministic approach. **Fracture Toughness and Fracture Energy, test methods for concrete and rocks**, edited by Mihashi, Takahashi and Wittmann, A.A. Balkema, pp. 383-394.

Schorn, H. (1986) Numerical simulation of composite material as concrete. **Fracture Toughness and Fracture Energy of Concrete**, edited by F.H. Wittmann, Elsevier.

Sentler, L. (1985) A stochastic model of concrete strength. **Mémoire AIPC, IABSE Proceedings**, pp. 125-140.

Vardar, O. and Finne, J. (1975) Analysis of Brazilian disk fracture test using the Weibull probabilistic treatement of brittle strength. **International Journal of Fracture**, vol. 11, n° 3, pp. 495-508.

84 SPLITTING FAILURE OF REINFORCED CONCRETE AT VARIOUS TEMPERATURES

C. van der VEEN
Department of Civil Engineering, Delft University of Technology,
The Netherlands

Abstract
The magnitude of the ultimate bond stress of deformed bars depends upon the failure mode. Splitting failure usually determines the bond strength in the case of concrete covers commonly used in practice.

An analytical model is derived to predict the splitting failure of the concrete cover at various (low) temperatures. In the model presented we also take into account the softening behaviour of the concrete.

A lower cracking resistance relative to the tensile strength against splitting failure was found for a larger bar diameter. Furthermore, a minimum relative cracking resistance is calculated at -40 and -80°C.
Keywords: Concrete, Bond, Low Temperature, Splitting Failure, Bond Strength.

1 Introduction

It is generally accepted that bond forces (stresses) of deformed bars radiate out into the surrounding concrete at some inclination α from the bonding surface of the bar, see Fig. 1. The bond stresses can be divided into radial (σ_r) and tangential (σ_t) components. The radial compressive components of the bond forces are balanced by tensile stress rings in the concrete, according to Tepfers (1973). He stated that the radial component of the bond forces, $\tau_b \tan\ \alpha$, can be regarded as a hydraulic pressure acting on a thick walled concrete ring formed by the concrete surrounding the bar. The wall thickness of the ring is determined by the smallest possible dimension, the least thickness of the cover (c). For determination of the cracking resistance he applies three theories which, as regards the distribution of the stresses, refer to the elastic, partly cracked elastic and plastic stages.

The elastic and plastic stages give the lowest and highest cracking resistance respectively.

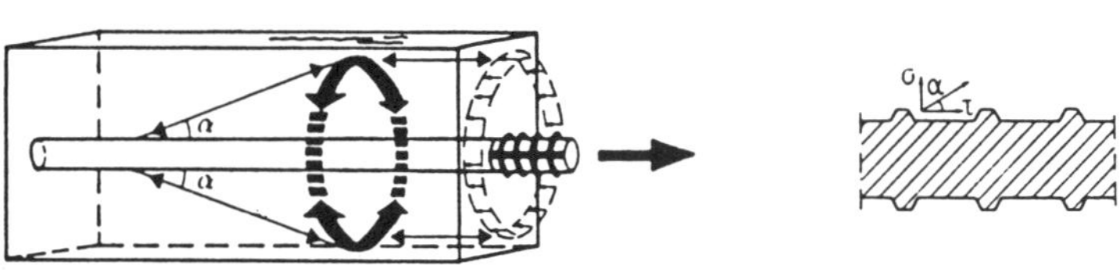

Fig. 1. Bond behaviour of deformed bars; the radial components of the bond stresses are balanced by tensile stresses in the uncracked concrete ring, Tepfers (1973).

Fracture Processes in Concrete, Rock and Ceramics Edited by J.G.M. van Mier, J.G. Rots and A. Bakker.
 Published by E. & F.N. Spon, 2–6 Boundary Row, London SE1 8HN. ISBN 0 419 15870 7

Tepfers (1973) showed that the load capacity of the concrete ring is not exhausted when cracking starts at the inner surface, but is reached later when the cracks penetrate some distance into the ring. In actual fact the elastic stage gives the load when cracking starts from the bond surface between the concrete and the reinforcing bar, while the partly cracked elastic stage gives the load when the longitudinal crack goes right through the concrete cover. However, the softening behaviour of the concrete was disregarded in the model. Consequently, no tensile stresses were allowed to transfer in the crack.

In the model presented here we also take into account the softening behaviour of the concrete according to Reinhardt et al. (1986), see formula (1).

$$\frac{\sigma}{f_{ct}} = 1 - \left(\frac{\delta}{\delta_o}\right)^k \tag{1}$$

in which δ = crack opening displacement
δ_o = stress-free crack opening displacement
f_{ct} = tensile strength

Furthermore, the effect of low temperature on the failure load is examined.

2 Model for the splitting failure mechanism

A model is derived for the tangential stress distribution over the cross section of a thick-walled concrete ring taking into account the softening of the concrete after cracking. The radial stress $\tau_b \tan\,\alpha$ due to bond action on the concrete is regarded as a hydraulic pressure against a thick-walled concrete ring. Consider the concrete ring with internal cracks, see Fig. 2.

The resistance of the uncracked part of the concrete ring was given by Tepfers (1973).

$$\tau_{br} \tan\,\alpha = f_{ct} \frac{2\,e}{d_s} \frac{(c + \frac{d_s}{2})^2 - e^2}{(c + \frac{d_s}{2})^2 + e^2} \tag{2}$$

in which e = crack depth and c = concrete cover.

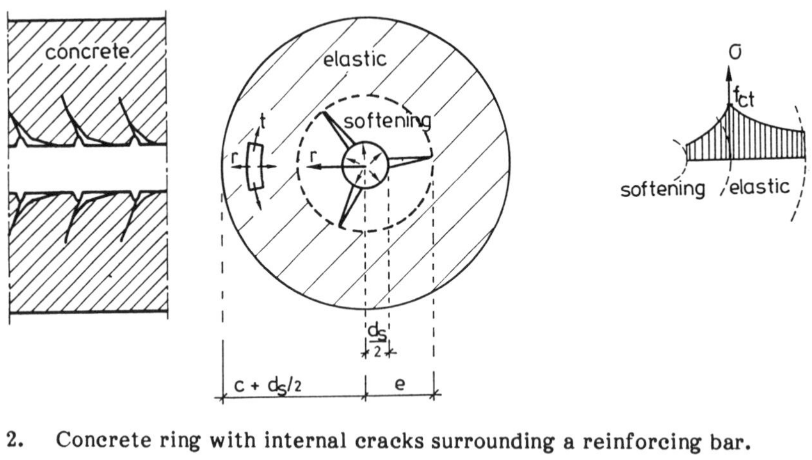

Fig. 2. Concrete ring with internal cracks surrounding a reinforcing bar.

Consider now the cracked part of the concrete ring. As a result of the cracks only a one-dimensional approach in the tangential (t) direction is applied. At a radial distance r the total tangential elongation δ_{tot} consists of an elastic part and n cracks. Thus the following deformation occurs.

$$\varepsilon_t 2\pi r + n\ \delta_o\ (1 - \frac{\sigma_t}{f_{ct}})^{\frac{1}{k}} = \delta_{tot} \tag{3}$$

in which n is the number of cracks.

At a radial distance e where the tensile strength has been reached and no cracks exist (in the elastic part) the total tangential elongation amounts approximately to:

$$\varepsilon_t\ 2\pi e \simeq \varepsilon_{cr} 2\pi e \quad \text{(no effect of Poisson's ratio and } \varepsilon_{cr} = \frac{f_{ct}}{E_c}\text{)} \tag{4}$$

The associated radial displacement amounts $u_r = \varepsilon_{cr}e$. To simplify the problem we assume a constant radial displacement for the cracked part $u_r = \varepsilon_{cr}e$. Thus also the total tangential displacement δ_{tot} is constant and is in fact imposed by the outer elastic concrete ring if no stresses exist in the cracked zone. Consequently, a rigid body translation in the radial direction $u_r = \varepsilon_{cr}e$ occurs for the cracked part, see Fig. 3.a. As a result of the elastic deformation due to the actual tangential stresses in the cracked part, the "gap" δ_{tot} is just closed at the crack tip (r = e) where the tensile strength is just reached, see Fig. 3.b.

However, even if it is assumed that for r < e the tangential stresses equal the tensile strength ($\sigma_t(r < e) = f_{ct}$) the associated elastic tangential deformation is still not able to close the gap δ_{tot}, particularly for small values of r. Thus, a crack opening always exists and this is the point of departure from which the model is developed.

Substituting expression (4) in equation 3 gives:

$$\varepsilon_t\ 2\pi r + n\ \delta_o\ (1 - \frac{\sigma_t}{f_{ct}})^{\frac{1}{k}} = \varepsilon_{cr}\ 2\pi e \tag{5}$$

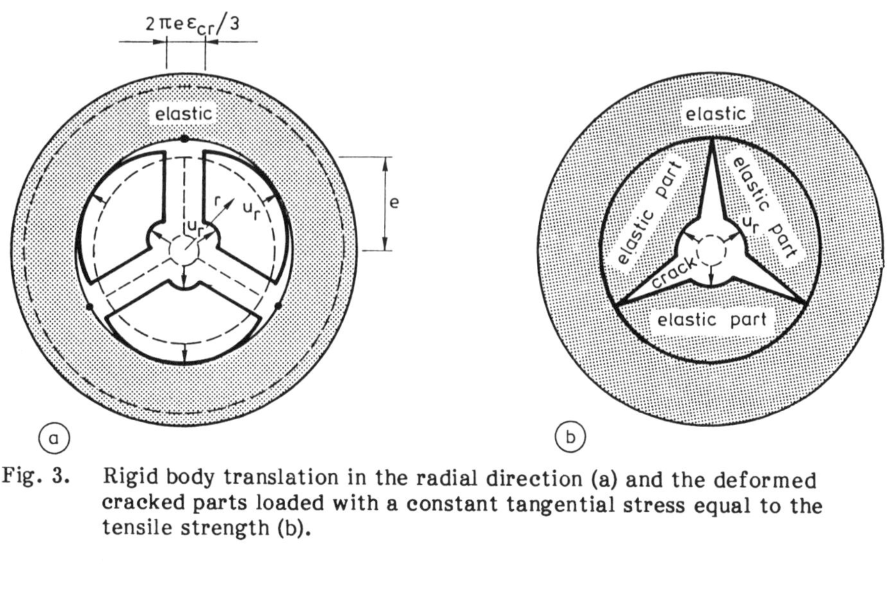

Fig. 3. Rigid body translation in the radial direction (a) and the deformed cracked parts loaded with a constant tangential stress equal to the tensile strength (b).

rearranging equation (5) gives:

$$\sigma_t = f_{ct} \left[1 - \left\{\frac{2\pi}{n\,\delta_o} (e\,\varepsilon_{cr} - r\,\varepsilon_t)\right\}^k\right] \tag{6}$$

Note: $\sigma_t = \sigma_t(r)$

It is assumed that $\varepsilon_t = \varepsilon_{cr}$ for $r \leq e$. This is (more or less) correct for $r = e$, but for smaller values of r the strain ε_t is overestimated. However, the expansion in tangential direction due to the radial compressive stress is not taken into account, which feature caused a smaller difference in the estimated ε_t and the real magnitude of ε_t. Furthermore, the stress is calculated from the stress-crack opening curve from which it follows that only small differences in stresses occur, see Fig. 4.

The force developed in the cracked zone (softening) follows from the integral:

$$\int_{r=\frac{d_s}{2}}^{r=e} \sigma_t dr = f_{ct} \left(e - \frac{d_s}{2}\right) \left[1 - \left\{\left(\frac{2\pi\ \varepsilon_{cr}}{n\ \delta_o}\right) \left(e - \frac{d_s}{2}\right)\right\}^k \frac{1}{k+1}\right] \tag{7}$$

Thus the cracking resistance developed by the softening amounts to:

$$\tau_{br} \tan\alpha = \frac{2\ f_{ct}}{d_s} \left(e - \frac{d_s}{2}\right) \left[1 - \left\{\left(\frac{2\pi\ \varepsilon_{cr}}{n\ \delta_o}\right) \left(e - \frac{d_s}{2}\right)\right\}^k \frac{1}{k+1}\right] \tag{8}$$

The total cracking resistance consists of the elastic part (formula 2) and the softening part (formula 8).

$$\tau_{br} \tan\alpha = \frac{f_{ct}2}{d_s} \left[\frac{e(c+\frac{d_s}{2})^2 - e^2}{(c+\frac{d_s}{2})^2 + e^2} + \left(e - \frac{d_s}{2}\right)\left[1 - \left\{\left(\frac{2\pi\ \varepsilon_{cr}}{n\ \delta_o}\right) \left(e - \frac{d_s}{2}\right)\right\}^k \frac{1}{k+1}\right]\right] \tag{9}$$

in which $\varepsilon_{cr} = \frac{f_{ct}}{E_c}$ (ultimate strain)

c = concrete cover

n = number of cracks

δ_o = 0.2 mm and $k = 0.248$

The number of cracks is an important variable in formula 9. If the number of cracks increases to infinite $n \to \infty$ the cracking resistance changes to the plastic case; cracks penetrating the concrete cover, thus $e = c + d_s/2$ and only softening exist.

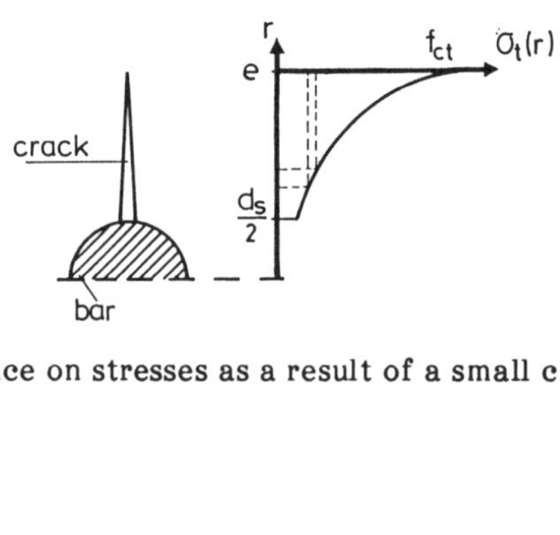

Fig. 4. Minor influence on stresses as a result of a small change in crack opening.

Thus

$$t_{br} \tan \alpha = \frac{2\ f_{ct}}{d_s} (e - \frac{d_s}{2}) = \frac{2\ f_{ct} c}{d_s} \text{ (plastic case, } n \rightarrow \infty)$$

A minimum is obtained if only a single crack (n = 1) exists, and this is the most important solution for practical applications, because this now gives us the lowest crack resistance.

2.1 Comparison of the model with results at room temperature

It is not possible to find a general applicable maximum for formula (9), because it depends not only on the crack depth e but also on the ultimate strain ε_{cr}, the concrete cover c and the bar diameter d_s. Consequently, a maximum can only be obtained if ε_{cr}, c and d_s are given. Calculations are made with 3 different bar diameters, 10, 20 and 40 mm respectively, 5 different concrete covers 0.5, 1.5, 2.5, 3.5 and 4.5 d_s respectively and an average ultimate concrete strain $\varepsilon_{cr} = 85 \times 10^{-6}$. The latter was calculated from $\varepsilon_{cr} = f_{ct}/E_c$.

For commonly used concrete ε_{cr} varies between 100 and 70 x 10^{-6} and an average value of 85 x 10^{-6} was used for the calculations. The results obtained with one crack (n = 1) are collected in Fig. 5 where the bond stress to the tensile strength of the concrete is plotted as a function of the cover to bar diameter ratio. For comparison purposes only, the angle α is assumed to be constant 45°. The model predicts values between the partly cracked elastic and plastic stage just as the experimental values. It is clearly shown that the cracking resistance (relative to the tensile strength) is higher for smaller bar diameters.

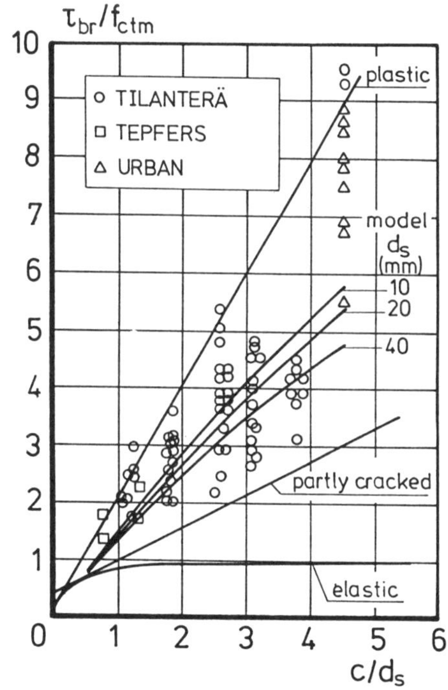

Fig. 5. Effect of thickness of concrete cover and bond capacity of the pull-out specimens on the occurrence of longitudinal cracking of the concrete cover.

To understand this feature one should realize that the crack opening depends on the elastic deformation of the elastic outer concrete ring. If the crack depth equals e.g. d_s (= e) then the crack opening equals $2\ \pi\ (d_s - r)\ \varepsilon_{cr}$. Thus at the bar surface ($r = \frac{d_s}{2}$) the crack opening equals $\pi\ d_s \varepsilon_{cr}$. Hence, for a smaller bar diameter d_s a smaller crack opening is found and a higher stress could be transferred. However, it should be noted that the differences are small (12%) in the practical area for a concrete cover c = 1.5 d_s. Thus it is difficult to find experimental evidence because of the large scatter of results.

The predicted results are influenced very much by the value of α, which is assumed to be 45° for comparison reasons only.

Probably the best way to estimate the angle α is to use the results of the pull-out tests. Based on the results of Urban (1980) an angle $\alpha = 35°$ is calculated for $c/d_s = 4.5$ and from the results of Tepfers an angle $\alpha = 40°$ is found for $1 \le c/d_s \le 2$. The latter angle is the most important one from a practical point of view.

2.2 Influence of the number of cracks

In Fig. 5 the effect of the bond capacity on the occurrence of longitudinal splitting cracking is given. However, it is evident that cracking of the cover does not imply that the ultimate bond stress has been reached. A 15% increase in load till failure is reported by Schmidt-Thrö (1987), whereas a 19% increase could be calculated from the results of Tepfers (1973).

As already discussed, the number of cracks is important to calculate the maximum (failure) load. The solution with one crack is compared with the experimental results in Fig. 5 and gives the lowest value. In order to calculate the ultimate load the number of cracks should be known. Schmidt-Thrö and Tepfers found that nearly always at least three cracks appeared.

When a fresh calculation was made with formula 9 and n = 3, an increase of about 5 to 10% was found for practical concrete covers, which seems to agree more or less with the experimental results. This feature is clear when we consider that the total tangential displacement remains constant as it is imposed by the still elastic outer concrete ring. If several cracks exist, each crack is not able to open as far as a single crack. Consequently, more stress can be transferred over the smaller crack openings.

2.3 Effect of low temperature on the failure load

Investigations have shown that in general the tensile splitting strength increases as the temperature is lower. This increase will be greater in concretes with higher moisture contents. The tensile splitting strength shows a marked increase for the water-saturated and sealed concrete in the temperature range from 0 to -40°C, see Fig. 6. In general a minimum value was observed at -80°C. At lower temperatures the increase in strength remained constant or even a small decrease was observed.

Only one investigation, Körmeling (1986) concerning the softening behaviour at low temperatures is known by the author. The stress-crack opening displacement curves show the same tendency at both temperatures, i.e. 20 and -170°C. The fracture energy increases by about 45% and the tensile strength by about 29% at -170°C. Consequently, the factor k/1+k should be multiplied by 1.45/1.29. Hence k = 0.315, which means that a higher relative cracking resistance can be expected. Nevertheless the same assumption holds true for room and low temperatures because of the lack of information for temperatures below zero.

Because the tensile stength increases more than the modulus of elasticity at lower temperatures, the ultimate tensile strain ε_{cr} increases too. A maximum value was found at -70°C by Bamforth (1987) which was about 100 to 185%

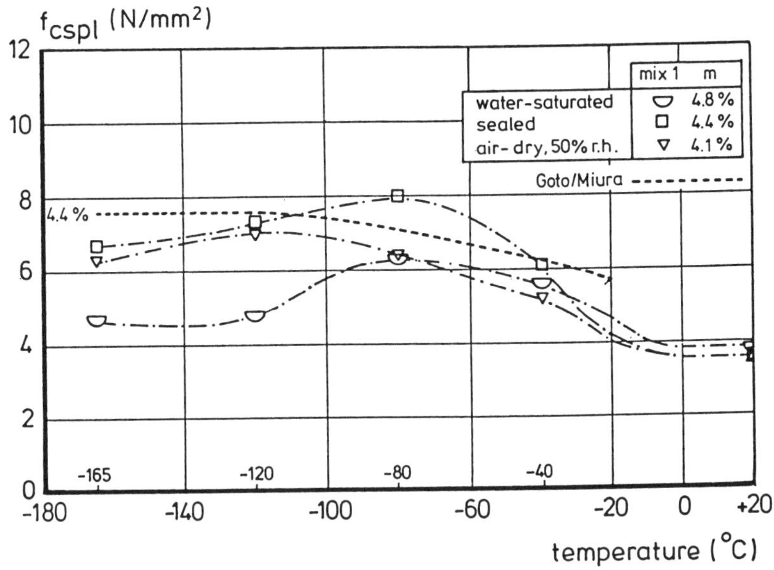

Fig. 6. Tensile splitting strength with different moisture content versus temperature

greater than at room temperature. This is due to the fact that the major part of the increase in the tensile (splitting) strength occurred in the temperature range from 0 to -60°C. Consequently, the crack opening (n = 1) given by the equation

$$\delta(r) = 2\pi(e - r)\,\varepsilon_{cr}(T)$$

will reach its maximum value at around -70°C and results in the lowest contribution of the softening part to the cracking resistance. Thus in general a lower (relative) cracking resistance will be found if $\varepsilon_{cr}(T)$ is larger than $\varepsilon_{cr}(20°C)$. This statement is in agreement with the experimental observations of Scheuermann (1987), who observed a 10% lower splitting failure load at -40 and -80°C.

3 Verification of the model at various temperatures

Experiments were conducted on ϕ 200 x 200 mm pull-out specimens with a centrically embedded steel bar of 20 mm in diameter. Thus the concrete cover amounts to 4.5 d_s (90 mm). The basic experimental programme comprised five temperatures, +20, -40, -80, -120 and -165°C respectively, two concrete grades and three curing conditions, namely water-saturated, sealed and air-dry at 50% r.h. and 20°C. Each experiment is described in detail in Van der Veen (1990).

The experimentally found bond stresses at which level splitting failure occurred are compared with the predicted values with the aid of formula 9 and n = 1. Thus the bond stress at the occurrence of cracking is predicted. First, the experimental results found at room temperature are assembled in Table 1. It is found that the ultimate load τ_{bu} is approximately 15% higher than the "cracking"

Table 1. Bond stress at splitting failure τ_{bu} and the occurrence of cracking τ_{br}

curing conditions	Mix 1 (w/c ratio 0.4) τ_{bu} (f_{ctm})	τ_{br} (f_{ctm})	$\frac{\tau_{bu}}{\tau_{br}}$ (.)	Mix 2 (w/c ratio 0.6) τ_{bu} (f_{ctm})	τ_{br} (f_{ctm})	$\frac{\tau_{bu}}{\tau_{br}}$ (.)
saturated	8.56	7.57	1.13	8.35	7.71	1.08
sealed	8.84	7.57	1.14	8.33	7.68	1.08
air-dry	9.28	7.55	1.23	8.98	7.37	1.22

load τ_{br}, which is in close agreement with the experimental results of Schmidt-Thrö (1987) and Tepfers (1973). When 3 cracks are assumed (n = 3) in formula 9 an increase in τ_{br} of about 14% is calculated which is again in close agreement with the experimental results.

Splitting failure at low temperatures was only observed for mix 1. The lowest relative crack resistance is found experimentally and theoretically at -40 and -80°C. The same phenomenon was found experimentally by Scheuermann (1987), who performed pull-out tests with a practical concrete cover of 2 d_s. The results of the derived theoretical formula 9 (n = 1, tan α = 40°) are compared with the results of Scheuermann, see Fig. 7. The calculated results were in close agreement with the experimental results of Scheuermann. Higher bond strength ratio's were calculated for the other temperatures which were slightly smaller than Scheuermann's ratio's. Thus a reasonable agreement is obtained for small concrete covers.

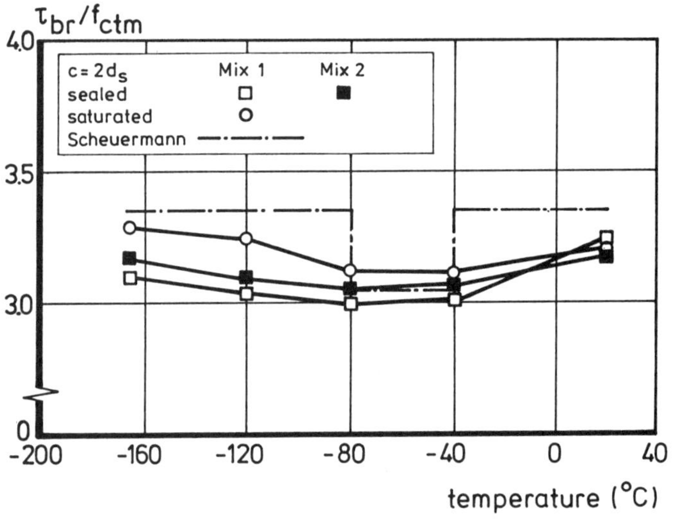

Fig. 7. The calculated (relative) splitting failure load for a concrete cover c = 2 d_s.

4 Conclusions

Splitting failure can be predicted theoretically with the model and the typical phenomenon (a lower bond strength to tensile strength ratio found at -40 and -80°C) is predicted as well.

If the bar diameter increases by 100% an approximately 10 to 15% lower relative cracking resistance against splitting failure is calculated.

5. Acknowledgement

These investigations have been carried out with the financial support of the Netherlands Technology Foundation (S.T.W. grant DCT 25.0298 to prof.dr.ir. A.S.G. Bruggeling and prof.dr.-ing. H.W. Reinhardt), which is gratefully acknowledged.

6 References

Bamforth, C.B. (1987) The structural permeability of concrete at cryogenic temperatures, **Dissertation** University of Aston, Birmingham.

Körmeling, H.A. (1986) Strain rate and temperature behaviour of steel fibre concrete in tension, **Dissertation** University of Delft.

Reinhardt, H.W., Cornelissen, H.A.W. and Hordijk, D.A. (1986) Tensile tests and failure analysis of concrete, **Journal of Structural Engineering**, 112, 2462-2477.

Scheuermann, J. (1987) Zum Einfluss tiefer Temperaturen auf Verbund und Rissbildung von Stahlbetonbauteilen, **Dissertation,** University of Braunschweig.

Schmidt-Thrö, G. (1987) Verbundverhalten von Rippenstählen im Beton unter besonderer Berücksichtigung einer einachsigen Querpressung, **Dissertation** Munchen.

Tepfers, R. (1973) A theory of bond applied to overlapped tensile reinforcement splices for deformed bars, **Report 73-2** Chalmers University of Technology Göteborg.

Urban, V. (1980) Anchorage length and bond stress calculation, **Research Report 458,** Technical University Prague.

Veen, C. van der (1990) Cryogenic bond stress-slip relationship, **Dissertation** University of Technology Delft.

85 BOND-SLIP BEHAVIOUR – A FRACTURE MECHANICS APPROACH

D.Z. YANKELEVSKY, D.G. ZEITOUN and
O. LEIBOWITZ
Faculty of Civil Engineering, Technion, Haifa, Israel

Abstract
This paper presents a first stage of a study aiming to develop the theoretical bond stress-slip relationship. The problem of a concrete cylinder in which a single steel bar is embedded, is considered. Energy and fracture considerations yield a nonlinear expression which is characterized by a linear part at the small slips range, followed by nonlinear relationship until a peak stress is reached. The model predicts the following descending branch as well.
Keywords: Bond Stress, Slip, Concrete, Steel, Fracture Mechanics.

1 Introduction

The behavior of a reinforced concrete element is based on the interaction between concrete and the steel bars. This interaction is governed by bond, which is the mechanism of load transfer from a steel bar to the surrounding concrete. In the case of deformed bars, load is mainly transferred through mechanical locking of the lugs. The load transferred per unit area of the bar circumference is commonly denoted as the bond stress. This load development is accompanied by some slip, which is a relative displacement between the steel bar and the concrete, as a result of frictional displacement as well as local cracking development at the contact between a lug and concrete and local crushing of concrete.

The interaction between a steel bar and concrete during a typical pull-out test may be characterized by the following four stages:

a. For low magnitude bond stresses the bond is mainly due to chemical bond and there is no slip.
b. For higher bond stresses exceeding the chemical bond, the steel bar lugs exert high magnitude local stresses on the concrete and inclined cracks are developed at the contact areas.
c. As the process continues the concrete against the bar lugs is crushed and slip is increasing.
d. Further slip develops lower resistance of a friction type which is less sensitive to the lugs shape.

In the present study, attention is given to the basic problem of a concrete cylinder and a centrally embedded single steel bar along the axis of symmetry.

Fracture Processes in Concrete, Rock and Ceramics Edited by J.G.M. van Mier, J.G. Rots and A. Bakker.
 Published by E. & F.N. Spon, 2–6 Boundary Row, London SE1 8HN. ISBN 0 419 15870 7

This problem has been treated intensively both theoretically and experimentally and the available information enables one to quantitatively define the global interaction, predict the primary cracks and the bond stress distribution along the steel bar.

A common approach to treat the bond problem is to define the bond stress-slip relationship as a property of the concrete-steel interface. This property is substituted into an equilibrium differential equation of the steel bar [Ciampi (1981), Edwards (1972), Eligehausen (1983), Giuriani (1981), Soroushian (1981), Yankelevsky (1985)]. Another approach assumes the bond stress distribution along the steel bar [Jiang (1984), Ngo (1967)].

Beyond the one-dimensional models, there exist several analytical axisymmetric models [e.g. Yankelevsky (1985)]. Finite element solutions may be subdivided into three main groups: The solutions based on interface representation by springs linking the steel bar to the concrete [Ngo (1967), Nilson (1984), Viwathanatepa (1979)]; solutions incorporating an interface layer [Bresler & Bertero (1963), Reinhardt (1984), Yankelevsky (1985)], and solutions introducing fracture to describe development at the lug root and its development [Igrafea (1984)]. Most of the Finite Element simulations consume considerable computer resources.

Most of the available models describing the steel bar-concrete interaction are based on empirical bond stress-slip relationships. This approach limit the physical understanding of the interaction. Moreover, recent experimental research has shown that the experimental bond stress-slip relationship is location dependent and is not constant along the bar, as most commonly is assumed. This local interaction, which depends on the location of a certain lug, has not yet been studied.

It is the purpose of the present research to study the local interaction of a lug with the surrounding concrete, in an attempt to obtain the theoretical bond stress-slip relationship which is location dependent.

2 Model description

Consider a concrete cylinder having an outer diameter $D=2.r_p$, an inner diameter d and length H, in which a central steel bar of diameter d is embedded (Figure 1). The steel bar is connected to the internal cylinderical surface along a short length lu located at a distance X_o from the cylinder end. A single circular lug is being in contact with the concrete at X_o. A tensile force F is applied to the steel bar end and its relative displacement is measured at its other end (Figure 1).

It is assumed that the lug height is small compared to the bar diameter and that the force acting on the lug may be presented by an equivalent bond stress τ acting along lu on the inner surface of the concrete cylinder. It is also assumed, at this stage of research, that the crack length is small compared with the bar diameter.

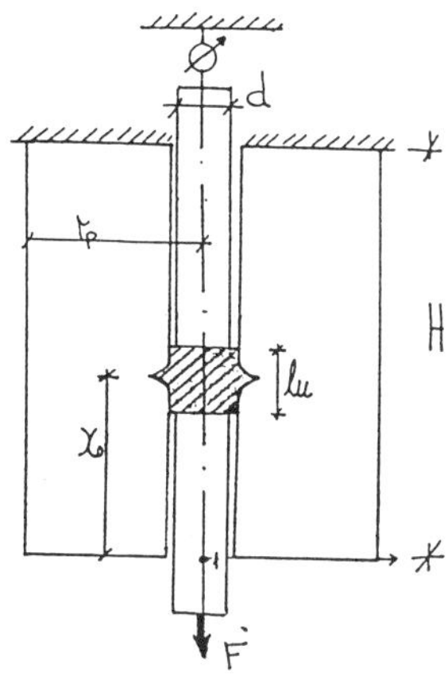

Fig. 1. The model specimen

3 Mathematical derivation

3.1 Energy balance considerations for a single lug

Assuming small deformations, energy balance will be considered for two cases: uncracked concrete and cracked state [Jeng (1986)].

For the uncracked case the energy balance expression is:

$$\int_{V_S} \sigma_{ij}\delta\epsilon_{ij}dV + \int_{V_C} \sigma_{ij}\delta\epsilon_{ij}dV = F\delta_S \tag{1}$$

where σ_{ij}, ϵ_{ij} are the stress and strain tensors components, respectively.
δ_s is the steel bar end displacement.
V_s and V_c are the steel bar and concrete volumes, respectively.

Energy balance for the cracked state is expressed [25]:

$$\int_{V_S} \sigma_{ij}\delta\epsilon_{ij}dV + \int_{V_C} \sigma_{ij}\delta\epsilon_{ij}dV + \delta\phi = F\,\delta_S \tag{2}$$

where $\delta\phi$ is term expressing energy release due to crack formation.

In both cases volume integrals may be converted into surface integrals:

$$\int_{V_C} \sigma_{ij}\delta\epsilon_{ij}dV = \int_{CES}\sigma_{ij}n_j\delta u_i ds \tag{3-a}$$

$$\int_{V_C} \sigma_{ij}\delta\epsilon_{ij}dV = \int_{SES}\sigma_{ij}n_j\delta u_i ds \tag{3-b}$$

where:
CES – concrete external surface.
SES – steel external surface.
n_j – the j-th component to the external normal to the surface.

These surface integrals will be now derived for the steel bar and the concrete cylinder:

a. Surface integral for the steel bar

The surface integral for the steel bar yields:

$$_{SES}\int \sigma_{ij} n_j \delta u_i ds = \pi d \tau \int_{\chi_o}^{\chi_o + l_u} w_s(z) dz + F\delta_s = F\delta_s + \tau I_s \qquad (4)$$

where $w_s(z)$ is the variable steel bar deformation.

Further development of equation (4) yields:

$$\tau I_s = F\Delta + \frac{F^2 lu}{6\pi E_s r^2} \qquad (5)$$

b. Surface integral for the concrete cylinder

In typical pull out tests [ASTM (1987), Yankelevsky (1985)], the circular concrete surface at the side of the applied load is stress free, while the surface on the opposite side is restrained to axial displacement. Assuming, at this stage, that the outer surface of the concrete cylinder is stress-free, the surface integral for the concrete cylinder will be:

$$_{CES}\int \sigma_{ij} n_j \delta u_i ds = \pi d \tau \int_{\chi_o}^{\chi_o + lu} W_c(r,z) dz = \tau I_c \qquad (6)$$

Substituting the expressions for I_s and I_c into the energy balance equations, one gets the fundamental bond stress-slip relationship for the uncracked and cracked states.

The bond stress-slip relationship for the uncracked state becomes:

$$\tau = \frac{\Delta}{R(\chi_o, lu)} \qquad (7)$$

where:

$$R(\chi_o, lu) = \int_{\chi_o}^{\chi_o + lu} \frac{K(r,z)}{lu\, G_c} dz - \frac{lu^2}{3E_s r} \qquad (7\text{-a})$$

The expression for the cracked state becomes:

$$\Delta = \tau\left[\int \frac{K(r,z)}{G_c\, lu} dz - \frac{lu^2}{3E_s r}\right] + \frac{\delta\phi}{F} \qquad (8)$$

3.2 Energy release in crack formation

The expression of the released energy due to crack opening depends on the way at which a crack develops at the contact area between the steel lug and the surrounding concrete. The lugs have an important role in crack development as was found by Goto (1979). Observation at experimental work shows that the crack, which starts at the lug, has a mixed shape including stress intensity factors K_I and K_{II}. Assuming that the crack is not affected by boundary conditions, the problem under discussion may be combined of four basic cases (Figure 3), where case (a) and case (b) describe crack opening (Mode I), and cases (c) and (d) describe a state of Mode II.

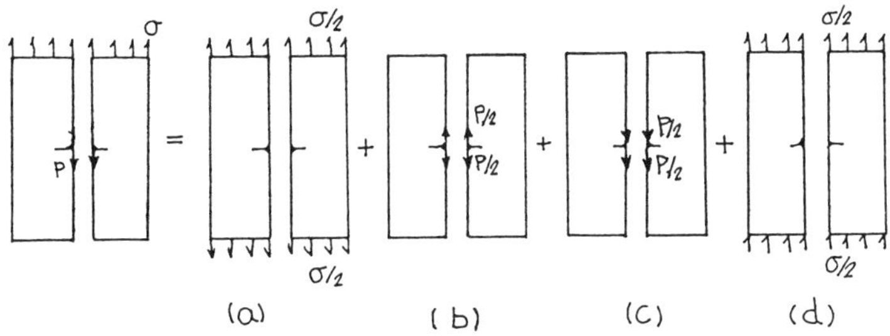

Fig. 2. The four basic cases of the model

Expressions for each of the cases have been developed and a criterion for the critical state was defined.

3.3 Dimensional analysis

It is important to carry dimensional analysis and present the bond stress-slip relationship in its general form which is applicable to various shapes.

The released energy in cases (a) and (d) is:

$$U_{ad} = \frac{\sigma^2 a^2}{E} \left[f_{1a}\left(\frac{r_p}{a}, \frac{g}{a}, \frac{g}{d}\right) + f_{1d}\left(\frac{r_p}{a}, \frac{g}{a}, \frac{g}{d}\right) \right] \pi d \qquad (9)$$

where:

f_{1a} and f_{1d} are nondimensional shape functions which depend on the geometry and type of loading.

g - is the radial length of a lug.

a - is the crack length at a measured displacement Δ.

The released energy may also be defined by the J integral which is defined in LFM:

$$J = G = \frac{\partial U}{\partial a} \qquad (10)$$

hence:

$$U = \frac{(1-\nu^2)}{E} \int_{c_1}^{c_2} (K_{Ia}^2 + K_{IId}^2) \pi d \, dr \qquad (11)$$

where K_{Ia} and K_{IId} are the stress intensity factors for cases a and d respectively and defined:

$$K_{Ia} = \sigma\sqrt{a}\; f_{2a}\left(\frac{r_p}{a}, \frac{g}{a}, \frac{g}{d}\right) \qquad (12\text{-a})$$

$$K_{IId} = \sigma\sqrt{a}\; f_{3d}\left(\frac{r_p}{a}, \frac{g}{a}, \frac{g}{d}\right) \qquad (12\text{-b})$$

f_{2a}, f_{3d} are nondimensional shape functions depending on the geometry and type of loading.

The released energy for cases (b) and (c) is:

$$U_{bc} = \frac{P^2}{Ea^2} \left[f_{1b}\left(\frac{r_p}{a}, \frac{g}{a}, \frac{g}{d}\right) + f_{1c}\left(\frac{r_p}{a}, \frac{g}{a}, \frac{g}{d}\right) \right] \pi d \qquad (13)$$

where P is a force magnitude.

f_{lb}, f_{lc} are nondimensional shape functions depending on the geometry and type of loading.

The released energy may be defined through the stress intensity factors:

$$U_{bc} = \frac{(1-\nu^2)}{E} \int_{c_1}^{c_2} (K_{Ib}^2 + K_{IIc}^2)\pi d \, dr \tag{14}$$

where K_{Ib} and K_{IIc} are the stress intensity factors for cases (b) and (c) respectively and may be defined by:

$$K_{Ib} = \frac{P}{\sqrt{a}} f_{2b}\left(\frac{r_p}{a}, \frac{g}{a}, \frac{g}{d}\right) \tag{15-a}$$

$$K_{IIc} = \frac{P}{\sqrt{a}} f_{3c}\left(\frac{r_p}{a}, \frac{g}{a}, \frac{g}{d}\right) \tag{15-b}$$

where f_{2b} and f_{3c} are shape functions depending on the geometry and type of loading.

3.4 Critical state

At this stage of the model development the crack length should be related to the mechanical properties of concrete, by introducing a crack growing criterion. The stress intensity factor reaches a certain magnitude at which the crack will grow. For Mode I the criterion may simply be $K_I = K_{Ic}$. For combined modes the following criterion may be assumed [Braek (1974), Parker (1981)]:

$$K_I^2 + 1.78\, K_{II}^2 = K_{Ic}^2 \tag{16}$$

Assuming that the following relationships exist:

$$\sigma = \alpha\, \tau \tag{17-a}$$

$$P = \beta\, \tau \tag{17-b}$$

where α, β are proportionality coefficients relating σ and P to the average stress τ.

Substituting equations (16,17) into equations (11,13) yields the expression for the released energy $\delta\phi$:

$$\delta\phi = \frac{\pi d}{\tau^2} K_{Ic}^4 \left[\frac{A+D}{\beta^2} + \frac{B+C}{\alpha^2}\right] \tag{18}$$

where:

$$A = \frac{f_{1a}}{F_{2a}^2\, E} \tag{19-a}$$

$$B = \frac{f_{1b}}{f_{2b}^2\, E} \tag{19-b}$$

$$C = \frac{f_{1c}}{(1.78\ f_{3c}^2)^2 E} \tag{19-c}$$

$$D = \frac{f_{1d}}{(1.78\ f_{3d}^2)^2 E} \tag{19-d}$$

4 Theoretical bond stress-slip relationship

Substituting the crack increase criterion and the expression for the released energy into the bond stress-slip relationship in the cracked state (equation 8), yields:

$$N_1 \tau^4 - \tau^3 + N_2 = 0 \tag{20}$$

where:

$$N_1 = \int_{X_o}^{X_o + lu} \frac{K(r,z)}{G_c\ lu}\, dz - \frac{lu^2}{3rE} \tag{21-a}$$

$$N_2 = \frac{K_{Ic}^4}{lu} \left[\frac{A+D}{\beta^2} + \frac{B+C}{\alpha^2}\right] \tag{21-b}$$

Qualitative examination of this equation yields the following observations:

a. The theoretical bond stress-slip relationship is location dependent as the constants A,B,C,D and K(r,t) depend on X_o.
b. The theoretical expression depends on material properties through the critical stress intensity factor K_{Ic}, and Young's modulus E.
c. As N_2 is a positive function it follows that equation (20) is valid for slips larger than a certain slip Δ_o. Experimental results of slips in the precracked state, support this result.
d. For the uncracked state at which $\Delta \leq \Delta_o$, the theoretical expression becomes linear:

$$\tau = \frac{\Delta}{N_1} \tag{22}$$

This finding is in accordance with the bond stress-slip behavior observed experimentally for small slips.

e. The theoretical relationship in equation (20) includes many geometrical parameters which had been found to affect the bond stress-slip relationship, such as: distance between lugs (lu), lug height, steel bar and concrete cylinder diameters, etc.
f. The shape of the theoretical bond stress-slip may be qualitatively analysed. It has already been mentioned that it starts as a linear relationship for small slips. Once a crack is developed N_2 is not zero and N_1 remains constant. Therefore, for some stresses slips will increase comapred to an uncracked state and this affects the nonlinear part of the curve until the peak is reached at $\Delta = \Delta^*$. When the crack increases a state is reached that the crack opens at a faster rate than it lengthens and, as

a result, N_1 value grows slower than N_2. Therefore, the stress values decrease and we observe the descending branch (Figure 3).

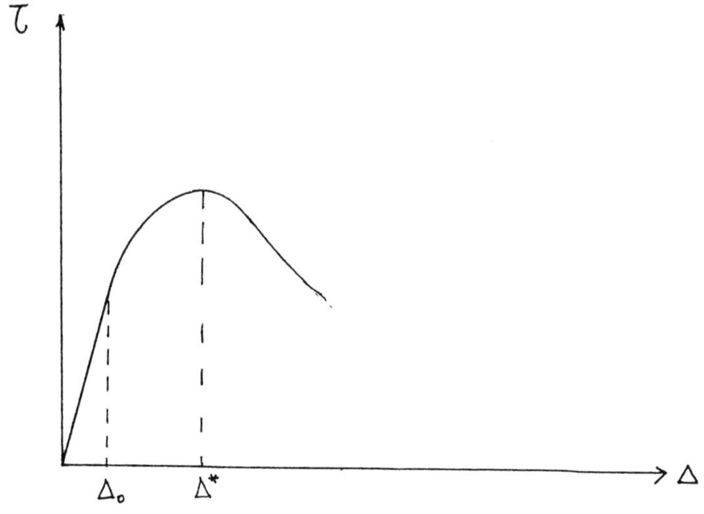

Fig. 3. Theoretical bond stress–slip relationship

5 References

ASTM, 1987, Annual book of Standards, Vol. 04.02, C234: Standard Test Method for Comparing Concretes on the Basis of the Bond Developed with Reinforcing Steel.

Alvarado, M.A., Shah, S.P. and John, R., "Mode I Fracture in Concrete Using Center Cracked Plate Specimen", **ASCE**, Mech. Eng., Vol. 115, No. 2, Feb., 1985, pp. 366–384.

Bresler, B. and Bertero, V., "Behavior of Reinforced Concrete Under Repeated Load", **ASCE**, Jour. of St. Div., Vol. 94, No. St6, June 1963, pp. 1567–1589.

Broek, D., "Elementary Engineering Fracture Mechanics", **Noordhoff Leiden**, 1974, 408 pp.

Ciampi, V. and Eligehausen, R., et al., "Analytical Model for Deformed Bar Bond Under Generalized Excitations", Proceedings, **IABSE** Colloquium on "Advanced Mechanics in Reinforced Concrete", Delft, June 1981, pp. 53–67.

Ciampi, V., Eligehausen, R., Bertero, V.V. and Popov, E.P., "Analytical Model for Concrete Anchorage of Reinforcing Bars Under Generalized Excitations", Report No. UCB/EERC–82/23, **Earthquake Engineering Research Center**, University of California, Berkeley, Nov. 1981, 111 pp.

Cowell, A.D., Popov, E.P. and Bertero, V.V., "Effects of Concrete Types and Loading Conditions on Local Bond Slip Relationships", Report No. UCB/EERC–82/17, **Earthquake Engineering Research Center**, University of California, Berkeley, Sept. 1982, 62 pp.

Edwards, A.D. and Picard, A., "Theory of Cracking in Concrete Members", **ASTM**, Jour. of St. Div., Vol. 98, No. ST12, 1972, pp. 2687-2699.

Eligehausen, R., Popov, E.P. and Bertero, V.V., "Local Bond Stress-Slip Relationships of Performed Bars Under Generalized Exciations", Report No. UCB/EERC-82/23, **Earthquake Engineering Research Center**, University of California, Berkeley, Oct. 1983, 169 pp.

Giuriani, Z., "Experimental Investigation on the Bond-Slip Law of Deformed Bars in Concrete", **Proceedings IABSE Colloquium** on "Advanced Mechanics on Reinforced Concrete", Delft, June 1981, pp. 121-142.

Goto, Y. and Otsuka, K., "Experimental Studies on Cracks Formed in Concrete Around Deformed Tension Bars", The Technology Reports of the Tohoku University, Vol. 44, No. 1, 1979, pp. 49-63.

Harajli, M.H. and Mukaddam, M.A., "Slip of Steel Bars in Concrete Joints Under Cyclic Loading", **ASCE**, Jour. of St., Div., Vol. 114, No. 9, 1988, pp. 2017-2035.

Hungspreng, S., "Local Bond Between a Steel Bar and Concrete Under High Intensity Cyclid Load", Ph.D. Thesis, Cornell University, 1981, 449 pp.

Ingraffea, A.R. et al., "Fracture Mechanics of Bond in Reinforced Concrete, **ASCE**, Jour. of St., Div., Vol. 110, No. 4, 1984, pp. 871-893.

Jeng, Y.S. and Shah, S.P., "Crack Propagation in Fiber Reinforced Concrete, **ASCE**, Jour. of St., Div., Vol. 112, No. 1, January 1986, pp. 19-34.

Jeng, V. and Shah, S.P., "Two Parameter Fracture Model for Concrete", **ASCE**, Eng. Mech., Vol. 110, No. 10, Oct. 1985, pp. 1227-1241.

Jiang, D.H., Shah, S.P. and Andorian, A.J., "Study of the Transfer of Tensile Forces by Bond", **ACI Journal**, Vol. 81, No. 3, 1984, pp. 251-259.

Morita, S. and Kaku, T., "Local Bond Stress-Slip Relationship Under Repeated Loading", Symposium: Resistance and Ultimate Deformability of Structures Acted On by Well Defined Repeated Loads, IABSE, Lisboa, 1973, pp. 221-227.

Nallathambi, P.N. and Karchaloo, B.J., "Determination of Specimen-Size Independent Fracture Toughness of Plain Concrete", **Magazine of Concrete Research**, Vol. 38, No. 135, June 1986, pp. 67-76.

Ngo, D. and Scordelis, A.C., "Finite Element Analysis of Reinforced Concrete Beams", **ACI**, Vol. 64, No. 3, March 1967, pp. 152-766.

Nilson, A.H., "Nonlinear Analysis of Reinforced Concrete by the Finite Element Analysis", **Materiaux et Construction**, Vol. 17, No. 100, 1984, pp. 311-320.

Parker, A.P., "The Mechanics of Fracture and Fatigue", London, **E.&F.N Spon Ltd.**, 1981, 165 pp.

Petersson, P.E., "Fracture Energy of Concrete: Practical Performance and Experimental Results", **Cement and Concrete Research**, Vol. 10, 1980, pp. 91-101.

Reinhardt, H.W. et al., "Prediction of Bond Between Steel and Concrete by Numerical Analysis", **Materiaux et Constructions**, Vol. 17, No. 100, 1984, pp. 311-320.

Sedov Continum Mechanics, Vol. I and Vol. II, Moscow Editions.

Somayaiji, S. and Shah, S.P., "Bond Stress Versus Slip Relationship and Cracking Response of Tension Members", **ACI Journal**, Vol. 78, No. 3, May/June, 1981, pp. 217-225.

Soroushian, P. and Choi, Ki-Bong, "Local Bond of Deformed Bars with Different Diameters in Confined Concrete", **ACI Struc. Journal**, Vol. 86, No. 2, 1989, pp. 217-222.

AUTHOR INDEX

SUBJECT INDEX

This index is compiled from the keywords provided by the authors, extended and revised by the Editors where necessary. The page numbers are those of the first page of the paper referred to.